Springer Geology

More information about this series at http://www.springer.com/series/10172

Mingguo Zhai
Editor

Precambrian Geology of China

Springer

Editor
Mingguo Zhai
Institute of Geology and Geophysics
Chinese Academy of Sciences
Beijing
China

ISSN 2197-9545 ISSN 2197-9553 (electronic)
Springer Geology
ISBN 978-3-662-47884-4 ISBN 978-3-662-47885-1 (eBook)
DOI 10.1007/978-3-662-47885-1

Library of Congress Control Number: 2015944154

Springer Heidelberg New York Dordrecht London

Printed on acid-free paper

Springer-Verlag GmbH Berlin Heidelberg is part of Springer Science+Business Media
(www.springer.com)

Preface

The Precambrian (4560–541 Ma) covers almost 90 % of the planet Earth's history. Precambrian continents experienced complex geological evolution, and carry important records of the secular changes in tectonics and metallogeny, including, at least, three important tectonic events that are the Neoarchean enormous crustal growth, the tectonic regime inversion from pre-plate tectonics to plate tectonics, and the Paleoproterozoic great oxygen event (GOE). Precambrian rocks are extensively distributed in China, not only in cratons, but also in Phanerozoic orogenic belts. The Chinese continent consists of several cratons, i.e., the North China Craton, the South China Craton and the Tarim Craton, and several Phanerozoic orogenic belts. These three cratons have different tectonic evolving history, and carry important records of crustal generation-growth-cratonization and interaction with circumjacent terrains. Some residual Precambrian rocks or micro-continental blocks also occur in Phanerozoic orogenic belts, such as in the Himalaya and Xing'an-Mongolia orogenic belts. There are abundant mineral resources in these cratons and micro-blocks, which were formed in different geological periods and recorded a common change in the pattern of metallogeny, mineral deposit character, spatial distribution, and genetic mechanisms that match well with the timings and styles of the major tectonic events. The assembly of the Chinese unified continent is considered to be amalgamated by several Triassic orogenic belts. Therefore, study of the three old lands in China will surely help deepen our understanding of China geology, as much as global continental tectonics and continental dynamics.

This book, *Precambrian Geology of China*, contains five parts and 12 chapters. Part I has only one chapter and is about the general Precambrian geology of China (by Mingguo Zhai and Yanyan Zhou; General Precambrian Geology in China). It summarizes the geology division of China, the general geology in the three cratons/blocks, and the assembly of the Chinese landmass.

Part II focuses on the Precambrian geology of the North China Craton. It contains three chapters. Formation and Evolution of Archean Continental Crust of the North China Craton (by Yusheng Wan) is about the Archean geology of the craton, and it mainly focuses on the Archean geological record of the craton back to >3.8 Ga, as well as the formation of the Archean basement. Paleoproterozoic Granulites in the North China Craton and their Geological Implications (by Jinghui Guo and others) concentrates on the Paleoproterozoic granulites and tectonic evolution of the craton. Late Paleoproterozoic–Neoproterozoic (1800–541 Ma) Mafic Dyke Swarms and Rifts in North China (by Peng Peng) narrates the late Paleoproterozoic to Neoproterozoic igneous records, sedimentation, and rift evolution in the craton.

Part III (by Yuan-sheng Geng) is about the Precambrian geology of the South China Craton. It has three chapters. Early Precambrian Geological Signatures in South China Craton focuses on the Archean geological records and their features, and the basement evolution during Paleoproterozoic; Mesoproterozoic Era of South China Craton is about the Mesoproterozoic magmatic events of the Yangtze and Cathaysian blocks/subcratons, and the tectonic evolution of the Jiangnan orogenic belt; and Neoproterozoic Era of South China Craton is dealing with the Neoproterozoic orogenic and rifting events in South China.

Part IV (by Bei Xu) summarizes the Precambrian Geology of the Tarim Craton. It divides into three chapters: Late Archean: Mesoproterozoic Geology of the Tarim Craton about the Archean to Mesoproterozoic formations and records; The Neoproterozoic Geology of the Tarim Craton about the Neoproterozoic formations and records; and The Precambrian Tectonic Evolution of the Tarim Craton about the Precambrian tectonic evolution of the Tarim Craton.

Part V is about the geology of micro-blocks in the Phanerozoic orogens, mainly concentrating on the Tibetan Plateau in the Himalaya Orogen (The Precambrian Geology of the Tibetan Plateau by Fulai Liu et al.) and the paleo-continents in the Xing'an-Mongolia orogenic Belt (Paleocontinents in Xing'an-Mongolia orogenic belt (XMOB) by Bei Xu).

Beijing, China — Mingguo Zhai
March 2015

Contents

Part I
General Precambrian Geology in China

General Precambrian Geology in China

Mingguo Zhai and Yanyan Zhou

Abstract The Earth has an evolving history ca. 4.6 Ga old. Precambrian continents experienced complex geological evolution since the Early Precambrian onward and carry important records of the secular changes in tectonics and metallogeny, including, at least, three important tectonic events that are Neoarchean enormous crustal growth, tectonic regime inversion from pre-plate tectonics to plate tectonics, and Paleoproterozoic great oxygen event (GOE). Without question, Precambrian geology is a key issue to understand continental geology. Precambrian rocks are extensively distributed in China, not only in cratons, but also in Phanerozoic orogenic belts. The Chinese continent consists of several cratons and surrounding Phanerozoic orogenic belts (folded belts). The main old lands in China include the North China Block (NCB), South China Block (SCB), and Tarim Block (TRB), all of which have individual tectonic evolving histories. The NCB experienced complex geological evolution since the Early Precambrian onward and carries important records from the old continental nuclei, giant crustal growth episode, and cratonization (stabilization), then to the Paleoproterozoic rifting–subduction–accretion–collision with imprints of the GOE, and to the Late Paleoproterozoic–Neoproterozoic multistage rifting representing North China platform tectonic features. The TRB has two-layer basement of the Early Precambrian metamorphic complexes and Neoproterozoic sedimentary sequences. Three till sheets have been reported. The SCB consists of the Yangtze Block (YZB) and Cathaysia Block (CTB) that were cohered in the Neoproterozoic. The YZB recorded tectonic processes of the Early Precambrian crustal growth, 1.0–0.9 and 0.8–0.6 Ga metamorphic–magmatic events, and two Neoproterozoic glaciations. The CTB consists of ca. 1.8, 1.0–0.9 Ga, and ca. 0.8 Ga granitic gneisses and metamorphic rocks, indicating there was a vast Precambrian basement. The Neoproterozoic sedimentary rocks overlie partly on the basement. That the YZB and CTB have a Neoproterozoic uniform cover layer illustrates the SCB should form, at least, during 1.0–0.9 Ga. The

M. Zhai (✉) · Y. Zhou
Institute of Geology and Geophysics, Chinese Academy of Sciences, Beijing 100029, China
e-mail: mgzhai@mail.igcas.ac.cn

M. Zhai
State Key Laboratory of Geodynamics, Northwest University Shanxi, Xi'an 710069, China

M. Zhai (ed.), *Precambrian Geology of China*,
Springer Geology, DOI 10.1007/978-3-662-47885-1_1

Nanhua Rift System developed in the SCB in ca. 0.8–0.7 Ga, which probably represented a global event indicating breakup of the Rodinia Supercontinent. The Central Chinese Orogenic System with high-ultra-high-pressure metamorphic rocks supports a suggestion that the above-mentioned three old lands were collided to assemble a unified Chinese continent during the Pangea orogenic period.

Keywords Precambrian rocks · Main old lands · Geological characteristics · Unified continent · China

1 Introduction

The Precambrian Era, from ca. 4.56 to 0.545 Ga, covers almost 90 % of this planet's history (Mclennan and Taylor 1982; Goodwin 1991). The growth of continental crust is generally believed that largely achieved in the Early Precambrian time, and ~75–80 % or more continental crust had formed before 2.5 Ga ago (e.g., Amstrong 1981; Dewey and Windley 1981; Condie 2000). The studies show that the oldest continental rocks in the Earth are tonalitic gneiss and the oldest zircons are detrital zircons in sedimentary rocks which were sourced from TTG protoliths (e.g., Wilde et al. 2001; Iizuka et al. 2006). Archean supercraton probably was similar in scale to the Pangea Supercontinent proposed by Rogers and Santosh (2003). After ca. 2.5 Ga global cratonization and ca. 2.0–1.9 Ga mobile belts, thick Late Paleoproterozoic–Neoproterozoic volcanic-sedimentary sequences, as cover of the craton, were extensively deposited on the early metamorphic basement, indicating the Earth had a stable lithosphere with underlying secular warm mantle that resulted in multistage magmatism and rifting during the Earth's middle age. Precambrian continents experienced complex geological evolution since the Early Precambrian onward and carry important records of the secular changes in tectonics and metallogeny, including, at least, three important tectonic events that are Neoarchean enormous crustal growth, tectonic regime inversion from pre-plate tectonics to plate tectonics, and Paleoproterozoic great oxygen event (GOE) (Zhai and Santosh 2011, 2013). However, our knowledge of this segment of time remains elusive, save for selective aspects. Many of the Archean terranes with the dictum "the present is key to the past" was born out of the necessity that specific geological problems or features in the past cannot be adequately explained or addressed due to limited information or outcrop exposure. However, present geological phenomena that we know and understand show some difference from the Precambrian phenomena, and questions have been brought, for example, why the oldest rock is felsic orthogneiss, how grand continental crust generates and grows, and when the plate tectonics starts to operate? Importance of Precambrian geology, went without saying, is more and more. Without question, Precambrian geology is a key issue to understand continental geology.

Precambrian rocks are extensively distributed in China, not only in cratons, but also in Phanerozoic orogenic belts, for example, relative big Precambrian micro-lithons in Xingmeng fold belt or Tibet. The Chinese continent consists of several cratons and surrounding Phanerozoic orogenic belts (folded belts), and Fig. 1 shows the Chinese regional tectonic map (Ren et al. 1990). The NCB, SCB, and TRB are three uppermost old lands. The SCB consists of two sub-blocks, including the YZB and CTB. It has been suggested that the SCB was formed through collision of the YZB and CTB with the joining of old magmatic arc complexes in the Neoproterozoic (Zhou et al. 2008; Li et al. 1995; Shu 2012). These three old continental blocks have individual tectonic evolving histories and carry important records of crustal generation–growth–stabilization and interaction with circumjacent terrenes. Some residual Precambrian rocks or micro-continental blocks also occur in Phanerozoic orogenic belts, such as in the Himalayan fold belts or Xing'anling fold belt.

In Precambrian period, the three old lands in China contain abundant mineral resources, which were formed in different geological periods and recorded a regular change in the pattern of metallogeny, mineral deposit character, spatial distribution, and genetic mechanisms that match well with the timings and styles of the major tectonic events. The assembly of Chinese unified continent is considered to be amalgamated by the three main continental blocks and several Triassic orogenic belts by Zhai and Santosh (2013), although more terrains and orogenic belts

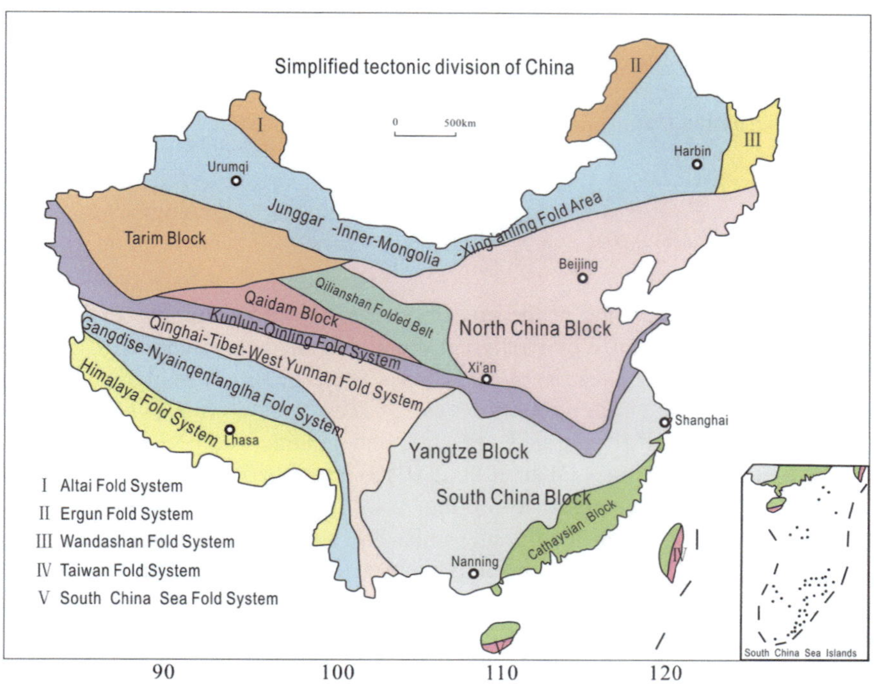

Fig. 1 Chinese map of regional geological units (modified from Ren et al. 1990)

probably joined in this tectonic process. Therefore, study of the three old lands in China will surely help deepen our understanding of China geology, so much as global continental tectonics and continental dynamics.

2 Division of Precambrian Complex and Stratigraphy in China

Early Precambrian rocks in China are commonly underwent strong metamorphism and deformation. Archean rocks consist of mainly felsic orthogneisses, supracrustal rocks, and gabbroic bodies. Supracrustal rocks are always associated with orthogneisses and gabbroic rocks, all of which folded and metamorphosed together. In some places, supracrustal rocks occur as residual slabs with different scale or lenses in orthogneiss "ocean." Therefore, Archean rocks are usually called "complex" that are consistent with international naming convention of Archean rocks. Paleoproterozoic supracrustal rocks in China distribute generally in mobile belts, although which were mostly metamorphosed to high-middle grade and intruded by granites. Meso–Neoproterozoic supracrustal rocks are mostly unmetamorphosed or weakly deformed. Paleo–Neoproterozoic rocks are traditionally named as Phanerozoic stratigraphy naming convention.

Archean and Paleoproterozoic metamorphic rocks extensively distributed in the NCB and locally occur in the SCB and TRB. Meso–Neoproterozoic (1800–700 Ma) thick sedimentary sequences with volcanic rocks are developed in the NCB, but Late Neoproterozoic rocks are less or absent in the NCB. In contrast, the Nanhua and Sinian (Ediacaran) systems are developed in the SCB, also in the TRB.

2.1 General Division

Figure 2 shows the division of the Precambrian rock and stratigraphy in China.

The Proto(Eo)archean rock-stratigraphy unit is called the Baijiafen Complex, which occurs only in Anshan area, NE China. The Baijiafen Complex consists of granodioritic gneiss with granitic veins. The granodioritic gneiss yields 3792 ± 12 Ma SHRIMP zircon U–Pb age, and granitic veins yield ca. 3.3 Ga SHRIMP zircon U–Pb ages (Wan et al. 2001; Wu et al. 2008). The granitic veins have been interpreted as crustal reworking.

The Paleoarchean rock-stratigraphy units are represented by the Chentaigou Complex in NE China and Caozhuang Complex in Yanshan area (eastern Hebei). The Caozhuang Complex can be divided into two parts that are the Caozhuang Group and the Huangboyü banded gray gneiss. The Caozhuang Group is a slab 1.9 km long and 400–500 m wide that is a complicated synformal fold structure.

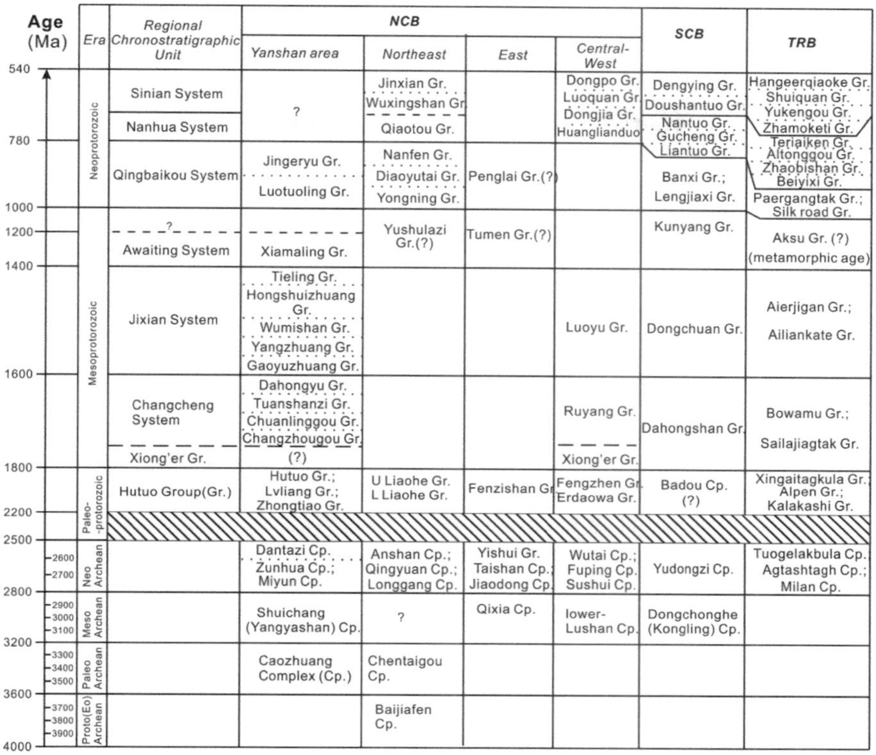

Fig. 2 Units of Precambrian rock and stratigraphy in China

The Huangboyü gray gneiss intruded into the supracrustal rocks of the Caozhuang Group prior to intense multistage deformations. Zircons from the Huangboyü gneiss have U–Pb ages of 3.0–3.3 Ga (Zhao 1993). The Caozhuang supracrustal rocks include various rock types including marble, diopsidite, amphibolite, biotite felsic gneiss, fuchsite quartzite, pyroxene–hornblende, and BIF (banded iron formation). These rocks were metamorphosed to high-amphibolite facies to granulite facies. Detrital zircons from the fuchsite quartzite show two prominent U–Pb age peaks of 3794 ± 15 and 3733 ± 17 Ma, and single-stage Hf modal ages between 3965 and 3633 Ma with a mean age of 3799 Ma (Wu et al. 2005). The fuchsite quartzite represents the oldest sedimentary unit in this suite, with a depositional age of probably between 3.7 and 3.3 Ga. The zircon dating of the amphibolite yielded concordant $^{207}Pb/^{206}Pb$ ages ranging from 3684 to 3354 Ma, and the later is represented metamorphic age (Liu et al. 2013). The Chentaigou Complex is composed of trondhjemitic-quartz diorite gneisses and granitic gneiss with subordinate metamorphosed supracrustal rocks. Orthogneisses and supracrustal were deformed together with complicated fold structure. The Chentaigou supracrustal rocks include quartz biotite schist, amphibolite, fine-grained felsic gneiss, and fuchsite quartzite. The detrital zircons from metasedimentary rocks have U–Pb ages from ca. 3.3 to ca. 3.8 Ga. The orthogneisses have U–Pb age peak of ca. 3.3 Ga.

Mesoarchean rock-stratigraphy unit is traditionally represented by the Shuichang Complex in eastern Hebei, Yanshan area, although it is controversial to its forming ages. Previous studies provided a lot of Sm–Nd, Rb–Sr, and zircon U–Pb ages of ca. 2900–3300 Ma. The recent SHRIMP zircon U–Pb ages are mostly focused on ca. 2.5–2.6 Ga with a few of ca. 2.7–2.9 Ga. However, Zhai (2011, 2014) suggested the Shuichang Complex is Mesoarchean unit based on its multiple phases of deformation, complicated metamorphism, and relationship with Neoarchean complex. The Shuichang Complex includes three parts, which are banded TTG gneisses, supracrustal rocks, and intrusive bodies of gabbro and granite. All of these rocks were metamorphosed to granulite facies. The TTG gneisses have superimposed fold structure with earlier tight fold and middle and later folds. Granitic gneisses are commonly intruded into TTG gneisses as sills and bodies, although they have consonant gneiss foliations. Migmatization is extensive and shows strong characteristics of partial melting and intruded by webbed veins of granite, pegmatite, and alpite in multistages. The supracrustal rocks consist of BIF, mafic granulite, (pyroxene) amphibolite, felsic gneiss, biotite quartz schist, and mica schists. Metagabbros (pyroxene amphibolites) and hypersthene-bearing granites intruded into orthogneisse and supracrustal rocks. The metagabbros and hypersthene-bearing granites have ca. 2.52–2.54 Ga zircon U–Pb ages, representing metamorphism of granulite facies. Another two Mesoarchean units are the Qixia Complex in east Shandong and the Lower Lushan Complex in Henan, southern-central west. They have similar rocks association. The Dongchonghe (Kongling) Complex occurs in the SCB. The Kongling Complex consists of metamorphosed sedimentary rocks and TTG gneisses. The metamorphosed sedimentary rocks mainly comprise banded biotite leptynite with or without graphite. The rocks also include garnet–sillimanite quartzite, garnet–sillimanite–biotite gneiss, graphite schist, marbles, calcium silicate, quartzite, and magnetite quartzite. Recent study reported the oldest forming age (2.95–2.9 Ga) of the Kongling TTG using SHRIMP U–Pb zircon dating (Qiu et al. 2000; Gao et al. 2001) and also obtained detrital zircon ages of 3.28–2.87 Ga from nearby Archean metapelites with metamorphic age of 2.01–1.95 Ga (Liu et al. 2008).

Neoarchean rocks are developed in the NCB, also in the SCB and TRB (Fig. 2). Two kinds of the Neoarchean units in the NCB are of greenstone-granite belt (GGB) type and high-grade gneiss region (HGR) type, respectively. A typical GGB unit is the Wutaishan Complex in central-western NCB. The BIF-bearing volcanic-sedimentary sequences here are developed on a granitic gneiss basement with magmatic zircon U–Pb ages of 2530–2565 Ma. The supracrustal rocks can be subdivided into three formations as follows: the Jingangku Formation (JGK) of amphibolite facies, the Wutai Formation (WT) of greenschist facies, and the sub-greenschist facies turbidites (Wang 2009). The age of the WT is constrained by SHRIMP zircon dating of the intermediate and felsic arc volcanic rocks comprising meta-andesite, metadacite, metarhyodacite and metarhyolite at ca. 2530–2515 Ma

(Wilde et al. 2004a). No reliable magmatic zircon ages are currently available for either the N–JGK or S–JGK. One detrital zircon in garnet quartzite is dated at 2510 ± 36 Ma (Wang et al. 2009). SHRIMP zircon dating of the metarhyolite dike, which intruded into the subgreenschist facies turbidites, yielded an age of 2528 ± 6 Ma (Wilde et al. 2004a). BIFs are interlayered with amphibolites and greenschists which represent metamorphosed basaltic rocks. The Zunhua Complex in eastern Hebei is also a typical Neoarchean GGB-type unit. Its main characteristic is that all rocks underwent amphibolite–granulite facies metamorphism and partly migmatization. The Taishan Complex is a HGR-type unit, which is dominately composed of orthogneisses with a minor amount supracrustal rocks as lenses. The high-Na and high-K granite gneisses occur in zoning in Taishan-Mengyin area, western Shandong (Wan et al. 2012). The Late Neoarchean granitoids in Taishan-Mengyin area have SHRIMP zircon U–Pb ages of 2495–2550 Ma. They can be further subdivided into quartz diorite–granodiorite and monzogranite–potassium granites. They occur as two belts: a Late Neoarchean (2525–2490 Ma) crust-derived granite belt (monzogranite and potassium granite) in the northeast and a Late Neoarchean (2550–2500 Ma) ocean-derived juvenile crustal rock belt in the southwest. An Early Neoarchean (2.75–2.60 Ga) TTG belt is situated as a central zone between the above-mentioned belts. Several gabbroic bodies with ca. 2.5 Ga also occur in the area. The Tuogelakbulak Complex in the TRB is composed of black-gray amphibolite, amphibole schist, gray felsic gneiss, garnet biotite schist, actinolite mica quartz schist, biotite quartz schist, migmatite, and some discontinuous marble lens interlayer. Schists and marble lenses are metamorphosed with gneiss foliations along the strike. The Yudongzi Complex in southern Qinling Mountains is believed to represent Archean basement of the SCB. The main rocks are the Yudingzi supracrustal rocks and granitic gneisses. The forming ages of granitic gneisses are constrained by LA–ICP–MS zircon dating at ca. 2660–2703 Ma (Zhang et al. 2010). The Yudingzi supracrustal rocks include greenschist, mica schist, fine-grained felsic gneiss, and BIF. Detrital zircons from BIF yield ca. 2645 Ma age and ca. 2527 age, and the later age has been suggested as metamorphic age (Wang et al. 2011).

The typical Paleoproterozoic stratigraphical unit (ca. 2250–1900 Ma) is the Hutuo Group and Fengzhen Group. These two groups have different rock associations and metamorphosed to different grades. The Hutuo Group is distributed in Wutaishan area in the central NCB and is composed of basic–acid volcanics and sedimentary rocks metamorphosed to low-grade amphibolite–greenschist facies. The volcanic rocks display bimodal features and geochemical characteristics suggestive of within-plate or island-arc-type affinity. The similar stratigraphical units to the Hutuo Group are the Lüliang Group and Zhongtiao Group in the central NCB, the Erdaowa Group in the western NCB, the Erdaowa Group in the northwetsern NCB, and Lower Liaohe Group in northeast NCB. The Fengzhen Group includes argillaceous schists, siltstones, and marbles, with the common presence of graphite, and has been metamorphosed to granulite facies under temperatures of 800–850 °C (HT). The similar stratigraphical units to the Fengzhen Group are the Upper Liaohe

Group in the northeastern NCB, the Fengzishan Group in the eastern NCB, upper layers of the Lüliang Group in the central NCB, and upper layers of the Upper Lushan Group in the southern NCB. The detrital–argillite–carbonate sedimentary rocks of the Upper Liaohe Group contain a large volume of magnesite deposit. A minor amount of metamorphosed sedimentary rocks have been reported in the TRB and SCB, although they locally occur and outcrop discontinuously (Fig. 2).

Thick Late Paleoproterozoic–Neoproterozoic sedimentary sequences were developed on the basement of the NCB under a stable tectonic setting and experienced minor deformation and metamorphism. The main sedimentary basins include the Xiong'er aulacogen system in the southern-central NCC, the Yan-Liao aulacogen system in the northern-central NCC, the northern marginal rift system in the northwestern NCC, and the eastern marginal rift system in the eastern NCC and the Korean Peninsula. All the rift systems occur separately in present, although they might have been linked each other during their evolution (Zhao 1993). The Yanshan is a typical area, where the Late Paleo- and Mesoproterozoic sedimentary sequences are assigned as the type section of the Stratigraphic Lexicon of China. The lower boundary age of the Changcheng System is traditionally approved to be 1.8 Ga. The boundary ages between the Jixian, Await Name, and the Qingbaikou systems are 1.6, 1.4, and 1.0 Ga, respectively (Cheng et al. 2009; Gao et al. 2009). Volcanic rocks of the Xiong'er Group in the southern NCB have magmatic zircon U–Pb ages of ca. 1.78 Ga, possibly representing the oldest horizon of Mesoproterozoic sequences. The upper boundary age of the Qingbaikou System remains unknown. Neoproterozoic sequences are mostly missing in the Yanshan area, only possibly preserved in NE China and the southern margin or eastern margin of Korean Peninsula. Neoproterozoic strata include the Qingbaikou (1000–780 Ma, $Pt_{3\text{-}1}$), the Nanhua (780–635 Ma, $Pt_{3\text{-}2}$), and the Sinian systems (635–542 Ma, $Pt_{3\text{-}3}$) (cf. Fig. 2). On the contrary, Neoproterozoic sequences are well distributed in the SCB, and Mesoproterozoic rocks were uncommonly reported. The Nanhua and Sinian systems have been suggested formed during ca. 780–540 Ma in Fig. 2, although a great debates remains. The Nanhua System mainly comprises the Liantuo Formation, Nantuo Formation, and equivalent strata, as well as interlayered strata of the Datangpo and Gucheng formations between the Liantuo and Nantuo sedimentary rocks in some regions. There exist two epochs of glacial sedimentary event in the Nanhua System, and they are the lower Chang'an Glaciation (including Gucheng and Jiangkou Glaciation) and the upper Nantuo Glaciation with the Datangpo Interglaciation (or equivalent manganese-bearing strata) between the two glacial epochs. The Sinian System disconformably overlies the Nanhua System. The Sinian standard section in the Three Gorges area is overlain by the Cambrian strata. The top and bottom boundary layers of the Sinian System are, respectively, the Lower Cambrian Shuituojing Formation and the Neoproterozoic Doushantuo Formation. Meso–Neoproterozoic rocks of TRB occur mainly in the northern and eastern margin of the Tarim Basin.

2.2 Distribution of the Precambrian Rocks in China

The three old continental blocks have individual tectonic evolving histories and carry important records of crustal generation–growth–stabilization and interaction with circumjacent terrenes. Figures 3, 4, and 5 show distributions of the Precambrian rocks formed in different geological periods in China.

The Archean rocks are mainly distributed in the NCB and Korean Peninsula. They constitute crystalline basement of the NCB or termed as the Sino-Korea Craton. The Archean rocks are sporadically located in the SCB. Recently, Archean detrital zircons yield ages from ca. 2.5 to ca. 3.8 Ga in the Yangzte and Cathaysia blocks, indicating Archean rocks are possibly covered by Neoproterozoic and Phanerozoic rocks. Some Archean rocks have been discovered in the southern and northern margins of the TRB, especially in eastern Hami to Beishan in Qinghai. A minor amount of Archean rocks occur in the northern margin of the Ili Block.

The Paleoproterozoic rocks occur in the NCB as mobile belts, i.e., the Jinyu, Fengzhen, and Jiaoliaoji mobile belts (refer to Fig. 8). The Paleoproterozoic rocks in the NCB were metamorphosed to amphibolite to granulite facies. The Paleoproterozoic mobile belt event took place in the NCB, termed the Hutuo Movement. This event has been interpreted to represent cratonic reworking characterized by rifting–subduction–collision processes. The NCB then evolved into a stable platform or para-platform tectonic setting. The Paleoproterozoic rocks mainly occur along the western or northwestern margins of the SCB, and along the

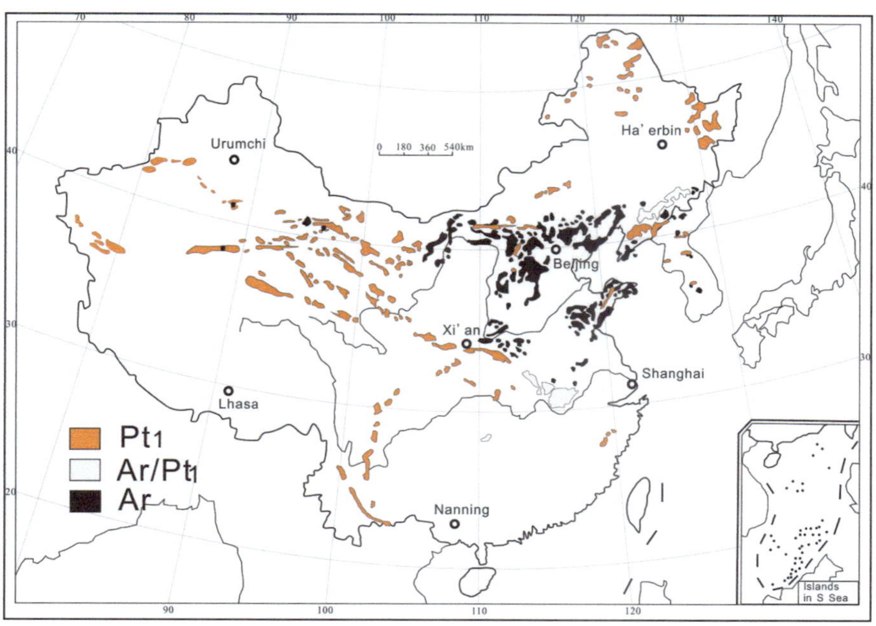

Fig. 3 Distribution of the Early Precambrian rocks in China

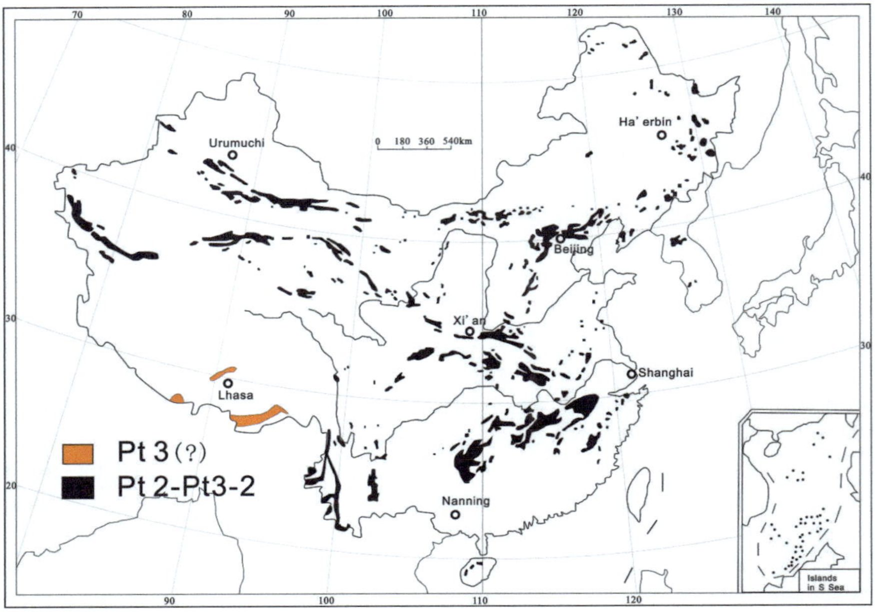

Fig. 4 Distribution of the Meso–Neoproterozoic (Nanhua) rocks in China

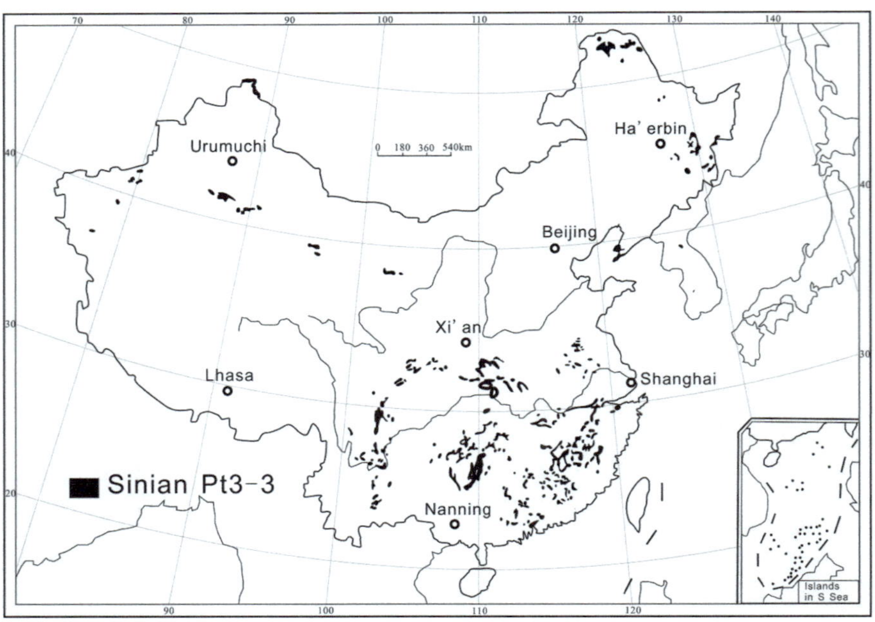

Fig. 5 Distribution of Sinian rocks in China

northern margin of the TRB and west-northern margin of the Ili Block. Many outcrops of the Paleoproterozoic rocks have been marked in Fig. 3. However, their occurrences, rock associations, grades of metamorphism and deformation, and isotopic ages are still not clear.

Meso–Neoproterozoic Nanhua System rocks (Pt_2–$Pt_{3\text{-}1}$) are extensively distributed in China (Fig. 4). Typical Mesoproterozoic rocks occur in the NCB. The Qingbaikou System rocks ($Pt_{3\text{-}1}$) are distributed in the NE and southern NCB, as well in the SCB. The Nanhua and Sinian sequences ($Pt_{3\text{-}2}$ and $Pt_{3\text{-}3}$) are well developed in the SCB (Figs. 4 and 5). Recent studies demonstrate that the $Pt_{3\text{-}2}$ and $Pt_{3\text{-}3}$ rocks occur in the northern margin of the TRB, and three or four tillite layers in the TRB seemly can be corresponding to the tillites in the Yangtze Block (Xu et al. 2003).

3 The North China Block

The North China Block consists of the Precambrian metamorphic basement, Late Paleoproterozoic–Paleozoic sedimentary cover, and Mesozoic intrusions. It is traditionally termed the North China Craton or Sino-Korea Craton. The NCB covers over 300,000 km^2, is one of the oldest cratons in the world with the old crust up to ca. 3.8 Ga, and has a complicated evolutionary history that experienced multistage tectonic evolving evolution (Zhao 1993; Goodwin 1991; Windley 1995; Liu et al. 2002; Rogers and Santosh 2003; Zhao et al. 2005; Kusky et al. 2007; Zhai 2011) and recorded imprints of almost all major Precambrian geological events (Zhai and Santosh 2011). The NCB experienced extensive lithosphere thinning and craton destruction, particularly in the eastern and central domains during the Early Cretaceous (Fan and Hooper 1991; Fan and Menzies 1992; Menzies et al. 1993; Zhai et al. 2002; Zhai 2008). The two scientific issues have attracted attentions of the geologists all over the world, i.e., (1) discovery of the Paleoproterozoic high-pressure granulites and high-ultra-high-temperature granulites and (2) the crustal growth–cratonization and tectonic regime at boundary time bound between the Archean and the Proterozoic. Figure 6 is a diagram showing the Precambrian crust growth and important geological events in the NCB and their relationship with global events. From this diagram, the NCB shows its similarity in early tectonic evolution to other cratons in the world, as well as its dissimilarity in some places. The major geological events and processes in the NCB can be grouped into the following distinct tectonic episodes: the Neoarchean crustal growth and construction of the major tectonic framework of the craton; the Paleoproterozoic rift–subduction–collision and operation of primitive plate tectonics; the GOE event marking a sudden change in Earth's environment from oxygen-poor to oxygen-rich in the end of Neoarchean to Paleoproterozoic; and the Late Paleoproterozoic–Neoproterozoic extensional tectonics and multistage rifting. These major geological events are accompanied by the major metallogenic events and the formation of potential ore deposits. For example, a huge amount of Algoma-type banded iron

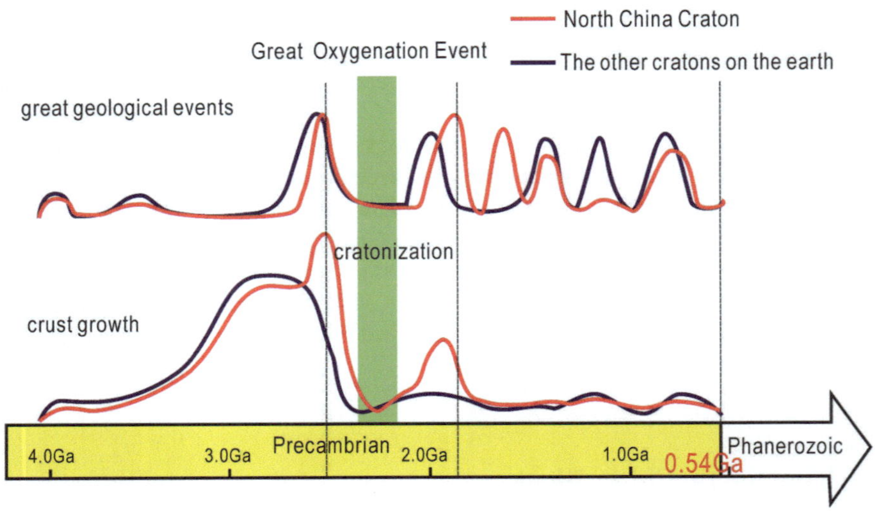

Fig. 6 Diagram showing Precambrian crust growth and important geological events in the NCB and their relationship with global events (after Zhai and Santosh 2013)

formation (BIF) is awarded industrial ore in China, REE deposit in Bayan Obo and magnesite deposit are the largest in the world, and Boron and Pb–Zn deposits are also preponderant. On the other hand, Superior-type BIF, greenstone-type gold deposit, and conglomerate-type U–Au deposit that are developed in the some cratons in the world are poor in the NCB. This "suddenly rich and suddenly poor" metallogenic characteristic is controlled by early tectonic processes in the NCB (Zhai 2010; Zhai and Santosh 2013).

3.1 Early Crustal Growth and Cratonization (Add Multistage)

The NCB contains several old continental nuclei. The Tiejiashan nucleus in NE China and Caozhuang nucleus in eastern Hebei are two of the oldest (Liu et al. 1992, 2007; Wan et al. 2001; Wu et al. 2008). The former is composed of granodioritic–granitic gneisses with a few of supracrustal layers and lenses. Magmatic zircons from granodioritic gneiss yielded ca. 3.8 Ga U–Pb age. The latter is a set of supracrustal rocks and invaded by ca. 3.3 Ga gray gneiss. The main supracrustal rocks include amphibolite, serpentinized marble, fuchsitic quartzite, metamorphosed calc-silicate rock, banded iron formation, biotite gneiss, and sillimanite plagioclase gneiss. These rocks were metamorphosed to high-amphibolite facies to granulite facies. Zircons from the fuchsitic quartzite yielded four U–Pb age populations of 3.83–3.82, 3.8–3.78, 3.72–3.7, and 3.68–3.6 Ga. Recently, 3.7–3.8 Ga detrital zircons are in the succession reported from the Proterozoic–Paleozoic sedimentary rocks in the central, southern, and western parts of the NCB. A ca. 4.1 Ga detrital zircon U–Pb age is reported from the Paleozoic volcanic rock in southern margin of the NCB (Diwu et al. 2010).

The timings and mechanisms of the Earth's continental crust growth have been addressed in several studies. Some of the earlier popular models proposed that 80–90 % of the global felsic continental crust occurred during the Early Precambrian and mostly in the Neoarchean (2.7–2.5 Ga). About 80 % of crust formation in North China also occurred during the Early Precambrian and mostly in the Neoarchean (2.7–2.5 Ga). The metamorphosed mafic rocks and trondhjemitic-tonalitic-granidioritic gneisses (TTGs) with >3.0 Ga Sm–Nd T_{DM} ages from the NCB account for about 15 % of the total volume, whereas rocks with ages <2.5 Ga account for only about 7 %. Rocks with T_{DM} ages varying from 3.0 to 2.5 Ga account for 78 %. The two prominent age peaks at 2.9–2.7 Ga, and 2.6–2.5 Ga might correspond to the earlier events of major crustal accretion in the NCB (Zhai 2004). The Hf isotopic model ages mostly range from 2600 to 3000 Ma with a peak of ca. 2820 Ma. The old continental crust of the NCB is dominated by TTGs, with some mafic komatiitic volcanic rocks. All of these show that the NCB has similar characteristics to most cratons in the world on early crustal generation and growth. The 2.9–2.7 Ga crustal growth event is suggested to be related to a superplume event (Condie and Kröner 2008). The Yanlingguan greenstone belt in western Shandong contains komatiites with spinifex textures. The metabasaltic rocks interlayered with komatiites yield ca. 2.7 Ga zircon U–Pb ages (Polat et al. 2006). Some ultramafic–mafic rocks, yielding ca. 2.7 Ga zircon Hf model ages and having similar geochemical compositions to komatiitic rocks but without spinifex textures in the NCB, are believed to be komatiites and metamorphosed to high-amphibolite–granulite facies. Therefore, ca. 2.7 Ga greenstone belts could be extensively distributed in the NCB (Zhai 2011). The diagram of ε_{Hf}-age ($^{207}Pb/^{206}Pb$) for zircons from the Early Precambrian rocks in the NCB shows that the most important crust growth occurred during 2.9–2.7 Ga (Fig. 7), in accord with global crust growth (Zhai and Bian 2001; Zhai and Liu 2003). The metamorphic zircon grains from the widespread mafic rocks with Nd and Hf model ages ranging

Fig. 7 Diagram of ε_{Hf} versus $^{207}Pb/^{206}Pb$ ages (after Zhai and Santosh 2011)

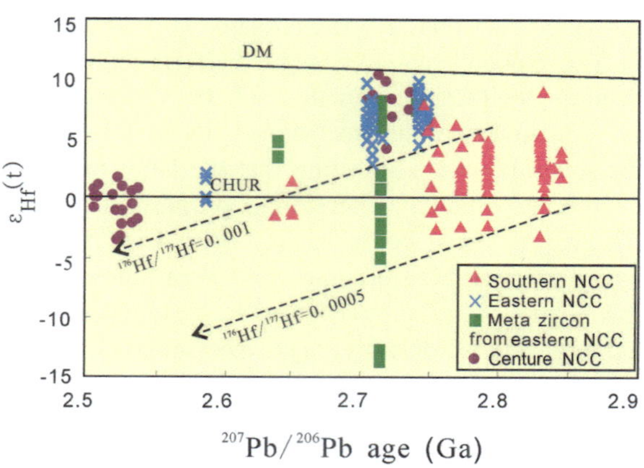

from 2.9 to 2.7 Ga yield the age of 2.6–2.5 Ga. Furthermore, the 2.6–2.5 Ga TTGs have zircon Hf model ages varying from 2.9 to 2.7 Ga, and they are mostly attributed to partial melting of 2.9–2.7 Ga mafic rocks according to previous experimental petrology data. Therefore, the NCB has a distinct characteristic of multistage crust growth, i.e., 2.9–2.7 Ga major growth event and 2.6–2.5 Ga crust reworking–growth event which is stronger than other cratons in the world. The crustal growth model in the NCB is generally proposed to be that crust is accreted surrounding several ancient cratonic nuclei and finally formed ten to five micro-blocks (Wu et al. 1998; Deng et al. 1999). Until now, at least seven micro-blocks have been identified including the Jiaoliao Block (JL), the Qianhuai Block (QH), the Ordos Block (OR), the Jining Block (JN), the Xuchang Block (XCH), the Xuhuai Block (XH), and the Alashan Block (ALS) (Zhai 2011).

The ca. 2.5 Ga is the time boundary between the Archean and Proterozoic. However, although the ca. 2.5 Ga geological records are weak in some cratons, the record of Earth's environment change is abrupt and enormous in all cratons in the world, which was followed by an "silent period" without tectonic-thermal action lasting 150–200 Ma (from 2.5 to 2.3 or 2.35 Ga), and then followed by the GOE (Condie and Kröner 2008). It is worth pointing out that the ca. 2.5 Ga geological event in the NCB is strong with abundant geological phenomena, mainly as follows: (1) 2.5–2.6 Ga TTG gneisses are extensively distributed in the NCB and usually contain layers or lenses of amphibolites and mafic granulites; (2) ca. 2.5 Ga greenstone belts are located in the northeastern, northwestern, eastern, central, and southern NCB, such as the Hongtoushan, Anshan, Xuchang, Zunhua, and Wutai Mountains. The ca. 2.7 Ga greenstone belts in the NCB, such as the Yanlingguan, have been metamorphosed and deformed at ca. 2.5 Ga. All greenstone belts surround old micro-blocks (high-grade regions) and constitute a tectonic pattern marked by the coexisting of greenstone belt and high-grade region; (3) a large number of granites derived from crustal partial melting at 2.52–2.49 Ga and intruded into both greenstone belts and high-grade regions; and (4) all the Archean rocks in the NCB underwent a strong ca. 2.5 Ga metamorphism from granulite facies to amphibolite facies (Zhai 2004, 2010). To summarize the above mentioned, I have suggested that the NCB formed through amalgamation of micro-blocks at ca. 2.5 Ga (Zhai 2004, 2011; Zhai and Bian 2001).

The Neoarchean greenstone belts probably represent arc–continent collision resulting in the amalgamation of the various micro-blocks. However, the metamorphic grades of the greenstone belts are lower than those of the complexes within the micro-blocks, suggesting that the latter might have developed under a higher geothermal gradient. Therefore, I propose that the various micro-blocks were surrounded by small ocean basins in the Late Neoarchean, whereas both old continental crust and the oceanic crust were hotter. The subduction and collision were much smaller in scale compared to those in the Phanerozoic plate tectonic regime, although the tectonic style and mechanisms are more or less similar (Zhai 2010). The formation of a large volume of granites by crustal melting is an important process of cratonization, generating the stable upper and lower crust layers. This process would lead to an upper crust of more granitic composition and a lower crust with molten residual materials together with underplated material of gabbroic composition. The underlying mantle drives the cratonization process through thermal and material input. A part of asthenosphere mantle was transformed into the lithosphere mantle through magma extraction, resulting in crust–mantle coupling (Zhai 2004, 2011).

BIFs occur extensively throughout the Archean units in the NCB, and four major iron metallogenic provinces have been identified: Anshan in NE China; E Hebei and Wutai in northern NCB, and Xuchang–Huoqiu in southern NCB. The BIFs have been metamorphosed to granulite facies and display strong deformational features. The BIFs occur not only in the greenstone belts, but also in high-grade regions within the all micro-blocks, mostly in association with mafic metavolcanics suggesting the imbrication of oceanic crust and pelagic sediments onto the continent during early accretionary growth of the NCB (Zhai and Windley 1990). The Archean high-grade domains in the NCB consist of upper amphibolite to granulite facies TTG gneisses, amphibolites, metagabbros, and minor supracrustal rocks. Lenses and boudins of the BIFs of different sizes occur as enclaves within the orthogneisses with or without associated metasupracrustal rocks. The host orthogneisses and associated supracrustal rocks yield ca. 3.3–2.55 Ga zircon U–Pb ages, suggesting that BIFs formed during Meso- to Neoarchean.

3.2 *Paleoproterozoic Geology in the NCB*

A Paleoproterozoic rift tectonic event took place closely following 2500–2300 Ma "silent period" without tectonic-thermal action in the NCB, corresponding to the global Paleoproterozoic breakup event of the Archean supercraton suggested by Condie and Kröner (2008). As a result of rifting event, three main mobile belts formed (Fig. 8), which are the Jiaoliao Mobile Belt in the eastern NCB, the Jinyu Mobile Belt in the central-western NCB, and the Fengzhen Mobile Belt in the northwestern NCB. The main volcanic-sedimentary sequences are called the Liaohe Group and the Fengzishan Group in the Liaoji Belt, the Lüliang Group, the Hutuo Group, and Zhongtiao Group in the Jinyu Belt, and the Fengzhen Group and Erdaowa Group in the Fengzhen Belt (Sun and Hu 1993; Li et al. 1998; Miao et al. 1999; Yu et al. 1999; Wan et al. 2000; Geng et al. 2003). The abundant carbonates and evaporates in the Liaohe Group in northeastern (NE) China include thick layers of marble–magnesites with boron ore-bearing fine-grained felsic gneiss. A unique, worldwide 2.33–2.06 Ga positive δ^{13} Ccarb excursion has been correlated with the GOE. Studied samples of dolomite–marble with 0.91 ± 0.03 CaO/MgO (mol) ratios show higher δ^{13}CPDB values of 0.6–1.4 ‰ (av. 1.2 ± 0.3 ‰) than those of normal marine carbonates around the globe. However, they display lower δ^{18}O SMOW values of 16.4–19.5 ‰ (av. 18.2 ± 1.1 ‰) compared to their contemporaneous counterparts, suggesting that the primary carbonates in the Liaohe Group possess a positive δ^{13}C anomaly (possibly 4.2 ‰), reflecting the impact of the GOE. The δ^{13}C and δ^{18}O have been depleted in post-sedimentation diagenesis and/or regional

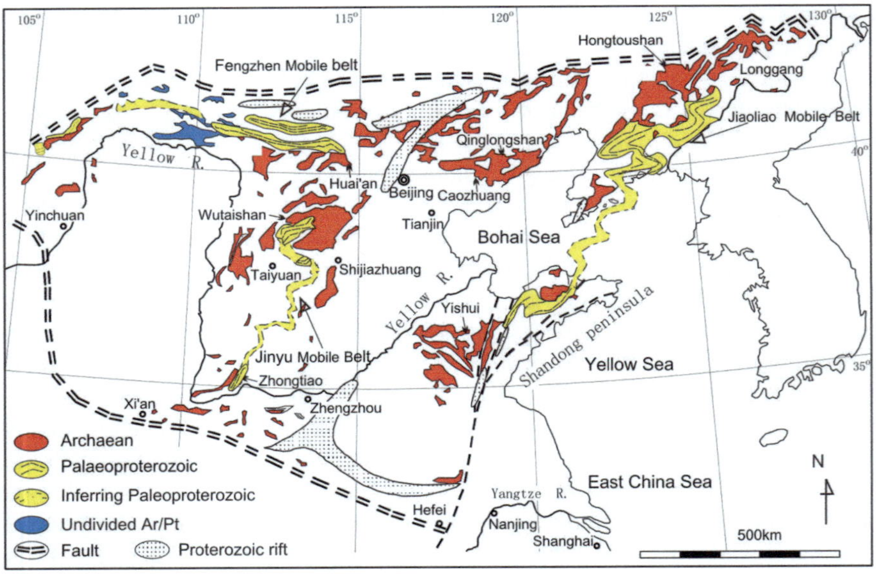

Fig. 8 Sketch map showing the Archean rocks, Paleoproterozoic mobile belts, and Proterozoic rifts

metamorphism. The >550 m-thick magnesite layer in the studied section has CaO/MgO ratios of 0.01–0.23 (av. 0.12 ± 0.1, $n = 6$). These rocks show $\delta^{13}C$ and $\delta^{18}O$ values of 0.1–0.6 ‰ and 9.2–12.7 ‰, with average values of 0.4 ± 0.2 ‰ and 10.9 ± 1.4 ‰, respectively (Tang et al. 2009). A large amount of graphite deposits occur in the Fenzishan Group in the Jiaoliao Belt, the Jining Group in the Fengzhen Belt, and the Upper Lushan Group in the Jinyu Belt. Their protoliths are carbonaceous shales. All B-bearing rocks, Pb–Zn-bearing rocks, magnesites, and graphite rocks have specific geochemical and isotope characteristics and thus constitute a Paleoproterozoic GOE metallogenic system.

Paleoproterozoic BIFs, widely developed in various parts of the world, are important sources of iron ores, commonly termed as Superior-type BIF. The Paleoproterozoic BIFs are regarded as a mark of GOE. However, there are only minor volumes of Paleoproterozoic BIF deposits in the NCB, and possible examples include those in the Lanxian Group (Yuanjiacun iron deposit) in the Jinyü Belt and in the Upper Huoqiu Group in Anhui Province, southern NCB. Zhai (2010) proposed that the depth of the sea water in 2.35–2.0 Ga in rift basins of the NCB possibly was shallower than that in those of the Canadian Shied and the west Australian Shield. The possible sedimentary setting in the ca. 2.35–2.0 Ga NCC is shallow sea-lagoon with depth of <200 m, in which huge thickness of magnesites and dolomites was precipitated instead of BIFs.

The rock sequences in these three mobile belts are typically divided into two formations. The lower formation is composed of basic–acid volcanic and sedimentary rocks metamorphosed to low-grade amphibolite–greenschist facies, where they form linear fold belts and have been affected by intense deformation and metamorphism. The volcanic rocks display bimodal features and geochemical characteristics, suggestive of within-plate or island-arc-type affinity. The upper formation is composed of layered detrital–argillite–carbonate sedimentary rocks. Zhai and Santosh (2011, 2013) and Zhai (2008, 2010) have proposed that the Paleoproterozoic rifts were widened into ocean basins, following which subduction was initiated. Subduction–accretion–collision processes built three major collisional sutures (mobile belts) during the Late Paleoproterozoic. These linear fold belts show imprints of intense deformation, and their high-grade metamorphic rocks record high-pressure and high-temperature P–T conditions. The sutures are distinct from the Archean granite-greenstone belts and display similar characteristics to Phanerozoic orogenic belts, including the preservation of volcano-sedimentary strata, deformation style, and degree of metamorphism. Therefore, the primitive plate tectonics could begin to operate in Paleoproterozoic with a similar principle, although much smaller in scale compared to modern-style plate tectonics (Zhai 2010, 2011). The Paleoproterozoic orogenic-like metallogenic systems mainly include Cu, Cu–Mo, Cu–Co, and Cu–Pb–Zn deposits. These ore deposits formed in rift and subduction environments and have characteristics of orogenic metallogeny.

Some high-pressure mafic granulites (HPG) and retrograded eclogites occur as lenses in the Archean basement orthogneisses of NCB. Their protoliths are considered as mafic dikes. The typical mineral assemblages indicate that they are metamorphosed to HPG facies and some to eclogite facies (Zhai et al. 1992, 1995). The garnets are commonly surrounded by a symplectite of fine-grained minerals, indicating a decompressional process. Some ultra-temperature metamorphic (UHT) minerals have been discovered in pelitic metamorphic rocks, such as sapphire (Guo et al. 2006; Santosh et al. 2007). According to HP and UHT granulites, some researchers have proposed the Early Precambrian plate tectonics operated in the NCB (Zhao et al. 1999, 2005; Kusky and Li 2003). Zhai et al. (1992, 1995) firstly suggested that a high-pressure granulite zone is located along the northern Hebei–Shanxi via middle Inner Mongolia to Southern Liaoning, representing a collisional belt between the Fuping continental massif and Huai'an continental massif. Zhao et al. (2005) suggested subduction–collision models for the Trans-North China Orogen, whereas Kusky and Li (2003) suggested collision along the northern Hebei–Inner Mongolia Orogenic Belt, and Santosh et al. (2007) suggested a double-sided subduction model. However, detailed geological investigation shows that these high-grade granulites are extensively distributed in area rather than into zones, and their metamorphic temperature–pressure gradients are of moderate pressure metamorphic facies system. Therefore, their tectonic significance requires further study and discussion (Zhai 2004, 2011).

Recently, Zhai (2011) proposed a Paleoproterozoic rift–subduction–collision orogenic-like event and emphasized that subduction depth in the Paleoproterozoic mobile belts of the NCB was limited. The Paleoproterozoic orogenic-like event was finished before ca. 1800 Ma. After that, an extensional tectonic event with mantle upwelling closely followed the collision event. Between 1.78 and 1.65 Ga, rift-related anorogenic magmatism and mafic dike swarm emplacements took place

(Peng et al. 2008), possibly associated with the global disruption of the Columbia supercontinent, subsequent to which the NCB entered into a platform regime (Zhai and Liu 2003; Peng et al. 2010a).

3.3 Late Paleoproterozoic–Neoproterozoic Multiple Rifts (Add Discussion)

During the latest Paleoproterozoic–Neoproterozoic, the NCB was a stable platform or para-platform on which vast, thick sedimentary sequences were deposited. The major sedimentary basins are the Xiong'er "aulacogen" in the southern-central NCB, the Yanshan "aulacogen" in the northern-central NCB, the northern marginal rifts in the northwestern NCB, and the eastern margin of Korean rifts in the eastern NCB. The oldest rocks are 1800–1780 Ma (Zhao et al. 2001, 2004b) bimodal volcanic rocks in the bottom of the Xiong'er Group, and upward sequences are successively the Ruyang Group and Luoyu Group in the Xiong'er "aulacogen" (Fig. 9). The main strata in the Yanshan "aulacogen" are the Changcheng System, Jixian System, and Qingbaikou System from bottom to top. The Changcheng System can be subdivided into two subsystems of the Changcheng subsystem (the Chuanlinggou Group, Chuanlinggou Group, and Tuanshanzi Group) and Nankou subsystem (the Dahongyu Group and Gaoyuzhuang Group). Volcanic rocks occur mainly in the Dahongyu Group and also in the Tuanshanzi Group. The magmatic zircons from the Dahongyu volcanic rocks yield U–Pb ages of 1680–1620 Ma, and zircons from anorogenic magmatic intrusive bodies related to volcanic rocks (rapakivi granites, gabbro–anorthosites, and porphyritic granites) yield U–Pb ages of 1700–1670 Ma (Zhai and Liu 2003; Zhao et al. 2004a). In recent years, mafic sills were distinguished from the Xiamaling Group that was believed to belong to the Qingbaikou System in previous study, and their U–Pb ages of zircon and baddeleyite are 1370–1320 Ma (Gao et al. 2007; Li et al. 2009; Zhang et al. 2009). Therefore, a new stratigraphical division has been proposed by the Chinese Commission on Stratigraphy (CCOS), i.e., establishing an Unnamed System to put under the Qingbaikou System (Gao et al. 2011). Therefore, the new Proterozoic strata division in the Yanshan "aulacogen" is, from bottom to top, the Changcheng System (Ch, 1800–1600 Ma, $Pt_{2\text{-}1}$); the Jixian System (Jx, 1600–1400 Ma, $Pt_{2\text{-}2}$); and the Unnamed System (Xm, 1400–1200 Ma, $Pt_{2\text{-}3}$). Some of Neoproterozoic strata (1200–1000 Ma, $Pt_{2\text{-}4}$) are absent in the Yanshan "aulacogen," which occur in the eastern and southern part of the NCB. The Neoproterozoic strata include the Qingbaikou System (Qb, 1000–780 Ma, $Pt_{3\text{-}1}$), the Nanhua System (Nh, 780–635 Ma, $Pt_{3\text{-}2}$), and the Sinian System (Z, 635–542 Ma, $Pt_{3\text{-}3}$). The boundary between the Unnamed System and the Qingbaikou System is still marked by the Qinyu Uplift unconformity interface. Recent studies (Peng et al. 2011a, 2011b) also distinguish ca. 900 Ma mafic dikes in the central and eastern NCB and the northern Korea. Moreover, ca. 820 Ma volcanic rocks are discovered in the Zhaertai Group

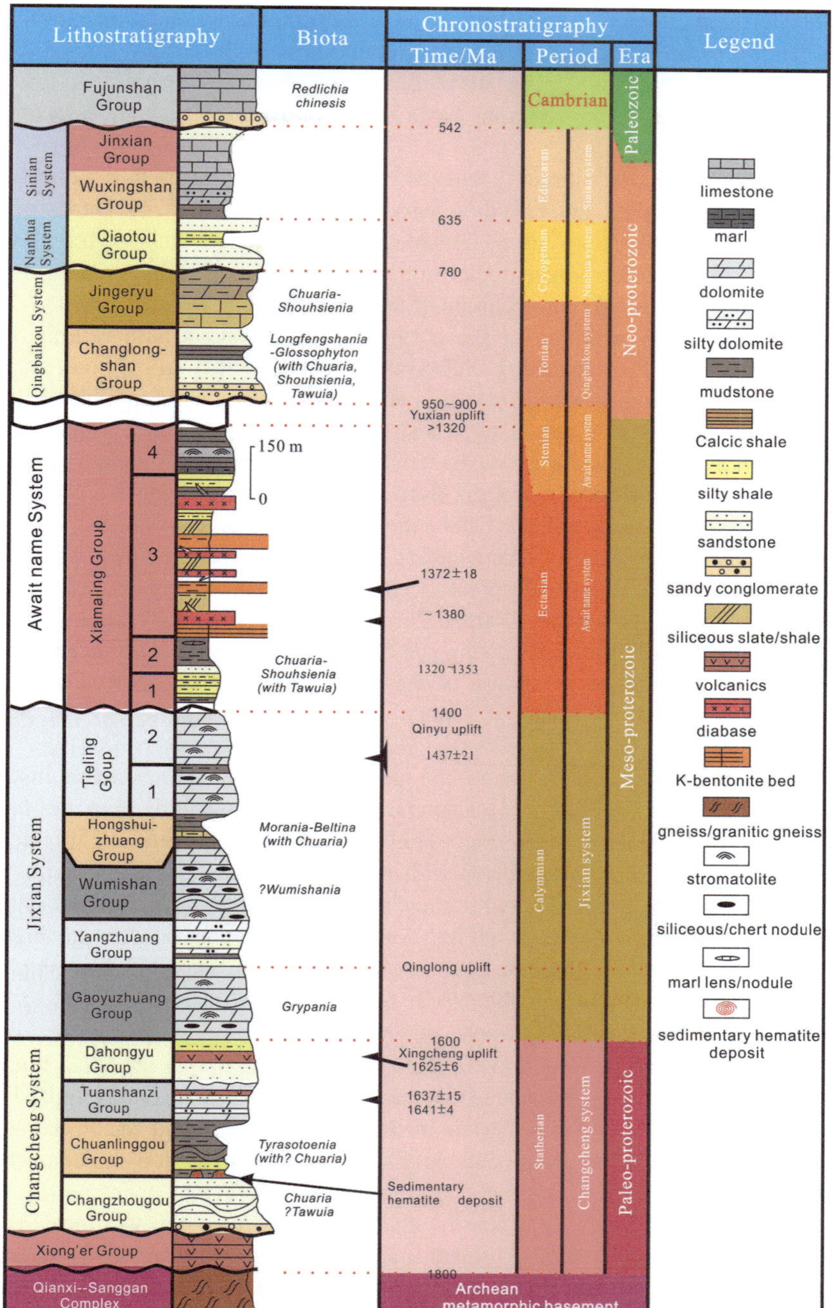

Fig. 9 Late Paleoproterozoic–Neoproterozoic stratigraphic chart in the NCC

in the northern marginal rifts (Peng et al. 2010), and detrital zircons from the sedimentary rocks in the eastern margin rifts obtained U–Pb ages of 800–900 Ma and ca. 1400 Ma (Hu et al. 2012). The study of Proterozoic strata confirmed that the northern marginal rifts have similar basin evolution to the Yanshan "aulacogen" (Meng et al. 2011).

The Proterozoic depositional sequences in the northern margin rift basin are uninterrupted, providing a good sedimentary record for basin analysis. Depositional sequences of the two groups are both characterized by gradual changes from lower siliciclastic to upper carbonate rocks, but the Changcheng Subsystem is much thicker than the overlying Nankou Subsystem (Meng et al. 2011). The tectono-sedimentary evolution of the northern North China Craton exhibits two distinct stages. The Changcheng Subsystem was deposited in rift basins that experienced strong subsidence in association with volcanism, and basin fills include fluvial, deep marine, and carbonate facies. Widely distributed shallow marine siliciclastics and carbonates are the characteristics of the Nankou Subsystem, thereby indicating a slow subsidence process over a broad area. A regional diachronous unconformity exists between the Changcheng and Nankou subsystems, with the Dahongyu arenite and Gaoyuzhuang carbonate deposited over the units of different ages. Slow subsidence and widespread occurrence of shallow marine facies imply that the Nankou Subsystem was formed during a post-rift period. The transgressive unconformity between the Changcheng and Nankou subsystems might be generated as a result of a complete separation of the North China Block from the adjacent continent. In other word, it can be defined as a "breakup unconformity" in origin (Fig. 10).

This paper emphatically points out the fact that the NCB experienced multistage rifting processes and landed in extensional setting for long term from the latest Paleoproterozoic to Neoproterozoic. This has been interpreted as a cratonic reworking event (Hutuo Movement) with rifting–subduction–collision processes, after which the NCC evolved into a stable platform or para-platform tectonic setting in Earth's middle age period extending longer than ca. 1.0 Ga. Any geological imprint of the Grenville orogenic event is not recorded in the NCB, but only ca. 900 Ma granites occur along the North Qinling Orogenic Belt relative to the southern margin of the NCB, whose geological significance is not clear. Some sedimentary rocks of the Nanhua Period are locally distributed in the NCB, but the sedimentary rocks of the Ediacaran System (Sinian) are not assured. Some borderline tillites in the Luoquan Group in the southern NCB and the Pirangdong Group in North Korea are debatable and also important for understanding Neoproterozoic evolution in the NCB. Combining geological characteristics of adjacent blocks to the NCB, it is possible that the Proterozoic NCC was located at a remote edge of the Nuna supercontinent if such a supercontinent existed. Or, there is another case, i.e., the Earth's middle age represents a particular tectonic evolution period, during which the Earth had a stable lithosphere with underlying secular warm mantle that resulted in multistage magmatism and rifting from the Late Paleoproterozoic to Neoproterozoic (Zhai et al. 2014a, b).The lithosphere should have become more stable and rigid throughout secular warm mantle–crust

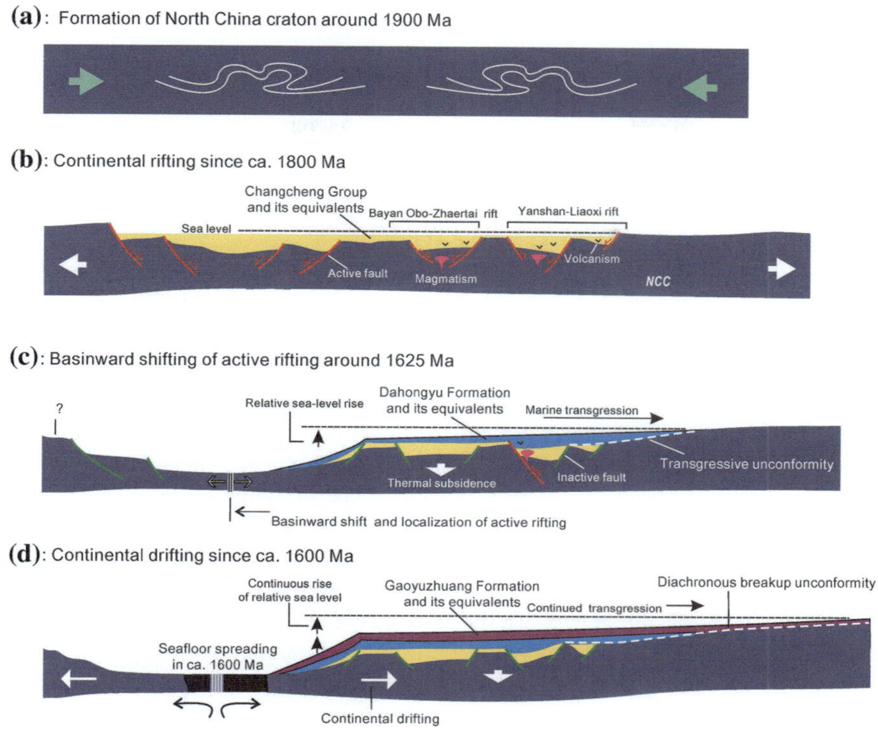

Fig. 10 Basin evolution model of the Changcheng system (after Meng et al. 2011)

interaction in the Earth's middle age. We propose that modern plate tectonics began operating after breakup of the Rodinia supercontinent, which contrasted with Paleoproterozoic small-scale primitive plate tectonics although they share similarity in dynamic principle (Zhai et al. 2014b).

In addition, Geng et al. (2006) argued that the Alashan Block should belong to the SCB rather than the NCB. The main evidence is discovery of the Neoproterozoic volcanic-sedimentary rocks with similar age to the Nanhua System. Recently, more and more Neoproterozoic volcanic rocks and dikes in the NCB are reported, which are helpful to understand tectonic history of the NCB.

The major ore deposits related to Late Paleoproterozoic–Neoproterozoic multiple rifting events are sedimentary hematite as occurring as discontinuous layers and lenses in the Chuanlinggou Group. Another iron ore deposit is the titanic magnetite related to anorogenic gabbro–leucogabbro association in the Damiao and Miyun in Yanshan aulacogen. The REE deposit, associated with Nb-bearing iron ores in Bayan Obo, constitutes another important ore deposit formed at this time. The middle Bayan Obo Group includes a series of clastic rock–sandstone–argillaceous slate–shale and black shale–dolomite. Several layers of dolomites have been distinguished. The REE mineralization is believed to be closely related with the carbonatite dikes (Fan et al. 2001; Yang et al. 2011a).

3.4 Metallogeny in the NCB

As mentioned above, the NCB experienced complex geological evolution since the Early Precambrian onward and carries important records of the secular changes in tectonics and metallogeny. Here, we synthesize the salient geological and tectonic features of the evolution and destruction of the NCB vis-à-vis the major metallogenic events and the formation of potential ore deposits. We identify a close relationship between the major geological events in the NCB with those elsewhere on the globe. We trace the records of a regular change in the pattern of metallogeny, mineral deposit character, spatial distribution, and genetic mechanisms which match well with the timings and styles of the major tectonic events in this craton.

The NCB went through five major tectonic cycles: (1) the Neoarchean crustal growth and stabilization, (2) Paleoproterozoic rifting–subduction–accretion–collision with imprints of the GOE, (3) Late Paleoproterozoic–Neoproterozoic multistage rifting, (4) Paleozoic orogenesis at the margins of the craton, and (5) Mesozoic extensional tectonics associated with lithosphere thinning and decratonization. Coinciding with these major geological events, five major metallogenic systems are identified as follows: Archean BIF system, Paleoproterozoic Cu–Pb–Zn and Mg–B systems, Mesoproterozoic REE–Fe–Pb–Zn system, Paleozoic orogenic Cu–Mo system, and Mesozoic intracontinental Au and Ag–Pb–Zn and Mo systems. The ore deposit types in each of these metallogenic systems show distinct characteristics and tectonic affinities (Zhai and Santosh 2013).

From Early Precambrian through Late Precambrian to Paleozoic and Mesozoic, the NCB records a transition from primitive- to modern-style plate tectonics. Evidence for imbricated oceanic plate stratigraphy in a subduction–accretion setting and collisional orogenesis along at least three major zones of ocean closure are documented. The major transitions in tectonic style and surface environmental changes in our planet are clearly recorded in the geological history and metallogenic events in the NCC. The large-scale gold deposits formed through intraplate tectonics during Mesozoic provide important insights into mantle dynamics and crust–mantle interaction associated with lithospheric thinning and craton destruction. The NCB provides one of the best examples to address secular changes in geological history and metallogenic epochs in the evolving Earth.

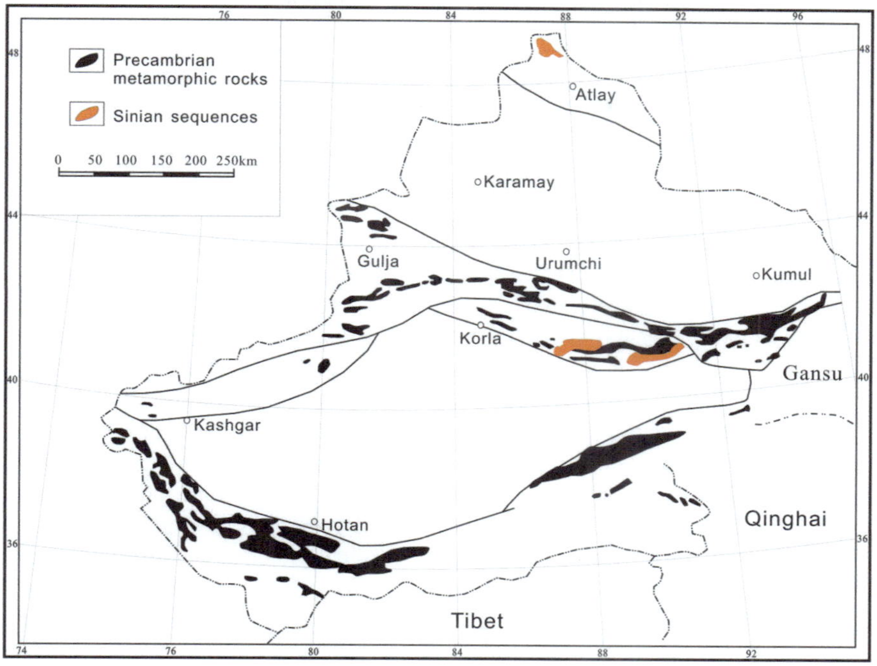

Fig. 11 Sketch map of the Precambrian rocks in Xinjiang (modified from Xinjiang geologic map and related data)

4 The Tarim Block

The Precambrian rocks in Xinjiang Autonomous Region are exposed mainly around the Tarim Basin and extend from eastern Hami to Beishan area (Fig. 11). Recently, more studies indicate clearly that the Early Precambrian basement rocks are underlain by the Neoproterozoic cover sequences in the Tarim Basin (Gao et al. 1985; Xinjiang Bureau of Geology and Mineral Resources 1991, 1999). The Tarim Block occupies about 600,000 km^2, whereas most areas were covered by the desert. The basement rocks are exposed discontinuously in the southwestern, northern, and northeastern Tarim Basin (Guo et al. 2003; Zhang et al. 2012; Long et al. 2011a, 2012). However, it is the least well studied among the three significant Precambrian blocks in China.

4.1 The Archean Basement Rocks in the Tarim Block

The Archean basement rocks are mainly composed of complex restricted to Neoarchean TTG gneisses, Paleoproterozoic medium to high-grade metamorphosed supracrustal rocks, and Early Meso–Neoproterozoic sedimentary sequences of greenschist facies. They are unconformably overlain by the younger unmetamorphosed Neoproterozoic sedimentary sequence containing some tillites and volcanic rocks.

Kuluktage area, located at the northern margin of the Tarim Basin, is the most significant outcrop area for studying basement rocks. It shows an east–west trending distribution and connects with the South Tianshan Orogenic Belt to the north. Previous studies have revealed the Archean rocks in the Kuluktage (Guo et al. 2000; Hu and Wei 2006; Long et al. 2011b). The oldest Precambrian crystalline basement rocks in this area are constituted mainly by TTG gneisses, which are a part of the Tuogelakbulak Complex and Xingditag Group. The Tuogelakbulak Complex is exposed in the south of Singer and is composed of black-gray amphibolite, amphibole schist, gray felsic gneiss, garnet biotite schist, actinolite mica quartz schist, biotite quartz schist, migmatite, and some discontinuous marble lens interlayer. Schists and marble lenses are metamorphosed with gneiss foliations along the strike. This complex is intruded by younger pink dimicaceous gneissic granite. In the southern Singer village, a complete profile without the lowest lithohorizon, which outcrops up to 800–1000 m, can be observed along the gully. This is equal to the Early Archean rocks in the 1:200,000 Singer and Otura-Tuogelakbulak geologic maps. It is unconformably overlain by the Paleoproterozoic Xinditag Group and thereon Mesoproterozoic sequences. The Sinian strata are distributed widely in the region. Some Cambrian and Ordovician exposures are also identified.

Magmatic rocks are the Precambrian mafic to felsic rocks. The Neopaleozoic magmatism develops in the north of Singer fault. Most of the Tuogelakbulak Complex is gray felsic gneiss and actinolite mica quartz schist, which have TTG chemical composition. They contain dark-gray amphibolite or amphibole schist inclusions. These rocks can also be found in the northeastern Korla. The TTGs are composed mainly of gray muscovite plagiogneiss, dimicaceous plagiogneiss, and biotite plagiogneiss. They often have banded structures, and part of them has been reformed into crystalline schist as the result of strong metamorphism and deformation. Zircon SIMS U–Pb dating of gray gneisses from Singer is 2565 ± 18 Ma, whereas the Sm–Nd isochron age of residual amphibolites is 3263 ± 126 Ma and $\varepsilon_{Nd}(t) = +3.2$ to ± 0.7 (Hu and Wei 2006), indicating they should be the Archean Complex. Some zircons from the Singer gray gneiss also have U–Pb ages of ca. 2300 Ma, 2100–1800 Ma (Long et al. 2012). The gneissic granite that intruded into the Singer TTGs is ca. 2000 Ma, and Tiemenguan amphibolite is ca. 1800 Ma (Guo et al. 2003), recording the Paleoproterozoic tectono-thermal events. Zircon U–Pb data confirm that the TTG gneisses near Korla are ca. 2.65 Ga (Long et al. 2011b). Zircons from these TTGs yield $\varepsilon_{Hf}(t)$ values between −5 and +1 with corresponding $T_{Hf(DM)}$ ages between 3.0 and 3.3 Ga, indicating that regional Neoproterozoic basement rocks are derived from the partial melting of the Paleo–Mesoarchean juvenile crustal material. Furthermore, >3.3 Ga continental crust should not exist in Kuluktage area. However, single zircon U–Pb age of 3605 ± 43 Ma reveals the existence of ancient crust in Agtashtagh, Altun Mountains (Lu and Yuan 2003).

Meanwhile, the chronologic framework of the Precambrian magmatic rocks in this area is constructed briefly, i.e., quartz monzonite (and/or veins): 1825 ± 23 Ma, T_{DM} = 2920 Ma; trondhjemitic granite (and/or gneissic): 2374 ± 10 Ma, T_{DM} = 3460 Ma; tonalite (and/or gneissic): 2604 ± 102 Ma, T_{DM} = 3063 Ma; and monzogranite granite (and/or gneissic): 3605 ± 43 Ma, T_{DM} = 3525 Ma,demonstrating the oldest crust is exposed in Aqtashtagh in the Tarim Block. The oldest terrain in Tarim may occur in the north slope of the Altun Mountains and it stretched to Kuluktage in the Late Neoproterozoic. Eventually, a coherent Archean basement at scale is established (Long et al. 2011a, 2011b, 2012).

Four major material sources of the sedimentary rocks in the Tarim Basin can be identified from detrital zircon studies, which are the Ordovician, Early Neoproterozoic, Early Mesoproterozoic, and Middle Paleoproterozic, respectively. The detrital zircon populations have similar age clusters between layers in a given area, whereas they are distinguished differences in other areas in the Tarim Block. Geochronological data show a distinct evolution history between the northern and southern Tarim Blocks, which need more detailed studies (Wu et al. 2010).

4.2 The Paleoproterozoic Metamorphic and Magmatic Rocks in the Tarim Block

The low-grade metamorphosed and strongly deformed sedimentary sequences are distributed in the Tieklike Uplift zone, southwestern Tarim. They are overlain by the Neoproterozoic volcanic-sedimentary rocks. The detrital zircons have U–Pb ages from Late Archean to Mesoproterozoic, in which the youngest ones are ca. 1.3 Ga. The metamorphic zircons from metamorphosed sedimentary rocks with light color granitic dikes yield ages from 0.9 to 0.8 Ga. This sedimentary sequence is considered to be contemporary with the Changcheng–Jixian systems in the NCB (Zhang et al. 2012).

The prominent Aksu blueschists are exposed in the Keping Uplift zone, northwestern Triam Basin. They are distributed continuously and do not show characteristics of tectonic mélange (Xiao et al. 1992). The south part of the Aksu blueschists is unconformably overlain by the Neoproterozoic upper Sinian strata, without the lower Sinian strata. The conglomerate layer, located at the bottom of the upper Sinian strata (i.e., upon the unconformity), contains gravels from both the underlain Aksu bluschists and the unmetamorphosed diabase dikes. These diabase dikes intrude into the Aksu Group but do not intrude into the overlying Sinian strata. The Aksu blueschists are considered to be one set of blueschist–greenschist facies rocks and the peak metamorphic temperature ranges from 300 to 400 °C. They are constituted mainly by strongly foliated chlorite-stilpnomelane graphite schist, stilpnomelane-phengite mica schist, greenschist, blueschist and minor quartzite and metaferruginous rocks. The Aksu blueschists are of intense folding deformation and show a northeast-southwest-trending belt style with the width of

20 km and the length of 40 km. The metamorphic period of the Aksu blueschists is 862 to 760 Ma (from mineral $^{40}Ar/^{39}Ar$ and whole-rock Rb-Sr isochron ages (Chen et al. 2004). Zircons from unmetamorphosed mafic dike swarm yield U–Pb ages from 803 to 780 Ma (Zhan et al. 2007; Zhang et al. 2008a, b, c). Detrital zircons from the metasedimentary rocks of the Aksu Group give peak ages from 830 to 780 Ma. It is interpreted that the depositional time is younger than ca. 830 Ma, and the metamorphism occurred after ca. 780 Ma. Briefly, the thermal evolution history of the blueschists is reconstructed according to the fission-track thermal history simulating. (1) A rapid exhumation occurred after the formation (872–862 Ma). They may be exhumated in the Early Sinian and started deposition in the Late Sinian. (2) The stratigraphic sequences of the Late Sinian were continuous. The mid-upper Silurian and mid-low Carboniferous series are absent in the integrated Paleozoic stratigraphic column. At the end of the Paleozoic era, the thickness of the Early Sinian and Paleozoic strata were approximately ten thousands meters. The blueschists annealed thoroughly, and the isotope system was reset and the timer restarted. (3) Regional uplift occurred during the Late Mesozoic (Zhang et al. 2008a, b, c). The metamorphic pressure of the south part calculated from phengite manometer is higher than that of the north part (Huang et al. 2009). It is inferred that the Aksu blueschists could be a high-pressure metamorphic belt resulted from the ancient oceanic plate southbound subduction downward the Paleo-Tarim massif. Moreover, some geologists are interested in the metamorphic rocks and possible ophiolite located along the northern and southern margins of the Altun Mountains. Considering the ophiolites from northern and southern margins formed at 829 ± 60 Ma and 1449 ± 270 Ma, respectively, they concluded the Tarim Block was amalgamated by various old terrains during various periods (Guo et al. 2000; Liu et al. 2009).

Most of the Early Neoproterozoic granites in the Tarim Block formed during 820–800 Ma, and a minority of them formed in ca. 760 Ma (Wu et al. 2010). Ultramafic and mafic complexes and mafic dikes are divided into two phases at ca. 820 Ma and 780–760 Ma, respectively. These rocks can be observed in both the southern and the northern margins of the Altun Mountains. The 740–735 Ma bimodal volcanic rocks and 735–650 Ma mafic dike swarm from the northern margin are probably related to the continental rifting (Zhang et al. 2012; Long et al. 2010).

4.3 Neoproterozoic Sedimentary Rocks and Glacial Events in the Tarim Block

Kuluktage is the most significant area to study the Neoproterozoic glacial events in the Tarim Block. The upper Proterozoic sequences in this area are continuous and complete (Fig. 12). From east to west, the rocks are exposed in Saimashan, Singer, Xingdi, Xishankou, etc. The upper Neoproterozoic series unconformably overlain

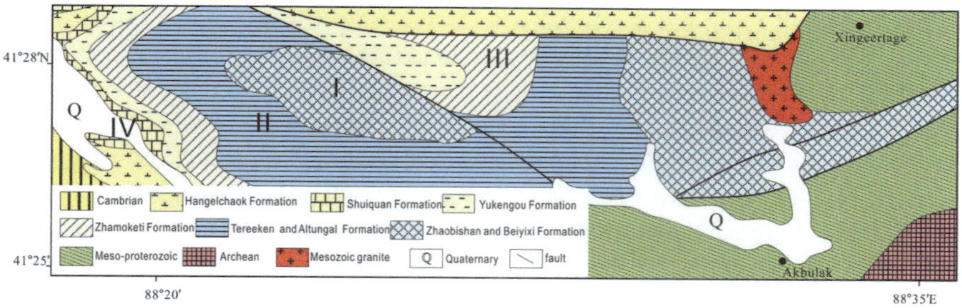

Fig. 12 Sketch map of the Precambrian rocks in the Kuluktage area (After Xu et al. 2008). I, II, III, IV show the four glacial events

by the Mesoproterozoic metamorphic rocks and is parallel-unconformably overlain by the Cambrian strata. The upper Neoproterozoic series can be divided into the Beiyixi, Zhaobishan, Aletonggou, Teriaiken, Zhamoketi, Yukengou, Shuiquan, and Hangeerqiaoke groups (Gao et al. 1985). The top and the bottom of the Beiyixi Group are composed of volcanic rocks, and the middle is composed of anagenite and glutenite deposits. The Lower Zhaobishan Group consists of quartz sandstone and siltstone. The Upper Zhaobishan Group is the interbedding of shale and sandstone. The Lower Aletonggou Group is composed of sandstone, siltstone, and mudstone containing anagenite formation, whereas the upper is mainly of shale and sandstone, containing 2–3 layers of volcanic rocks or tuffs. The Terianken Group is constituted by mega-thick anagenite bearing minor shale and dolomite with ~5 m beige dolomite on the top. The Lower Zhamoketi Group is composed of flysch deposits characterized by incomplete Bouma sequence. The Upper Zhamoketi Group is of thick volcanic rocks. The deposition of the Yukengou Group is simple, which is composed of grayish-green silty mudstone and shale. The Shuiquan Group is mainly of thin-layered dolomite and limestone deposits with the upper containing silver gray siltstone and mudstone and the top of several meters volcanic rocks. The top sequence of the Hangeerqiaoke Group is metathick beige celadon tillite association and finished by ~4 m gray dolomite on the top. The four layers of anagenite in the Neoproterozoic sedimentary rocks have been distinguished. The depositional feature, stratinomy, conglomerate origin, and C-isotopic characteristics have been reported in detail. Most conglomerates from the bottom of the Beiyixi Group show positive $\delta^{13}C$, possibly indicating the starting of continental rift. However, the conglomerate $\delta^{13}C$ curves of the other three phases demonstrate significant variations. First, the $\delta^{13}C$ of the Beiyixi Group varies from positive value of the bottom to ~−5 ‰ of the top. Second, the $\delta^{13}C$ varies from 0 to −3 ‰ of the Aletonggou Group, upper seated of the lower Sinian System, from −3.4 to −14.4 ‰ of the Teriaiken Group. Third, the $\delta^{13}C$ varies from positive value of the Zhamoketi Group, lower seated of the upper Sinian System, to negative value of the glacial laminated clay from the top of Hangeerqiaoke Group. Therefore, from bottom to top, they are known as Beiyixi, Teriaiken, and Hangerqiaoke glacial events, respectively (Song et al. 2002; Xu et al. 2003; Xiao et al. 2004).

The snowball Earth has a crucial issue on the synchroneity of the global glacial events, which should be supported by vast isotopic geochronological data. Four significant Neoproterozoic glacial events can be identified around the world, i.e., the Kaigas, Sturtian, Marinoan, and Gaskiers. The Marinoan glacial event ended at ca. 635 Ma, and the Sturtian glacial event possibly started during 720–710 Ma and ended before ca. 670 Ma. The younger Gasliers glacial event occurred during 590–580 Ma. Magmatic zircons from volcanic rocks of the top Beiyixi Group yield SHRIMP U–Pb age of 732 ± 7 Ma, suggesting the minimum occurrence time of the Beiyixi glacial event. Given the dating result of the volcanic rocks under the anagenite layer, the Beiyixi glacial event is restricted from 740 to 732 Ma. The four Neoproterozoic groups containing anagenite in Kuluktage were deposited during three major periods, i.e., Beiyixi Group in 740–732 Ma, the Aletonggou and Teriaiken groups in 732–615 Ma, and Hangeerqiake Group in 615–542 Ma, respectively. They also represent the three regional Neoproterozoic glacial epochs (Xu et al. 2003) and can be compared with globe glacial events. By contrast, the Beiyixi is equivalent to the Kaigas glacial event, meanwhile the Aletonggou and Teriaiken are possible equivalent to the Sturtian and Elatina glacial events, and the Hangeerqiake is equivalent of the Gaskiers glacial event.

Definitely, the Early Precambrian basement rocks are exposed in the Tarim Block where Paleoproterozoic magmatism and metamorphism occurred during 2000–1800 Ma. The distribution of the Proterozoic rocks is limited but represents affluent geological phenomenon. Some researchers suggested the Tarim Block has affinity with the Yangtze Block in view of the similar Neoproterozoic glacial events. Other researchers paid much attention to the Early Precambrian evolution between the North China and Tarim Blocks, especially the Dunhuang terrain stretched from the eastern margin of the Tarim (Liu et al. 2009; Zhang et al. 2013). Hypothetical affinity with the North China Block is inferred when considering the age peaks of 2.6–2.5 Ga and 1.9–1.8 Ga revealed by the zircons from TTG gneisses and metamorphic rocks. Further studies should be proceeded to gather more evidence.

5 The South China Block

The South China Block is bounded by the Qinling–Dabie Orogenic Belt to the north and the western Sichuan Plateau and Hengduan Mountains to the east (Ren 1989). Conventionally, the South China Block is subdivided into the Yangtze Block in the northwest and the Cathaysia Block in the southeast, separated by the Neoproterozoic Jiangnan Orogenic Belt (Fig. 13). Some researchers also suggest that the South China Orogen is the Paleozoic or Early Mesozoic orogeny wrapped with the Precambrian residual fragments. Ma (2006) proposed that the South China was a marginal basin tectonically captured by the Yangtze active continental margin during the Neoproterozoic to the Early Paleozoic. At the end of the Early Paleozoic, the Fujian coast arc collided with the Wuyi arc, and then, both of them collided

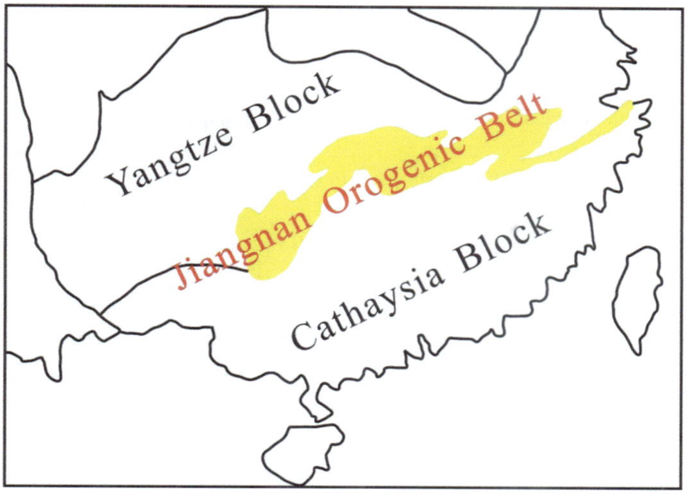

Fig. 13 Main geological units in the SCB (after Zhou et al. 2008)

together with the active continental margin of the Yangtze Block, forming the South China orogenic system which induced the metamorphism of the marginal basin and intense tectonic deformation of the SCB, and finally formed the Early Paleozoic orogenic belt. Some studies later suggest that the SCB could be correlated with the Rodinia supercontinent (Li et al. 2008a, b, c) and consider it to be a missing link between Laurentia and Australia. The SCB was presented to be above a super mantle plume center in the Rodinia supercontinent. Significant superplume activity induced a series of tectono-thermal events and finally resulted in the breakup of the Rodinia in the Neoproterozoic. Published a batch of papers from approvers and opponents promoted evolutional research processes on the South China Block (Yan et al. 2002; Li et al. 2002b; Wang and Pan 2009; Zheng 2004; Zheng et al. 2008). The SCB has attracted considerable attention in terms of its widespread and multi-episodes of the Phanerozoic magmatism and related poly-metallic mineral resources including W, Sn, U, and REE deposits. The Precambrian tectono-magmatic processes have constructed the structural framework of the South China and also become the research basis of the large-scale mineralization.

5.1 Yangtze Block

The Yangtze Block was called the Yangtz Platform because of its scarce outcrops of metamorphic crystallization basement but widespread Phanerozoic supracrustal sediments and deformation structures. The Yangtze Block comprises double basements including middle-level metamorphic basement in the lower part and upper-level metamorphic fold basement in the upper part. The Banxi Group and its equivalent strata are classified as parts of basement, whereas the Liantuo sandstones and above strata to be sedimentary cover (Ren et al. 1990). The Proterozoic Jinning Orogenic Belt led to the formation of the cratonic fold basement.

5.1.1 The Crystalline Basement in the Yangtze Block

The middle-level metamorphic basement in the Yangtze Block mainly comprises the Kongling Complex in the eastern Yangtze Block, the Houhe Complex in the northwestern margin of the Yangtze Block, and the Dahongshan Complex in Yunnan Province, the southwestern Yangtze Block. The Kongling Complex (high-grade metamorphic terrane) is distributed in the areas Yixingshan, Zhugui, and Huangling, Yichang City, and displays a dome-like occurrence with an outcrop area of 360 km^2 (Fig. 14). It was intruded by the Paleoproterozoic (ca. 1.85 Ga) Quanqitan granite in the north and extensively by the Early Neoproterozoic (820–750 Ma) Huangling intrusions in the south (Jiao et al. 2009). The traditionally called Kongling Group is a typical Archean high-grade metamorphic complex mainly consisting of TTG gneisses, granitic gneisses, and metamorphosed supra-crustal rocks including minor basic granulites. The metamorphosed sedimentary rocks mainly comprise banded biotite leptynite with or without graphite. The rocks also include garnet–sillimanite quartzite, garnet–sillimanite–biotite gneiss, graphite schist, marbles, calcium silicate, quartzite, and magnetite quartzite. A spate of garnet and sillimanite are widely developed in metamorphosed sedimentary rocks. Dehydration melting of biotite and local presence of basic granulite can be observed, indicating that the peak metamorphic temperature is >750–900 °C and metamorphic pressure is 0.55–1.1 Gpa (Gao et al. 2001). Qiu et al. (2000) and Gao et al. (2001) firstly reported the oldest forming age (2.95–2.9 Ga) of the Kongling TTG using SHRIMP U–Pb zircon dating and also obtained detrital zircon ages of 3.28–2.87 Ga from nearby Archean metapelites with metamorphic age of 2.01–1.95 Ga (Liu et al. 2008). Jiao et al. (2009) reported the forming age (3218 ± 13 Ma) of magmatic zircons from biotite plagioclase gneiss in Huangliangcun, northeastern Kongling area. All the zircons have $\varepsilon_{Hf}(t)$ values of −2.33 ± 0.51 with $T_{DM}{}^{C}$ of 3679 ± 49 Ma, indicating that they likely originated from partial melting of much older (>3.6 Ga) Paleoarchean crustal materials (Zhang et al. 2006a; Zheng et al. 2006). This result is consistent with the ancient detrital zircons (>3.5 Ga, up to 3.8 Ga) selected from sedimentary rocks in the Yangtze Block (Zhang et al. 2006b), implying that Paleoarchean crustal materials exist in the Yangtze Block. Both magmatic (ca. 2.9 Ga) and residual zircons (ca. 3.2 Ga) from the Kongling TTG gneisses yield consistent $\varepsilon_{Hf}(t)$ values and $T_{DM}{}^{C}$ ages with biotite plagioclase gneiss, indicating that the two types of gneisses may come from similar sources. Metamorphic zircons selected from metamorphic rocks of the Kongling Complex yield ages varying from 2.0 to 1.9 Ga (Qiu et al. 2000), consistent with high-K granite intruding into the Kongling Complex, marked the end of Paleoproterozoic tectono-thermal events in the Yangtze Block (Li et al. 2012). The Houhe Complex in the South Qinling Mountains mainly comprises tonalitic

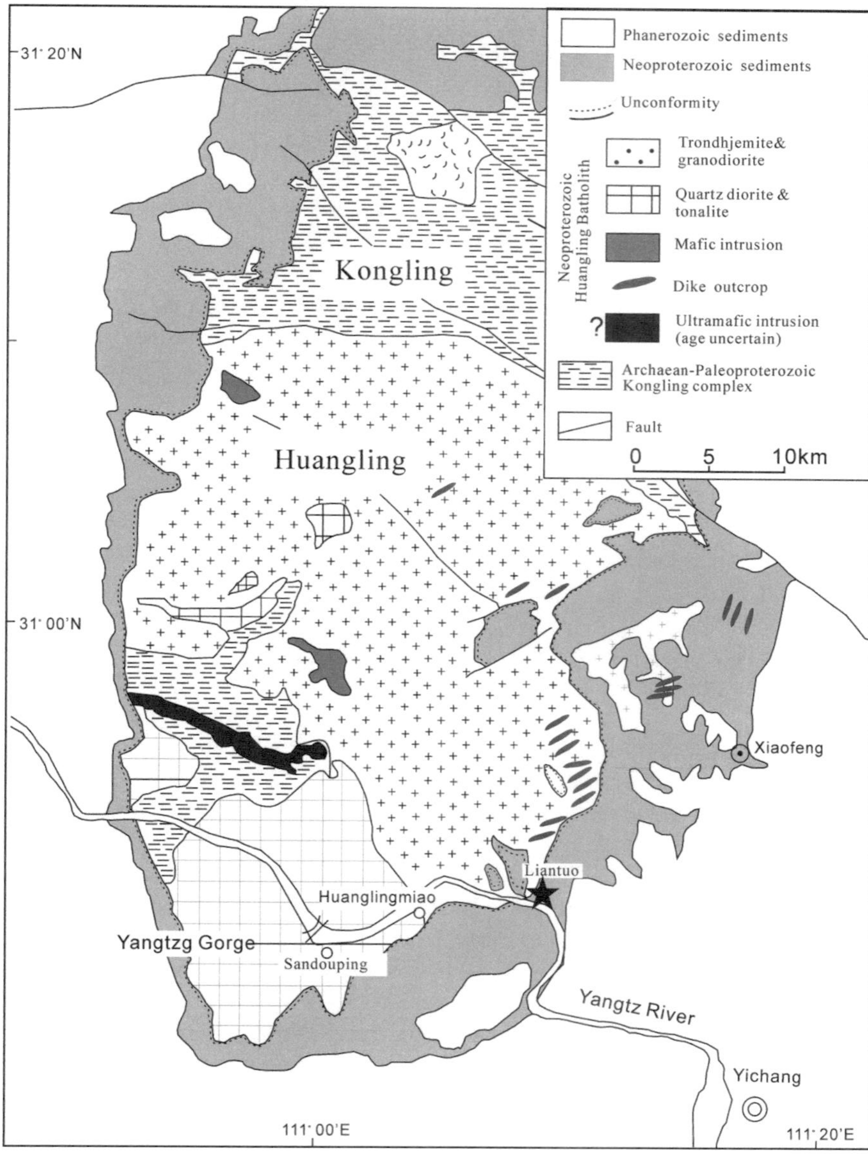

Fig. 14 Generalized geological map for the Kongling Complex (after Yu et al. 2012)

gneisses, minor amphibolite and marbles, and experienced high-amphibolite facies metamorphism and migmatization. Magmatic zircon crystallization U–Pb age is ca. 2.08 Ga (Wu et al. 2012). The Dahongshan Group underwent metamorphism of low-amphibolite to high-greenschist facies, and the volcanic rocks inside erupted at ca. 1.68 Ga (Greentree and Li 2008), indicating that the exposed metamorphic basement in the western Yangtze Block is much younger.

The upper-level metamorphic basement in the Yangtze Block also can be subdivided into two parts. The lower part consists of the Lengjiaxi Group and equivalents strata including the Fanjingshan Group, Sibao Group, Shuangqiaoshan Group, and Shangxi Group. The rocks were metamorphosed mainly to low-greenschist facies, which are of phyllite-slate series, and exposed as a NNE fold system. Its upper part is called as the Banxi Group which is a suite of sedimentary sequences with volcanic rocks (Ren et al. 1990). Liu et al. (1991) considered that there was a southward sedimentary phase transition in the Three Gorges area from continental sandstones of the Sinian Liantuo Formation to marine deposits of the Banxi Group. It has been debated whether the Banxi Group is equivalent to the Liantuo Formation, especially when the youngest ages of ca. 720 Ma from the top of the Banxi Group was obtained (Zhang et al. 2009). The debate will remain up to now because discontinuously outcropped strata occur in various areas in the South China and respective geological investigations by different researchers. This difference of opinion also affects the division of the Nanhua System and its geological implication.

5.1.2 The Early Neoproterozoic Volcanic-Sedimentary Rocks and Intrusions

In the strata correlation diagram (Fig. 15), the lower boundary age of the Nanhua System and the upper boundary age of the Qingbaikou System are both limited at ca. 780 Ma according to the demarcation ages of the Liantuo and Chang'an formations. The age limit is inferred to be the onset of the Liantuo Glaciation (Gao et al. 2011). Some researchers favor that the Nanhua rift began to form at ca. 820 Ma, denying that the Banxi Group belongs to basement but the lower part of the cover. They emphasize that above the Jining-Sibao Unconformity, there exist wedge-shaped strata with characteristics of discontinuous lateral extension, various strata thickness and sedimentary lithofacies, and anisochronous bottom boundaries. The strata are distributed mainly in the Hunan–Guangxi and North Zhejiang sub-basins. Accompanied with the Upper Yangtze massif, Jiangnan ancient uplift, and Cathaysia massif, they constitute the Neoproterozoic paleogeographic pattern characterized by an alternation of grabens and horsts in the South China (Wang and Pan 2009). Some workers documented that the Neoproterozoic "wedge-shaped strata" are the products of rift which started in Late Qingbaikou Period (820–815 Ma) and died out at end of Qingbaikou Period (ca. 780 Ma). It consists of Anhui–Zhejiang–Jiangxi grabe–horst belt, Hunan–Guangxi step slope belt, and South China rift oceanic basin. The graben–horst belt is composed of "double upwarpings and double balleys" accompanied with intensive volcanic activities. According to the above-mentioned opinion, the Nanhua Rift System is composed of the Banxi Group and equivalent suite. The subsequent Liantuo Formation deposited in residual trough parallel-unconformably overlies the Banxi Group (Yang et al. 2012).

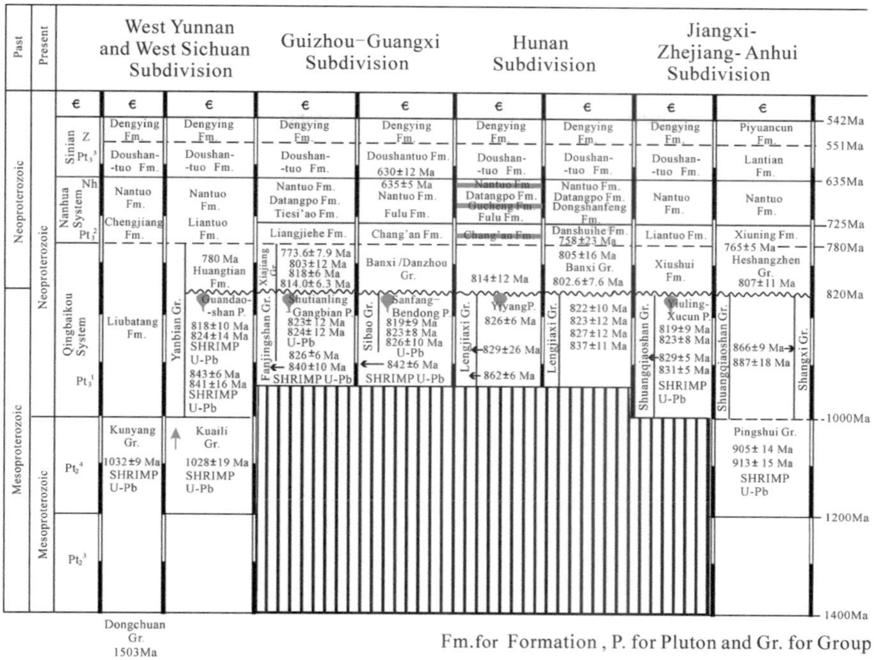

Fig. 15 Diagram of the Neoproterozoic strata correlation in the Yangtze Block (after Gao et al. 2011)

According to the evolutional characteristics, a rift is subdivided into rift base, rift body, and rift cover (Wang and Pan 2009). The Jining-Sibao Unconformity and the underlying metamorphic strata consist of the rift base, wedge-shaped strata make up the rift body, and the widespread Sinian System in the Nanhua Basin constitutes the rift cover. Preliminary studies indicate that sedimentary associations representative of the early stage of rift basin include alluvial and diluvium facies association, continental (or marine) volcano-clastic and pyroclastic facies association, littoral to shallow marine facies association, and finished with drowned carbonate-platform to starved-basin carbonaceous shale facies association, reflecting a complete sedimentary process from continent to marine. The lowest layer of the first depositional stage represents the beginning of Neoproterozoic sedimentary sequence after the Sibao Orogeny in the South China, such as the Cangshuipu Formation in the North Hunan, the Baizhu Formation in the North Guangxi, and the Shiqiaopu Formation in the Northwest Hunan. The formation age of igneous rocks (819–812 Ma) from the bottom of all above strata represents the earliest depositional time in the Neoproterozoic after the Sibao Orogeny in the SCB (Li et al. 2005, 2008a, b, c, d; Zhou et al. 2002a, b; Ling et al. 2003; Zhang et al. 2008a, b, c).

The distribution of the Middle Neoproterozoic intrusions and volcanic-sedimentary rocks in the South China Block are shown in Fig. 11 (Li et al. 2012). The Neoproterozoic basin in the SCB formed accompanied by numerous of volcanic–magmatic activities. The geochemical data from volcanic–pyroclastic rocks have significant implications to better understanding of geodynamic environments. Li et al. (1999, 2002a, 2003, 2007) and other scholars suggest that the granitoids, mafic intrusions, and dikes (830–820 Ma) overlapped by the Neoproterozoic wedge-shaped strata are the products of mantle plumes. Geochemical features of volcanic rocks from the Neoproterozoic Suxiong Formation also reveal similar characteristics. Igneous rocks from the Tiechuanshan, Mamianshan, Hongchicun, and Yingyangguan formations related to sedimentary overlap of the wedge-shaped strata after the rift basin opening show affinity with rift magmatism. Moreover, many intrusions overlapped by wedge-shaped strata also formed in a rift setting, such as mafic–ultramafic rocks in the Baotan-Yuanbaoshan area and spilite and mafic–ultramafic rocks in the Longsheng area in North Guangxi, and granites in Guandaoshan area of West Sichuan. Consequently, all lines of evidence are supportive of a rift setting induced by mantle plume in the South China in Neoproterozoic (Ren et al. 1990; Wang and Pan 2009; Yang et al. 2012). Based on geochemical characteristics of granitoid and mafic–ultramafic intrusions in the South China, Zhou et al. (2002a, b) consider that the southeastern, western, and northern margins of the Yangtze Block were active continental margin with arc–basin system in Middle Neoproterozoic. The volcanic magmatism of this stage formed, which is related to subductions, collision, and post-collisional collapse (Zheng et al. 2007).

The Nanhua System mainly comprises the Liantuo Formation, Nantuo Formation, and equivalent strata (see Fig. 16, in light of CCOS), as well as interlayered strata of the Datangpo and Gucheng formations between the Liantuo and Nantuo sedimentary rocks in some regions. There exist two epochs of glacial sedimentary event in the Nanhua System, and they are the lower Chang'an Glaciation (including Gucheng and Jiangkou Glaciation) and the upper Nantuo Glaciation with the Datangpo Interglaciation (or equivalent manganese-bearing strata) between the two glacial epochs. Great debates remain on whether the Nantuo Glaciation is correspondent to Sturtian or Marinoan Glaciations because of inaccurate Rb–Sr and Sm–Nd isotopic ages obtained from the Datangpo Interglaciation. However, the onset and termination ages of the Chang'an Glaciation, which is correspondent to the Sturtian Glaciation, recently have been more accurately defined at 750–725 Ma. The Chang'an Glaciation is characterized by a set of marine glacial deposits, including marine water gravity flow deposit or low-density debris flow deposit. The Nantuo Glaciation correspondent to the Marinoan Glaciation is limited at approximately 660–635 Ma. Its sedimentary types are complex and might include marine glacial deposit, continental glacial deposit, glacial debris flow deposit, ice water gravity flow deposit, or debris flow deposit (Yin et al. 2003; Chu 2004; Huang et al. 2007; Zhang et al. 2008b, 2009a, b). The SCB yields a paleolatitude of 38 ± 8 °N at ca. 750 Ma. The Liantuo Formation produced similar paleolatitudes of 33.6 ± 1.7 °N and $37.7 + 7.6/- 6.5$ °N or, briefly, intermediate of 30–40 °N (Li et al. 2008a, b, c). There are no paleomagnetic data from the Chang'an Glaciation. Thus, the related paleolatitude of South China Block during the Gaskiers Glaciation is also likely intermediate. However, the Gaskiers

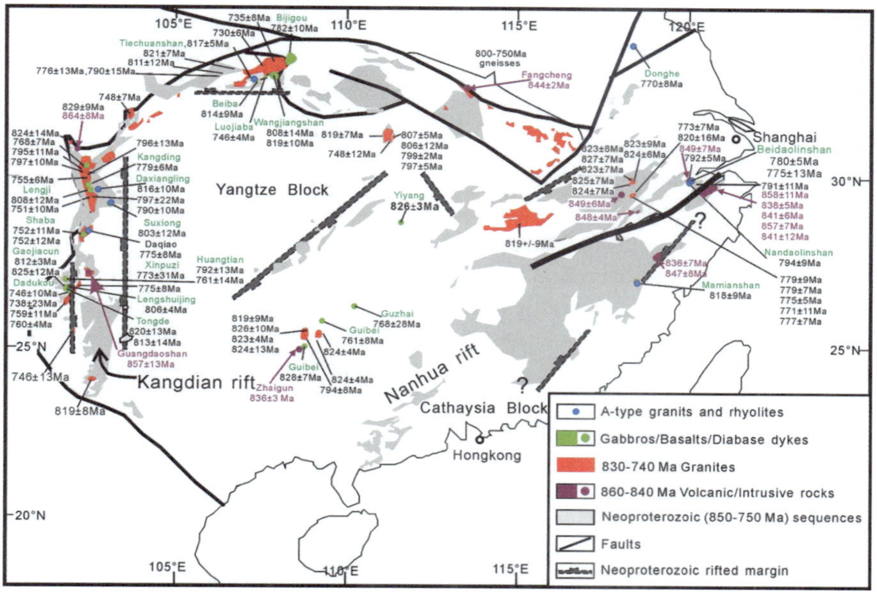

Fig. 16 The distribution of the Middle Neoproterozoic intrusive rocks and volcanic-sedimentary rocks in South China (after Li et al. 2012)

Glaciation has no glacial sedimentary records in the South China Block, and only isotopic and chemical data indicate that the Kaigas was likely present as cold paleoclimate and/or mountain glaciation (Zhang et al. 2009a, b). It is still controversial as to the formation ages of the two glacial epochs, especially the onset and termination ages of the lower Sturtian Glaciation. Thus, further detailed geochronological study on the Gucheng, Tiesiao, Chang'an, and Jiangkou formations is significant for limiting the duration of the Sturtian Glaciation and onset ages of the Cryogenian and Nanhua systems.

5.1.3 Sinian System

The Sinian System disconformably overlies the Nanhua System. The Sinian standard section in the Three Gorges area is overlain by the Cambrian strata. The top and bottom boundary layers of the Sinian System are, respectively, the Lower Cambrian Shuituojing Formation and the Neoproterozoic Doushantuo Formation. The Sinian System, corresponding to the Ediacaran, yields a boundary age limited at ca. 635 Ma between the bottom of the Liantuo Formation and the underlying Nanhua System (Fig. 10). A complete depositional sequence is recognized from the Doushantuo to Dengying formations in the Three Gorges area. The basal siliciclastic facies are interpreted as deposits in a shelf setting. Mixed clastic and carbonate facies characterize the lower portion of the sequence, whereas platform carbonate facies dominate the middle and upper portions. The vertical facies variation might have been related to sea level rise after glacial epoch at the end of the pre-Sinian Period (Wang et al. 2002; Liu et al. 2003; Yang et al. 2012). The Doushantuo Formation experienced initial low-water deposition and the following rapid and extensive transgressive deposition. From bottom to top, the Doushantuo Formation consists mainly of supratidal–intertidal sabkha with evaporates, peritidal carbonate, and siliciclastics of the inner ramp subfacies. However, from middle to upper parts, sedimentary phase gradually changed to be deepening and thinning-upward intertidal via subtidal to outer ramp subfacies. At this stage, the sea lever rose rapidly with high amplitude and low frequency and thus formed a series of deepening and thinning-upward sedimentary cycle and overall retrogressive sedimentary sequences. The Dengying Formation shows that sea lever changed to be low amplitude and high frequency trend. The sedimentation of the Dengying Formation is based on aggradation and progradation. Meanwhile, topographical structure of basin basement became wider, and sedimentary rocks of various typical inner ramp subfacies were developed, which are called carbonate beach bar deposit with pattern of double sandy bars. The Hamajing Member at the bottom is of oolitic carbonate bar from intersubtidal subfacies. The Shibantan in the middle is of intertidal lagoon deposits with shallowing and thinning-upward sequences. The Baimatuo on the top is mainly of beach oolitic and peritidal carbonates. Isotopic researches on carbonate of the Sinian System (Wang et al. 2002) prove that the $\delta^{13}C$ values of the "carbonate cap" vary from −3.1 to −3.7 ‰ at the base of the Doushantuo Formation and reach a maximum of +6.7 ‰ at the sequence III(DS3) with the positive upward change. At the top of the formation (DS6), $\delta^{13}C$ change to be negative of −2.5 ‰, forming a complete circle of change. Meanwhile, $^{87}Sr/^{86}Sr$ ratios show an upward increase trend with progressive increase from 0.707743 at the bottom of the Doushantuo Formation (DS1) to 0.707965 at the top of the Doushantuo Formation (DS6). The $\delta^{13}C$ values of the Hamajing Member of the Doushantuo Formation (DS7) vary from −2.5 to positive, and reach a maximum of +5 ‰ in DS8 and then decrease to +1.7 ‰ in the middle part of the member (DS1O), and then keep a smooth change ranging from +1.7 ‰ to +3.6 ‰. The $\delta^{13}C$ values at the top of Baimatuo Member go down to + 0.5 ‰ and show a decrease to negative trend. $^{87}Sr/^{86}Sr$ ratios (0.708244 to 0.708993) still show an upward increase trend from bottom to top of the Dengying Formation.

The Sinian System is well outcropped in the Guizhou and adjacent areas in the Upper Yangtze Region and shows regular sedimentary sequence changes from northwest to southeast (Mei et al. 2006). The Dengying Formation is characterized by very thick dolomite (the thickness varies from 400 to 600 m) within shallow-water environment in the northwest area and contains particle, clay grain, and oncolite dolomite in the central part. Correspondingly, a black, shaly condensed section with basin facies developed in the Doushantuo Formation. However, the thickness of the Sinian System decreased significantly (tens to hundreds meters) within deepwater environment in the southeast area. The Doushantuo Formation is mainly silicalite with thickness increasing, while the Dengying Formation is typical of dolomicrite with thickness decreasing sharply. Consequently, it is hard to divide it into three intervals but one condensed sequence. More specifically, the Dengying Formation only has dolomite from the shallow to the deep, extremely different from the feature of the Cambrian Loushanguan Group that dolomite changed to be limestone from northwest to southeast (Feng et al. 2001), possibly indicating a particular different dolomitization in the Sinian from the Phanerozoic. An overall upward shoaling sedimentary sequence from the Doushantuo Formation to the Dengying Formation represents the formation and development of a carbonate ramp platform. Transgression during the Doushantuo stage provides a living environment for biological diversity represented by Doushantuo biota (Yuan et al. 2002). Rapid transgression represented by typical drowned unconformity surface between the Sinian and Cambrian is the prelude of the Cambrian explosion (Gao 1999) and reflects the complex relationships and organic connection between biological and sedimentary environment changes.

5.2 Cathaysia Block

Ever since the term "Cathaysia Old Land" was proposed by Grabau in 1924 based on the Devonian sandstone unconformably covering the metamorphic basement in the southeastern coastal areas, the Chinese and foreign geologists discussed the geological structures and evolutional history of the Cathaysia Block from different perspectives and spatial–temporal scales (Chen 1956; Hsü et al. 1987, 1990; Shui et al. 1988). Since then, more and more evidence states that the Precambrian basement rocks are widely distributed in the Cathaysia Block, such as metamorphic series at high-greenschist to amphibolite facies in the northwestern Fujian–southwestern Zhejiang, Yunkai, and coastal areas (Ma and Chen 2000). In general, the formation and evolutional features of the Precambrian basement are inconsistent between the Cathaysia and Yangtze blocks (Gao 1999). It has been identified that the oldest rocks exposed in the southern Zhejiang and northern Fujian in the Cathaysia Block are Paleoproterozoic metamorphic igneous rocks (Hu 1994; Gan et al. 1995; Li 1997), in some of which the Neoarchean detrital (or inherited) zircons are also present (Xu et al. 2007; Zheng et al. 2011), implying the Cathaysia Block is either underlain by Archean crust or once developed on the margin of an Archean block (Fig. 17). Furthermore, recent studies obtained that the Guzhai granodiorite and the Longchuan granitic gneiss in the eastern Guangdong Province contain a few inherited/detrital zircons with ages of 3.1–3.0 Ga (Yu et al. 2007), implying that the Cathaysia Block may also have very ancient evolutionary history. Yu et al. (2007) reported U–Pb dating from fifty-six detrital zircons from a paragneiss in Nanxiong area, northern Guangdong Province, and suggested that the Late Neoproterozoic sediments in the Cathaysia hinterland are composed of numerous Genvillian and Neoarchean clastic materials, as well as some Mesoproterozoic detritus. Minor Paleoarchean (ca. 3.76 Ga) and Mesoarchean (3.2–3.0 Ga) zircons are also found in sediments, suggesting that the Cathaysia Block may contain very old rocks. The Hf isotope compositions from thirty-seven

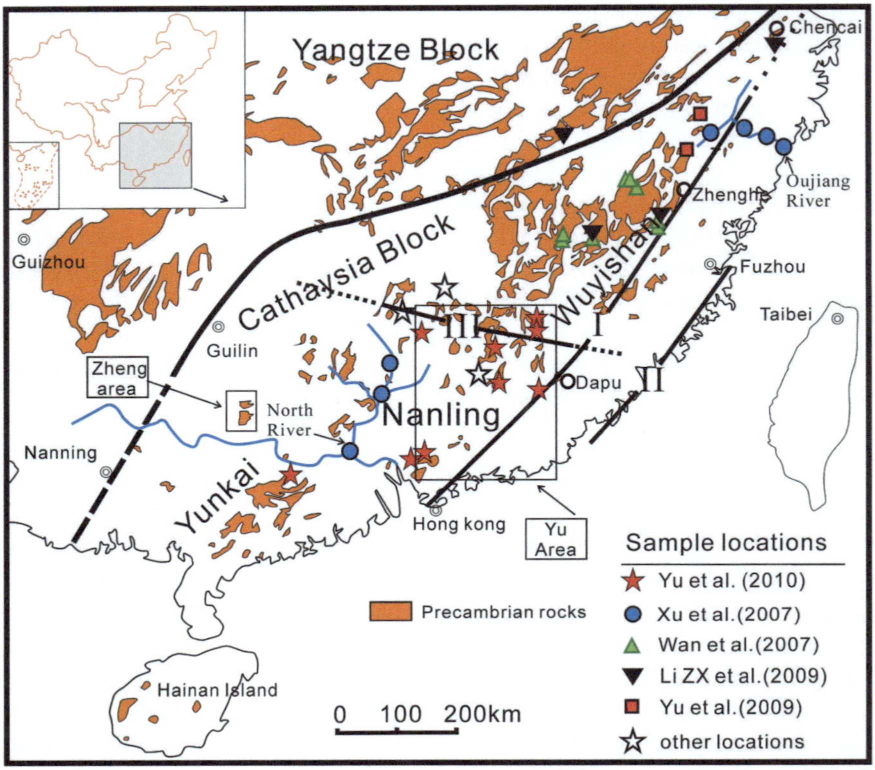

Fig. 17 Sketch geological map of the Precambrian rocks in the Cathaysia Block (after Yu et al. 2012; Zheng et al. 2011)

zircons reveal that these clastic materials are derived from different sources. Minor zircons crystallized from magma generated from juvenile crust, while the parental magma of most zircons derived from ancient crust. Zircon U–Pb and Hf isotope analysis shows that the generation of juvenile crust in the Cathaysia Block occurred mainly at 2.6–2.5 Ga. Mesoarchean (3.3–3.0 Ga), Late Paleoproterozoic (ca. 1.8 Ga), and Paleoarchean (ca. 3.7 Ga) are also important episodes of crustal growth (Fig. 18). Meso–Neoproterozoic magmatism was extremely intense, but it mainly involved recycling of ancient crustal components with little juvenile crust formation. The marked presence of ca. 2.1 Ga Hf model ages and the absence of contemporary zircons indicate that the parental magma of many zircons was derived from sources mixed with Neoarchean and Late Paleoproterozoic materials.

The oldest rocks of the Cathaysia Block, preserved in the southern Zhejiang and in Wuyishan area of western Fujian, comprise Badu Complex and related Paleoproterozoic granites (Yu et al. 2010, 2012) (Fig. 17). Metamorphic rocks consist mainly of mica–quartz schist, epidote amphibolite, actinolite schist, fine-grained biotite gneiss, magnetite-bearing quartzite, and marbles. Many granites

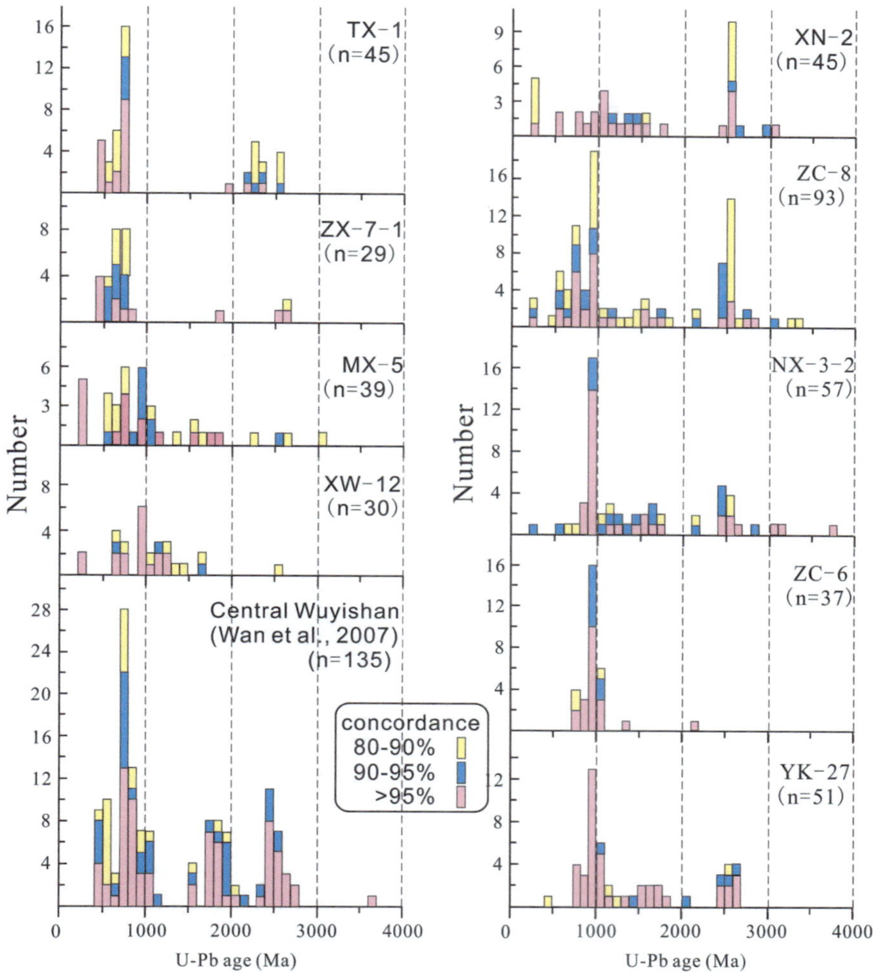

Fig. 18 Detrital zircons U–Pb ages in the Cathaysia Block (after Yu et al. 2007)

intruded into the Badu Complex before ca. 1888 Ma (Yu et al. 2009). The above-mentioned Precambrian rocks are intruded by the Early Paleozoic rocks and covered by the Paleozoic and Early Mesozoic sediments. Inherited cores of zircon grains from metamorphic rocks predominantly yield ages at ca 2.5 Ga, while overgrowth rims reflect two episodes of granulite facies metamorphism possibly related to collisional orogeny at 1.89–1.88 Ga and 252–234 Ma. The unimodal age distribution (ca. 2.5 Ga) of detrital zircons and the positive $\varepsilon_{Hf}(t)$ of most Neoarchean zircons reveal that the detritus of these sedimentary protoliths from the Badu Complex came from Late Archean magmatic rocks. In addition, minor zircons formed at 3.5–3.3 Ga are indicative of the existence of older source materials.

Recently, master's thesis from Zhao (2012) reported three types of paragenetic metamorphic rocks in the Badu Complex, western Zhejiang, and they are sillimanite–garnet–biotite plagioclase gneiss, garnet amphibole gneiss, and garnet two-pyroxene granulite. The rocks occur as lenses in granitic migmatites. Their protoliths are sedimentary and magmatic rocks. Samples are selected from Suichang–Datuo areas in the northern Longquan. The exposed rocks are mainly granitic gneisses consisting of biotite two-feldspar gneiss, hornblende plagioclase gneiss, and garnet hornblende gneiss, interlayered with sillimanite–garnet–biotite gneiss, impure marble, and dike with unequal thickness. Felsic gneiss has a coarse-grained blastic texture with minor garnets. Sillimanite–garnet–biotite gneiss has consistent mineral assemblage and structure with pelitic granulite and contains minor antiperthite. The exsolution of acicular rutile is observed in some biotites. Garnets from pelitic granulite develop fine-grained symplectite and mainly comprise plagioclase, biotite, and quartz. P–T conditions of the peak stage for pelitic granulites are constrained at 0.6–0.7 Gpa and 800–850 °C. Garnets in garnet amphibole gneiss are obvious, locally zoned with green hornblende and fine-grained mineral inclusions. Hornblende is usually brown in the matrix, with high titanium content, belonging to metamorphic hornblende at granulite facies. The third stage of green metasomatic garnet and brown hornblende is also observed. The above mineral changes indicate a metamorphic process from prograde to peak to retrograde. P–T conditions for the peak stage are constrained at 0.92–1.1 Gpa and 850–900 °C. Metamorphic zircons reveal SHRIMP U–Pb metamorphic age varying from 1880 to 1850 Ma and inherited age of 2800–2000 Ma.

Zheng et al. (2011) reported the Archean basement in the west Cathaysia Block (Fig. 18 Zheng area). Lamproite diatremes are emplaced into three Proterozoic outcrop areas (i.e., Ningxiang, Jingshan, and Zhenyuan). Xenocrystic zircons yield age populations of 2.9–2.8 and 2.6–2.5 Ga, and Hf model ages (T_{DM}) vary from 2.6 to 3.5 Ga. Zircons also record a thermal event at ca. 2020 Ma, which was integrated to be products of older crustal remelting. Zircons aged from 1000 to 850 Ma are integrated to be additions of the Neoproterozoic magmatism and juvenile crustal materials. Thus, a coherent outline of the Cathaysia Block framework and Precambrian crustal evolution has been constructed: Before ca. 2.5, ancient crustal materials have already existed in the Cathaysia Block. At 2.8–2.5 Ga, an important continental growth event occurred. However, crustal generation process mainly took place at ca. 2.5 Ga, and further metamorphic thermal events occurred at 1.9–1.8 Ga. At 1.1–0.85 Ga; the addition of mantle-derived materials has been an important influence on the genesis of the Neoproterozoic migmatite.

According to gravity and seismic data, Liu et al. (2007a, b) put forward new descriptions for the basement of the Cathaysia Block: (1) The widespread ancient metamorphic basement in the Cathaysia Block may extend to the Taiwan island and Kyushu strait and is even bigger than we observed; (2) there exist several old continental nuclei in South China, and the Cathaysia Block may be formed by the accretion of many remaining continental nuclei.

5.3 *The Formation of the Jiangnan Orogenic Belt and Formation of the South China Block*

There are two main issues remaining to be debated about the formation of the South China Block: Whether Cathaysia Block is an ancient continental block or not, and the nature and timing of the Jiangnan Orogenic Belt.

As mentioned previously, there is no doubt on the existence of ancient metamorphic rocks and the Neoproterozoic sedimentary cover in the Cathaysia Block. However, the nature of Cathaysia is still controversial, with three major viewpoints, i.e., (1) Cathaysia Block preserved the Precambrian crystallized basement and cover; therefore, it is an ancient continental block (Zhou et al. 2009a; Li et al. 2012; Yu et al. 2012; Wang et al. 2012a); (2) the Cathaysia contains ancient materials which possibly came from a nearby block, but the Cathaysia in nature belongs to a Paleozoic orogenic belt because it experienced an orogenic process indicated by its characteristics of deformation and metamorphism (e.g., Ma 2006); and (3) there is a pre-Nanhua crystalline basement in the Cathaysia, which was separated into several small blocks during the ca. 800 Ma byan extensional or breakoff event and then collided again onto the Yangtze block in the Silurian (Shu 2012).

Although most researchers agree on the existence of Jiangnan Orogenic Belt, its nature and age is still under debate: (1) The Cathaysia Block is likely to be a part of the southern Laurentian continent with the Middle Proterozoic anorogenic magmatic rocks, which can be indicated by the 1.3–1.0 Ga Grenville (Sibao) metamorphism (Li et al. 1999, 2008c) and a lot of Mesoproterozoic detrital zircons in the ca. 1.43 Ga granite and slate. The Tianli schist, the only metamorphic rock of Sibao Era, is located in the southeastern margin of the Yangtze Block and deposited at 1.53–1.04 Ga with metamorphic age of 1042–1015 Ma. The unconformity between the Tianli schist and unmetamorphosed sedimentary cover records the >0.9 Ga compression deformation and ca. 830 Ma extensional structure and has been considered as the powerful evidence for the Sibao Orogenic Movement. In addition, there are also a large number of 1.2–0.96 Ga granites and magmatic detrital zircons from and sedimentary rocks in the Cathaysia and Yangtze Blocks (Li et al. 2012). The occurrence of ophiolite debris around the Yangtze Block is also controversial. Some of them, with relative younger age, e.g., Longsheng ophiolitic gabbro (760 Ma) (Ge et al. 2000), may be produced under rifting background. Some of them may be derived from back-arc basin (Li et al. 2006), e.g., the positive εNd (*t*) (ca. +5) and depleted REE-trace element coefficiency pattern of the northeast Jiangxi Ophiolite proves its SSZ origin from small back-arc oceanic basin (Li et al. 1997). The mineral-whole-rock Sm–Nd isochron age of Zhangshuduan mafic–ultramafic rocks and SHRIMP zircon U–Pb age of Xiwan adakites are 1.0 Ga (Chen et al. 1991) and 0.97 Ga (Gao et al. 2009), respectively, both of which are similar to the age of the Tianli Schist. Similarity also occurred to the "Fuchuan Ophiolite in southern Anhui Province" (Ding et al. 2002) and "Huanglingmiaowan Ophiolite" (Peng et al. 2012). Therefore, the Sibao Orogenic Movement could happen between 1.1 and 0.9 Ga. The Cathaysia Block, as the subducted part, initially collided to the southwestern margin of the Yangtze Block, and its clastics contribute greatly to the sedimentary rocks in the foreland basin (the Kunyang Group). The subducted oceanic crust along the southeastern margin of the Yangtz Block led to formation of the high-greenschist facies, Tianli schists, and the sequent Shuangxiwu magmatic arc and back-arc basin in northern Jiangxi. The subduction of back-arc basic induces the formation of adakitic granite. The appearance of abducting ophiolite and subduction-generated granite represents the final collision and formation of a unified SCB at ca. 0.88 Ga (Li et al. 2012).

(2) In addition, some researchers have speculated that it should not be a simple process of continent–continent collision between the Yangtze and Cathaysia Blocks (Zheng et al. 2007; Zhou et al. 2009a; Wang et al. 2012a, 2012b). The Shuanxiwu Complex is not representative of a continental magmatic arc, but an oceanic arc. Thus, the Yangtze and Cathaysia blocks were finally amalgamated by continent–arc–continent collision. The magmatism of 930–820 Ma records complicated orogenic process, and by contrast, the post-collisional magmatism is younger ca. 820 Ma. Zhou et al. (2009a) emphasized that isotopic dating of blueschist (866 ± 14 Ma) in the east Jiangxi can further constrain the tectonic process of the Jiangnan Orogenic Belt (Shu et al. 1995): (1) An oceanic lithosphere was northwestward subducted beneath the southeastern margin of the Yangtze Block (before 866–835 Ma, with peak subduction at ca. 866 Ma); (2) arc magmatism (from 878 to 822 Ma) includes arc affinity ellipsoidal lavas, komatiitic basalts, and keratophyres associated with basement sedimentary strata; (3) sedimentation in back-arc foreland basin (from 872 to 835 Ma), forming the sedimentary rocks in the basement of the Jiangnan Orogenic Belt; (4) collision between the Yangtze and Cathaysia blocks (from 835 to 820 Ma), forming the early stage of S-type granodiorite; and (5) post-collisional extension (after ca. 820 Ma), forming the late stage of S-type granite at 804–771 Ma and mafic–ultramafic rocks with two types of geochemical features at ca. 760 (Zhou et al. 2009b). Wang et al. (2012c) present LA–ICP–MS zircon U–Pb data carried out for the Shijiao–Huangshan intrusions from Zhuji area in Zhejiang Province, which is also tectonically adjacent to the Jiangshan–Shaoxing fault in eastern Jiangnan Orogenic Belt. The data set indicates that ultramafic rocks (including orbicular pyroxenite) crystallized at 844 ± 3 Ma, but diorites crystallized at ca. 816 ± 6 Ma. Combined with other available dating results, Wang et al. (2012a) suppose that the Shuangxiwu oceanic magmatic arc along the Jiangshan–Shaoxing fault might have been formed during the span of 930–810 Ma, and the continent–arc collision happened at 870–840 Ma, representative of the closure of back-arc basin. The volcanic activity related to post-collisional extension setting occurred at 800–760 Ma.

(3) There are other opinions. For example, Wang et al. (2012a, b, c) suggested most metamorphic basement rocks from the "Jiangnan Orogenic Belt" were younger than ca. 900 Ma, according mainly to regional stratigraphic correlation, tectonic characteristics of volcanic-sedimentary sequences, distribution and tectonic settings of leucogranite, and cordierite bearing in granodiorite. The unconformity surface, which represents the "Sibao Movement," is constrained during 840–820 Ma. Therefore, formation and evolution of the Jiangnan Orogenic Belt has no

relation to 1.1–0.9 Ga Greenville orogenesis. It is an accretionary edge of the Mid–Late Neoproterozoic Yangtze old block, which resulted by the continental marginal deposition related to an 830–780 Ma collision and post-collision processes after the Rodinia breakoff. The bottom boundary age of the Nanhua System is constrained to be ca. 780 Ma. Another study proposed there was a Paleo-Banxi ocean and further suggested the South China Block was formed by the collisional orogenesis during the Early Paleozoic or Early Mesozoic (Yang et al. 1995; Hsü et al. 1990). However, these views are not widely accepted because of the absence of the Paleozoic or Mesozoic ophiolites and typical rock association from active continental margin. The continuous sedimentary facies or biophases both in Yangtze and Cathysia blocks also do not support above-mentioned views (Chen et al. 1995). This paper prefers that the South China Block underwent an intracontinental tectonics during the Paleozoic.

In conclusion, many lines of evidence indicate the existence of the Cathysia old land, and after that Cathysia old land was multiple superimposed by Paleozoic tectonics. The Jiangnan Orogenic Belt separated the Yangtze and Cathysia blocks and represents, as a collisional belt, final assembly of the Yangtze and Cathysia blocks. Previous studies suggest that the united South China Block may form before 820–800 Ma.

It needs to emphasize that as old landmasses, the crystalline basement formation of the Yangtze and Cathysia blocks could reflect their cratonization time. It can be inferred that they have stabilized before 1.9–1.8 Ga even before ca. 2.5 Ga. Previous studies misunderstood the cratonization of the South China Block is represented by the Sibao–Jinning Movement or Nanhua cover sequences. The amalgamation of blocks (plates) represents assembly of various continents rather than cratonization; e.g., the Triassic amalgamation of North China and Yangtze blocks along the Dabie–Sulu metamorphic belt represents continental collision in the Pangea cycle rather than cratonization of the NCB or Yangtze Block. Similarly, the Jiangnan Orogenic Belt represents an amalgamation event of the Yangtze and Cathysia blocks rather than cratonization of the Yangtze or Cathysia blocks.

6 Assembly of Chinese Unified Continent

The assembly of Chinese unified continent is considered to be amalgamated by the three main continental blocks and several Triassic orogenic belts by this paper, although more terrains and orogenic belts probably joined in this tectonic process. Some researchers concern the Neoproterozoic Paleo-Qinling Orogenic Belt between the North China and Yangtze blocks and the Jiangnan (Jinning or Sibao) Orogenic Belt between the Yangtze and Cathysia blocks (Huang et al. 1977, 1980; Ren et al. 1990; Bai et al. 1993). They suggest the amalgamation between Yangtze and North China blocks, even the Tarim Block together, possibly occurred by terrains and orogenic belts during Jinning Period (Neoproterozoic in 1000–800 Ma). Hence, a modernized Neoproterozoic mainland (or Precambrian para-platform) was formed. Some people also presented a complex multistage orogenesis model for the

Paleo-Qinling Orogenic Belt and suggested the southern margin of the Tarim–North China blocks is a multiple stage orogenic belt which had been multistage reactive during Meso–Neoproterozoic (Cheng 1994). More attention has been paid to the petrogenesis and geological significance of the Neoproterozoic granites that are exposed along a long regional zone from the Altun–Kunlun Mountains to Dabie Mountains, even to the Korean Peninsula; as such, the Neoproterozoic Altun–Kunlun–Qilian–Qinling–Dabie–Sulu granitic zone could be extended eastward to Korea during 800–900 Ma.

However, studies of the ultrapressure metamorphism and geodynamics in Sulu–Dabie orogen demonstrated that the collision between the North China and Yangtze blocks occurred during the Indo-Chinese epoch. It supports the final time of the amalgamation of main old lands in China. Recently, new studies reveal the metamorphism in the Central Orogenic System including the Kunlun, Altun, Qilian, and Qinling Orogenic Belt. Data from the ophiolites, regional tectonics, and chronology reflect a complex multistage orogenesis. The Central Orogenic System (Xu et al. 2006; Yang et al. 2010) is nominated, which records the tectono-thermal events related to the Neoproterozoic rifting via Early Paleozoic to Triassic tectonic processes, even Mesozoic superimposed structure. It undergoes a long time activity and evolution history for >600 Ma since the Neoproterozoic. The northern part of the Central Orogenic System (including the west Kunlun, Altun, Qilian, east Kunlun, and North Qinling Orogenic Belts) formed during Early Paleozoic and the collision finished in Devonian. The southern part (including the south part of east Kunlun, south Qinling, Dabie, and Sulu Orogenic Belts) underwent the subduction and collision during the end of Late Paleozoic–Triassic, and this collision completed during the end of Late Triassic–Early Jurassic. These two parts formed in various periods constitute the composite Central Orogenic System. In conclusion, at least such studies show the final amalgamation of the three main old lands (blocks) and the formation of the united mainland in China occurred during the Indo-Chinese epoch, which is simultaneous with the Pangea supercontinent .

Acknowledgements This study represents the research results of a project supported by the 973 program (Grant No. 2012CB4166006), and research programs (Grant Nos. 41030316 and 41210003) supported by the National Nature Science Foundation of China. I express my gratitude to Xiyan Zhu, and Prof. Qingreng Meng, and also Dr Wang W, Wang F, and doctoral students Yan Zhong, Haozheng Wang, and Lei Zhao for their discussion and drawing figures. I also thank Drs. Jinghui Guo, Yong Sun, Taiping Zhao, Neng Jiang, Peng Peng, Huafeng Zhang, Bo Hu, and Guochun Zhao for their helpful suggestions.

References

Armstrong R-L (1981) Radiogenic isotopes: the case for crustal recycling on a nearsteady-state no-continental-growth Earth. Phil Trans R Soc Lond A 301:443–472

Bai J, Huang X-G, Dai F-Y, Wang H-C, Guo J-J, Yan Y-Y, Xiu Q-Y, Dai F-Y, Xu W-Z, Wang F-G (1993) The Precambrian evolution of China. Geological Publishing House, Beijing, pp 199–203 (in Chinese)

Chen G-D (1956) The example of "reworking area" in China platform and discussion of "the Cathaysia old land". Acta Geol Sin 36(3):239–271 (in Chinese)
Chen J, Foland K-A, Xing F-M, Xu X, Zhou T-X (1991) Magmatism along the southeastern margin of the Yangtze block: precambrian collision of the Yangtze and Cathaysia block of China. Geology 19:815–818
Chen X, Rong J-Y, Rowley D-E, Zhang J, Zhang Y-D, Zhan R-B (1995) Is the early Paleozoic Banxi Ocean in South China necessary? Geol Rev 4:389–400 (in Chinese with English abstract)
Chen Y, Xu B, Zhan S, Li Y-A (2004) First mid-Neo Proterozoic paleomagnetic results from the Tarim Basin (NW China) and their geodynamic implications. Precambr Res 133:271–281
Cheng Y-Q (Edit). (1994) Regional Geology in China. Geological Publishing House, Beijing, pp 215–217 (in Chinese)
Cheng Y-Q, Wang Z-J, Huang Z-G (eds) (2009) Stratifical lexicon of China. Geological Publishing House, Beijing, pp 39–47
Chu X-L (2004) Snow ball earth during the Neoproterozoic. Bull Mineral Petrol Geochem 23(3):233–238 (in Chinese with English abstract)
Condie K-C (2000) Episodic continental growth models: afterthoughts and extensions. Tectonophys 322:153–162
Condie K-C, Kröner A (2008) When did plate tectonics begin? Evidence from the geologic record. Geol Soc Ame Spe Pap 440:281–294
Deng J-F, Wu Z-X, Zhao G-C, Zhao Z-H, Luo Z-H, Mo X-X (1999) Precambrian granitic rocks, continental crustal evolution and craton formation of the North China platform. Acta Petrol Sin 15(2):190–198 (in Chinese with English abstract)
Dewey J-F, Windley B-F (1981) Growth and differentiation of the continental crust. Phil Trans Soc Lond A301:189–206
Ding B-H, Shi R-D, Zhi X-C, Zheng L, Chen L (2002) Neoproterozoic (~850 Ma) subduction in the Jiangnan orogen: evidence from the SHRIMP U-Pb dating of the SSZ-type ophiolite in southern Anhui Province. Acta Petrol Miner 29:375–388 (in Chinese with English abstract)
Diwu C-R, Sun Y, Dong Z-C, Wang H-L, Chen D-L, Chen L, Zhang H (2010) In situ U-Pb geochronology of Hadean zircon xenocryst (4.1-3.9 Ga) from the western of the Northern Qinling orogenic belt. Acta Petrol Sin 26:1171–1174 (in Chinese with English abstract)
Fan Q-C, Hooper P-R (1991) The Cenozoic basalti rocks of Eastern China: petrology and chemical composition. J Petrol 32:765–810
Fan W-M, Menzies M-A (1992) Destruction of aged lower lithosphere and asthenosphere mantle beneath eastern China. Geotect Metall 16:171–179
Fan H-R, Xie Y-H, Wang K-Y, Yang X-M (2001) Carbonatitic fluids and REE mineralization. Earth Sci Front 8:289–295 (in Chinese with English abstract)
Feng Z-Z, Peng Y-M, Jin Z-K, Jiang P-L, Bao Z-D, Luo Z, Ju T-Y, Tian H-J, Wang H (2001) Lithofacies Palaeogeography of the Cambrian and Ordovician in South China. Geological Publishing House, Beijing, pp 1–221 (in Chinese)
Gan X-C, Ji H-M, Sun D-Z, Jin W-S, Zhao F-Q (1995) A geochronological study on early Proterozoic granitic rocks, southwestern Zhejiang. Acta Petrol Miner 14(1):1–8 (in Chinese with English abstract)
Gao J, Klemd R, Long L-L, Xiong X-M, Qian Q (2009) Adakitic signature formed by fractional crystallization: an interpretation for the Neo-Proterozoic meta-plagiogranites of the NE Jiangxi ophiolitic melange belt, South China. Lithos 110:277–293
Gao L-Z (1999) Neoproterozoic and early Paleozoic biological explosion and earth's evolutionary rhythms. Geosciences 13(2):234–235 (in Chinese with English abstract)
Gao LZ, Ding XZ, Pang WH, Zhang CH (2011) New geological time scale of Meso- and Neoproterozoic of China and geochronologic constraint. J Stratigr 35:1–7 (in Chinese with English abstract)
Gao L-Z, Zhang C-H, Shi X-Y, Zhou H-R, Wang Z-Q (2007) Zircon SHRIMP U- Pb dating of the ash-bed from Xiamaling Formation, Qingbaikou group in North China. Geol Bull China 26(3):249–255 (in Chinese with English abstract)

Gao S, Qiu Y-M, Li W-L (2001) Study of single zircon SHRIMP dating on Kongling high-grade metamorphic terrane: discovery of >3.2 Ga material in the Yangtze Craton. Sci China Ser D-Earth Sci 31:27–35

Gao Z-J, Wang W-Y, Peng W-C, Li Y, Xiao B (1985) Sinian System in Xinjiang. Xinjiang People Press, Urumqi, pp 5–78 (in Chinese)

Ge W-C, Li X-H, Li Z-X, Zhou H, Wang J, Lee C-Y (2000) Longshan ophiolite in North Guangxi revisited. Acta Petrol Sin 16:111–118 (in Chinese with English abstract)

Geng Y-S, Wan Y-S, Yang C-H (2003) The Paleoproterozoic rift-type volcanism in Luliangshan area, Shanxi province, and its geological significance. Acta Geosci Sin 24:97–104 (in Chinese with English abstract)

Geng Y-S, Wang X-S, Shen Q-H, Wu C-M (2006) Redefinition of the Alxa Group-complex (Precambrian metamorphic basement) in the Alxa area, Inner Mongolia. Geol China 33(1):138–145 (in Chinese with English abstract)

Goodwin A (1991) Precambrian geology. Academic Press, London, p 666

Greentree M-R, Li Z-X (2008) The oldest known rocks in south-western China: SHRIMP U-Pb magmatic crystallization age and detrital provenance analysis of the Paleoproterozoic Dahongshan group. J Asian Earth Sci 33:289–302

Guo J-H, Chen Y, Peng P, Liu F, Chen L, Zhang L-Q (2006) Ultra-high pressure (UHT) metamorphism of sapphirine-bearing granulite with ~1.8 Ga age in Daqingshan, Inner Mongolia. In: Abstract Volume of 2006 National Symposium of Petrology and Geodynamics, Nanjing University, Nanjing, pp 215–218 (in Chinese)

Guo Z-J, Zhang Z-C, Jia C-Z, Wei G-Q (2000) The tectonic framework of Precambrian basement of the Tarim Craton, NW China. Sci China Ser D-Earth Sci 30(6):568–575 (in Chinese with English abstract)

Guo Z-J, Zhang Z-C, Liu S-W, Li H-M (2003) U-Pb geochronological evidence for the Precambrian complex of the Tarim Craton, NW China. Acta Petrol Sin 19:537–542 (in Chinese with English abstract)

Hsü K-J, Li J-L, Chen H-H, Wang Q, Sun S, Sengör A-M-C (1990) Tectonics of South China: key to understanding West-Pacific geology. Tectonics 183:9–39

Hsü K-J, Sun S, Li J-L (1987) It is South China Orogen and no South China Platform. Sci China, Ser B 10:1107–1115 (in Chinese with English abstract)

Hu A-Q, Wei G-J (2006) On the age of the Neo-Archean Qingir Gray Gneisses from the Northern Tarim Basin, Xinjiang, China. Acta Geol Sin 80:126–134 (in Chinese with English abstract)

Hu B, Zhai M-G, Li T-S, Li Z, Peng P, Guo J, Kusky T-M (2012) Mesoproterozoic magmatic events in the eastern North China Craton and their tectonic implications: geochronological evidence from detrital zircons in the Shandong Peninsula and North Korea. Gondwana Res 22:828–842

Hu J-X (1994) Geochronology of the late Proterozoic Badu Group in southwestern Jejiang Province. Geochimica 23(supl):18–24 (in Chinese with English abstract)

Huang J-Q, Ren J-S, Jiang C-F, Zhang Z-M, Xu Z-Q (1977) Tectonic framework of China. Acta Geol Sin 2:117–135 (in Chinese with English abstract)

Huang J-Q, Ren J-S, Jiang C-F (1980) Geotectonics and evolution in China. Scientific Press, Beijing (in Chinese)

Huang J, Chu X-L, Zhang Q-R, Feng L-J (2007) Constraints on the age of Neoproterozoic global glaciations. Earth Sci Front 14(2):249–256 (in Chinese with English abstract)

Huang W-T, Yu J-J, Zheng B-H, Chen X-Y, Zhu W-B (2009) Study on Phengite in Aksu Precambrian Blueschists, Xinjiang. Acta Mineral Sin 29(3):338–344 (in Chinese with English abstract)

Iizuka T, Horie K, Komiya T, Maruyama S, Hirata T, Hidaka H, Windley B-F (2006) 4.2 Ga zircon xenocryst in an Acasta gneiss from northwestern Canada: evidence for early continental crust. Geology 34(4):245–248

Jiao W-F, Wu Y-B, Peng M, Wang J (2009) The oldest basement rock in the Yangtze Craton revealed by zircon U-Pb age and Hf isotope composition. Sci China Ser D-Earth Sci 52(9):1393–1399

Kusky T-M, Li J-H (2003) Paleoproterozoic tectonic evolution of the North China Craton. J Asian Earth Sci 22:383–397

Kusky T-M, Windley B-F, Zhai M-G (2007) Tectonic evolution of the North China Block: from orogen to craton to orogen. In: Zhai MG, Windley BF, Kusky T, Meng QR. (eds), Mesozoic sub–continental thinning Beneath Eastern North China, vol 280. Geological Society Special Publication, London, pp 1–34

Li H-K, Lu S-N, Li H-M, Sun L-X, Xiang Z-Q, Geng J-Z, Zhou H-Y (2009) Zircon and beddeleyite U- Pb precision dating basic rock sius intruding the Xiamaling Formation, North China. Geol Bull China 28(10):22–29 (in Chinese with English abstract)

Li S-Z, Liu Y-J, Yang Z-S, Ma R (1998) Relations between deformation and metamorphic recrystallization in metapelite of Liaohe Group. Acta Petrol Sin 14:351–365 (in Chinese with English abstract)

Li W-X, Li X-H, Li Z-X (2005) Neoproterozoic bimodal magmatism in the Cathaysia Block of South China and its tectonic significance. Precambr Res 136:51–66

Li W-X, Li X-H, Li Z-X (2008a) Middle Neoproterozoic syn-rifting volcanic rocks in Guangfeng, South China: Petrogenesis and tectonic significance. Geol Mag 145(4):475–489

Li X-H, Li W-X, He B (2012) Building of the South China Block and its relavance to assembly and breakup of Rodinia supercontinent: observations, interpretations and test. Bull Mineral Petrol Geochem 31:543–559 (in Chinese with English abstract)

Li X-H, Li Z-X, Ge W-C, Zhou H-W, Li W-X, Wingate M-T-D (2003) Neoproterozoic granitoids in south China. Crustal melting above a mantle plume at ca. 825 Ma? Precambr Res 122:45–83

Li X-H, Li Z-X, Sinclair J-A, Li W-X, Carter G (2006) Revisiting the "Yanbian Terrane": implication for Neoproterozoic tectonic evolution of the western Yangtze Block, South China. Precambr Res 151:14–30

Li X-H, Li Z-X, Sinclair J-A, Li W-X, Carter G (2007) Understanding dual geochemical characters in a geological context for the Gaojiacun intrusion: response to Munteanu and Yao's discussion. Precambr Res 154:164–167

Li X-H, Li Z-X, Zhou H-W, Liu Y, Kinny P-D (2002a) U-Pb zircon geochronology, geochemistry and Nd isotopic study of Neoproterozoic bimodal volcanic rocks in the Kangdian Rift of South China: implications for the initial rifting of Rodinia. Precambr Res 113:135–154

Li X-H, Zhao J-X, McCulloch M-T, Zhou G-Q, Xing F-M (1997) Geochemical and Sm-Nd isotopic study of late Proterozoic ophiolite from southeast China: implication for ophiolite petrogenesis and tectonic evolution. Precambr Res 81:129–144

Li X-H, Zhou H-W, Li Z-X, Liu Y (2002b) Petrogenesis of Neoproterozoic bimodal volcanics in western Sichuan and its tectonic implications: Geochemical and Sm-Nd isotopic constraints. Chin J Geol 37(3):264–276 (in Chinese with English abstract)

Li X-H (1997) Timing of the Cathaysia block formation: constraints from SHRIMP U-Pb zircon geochronology. Episodes 20:188–192

Li Z-X, Bogdanova S-V, Collins A-S, Davidson A, De Waele B, Ernst R-E, Fitzsimons I-C-W, Fuck R-A, Gladkochub D-P, Jacobs J (2008b) Assembly, configuration and breakup history of Rodinia: a synthesis. Precambr Res 160:179–210

Li Z-X, Li X-H, Kinny P-D, Wang J (1999) The breakup of Rodinia: did it start with a mantle plume beneath South China? Earth Planet Sci Lett 173:171–181

Li Z-X, Li X-H, Li W-X, Ding S-J (2008c) Was Cathaysia part of Proterozoic Laurentia? -new data from Hainan Island, South China. Terra Nova 20:154–164

Li Z-X, Li X-H, Zhou H-W, Kinny P-D (2008d) Grenvillian continental collision in south China: new SHRIMP U-Pb zircon results and implications for the configuration of Rodinia. Geology 20:154–164

Li Z-X, Zhang L, Powell C-M (1995) South China in Rodinia: part of the missing link between Austrslia-East Antarctica and Laurentia? Geology 23:407–410

Ling W-L, Gao S, Zhang B-R, Li H-M, Liu Y, Cheng J-P (2003) Neoproterozoic tectonic evolution of the northwestern Yangtze craton, South China: implications for amalgamation and break-up of the Rodinia Supercontinent. Precambr Res 122:111–140

Liu D-Y, Nutman A-P-W, Compston W, Wu J-S, Shen Q-H (1992) Remmants of ≥3800 Ma crust in the Chinese part of the Sino-Korean Craton. Geology 20:339–342
Liu D-Y, Wan Y-S, Wu J-S, Wilde S-A, Zhou H-Y, Dong C-Y, Yin X-Y (2007) Eoarchean rocks and zircons in the North China Craton. Developments in Precambrian Geology, 'Earth's Oldest Rocks', vol 15. Elsevier, Amsterdam, pp 251–273
Liu H-Y (1991) China Sinian system. Scientific Press, Beijing pp 1–388, 112 (in Chinese)
Liu S-W, Pan Y-M, Li J-H, Li Q-G, Zhang J (2002) Geological and isotopic geochemical constraint on the evolution of the Fuping complex, North China Craton. Precam Res 117:41–56
Liu X-M, Gao S, Diwu C-R, Ling W-L (2008) Precambrian crustal growth of the Yangtze Craton as revealed by detrital zircon. American J Sci 308:421–468
Liu Y-H, Yu X-Z, Li J (2007b) Geophysical field information of Cathaysia. Geophys Geochem Explor 31:514–517 (in Chinese with English abstract)
Liu Y-Q, Liu Y-Q, Gao L-Z (2003) Advances in the study of Sedimentary Facies of the Sinian Candidate Stratotype in the Eastern Areas of the Three Gorges, Hubei. Geol Rev 49:187–194 (in Chinese with English abstract)
Liu Y-S, Yu H-F, Xin H-T, Lu S-Y, Xiu Q-Y, Li Q (2009) Tectonic units division and Precambrian significant geological events in Altyn Tagh Mountain, China. Geol Bull China 28(10):1430–1438 (in Chinese with English abstract)
Liu S-J, Wan Y-S, Sun H-Y, Nutman A-P, Xie H-Q, Dong C-Y, Ma M-Z, Liu D-Y, Jahn B-M (2013) Paleo- to Eoarchean crustal evolution in eastern Hebei, North China Craton: New evidence from SHRIMP U-Pb dating and in-situ Hf isotopic study of detrital zircons from paragneisses. J Asian Earth Sci 78:4–17
Long X-P, Sun M, Yuan C, Kröner A, Hu A-Q (2012) Zircon REE patterns and geochemical characteristics of Paleoproterozoic anatectic granite in the northern Tarim Craton, NW China: implications for the reconstruction of the Columbia supercontinent. Precambr Res 222–223: 474–487
Long X-P, Yuan C, Sun M, Kröner A, Zhao G-C, Wilde S, Hu A-Q (2011a) Reworking of the Tarim Craton by underplating of mantle plume-derived magmas: evidence from Neoproterozoic granitoids in the Kuluketage area, NW China. Precambr Res 187:1–14
Long X-P, Yuan C, Sun M, Xiao W-J, Zhao G-C, Zhou K-F, Wang Y-J, Hu A-Q (2011b) The discovery of the oldest rocks in the Kuluketage area and its geological implications. Sci China Earth Sci 54:342–348
Long X-P, Yuan C, Sun M, Zhao G-C, Xiao W-J, Wang Y-J, Yang Y-H, Hu A-Q (2010) Archean crustal evolution of the northern Tarim Craton, NW China: Zircon U-Pb and Hf isotopic constraints. Precambr Res 180:272–284
Lu S-N, Yuan G-T (2003) Geochronology of Early Precambrian magmatic activities in Aketashitage, East Altyn Tagh. Acta Geol Sin 77(1):61–68 (in Chinese with English abstract)
Ma R-S (2006) New thought about the tectonic evolution of the South China: with discussion on several problems of the Cathaysian old land. Geol J China Univ 12(4):448–456 (in Chinese with English abstract)
Ma Z-D, Chen Y-J (2000) Geochemical discussion on Paleo-Mesoproterozoic basement crust of Yangzte and Cathaysia cratons in southern China: using traces as tracers. Geochimica 29(6):525–531 (in Chinese with English abstract)
Mclennan S-M, Taylor S-R (1982) Geochemical constraints on the growth of the continental crust. J Geol 90:347–361
Mei M-X, Zhou P, Zhang H, Chen Y-H, Nie R-Z (2006) Sequence stratigraphic framework and its palaeogeographical background for the Sinian of Upper Yangtze Region. J Paleogeogr 8(2):219–231 (in Chinese with English abstract)
Meng Q-R, Wei H-H, Qu Y-Q, Ma S-X (2011) Stratigraphic and sedimentary records of the rift to drift evolution of the northern North China craton at the Paleo- to Mesoproterozoic transition. Gondwana Res 20:205–218
Menzies MA, Fan WM, Zhang M (1993) Paleozoic and Cenozoic lithoprobes and the loss of >120 km of Archean lithosphere, Sino-Korean craton, China. London: Geological Society of London, Special Publication. 76:71–81

Miao P-S, Zhang Z-F, Zhang J-Z, Zhao Z-X, Xu S-C (1999) Discussion of Paleoproterozoic sedimentary sequences. Geol China 18(4):405–413

Peng P, Bleeker W, Ernst R-E, Soderlund U, McNicoll V (2011a) U-Pb baddeleyite ages, distribution and geochemistry of 925 Ma mafic dykes and 900 Ma sills in the North China craton: evidence for a Neoproterozoic mantle plume. Lithos 127:210–221

Peng P, Guo J-H, Zhai M-G, Bleeker W (2010a) Paleoproterozoic gabbronoritic and granitic magmatism in the northern margin of the North China craton: evidence of crust-mantle interaction. Precambr Res 183:635–659

Peng P, Zhai MG, Ernst RE, Guo JH, Liu F, Hu B (2008) A 1.78 Ga large igneous province in the North China craton: the Xiong'er volcanic province and North China dyke swarm. Lithos 101:260–280

Peng P, Zhai M-G, Li Q-L, Wu F-Y, Hou Q-L, Li Z, Li T-S, Zhang Y-B (2011b) Neoproterozoic (~ 900 Ma) Sariwon sills in North Korea: Geochronology, geochemistry and implications for the evolution of the south-eastern margin of the North China Craton. Gondwana Res 20:234–254

Peng R-M, Zhai Y-S, Wang J-P, Chen X-F, Liu Q, Lv J-Y, Shi Y-X, Wang G, Li S-B, Wang L-G, Ma Y-T, Zhang P (2010b) Discovery of Neoproterozoic acid volcanic rock in the south-western section of Langshan, Inner Mongolia. Chin Sci Bull 55(26):2611–2620

Peng S, Kusky T-M, Jiang X-F, Wang L, Wang J-P, Deng H (2012) Geology, geochemistry, and geochronology of Miaowan ophiolite, Yangtze craton: Implication for South China's amalgamation history with Rodinia supercontinent. Gondwana Res 21:577–594

Polat A, Li J, Fryer B, Kusky T-M, Gagnon J, Zhang S (2006) Geochemical characteristics of the Neoarchean (2800-2700 Ma) Taishan greenstone belt, North China Craton: evidence for plume-craton interaction. Chem Geol 230:60–87

Qiu Y-M, Gao S, McNaughton N-J, David I-G, Ling W-L (2000) First evidence of ≥3.2 Ga continental crust in the Yangtze Craton of south China and its implications for Archean crustal evolution and Phanerozoic tectonics. Geology 28:11–14

Ren J-S, Chen T-Y, Niu B-G, Liu Z-G, Liu F-R (1990) Continental lithospheric tectonic evolution and metallogenism of Eastern China and adjacent regions. Scientific Press, Beijing (in Chinese)

Ren J-S (1989) New knowledge of geotectonic evolution of eastern China and adjacent regions. Chin Reg Geol 4:3–14 (in Chinese with English abstract)

Rogers J-J-W, Santosh M (2003) Supercontinents in Earth history. Gondwana Res 6:357–368

Santosh M, Wilde S-A, Li J-H (2007) Timing of Paleoproterozoic ultrahigh-temperature metamorphism in the North China Craton: evidence from SHRIMP U-Pb zircon geochronology. Precambr Res 159:178–196

Shu L-S, Shi Y-S, Guo L-Z, Charvet J, Sun Y (1995) Plate-terrain tectonic evolution and kinematics of Collisional Orogen in the middle Jiangnan, Estaern China. Publishing House of Nanjing University, Nanjing, pp 1–174 (in Chinese)

Shu L-S (2012) An analysis of principal features of tectonic evolution in South China Block. Geol Bull China 31(7):1035–1053 (in Chinese with English abstract)

Shui T, Xu B-T, Liang R-H, Qiu Y-S (1988) Metamorphic basement in Zhe-Min of China. Scientific Press, Beijing (in Chinese)

Song B, Zhang Y-H, Wan Y-S, Jian P (2002) Mounting and analytical procedure and zircon SHRIMP dating. Geol Rev 48(supp):26–30 (in Chinese with English abstract)

Sun D-Z, Hu W-X (1993) The tectonic framework of Precambrian in Zhongtiao Mountains. Geological Publishing House Beijing, pp 165–167 (in Chinese)

Tang H-S, Wu G, Lai Y (2009) The C-O isotope geochemistry and genesis of the Dashiqiao magnesite deposit, Liaoning Province, NE China. Acta Petrol Sin 25:455–467 (in Chinese with English abstract)

Wan Y-S, Geng Y-S, Shen Q-H (2000) Khondalite series-geochronology and geochemistry of the Jiehekou Group in Luliang area, Shanxi province. Acta Petrol Sin 16:49–58 (in Chinese with English abstract)

Wan Y-S, Song B, Liu D-Y, Li H-M, Yang C, Zhang Q-D, Yang C-H, Geng Y-S, Shen Q-H (2001) Geochronology and geochemistry of 3.8-2.5 Ga Archean rock belt in Dongshan scenic park, Ahshan area. Acta Geol Sin 75:363–370 (in Chinese with English abstract)

Wan Y-S, Dong C-Y, Liu D-Y, Kröner A, Yang C-H, Wang W, Du L-L, Xie H-Q, Ma M-Z (2012) Zircon ages and geochemistry of late Neoarchean syenogranites in the North China Craton: A review. Precambr Res 222–223:265–289

Wang J, Pan G-T (2009) Neoproterozoic South China Palaeocontinents: an overview. Acta Sediment Sina 27:818–825 (in Chinese with English abstract)

Wang X-L, Li X-H, Li Z-X, Li Q-L, Tang G-Q, Gao Y-Y, Zhang Q-R, Liu Y (2012a) Episodic Precambrian crust growth: evidence from U-Pb ages and Hf–O isotopes of zircon in the Nanhua Basin, central South China. Precambr Res 222–223:386–403

Wang X-L, Shu L-S, Xing G-F, Zhou J-C, Tang M, Shu X-J, Qi L, Hu Y-H (2012b) Post-orogenic extension in the eastern part of the Jiangnan orogen: evidence from ca 800-760 Ma volcanic rocks. Precambr Res 222–223:404–423

Wang Z-Q, Gao L-Z, Ding X-Z, Huang Z-Z (2012c) Tectonic Environment of the Metamorphosed basement in the Jiangnan Orogen and Its Evolutional Features. Geol Rev 58:401–413 (in Chinese with English abstract)

Wang Z-Q, Yin C-Y, Gao L-Z, Liu Y-Q (2002) Chemostratigraphic characteristics of correlation of the Sinian Stratotype in the Eastern Yangtze Gorges Area, Yichang, Hubei Province. Geol Rev 48:408–415 (in Chinese with English abstract)

Wang H-L, Xu X-Y, Chen J-L, Yan Z, Li T, Zhu T (2011) Constraints from zircon U-Pb chronology of Yudongzi Group magnetite-quantzite in the Lueyang area, southern Qinling, China. Acta Geol Sinica 85:1284–1290 (in Chinese with English abstract)

Wang Z-H (2009) Tectonic evolution of the Hengshan–Wutai–Fuping complexes and its implication for the Trans-North China Orogen. Precambr Res 170:73–87

Wilde S-A, Cawood P-A, Wang K-Y, Nemchin A, Zhao G-C (2004a) Determining Precambrian crustal evolution in China: a case-study from Wutaishan, Shanxi Province, demonstrating the application of precise SHRIMPU–Pb geochronology. In: Malpas J, Fletcher CJN, Ali JR, Aitchison JC (eds.) Aspects of the tectonic evolution of China: Geological, vol 226. Society London Specal Publication, London, pp 5–26

Wilde S-A, Valley J-W, Peck W-H (2001) Evidence from detrital zircons for the existence of continental crust and oceans on the Earth 4.4 Ga ago. Nature 409:175–178

Windley B-F (1995) The evolving continents (3rd edn). Willy, New York, pp 2–4, 401–460

Wu F-Y, Zhang Y-B, Yang J-H, Xie L-W, Yang Y-H (2008) Zircon U-Pb and Hf isotopic constraints on the Early Archean crustal evolution in Anshan of the North China Craton. Precambr Res 167:339–362

Wu G-H, Sun J-H, Guo Q-Y, Tang T, Chen Z-Y, Feng X-J (2010) The distribution of detrital zircon U-Pb ages and its significance to Precambrian basement in Tarim Basin. Acta Geosci Sin 32(1):65–72 (in Chinese with English abstract)

Wu J-S, Gen Y-S, Shen Q-H, Wan Y-S, Liu D-Y, Song B (1998) Archaean geology characteristics and tectonic evolution of the China-Korea Paleo-continent. Geological Publishing House, Beijing, pp 192–211 (in Chinese)

Wu Y-B, Gao S, Zhang H-F, Zheng JP, Liu XC, Wang H, Gong HJ, Zhou L, Yuan HL (2012) Geochemistry and zircon U-Pb geochronology of Paleoproterozoic arc related granitoid in the Northwestern Yangtze Block and its geological implications. Precambr Res 200–203:26–37

Wu F-Y, Yang J-H, Liu X-M, Li T-S, Xie L-W, Yang Y-H (2005) Hf isotopes of the 3.8 Ga zircons in eastern Hebei Province, China: implications for early crustal evolution of the North China Craton. Chin Sci Bull 50:2473–2480

Xiao SH, Bao HM, Wang HF, Kaufman AJ, Zhou CM, Li GX, Yuan XL, Ling HF (2004) The Neoproterozoic Quruqtagh Group in eastern Chinese Tianshan: evidence for a post-Marinoan glaciation. Precambr Res 130:1–26

Xiao X-C, Tang Y-Q, Feng Y-M, Zhu B-Q, Li J-Y, Zhao M (1992) Tectonic evolution of the Northern Xinjiang and Adjacent Regions. Geological Publishing House, Beijing, pp 48–52 (in Chinese)

Xinjiang Bureau of Geology and Mineral Resources (1991) Regional Geological chronicles of Xinjiang Uygur Ethnic Autonomous. Geological Publishing House, Beijing (in Chinese)

Xinjiang Bureau of Geology and Mineral Resources. 1999. Litho-stratigraphic Sequences. Publishing House of Chinese Geological University, Wuha, pp 5–76 (in Chinese)
Xu B, Kou X-W, Song B, Wei W, Wang Y (2008) SHRIMP dating of the upper Proterozoic volcanic rocks in the Tarim Plate and constraints on the Neoproterozoic glaciation. Acta Petrol Sin 24(12):2857–2862 (in Chinese with English abstract)
Xu B, Zheng H-F, Yao H, Li Y-A (2003) C-isotope composition and significance of the Sinian on the Tarim Plate. Chin Sci Bull 48(4):385–389
Xu Z-Q, Yang J-S, Li H-B, Yao J-X (2006) The Early Paleozoic Terrene Framework and the Formation of the High-Pressure (HP) and Ultra-High Pressure (UHP) Metamorphic Belts at the Central Orogenic Belt (COB). Acta Geol Sin 80(12):1086–1793 (in Chinese with English abstract)
Xu X, O'Reilly S-Y, Griffin W-L, Wang X, Pearson N-J, He Z (2007) The crust of Cathaysia: age, assembly and reworking of two terranes. Precambr Res 158:51–78
Yan D-P, Zhou M-F, Song H-L, John M (2002) Where was South China located in the reconstruct of Rodinia? Earth Sci Front 9(4):249–256 (in Chinese with English abstract)
Yang J-S, Xu Z-Q, Ma C-Q, Wu C-L, Zhang J-X, Wang Z-Q, Wang G-C, Zhang H-F, Dong Y-P, Lai S-C (2010) Compound orogeny and scientific problems concerning the Central Orogenic Belt of China. Geol China 37(1):1–11 (in Chinese with English abstract)
Yang M-G, Zhu P-J, Xiong Q-H, Mao S-B (2012) Framework and evolution of early Neoproterozoic-Paleozoic South China Rift System. Acta Geol Sin 86(9):1367–1375 (In Chinese with English abstract)
Yang S-F, Chen H-L, Wu G-H, Dong C-W (1995) Discovery of Early Paleozoic is land-arc volcanic rocks in northern part of Fujian Province and the significance for tectonic study. Sci Geol Sin 30(2):105–116 (in Chinese with English abstract)
Yang KF, Fan HR, Santosh M, Hu FF, Wang KY (2011) Mesoproterozoic carbonatitic magmatism in the Bayan Obo deposit, Inner Mongolia, North China: constraints for the mechanism of super accumulation of rare earth elements. Ore Geol Rev 40:122–131
Yin C-Y, Liu D-Y, Gao L-Z, Wang Z-Q, Xing Y-S, Jian P, Shi Y-R (2003) Lower boundary age of the Nanhua System and the Gucheng glacial stage: evidence from SHRIMPII dating. Chin Sci Bull 48(16):1657–1662
Yu J-H, Wang S-Y, Lai M-Y, Chen S-X, Lu B-Q (1999) Re-division of the metamorphic facies zonation of Luliang Group in Shanxi province and its significance. Geol J China Univ 5:66–75 (in Chinese with English abstract)
Yu J-H, O'Reilly Y-S, Wang L, Griffin W-L, Jiang S, Wang R, Xu X (2007) Finding of ancient materials in Cathaysia and implication for the formation of Precambrian crust. Chin Sci Bull 52(1):13–22
Yu J-H, O'Reilly S-Y, Wang L, Griffin W-L, Zhou M-F, Zhang M, Shu L (2010) Components and episodic growth of Precambrian crust in the Cathaysia Block, South China: evidence from U-Pb ages and Hf isotopes of zircons in Neoproterozoic sediments. Precambrian Res 181:97–114
Yu J-H, O'Reilly S-Y, Zhou M-F, Griffin W-L, Wang L (2012) U-Pb geochronology and Hf-Nd isotopic geochemistry of the Badu Complex, Southeastern China: implications for the Precambrian crustal evolution and paleogeography of the Cathaysia Block. Precambr Res 222–223:424–449
Yu J-H, Wang L, O'Reilly S, Griffin W, Zhang M, Li C, Shu L (2009) A Paleoproterozoic orogeny recorded in a long-lived cratonic remnant (Wuyishan terrane), eastern Cathaysia Block, China. Precambr Res 174:347–363
Yuan X-L, Xiao S-H, Yin L-M, Andrew N, Zhou C-M, Mu X-N (2002) Doushantuo Fossils: life on the Eve of animal radiation. Press of University of Science and Technology of China, Hefei, pp 1–171 (in Chinese)
Zhai M-G (2004) Precambrian Geological events in the North China Craton. In: Malpas J, Fletcher CJN, Ali JR, et al (eds) Tectonic Evolution of China, vol 226. Geological Society Special Publication, London, pp 57–72

Zhai M-G (2008) Lower crust and lithosphere beneath the North China Craton before the Mesozoic lithospheric disruption. Acta Petrol Sinica 24:1604–2185 (in Chinese with English abstract)

Zhai M-G (2010) Tectonic evolution and metallogenesis of North China Craton. Mineral Deposits 39(1):24–36 (in Chinese with English abstract)

Zhai M-G (2011) Cratonization and the Ancient North China Continent: a summary and review. Sci China Earth Sci 54(8):1110–1120

Zhai M-G (2014) Multi-stage crustal growth and cratonization of the North China Craton. Geosci Frotiers 5:457–469

Zhai M-G, Bian A-G (2001) Amalgamation of the supercontinental of the North China craton and its break up during late–middle Proterozoic. Sci China Ser D-Earth Sci 43:219–232

Zhai M-G, Guo J-H, Li Y-G, Yan Y-H, Zhang W-H, Li J-H et al (1995) The discoveries of retrograde eclogites in North China craton in Archaean. Chin Sci Bull 40:1590–1594 (in Chinese with English abstract)

Zhai M-G, Guo J-H, Yan Y-H, Han X-L, Li Y-G (1992) Discovery and preliminary study of Archaean high-pressure basic granulites in North China. Sci China Ser B 36:1402–1408 (in Chinese with English abstract)

Zhai M-G, Yang J-H, Fan H-R, Miao L-C, Li Y-G (2002) A large–scale cluster of gold deposits and metallogenesis in the Eastern North China craton. Int Geol Rev 44(5):458–476

Zhai M-G, Liu W-J (2003) Paleoproterozoic tectonic history of the North China craton: a review. Precambr Res 122:183–199

Zhai M-G, Santosh M (2011) The early Precambrian odyssey of North China Craton: a synoptic overview. Gondwana Res 20(1):6–25

Zhai M-G, Santosh M (2013) Metallogeny of the North China Craton: link with secular changes in the evolving Earth. Gondwana Res 24(1):275–297

Zhai M-G, Hu B, Peng P, Zhao T-P (2014a) Meso-Neoproterozoic magmatic events and multi-stage rifting in the NCC. Earth Sci Front 21(1):100–119 (in Chinese with English abstract)

Zhai M-G, Hu B, Peng P, Zhao T-P, Meng Q-R (2014b) Late Paleoproterozoic-Neoproterozoic multi-rifting events in the North China Craton and their geological significance: a study advance and review. Tectonophysics, in press

Zhai M-G, Windley B-F (1990) The Archean and Early Proterozoic iron formation of North China: their characteristics, geotectonic relations, chemistry and implications for crustal growth. Precambr Res 48:267–286

Zhan S, Chen Y, Xu B, Wang B, Faure M (2007) Late Neoproterozoic paleomagnetic results from the Sugetbrak Formation of the Aksu area, Tarim basin (NW China) and their implication to paleogeograhpic reconstructions and the snowball Earth hypothesis. Precambr Res 154:143–158

Zhang C-L, Li H-K, Wang H-Y (2012) A review on Precambrian tectonic evolution of Tarim Block possibility of Interraction between Neoproterozoic plate subduction and mantle plume. Geol Rev 58(5):923–936 (in Chinese with English abstract)

Zhang J-X, Yu S-Y, Gong J-H, Li H-K, Hou K-J (2013) The latest Neoarchean-Paleoproterozoic evolution of the Dunhuang block, eastern Tarim craton, northwestern China: Evidence from zircon U-Pb dating and Hf isotopic analyses. Precambr Res 226:21–42

Zhang Q-R, Chu X-L, Feng L-J (2009a) Discussion on the Neoproterozoic glaciations in the South China Block and their related paleolatitudes. Chin Sci Bull 54(10):1797–1800

Zhang S-B, Zheng Y-F, Wu Y-B, Zhao Z-F, Gao S, Wu F-Y (2006a) Zircon isotope evidence for ≥3.5 Ga continental crust in the Yangtze craton of China. Precambr Res 146:16–34

Zhang S-B, Zheng Y-F, Wu Y-B, Zhao Z-F, Gao S, Wu F-Y (2006b) Zircon U-Pb age and Hf isotope evidence for 3.8 Ga crustal remnant and episodic reworking of Archean crust in South China. Earth Planet Sci Lett 252:56–71

Zhang S-H, Zhao Y, Yang Z-Y, He Z-F, Wu H (2009b) The 1.35 Ga diabase sills from the northern North China Craton: implications for breakup of the Columbia (Nuna) supercontinent. Earth Planet Sci Lett 288:588–600

Zhang S-H, Jiang G-Q, Dong J, Han Y-G, Wu H-C (2008a) New SHRIMP U-Pb age from the Wuqiangxi Formation of Banxi Group: implications for rifting and stratigraphic erosion associated with the early Cryogenian (Sturtian) glaciation in South China. Sci China Ser D Earth Sci 51(11):1537–1544

Zhang S-H, Jiang G-Q, Han Y-G (2008b) The age of the Nantuo Formation and Nantuo glaciation in South China. Terra Nova 20(4):289–294

Zhang Z-Y, Zhu W-B, Shu L-S, Wan J-L, Yang W, Su J-B (2008c) Tectonic evolution of Precambrian blueschsits in Aksu, Northern Xinjiang. Acta Petrol Sinica 24(12):2849–2856. (in Chinese with English abstract)

Zhang X, Xu X-Y, Song G-S, Wang H-L, Chen J-L, Li T (2010) LA-ICP-MS U-Pb dating and signicance of Yudongzi Group deformation granite from Lueyang area, western Qinling, China. Geol Bull China 29:510–517 (in Chinese with English abstract)

Zhao G-C, Wilde S-A, Cawood P-A, Lu L-Z (1999) Thermal evolution of two textural types of mafic granulites in the North China craton: evidence for both mantle plume and collisional tectonics. Geol Mag 136:223–240

Zhao G-C, Sun M, Wilde S-A, Li S-Z (2005) Late Archean to Paleoproterozoic evolution of the North China Craton: key issues revisited. Precambr Res 136:177–202

Zhao L (2012) Study on the early Precambrian metamorphic evolution and geochronology of the Badu Group in southern Zhejiang. Master's thesis (Geology). Chinese Academy of Geological Science (in Chinese with English abstract)

Zhao T-P, Chen F-K, Zhai M-G, Xia B (2004a) Zircon U-Pb ages of Damiao Proterozoic anorthosite and geological implication. Acta Petrol Sin 20:685–690 (in Chinese with English abstract)

Zhao T-P, Zhai M-G, Xia B, Li H-M, Zhang Y-X, Wan Y-S (2004b) Study on the zircon SHRIMP ages of the Xiong'er Group volcanic rocks: constraint on the starting time of covering strata in the North China Craton. Chin Sci Bull 9(23):2495–2502

Zhao T-P, Zhou M-F, Jin C-W, Guan H, Li H-M (2001) Discussion on the age of the Xiong'er Group in the Southern Margin of the North China Craton. Chin J Geol 36(3):326–334 (in Chinese with English abstract)

Zhao Z-P (Edit) (1993) Precambrian crustal evolution of the Sino-Korean Paraplatform. Science Press, Beijing, pp 389–390 (in Chinese)

Zheng J-P, Griffin W-L, Li L-S, O'Reilly S-Y, Pearson N-J, Tang H-Y, Liu G-L, Zhao J-H, Yu C-M, Su Y-P (2011) Highly evolved Archean basement beneath the western Cathaysia Block, South China. Geochim Cosmochim Acta 75:242–255

Zheng J-P, Griffin W-L, O'Reilly S-Y, Zhang M, Pearson N-J, Pan Y-M (2006) Widespread Archean basement beneath the Yangtze Craton. Geology 34:417–420

Zheng Y-F, Wu R-X, Wu Y-B, Zhang S-B, Yuan H-L, Wu F-Y (2008) Rift melting of juvenile arc-derived crust: Geochemical evidence from Neoproterozoic volcanic and granitic rocks in the Jiangnan Orogen, South China. Precambr Res 163:351–383

Zheng Y-F, Zhang S-B, Zhao Z-F, Wu Y-B, Li X-H, Li Z-X, Wu F-Y (2007) Contrasting zircon Hf and O isotopes in the two episodes of Neoproterozoic granitoids in South China: Implications for growth and reworking of continental crust. Lithos 96:127–150

Zheng Y-F (2004) Position of South China in configuration of Neoproterozoic supercontinent. Chin Sci Bull 49(8):751–753

Zhou H-W, Li X-H, Wang H-R, Li J, Li H-M (2002a) U-Pb Zircon Geochronology of Basic Volcanic Rocks of the Yingyangguan Group in Hezhou, Guangxi, and Its Tectonic Implications. Geol Rev 48(suppl):22–25 (in Chinese with English abstract)

Zhou J-C, Wang X-L, Qiu J-S (2008) Is the Jiangnan Orogenic belt a Grenvillian Orogenic Belt: some problems about the Precambrian geology of South China. Geol J China Univ 14:64–72 (in Chinese with English abstract)

Zhou J-C, Wang X-L, Qiu J-S (2009a) Some Neoproterozoic Geological Events Involved in the development of the Jiangnan Orogen. Geol J China Univ 15(4):453–459 (in Chinese with English abstract)

Zhou J-C, Wang X-L, Qiu J-S (2009b) Geochronology of Neoproterozoic mafic rocks and sandstones from northeastern Guizhou, south China: Coeval arc magmatism and sedimentation. Precambr Res 170:27–42

Zhou M-F, Yan D-P, Kennedy A, Li Y-Q, Ding J (2002b) SHRIMP U-Pb zircon geochronological and geochemical evidence for Neoproterozoic arc-magmatism along the western margin of the Yangtze Block, South China. Earth Planet Sci Lett 196:51–67

Part II
The North China Craton

Formation and Evolution of Archean Continental Crust of the North China Craton

Yu-Sheng Wan, Dun-Yi Liu, Chun-Yan Dong, Hang-Qian Xie, Alfred Kröner, Ming-Zhu Ma, Shou-Jie Liu, Shi-Wen Xie and Peng Ren

Abstract The North China Craton (NCC) has had a long geological history back to ca. 3.8 Ga ago. In the Anshan area, northeastern part of the craton, three distinct complexes with ages of 3.8–3.1 Ga (Baijiafen, Dongshan, and Shengousi) have been identified, along with widespread 3.1–2.5 Ga rocks of different origins and ages. In eastern Hebei Province, abundant 3.88–3.4 Ga detrital zircons were obtained from metasedimentary rocks of the Caozhuang Complex, and the oldest rock identified is a 3.4 Ga gneissic quartz diorite. The oldest zircons that may originally have been derived from the NCC are 4.1–3.9 Ga grains in Paleozoic volcano-sedimentary rocks in the northern Qinling Orogenic Belt bordering the NCC in the south. 3.0–2.8 Ga rocks occur in Anshan, eastern Hebei, eastern Shandong, and Lushan. ca. 2.7 Ga rocks of igneous origin are exposed in eight areas of the NCC, but ~2.7 Ga supracrustal rocks have so far only been identified in western Shandong. ca. 2.5 Ga intrusive and supracrustal rocks and associated regional metamorphism occur in almost all Archean areas of the NCC. Banded iron formations contain the most important ore deposit of the Archean in the NCC and mainly formed during the late Neoarchean. Ancient crustal records obtained from deep crust beneath the NCC are similar to those in the exposed areas, with the oldest ca. 3.6 Ga rock enclaves occurring in Xinyang near the southern margin of the NCC. This synthesis is based on the compilation of a large database of zircon ages as well as whole-rock Nd isotopic and Hf-in-zircon isotopic data in order to understand the formation and evolution of the early Precambrian basement of the NCC. Considering the craton as an entity, there is a continuous age record from 3.8 to 1.8 Ga, and two tectono-thermal events are most significant in the late Neoarchean to the earliest Paleoproterozoic and late Paleoproterozoic history, with age peaks at ~2.52 and ~1.85 Ga, respectively. Whole-rock Nd and Hf-in-zircon isotopic data show similar features, documenting the addition of juvenile material to the continental crust at 3.8–3.55, 3.45, 3.35–3.3, 2.9, and 2.85–2.5 Ga with the late

Y.-S. Wan (✉) · D.-Y. Liu · C.-Y. Dong · H.-Q. Xie · A. Kröner · M.-Z. Ma · S.-J. Liu
S.-W. Xie · P. Ren
Beijing SHRIMP Center, Institute of Geology, Chinese Academy of Geological Sciences, 26 Baiwanzhuang Road, Beijing 100037, China
e-mail: wanyusheng@bjshrimp.cn

M. Zhai (ed.), *Precambrian Geology of China*,
Springer Geology, DOI 10.1007/978-3-662-47885-1_2

Mesoarchean to early Neoarchean being the most important period. Crustal recycling began as early as 3.8 Ga and continued until 3.25 Ga and appears to have played a more important role than juvenile additions between 3.25 and 2.90 Ga. After outlining the general geological history of the NNC basement, we discuss several issues relating to Archean crust formation and evolution and arrive at the following major conclusions: (1) Similar to several other cratons, the late Mesoarchean to early Neoarchean was the most important period for rapid production of continental crust, and the most intensive and widespread tectono-thermal event occurred at the end of the Neoarchean. (2) In our new tectonic model, we define and outline three ancient terranes containing abundant 3.8–2.6 Ga rocks, namely the Eastern Ancient Terrane, Southern Ancient Terrane, and Central Ancient Terrane. (3) Vertical magmatic growth is seen as the main mechanism of crust formation prior to the Mesoarchean. We favor a multi-island arc model related to subduction/collision and amalgamation of different ancient terranes in the late Neoarchean. (4) The NCC may already have been a large crustal unit as a result of cratonic stabilization at the end of the late Neoarchean, probably due to magmatic underplating.

Keywords North China Craton · Archean · Magmatism · Metamorphism · Zircon dating · Nd–Hf isotopes

1 Introduction

The North China Craton (NCC) is one of the largest cratons in the eastern Eurasian continent with a total area of 300,000 km^2. It has a triangular shape and is surrounded by young orogenic belts in the north and south, and faces the Pacific Ocean in the east. The major tectonic structures are cut off by the craton boundary, and this suggests that the NCC is a fragment of a once larger craton (Li et al. 2000). It is one of a few areas on the Earth where >3.8 Ga rocks have been identified (Liu et al. 1992; Song et al. 1996; Wan et al. 2005a, 2012a). Rocks older than ca. 2.6 Ga occur widely in the craton (Fig. 1). The NCC is characterized by strong late Neoarchean tectono-thermal events (Shen et al. 2005; Wan et al. 2011a; Zhai and Santosh 2011), and this makes it different from several other cratons worldwide, providing a chance for better understanding the tectonic property of the Archean/Proterozoic boundary. Similar to other cratons, 80–90 % of the continental crust of the NCC may have formed in the Archean.

The Archean rocks are covered in many areas by Paleo- to Neoproterozoic sedimentary sequences deposited into major basins such as the Songliao basin, North China basin, and Ordos basin. Furthermore, several events since the Paleoproterozoic modified the original features of the Archean basement, including (1) Paleoproterozoic tectono-thermal events that resulted in strong reworking of Archean crust and overprinted the original geological relationships; (2) long-term

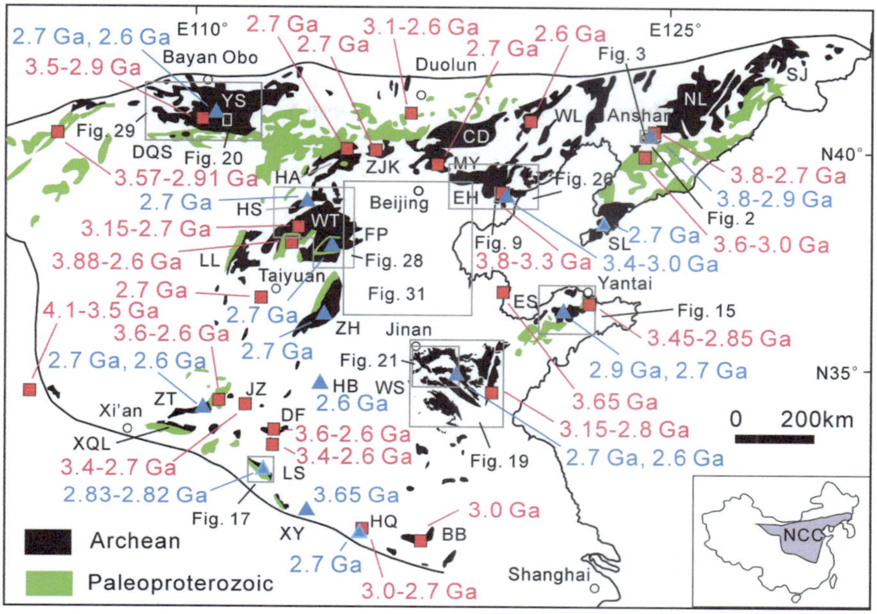

Fig. 1 Simplified sketch map showing distribution and zircon ages for the early Precambrian rocks in the North China Craton. Also shown are the locations of rocks and zircons of >2.6 Ga and Figs. 2, 3, 9, 15, 17, 19, 20, 21, 26, 28, 29, and 31. *YS* Yinshan; *HA* Huai'an; *ZJK* Zhangjiakou; *HS* Hengshan; *WT* Wutai; *FP* Fuping; *LL* Lüliang; *ZH* Zanhuang; *ZT* Zhongtiao; *JZ* Jiaozuo; *DF* Dengfeng; *XQL* Xiaoqinling; *LS* Lushan; *CD* Chengde; *MY* Miyun; *EH* eastern Hebei; *WL* western Liaoning; *SL* southern Liaoning; *NL* northern Liaoning; *SJ* southern Jilin; *WS* western Shandong; *ES* eastern Shandong; *HQ* Huoqiu; *BB* Bengbu. *Blue triangle* rock age; *red square* detrital or xenocrystic zircon age

erosion destroyed part of the Archean basement; and (3) craton destruction since the Mesozoic was mainly related to deep crustal processes but also influenced the shallow crust. All these modifications make it difficult to comprehensively understand the Archean geology of the NCC. Nevertheless, geoscientists carried out geological, petrological, geochronological, and geochemical studies in almost every Archean area of the craton. Research was also undertaken on rock samples recovered from drill cores that penetrate into basement beneath some sedimentary basins and on Archean enclaves brought up from the deep crust by young igneous rocks. In view of all these new data, the main features of the Archean basement of the NCC have become better understood although significant debates still occur as indicated by different tectonic models (e.g., Kusky and Li 2003; Wu et al. 1998; Zhai and Santosh 2011; Zhao et al. 2002, 2005; Zhao 2014).

The NCC has a long history of geological studies, and some specific case studies carried out in the last century laid important foundations for later work (e.g., Bai et al. 1986, 1996; Cao et al. 1996; Cheng et al. 1982; Li et al. 1986; Lu et al. 1996; Qian et al. 1985, 1994; Shen et al. 1990, 1992, 1994a, b, 2000; Sun et al. 1984;

Sun and Hu 1993; Wu et al. 1989, 1998; Zhang et al. 1988; Zhao et al. 1993). Since the beginning of this century, numerous papers were published on Archean rocks and their evolution, and in this chapter, we try to synthesize present knowledge by first outlining the general geological records of the Archean basement and then discussing several important issues relating to Archean crust formation and evolution.

2 Archean Geological Record

2.1 Eoarchean (>3.6 Ga)

Rocks older than 3.8 Ga have been discovered in only a few areas worldwide such as northern Canada (Bowring and Williams 1999; O'Neil et al. 2007), eastern Antarctica (Black et al. 1986), West Greenland (Kinny 1986; Nutman et al. 1996), and around Anshan city in the NNC (Liu et al. 1992; Song et al. 1996; Wan et al. 2005a, 2012a). Anshan is located southwest of the Anben (Anshan-Benxi) area where 2.5 Ga BIF-bearing supracrustal rocks (the Anshan "Group," we use the word group with quotation marks to indicate that it does not conform to modern stratigraphic terminology but constitutes a tectono-stratigraphic term) and granitoids (mainly syenogranites) occur widely (Fig. 2).

In Anshan, 3.8 Ga rocks occur within three complexes, namely the Baijiafen, Dongshan, and Shengousi complexes (Fig. 3). Baijiafen quarry was the site where 3.8 Ga rocks were first discovered (Liu et al. 1992). The complex is ~700 m long and ~50 m wide in exposure and extends in a NW–SE direction, in tectonic contact with the 3.3–3.1 Ga Chentaigou granite in the southwest and 3.36 Ga Chentaigou supracrustal rocks in the northeast (Wan et al. 2005a; Wan unpublished data). It is mainly composed of strongly mylonitized trondhjemitic gneiss of different ages with some biotite schist, gneissic monzogranite, and quartz diorite (Fig. 4). In an early study, all granitoids of the complex were considered to have formed at 3.8 Ga (Liu et al. 1992), but later studies indicated that these rocks formed at different times ranging from 3.8 to 3.1 Ga (Liu et al. 2008). The 3.8 Ga mylonitized trondhjemitic gneiss occurs in narrow layers such as sample A0518 (Fig. 5a) that is distinguished from the surrounding rocks (also trondhjemitic gneisses) by containing less biotite. The biotite schist (Fig. 5b) varies in thickness between a few centimeters and half a meter and alternates with igneous units, including trondhjemitic gneiss. There are two opinions on the origin of the schist, namely an altered mafic dike (Song et al. 1996) or a metasedimentary rock (Liu et al. 2008). The latter is based on the observation that the schist is interlayered with chert that is composed of recrystallized quartz with some banded biotite + epidote aggregates. However, at some localities, quartz veins occur together with biotite schist; therefore, the "chert" may, in fact, be a deformed quartz vein. Although it is uncertain whether the biotite schist is derived from a mafic dike or a supracrustal

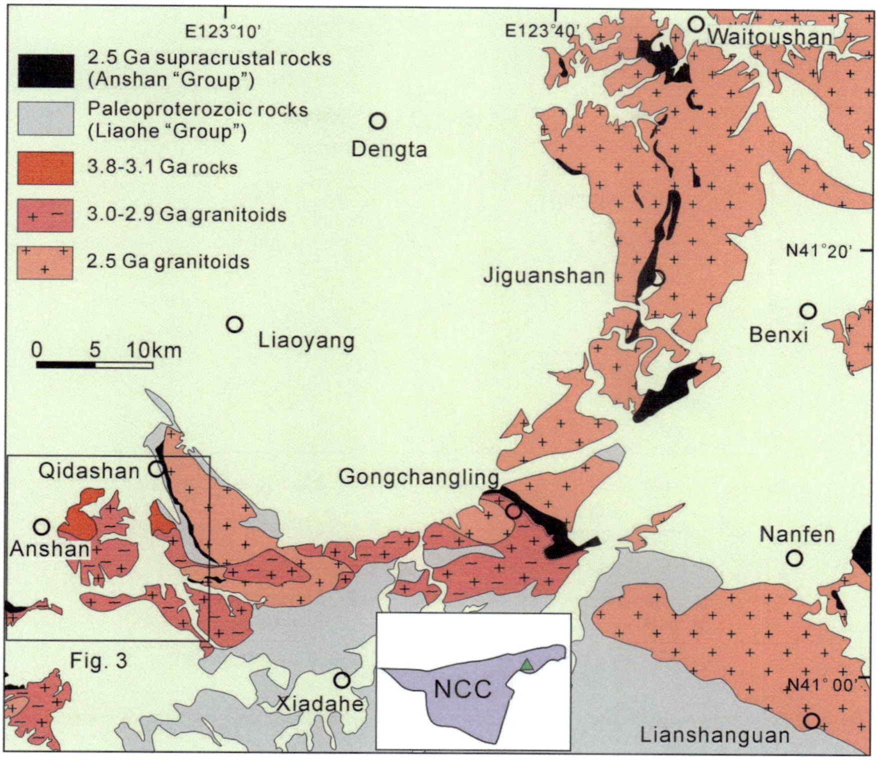

Fig. 2 Geological map of the Anben (Anshan-Benxi) area, North China Craton. Modified after LBGMRE (1975a, b, 1976); Wan (1993), and Wan et al. (2015a). *Triangle* in *inset* map shows location of Fig. 2 in the NCC

rock, it is considered to be older than 3.6 Ga because it is cut by 3.62 Ga trondhjemitic gneiss (Fig. 5b). Some younger magmatic rocks (samples A0405, A0517, A0521) contain xenocrystic zircons with ages of 3.8–3.6 Ga.

Rocks of the Dongshan Complex trend approximately WNW–ESE (Fig. 3), and the exposure is more than 10 m wide and up to 1000 m long. It occurs as a large enclave within the 3.14 Ga Lishan trondhjemite and is cut by veins derived from the Lishan trondhjemite intrusion. Different types of 3.8–3.1 Ga rocks have been identified in the complex, including banded trondhjemitic gneiss, gneissic trondhjemite, gneissic monzogranite, pegmatite, meta-ultramafic rock (komatiite?), amphibolite (metabasalts or gabbro), and meta-quartz diorite. These units are mostly structurally concordant to each other because of strong ductile deformation, and their original contact relationships are often difficult or impossible to recognize. Rocks older than 3.6 Ga were identified in three locations, including 3.81–3.68 Ga banded trondhjemitic gneisses and 3.79 Ga meta-quartz diorite. The 3.81 Ga banded trondhjemitic gneiss (sample Ch28, Fig. 5c) near a pavilion was first identified in the complex (Song et al. 1996). 3.79 Ga meta-quartz diorite (sample

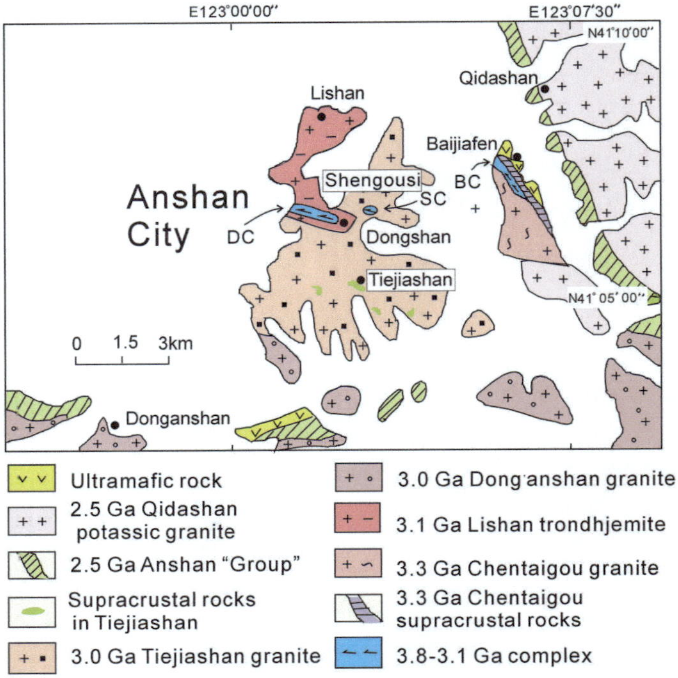

Fig. 3 Geological map of the Anshan area (Wan et al. 2012a). *BC* Baijiafen Complex; *DC* Dongshan Complex; *SC* Shengousi Complex

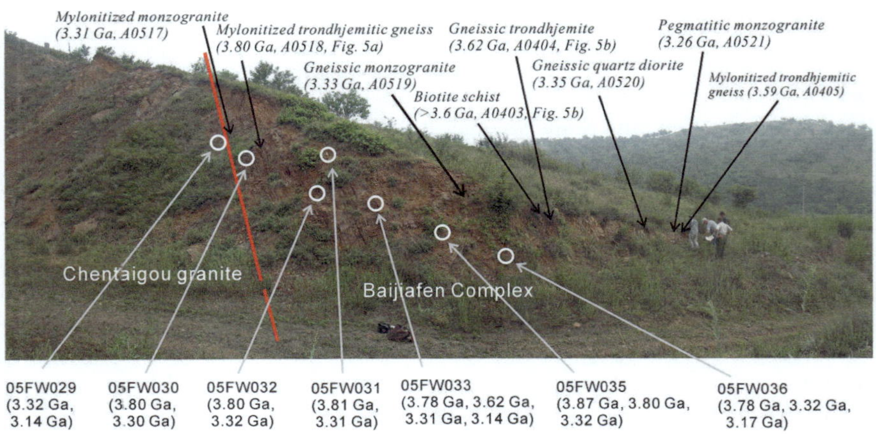

Fig. 4 Photographic mosaic showing studied section of the Baijiafen Complex with sampling sites marked (Liu et al. 2008). Samples with prefix 05FW are from Wu et al. (2008)

Fig. 5 Field photographs of Eoarchean rocks in the Anshan area. **a** 3.80 Ga mylonitized trondhjemitic gneiss (A0518), Baijiafen Complex; **b** biotite schist (A0403) intruded by 3.62 Ga trondhjemitic gneiss (A0404), Baijiafen Complex; **c** 3.81 Ga banded trondhjemitic gneiss (Ch28) intruded by trondhjemitic and pegmatitic veins, Dongshan Complex; **d** 3.79 Ga meta-quartz diorite (A9604) occurring in 3.79 Ga banded trondhjemitic gneiss (A0507), Dongshan Complex; **e** 3.68 Ga banded trondhjemitic gneiss (A0423) intruded by pegmatitic vein, Dongshan Complex; **f** 3.77 Ga banded trondhjemitic gneiss (A0512), Shengousi Complex

A9604) occurs as a small enclave within the 3.79 Ga banded trondhjemite (sample A0507) in an outcrop near the southeastern end of the Dongshan Complex (Fig. 5d). The identification of 3.68 Ga banded trondhjemitic gneiss (sample A0423, Fig. 5e) indicates that there are Eoarchean banded trondhjemitic gneisses of different ages in the complex although they are similar in field appearance and composition (Liu et al. 2008).

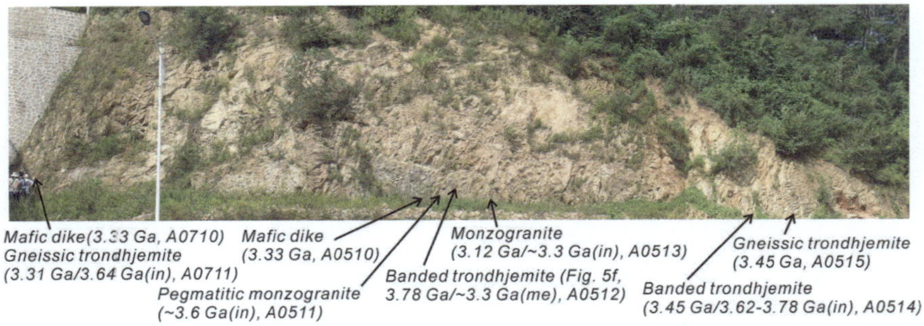

Fig. 6 Photographic mosaic showing the section of the Shengousi Complex in the Anshan area (Wan et al. 2012a)

The Shengousi Complex is located between the Dongshan and Baijiafen complexes, and its full extent is unknown due to poor exposure and is only readily apparent in a deep road cut (Wan et al. 2012a). The key locality is a ~50 m-long and up to ~10 m-high section on an elevated bench in this road cut (Fig. 6). The complex comprises polyphase migmatites, and despite a strong structural overprint through ductile deformation, there are small areas of lower strain where the relative chronologies between the different migmatite components are preserved. The youngest component on the basis of field relationships is a pegmatitic monzogranite dike. Eight samples were collected for zircon dating in this section, and these samples revealed a protracted tectono-magmatic history from 3.77 to 3.13 Ga with the oldest rocks being 3.77 Ga banded trondhjemitic gneisses (sample A0512, Fig. 5f). The complex mainly contains monzogranite and mafic metamorphic rocks besides trondhjemitic rocks of different ages (Fig. 6).

The above three complexes are similar in their geochronological record. Apart from >3.6 Ga gneisses, 3.45–3.0 Ga rocks have been identified. Many >3.6 Ga rocks notably contain younger (mainly 3.3 Ga) zircons. Some authors considered this as evidence that the trondhjemitic gneisses formed at 3.3 Ga and contained numerous 3.8 Ga zircon xenocrysts (Wu et al. 2008). However, based on field observations, zircon morphology, and cathodoluminescence (CL) images, a better explanation may be that these granitoid gneisses predominantly constitute strongly deformed igneous injection migmatites containing igneous components of several ages and/or are related to anatexis of 3.8 Ga rocks at 3.3 Ga (Liu et al. 2008; Nutman et al. 2009; Song et al. 1996; Wan et al. 2012a). There is no doubt that the 3.79 Ga quartz diorite identified in the Dongshan Complex is magmatic in origin, as indicated by its geological, petrological, and compositional features, although it also contains younger zircons (Wan et al. 2005a). Some of the <3.6 Ga rocks in the three complexes that do not show banded structures or strong deformation also contain old zircon xenocrysts (Fig. 7). All ~3.8 Ga gneisses are trondhjemitic in composition with high SiO_2 and Na_2O, and low ΣFeO ($FeO + 0.9 \times Fe_2O_3$), MgO, CaO, and K_2O. They have very low total rare earth element (REE) contents, positive or negligible Eu anomalies, and weakly fractionated REE patterns (Fig. 8a).

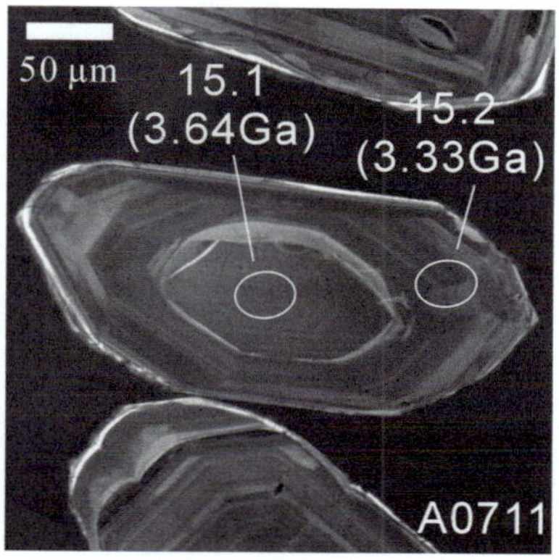

Fig. 7 3.64 Ga xenocrystic zircon core occurring in 3.33 Ga magmatic zircon from gneissic trondhjemite (A0711) in the Shengousi Complex, Anshan

Many >3.6 Ga detrital and xenocrystic zircons were also discovered in the Gongchangling and Waitoushan areas (Fig. 2) (Wan 1993; Wan et al. 2015a). This suggests that Eoarchean rocks may occur in a wider area of Anben than currently identified. The single zircon age of 4.17 Ga reported by Cui et al. (2013) from an amphibolite (metabasalt) of the 2.5 Ga Anshan "Group" in Waitoushan is questionable after rechecking the original LA-ICP-MS data (Diwu personal communication).

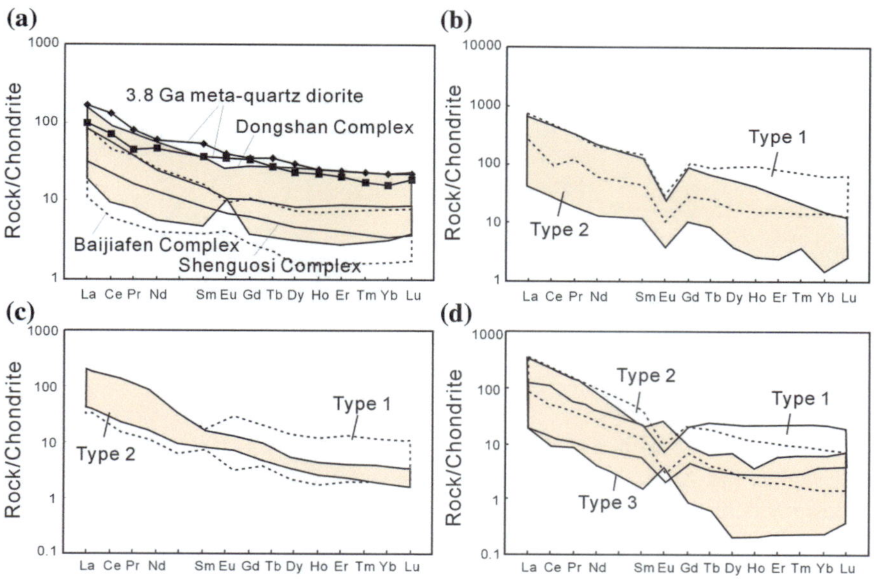

Fig. 8 Chondrite-normalized REE patterns for Archean rocks from the North China Craton. **a** 3.8 Ga trondhjemitic rocks in Anshan; **b** 3.0–2.9 Ga Tiejiashan K-rich granite in Anshan; **c** 2.75–2.7 Ga TTG rocks in the North China Craton; **d** 2.53–2.50 Ga syenogranite in the North China Craton. Data are from Jahn et al. (2008), Wan et al. (2005a, 2007, 2011b, 2012a, b, 2014a), Yang et al. (2013), and Zhu et al. (2013)

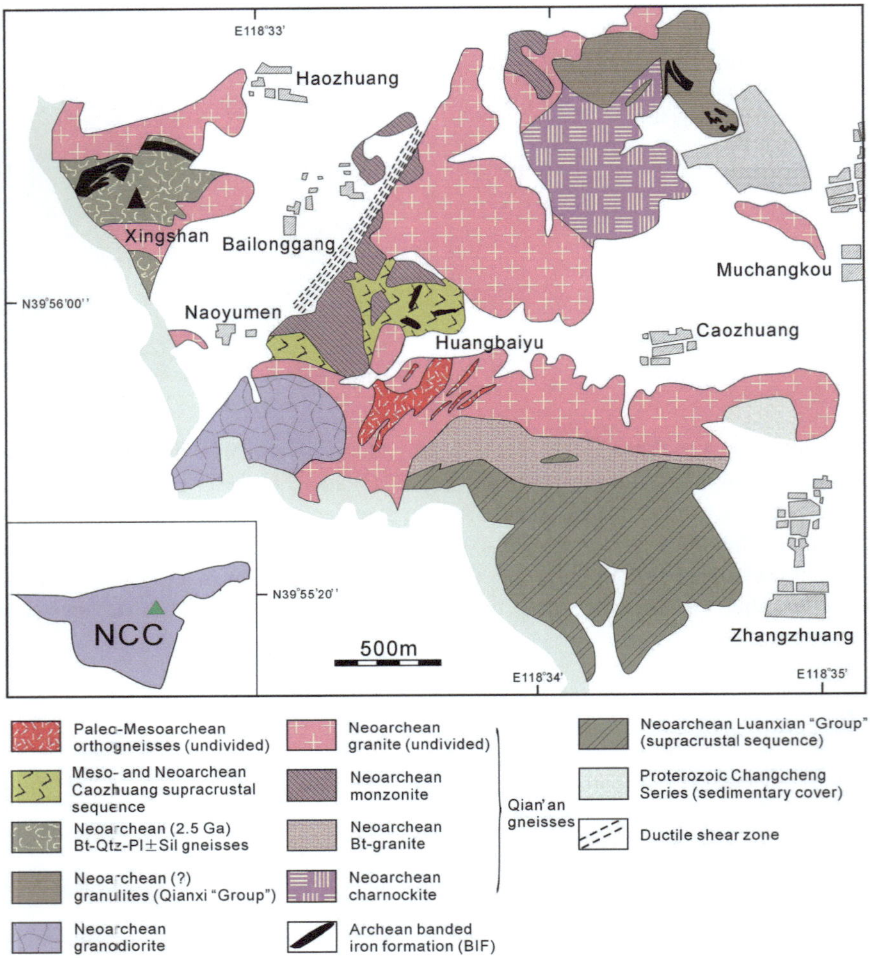

Fig. 9 Simplified geological map of the Caozhuang area, eastern Hebei. Modified after Chen (1988), Liu et al. (2013a), and Nutman et al.(2011). *Triangle* in *inset* map shows location of Fig. 9 in the NCC

Eoarchean rocks within relatively large areas in the NCC have so far only been identified in the Anben area. However, abundant Eoarchean zircons were discovered in rocks of the Caozhuang Complex, eastern Hebei Province. This complex is composed of amphibolite- to granulite-facies granitoids and a supracrustal sequence (Fig. 9). The supracrustal sequence is composed of Bt-Pl-Qtz ± Sil and Bt ± Grt gneisses, amphibolite, marble, calc-silicate, fuchsite quartzite, and banded iron formation (BIF) and was considered to have an age of 3.5 Ga on the basis of a Sm–Nd isochron age for amphibolites (Jahn et al. 1987). However, the Sm–Nd isochron was later interpreted as a mixing line (Nutman et al. 2011). The oldest rocks

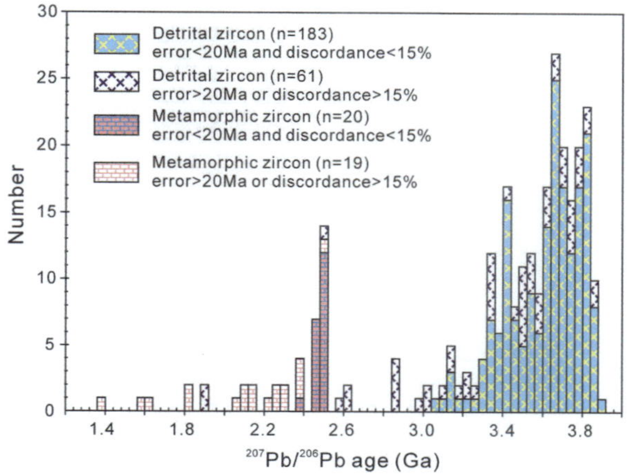

Fig. 10 Age histogram for zircons from metasedimentary rocks (including fuchsite quartzite, para-amphibolite, and metapelitic rock) in the Caozhuang area, eastern Hebei. Metamorphic zircon ages of 2.5 Ga have been identified only in para-amphibolite and metapelitic rock. Data are from Liu et al. (1992, 2013a), Nutman et al. (2011), Wilde et al. (2008), and Wu et al. (2005a)

identified in the area are 3.4–3.3 Ga granitoid gneisses (Nutman et al. 2011; Liu unpublished data). Detrital zircons with ages of 3.88–3.55 Ga were reported from the Caozhuang (fuchsite) quartzite (Liu et al. 1992; Nutman et al. 2011; Wilde et al. 2008; Wu et al. 2005a). Recently, Liu et al. (2013b) carried out SHRIMP dating on detrital zircons from para-amphibolite and Grt-Bt gneiss and obtained detrital core ages ranging from 3.8 to 3.4 Ga and metamorphic rim ages of ~2.5 Ga. Combining the detrital zircon data of the three rock types, age peaks at ~3.83, ~3.67, 3.55, and ~3.41 Ga can be recognized (Fig. 10), suggesting that detrital material was derived from crustal sources of different ages. Based on zircon morphology and inclusion studies, Nutman et al. (2014) suggested that the Caozhuang quartzite is most likely a <3.5 Ga immature sedimentary rock of local provenance. This interpretation strengthens the case for Eoarchean rocks with a substantial 3.88–3.80 Ga component occurring in eastern Hebei Province and indicates that the paragneiss protoliths of the Caozhuang Complex were deposited between 3.4 and 2.5 Ga ago (Liu et al. 2013b; Nutman et al. 2014).

Rocks and zircons older than 3.6 Ga have also been identified elsewhere in the NCC (Fig. 1), including 3.65 Ga felsic granulite enclaves in Mesozoic volcanic rocks of the Xinyang area (Zheng et al. 2004a) and ≥3.6 Ga detrital zircons in the Zhongtiao and Dengfeng areas (Liu et al. 2012a; Zhang et al. 2014a). Interesting is the discovery of three 4.1–3.9 Ga xenocrystic or detrital zircons, together with younger Archean and Paleoproterozoic grains, in Paleozoic volcano-sedimentary rocks of the northern Qinling Orogenic Belt (Diwu et al. 2010a, 2013, personal communication; Wang et al. 2007). One zircon has a 3.71 Ga overgrowth rim (Fig. 11), suggesting a tectono-thermal event at that time in the zircon source

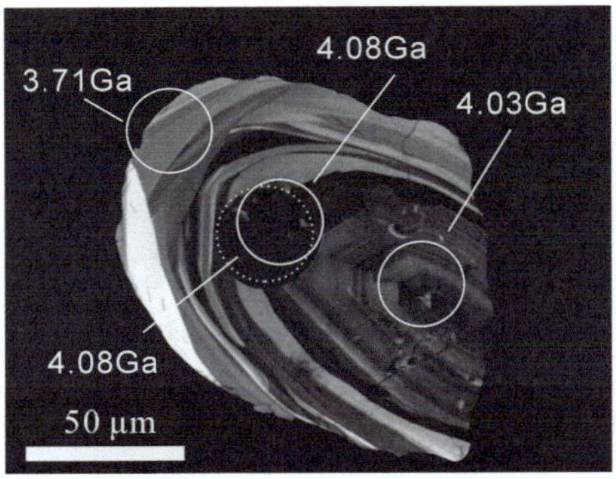

Fig. 11 Cathodoluminenscence image showing 3.71 Ga overgrowth rim on 4.08 Ga xenocrystic or detrital zircon from pyroclastic rock in the Paleozoic Caotangou Group in northern Qinling, near the North China Craton (from Diwu et al. 2010a, b). *Circle* and *dashed circles* refer to SHRIMP and LA-ICPMS dating locations, respectively

region. The volcanic rock is located near the southern margin of the NCC, and the old zircons are considered to be derived from the NCC basement (Diwu et al. 2013; Wan et al. 2009a).

2.2 *Paleoarchean (3.6–3.2 Ga)*

Paleoarchean rocks and components have been identified in many areas of the NCC and reflect the most important tectono-magmatic events in Anshan as indicated by their distribution and zircon age histogram (Fig. 12). The 3.36 Ga Chentaigou supracrustal rocks near Baijiafen (Fig. 3) are the oldest supracrustal rocks identified

Fig. 12 Age histogram of zircons from Archean rocks in the Anben area. Data are from Liu et al. (1992, 2008), Song et al. (1996), Wan et al. (2005a, 2007, 2012a, b, d, 2015a, unpublished data), Wu et al. (2008), Yin et al. (2006), and Zhou et al. (2007, 2009)

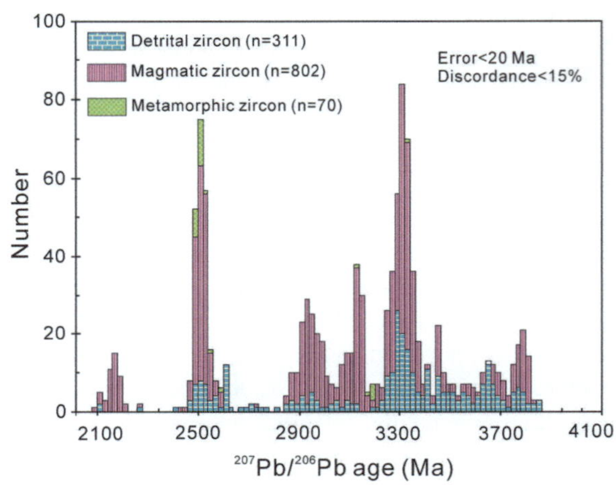

in the NCC. Their formation age is limited by 3.36 Ga felsic metavolcanic rocks in the sequence and 3.34–3.31 Ga trondhjemitic veins intruding the sequence (Fig. 13a, b) (Song et al. 1996; Wan unpublished data). The main rock types include amphibolite, meta-ultramafic rocks, felsic metavolcanic rock, meta-calc-silicate, fuchsite quartzite, and iron-bearing chert (Fig. 13c, d). The meta-ultramafic rocks occur over a relatively large area in the northeastern exposure of the sequence, and their protolith was uncertain (volcanic or intrusive) in early studies. They are intruded by 3.31 Ga trondhjemitic veins (Fig. 13b) and are considered to belong to the sequence. They are similar to komatiites in chemical compositions with MgO contents up to 38 %, but no spinifex textures have been identified because of strong modification. The amphibolite is tholeiitic in composition, showing a flat REE pattern with no large ion lithophile element (LILE) enrichment. This is the main reason for the speculative interpretation that these supracrustal rocks formed in an arc environment (Wan et al. 1997a).

The oldest Paleoarchean rock is a 3.45 Ga trondhjemitic gneiss in the Shengousi Complex (Fig. 13e). It is considered to have interacted with Eoarchean trondhjemitic rocks nearby to form an injection migmatite that subsequently underwent strong deformation to produce a rock with layers of different ages (Wan et al. 2012a). Igneous rocks of 3.35–3.2 Ga widely occur in the Anshan area, including the Baijiafen, Dongshan, and Shengousi complexes. The rock types are metagabbro, metadiorite, meta-quartz diorite, granite, as well as trondhjemitic gneiss (Dong et al. 2013; Liu et al. 2008; Wan et al. 2001, 2012a; Zhou et al. 2007, 2009). It is common that different rocks with similar or different ages occur within a single outcrop (Fig. 13f). The Chentaigou granite was considered to be the largest pluton formed at ~3.3 Ga (Song et al. 1996). However, Wu et al. (2008) obtained ~3.3 and ~3.1 Ga zircon ages for three Chentaigou granite samples near the Baijiafen Complex and thought the latter to represent the emplacement time of the pluton. Based on geological mapping and zircon dating, Li et al. (2013) suggested that the Chentaigou granite is composed of 3.3 and 3.1 Ga granitoid rocks. Trondhjemitic rocks in the Anshan area, between 3.45 and 3.3 Ga in age, show a significant variation in REE composition from very low ΣREE contents and weakly fractionated REE patterns to high ΣREE contents and strongly fractionated REE patterns (Wan unpublished data).

In Caozhuang, eastern Hebei Province, 3.4 Ga quartz dioritic gneiss and 3.3–3.2 Ga tonalitic gneiss (Fig. 13f, g) occur as enclaves within 2.5 Ga granite (Nutman et al. 2011; Liu unpublished data). This is the second area where Paleoarchean igneous rocks were identified in the NCC. Based on the youngest detrital zircons at 3.55 Ga from a fuchsite quartzite and the oldest TTG rock at ~3.3 Ga in the area, Wan et al. (2009a) suggested that deposition of the Caozhuang Complex supracrustal rocks occurred between 3.55 and 3.3 Ga. However, the relationship between the supracrustal rocks and the ~3.3 Ga TTGs is uncertain. Based on the metamorphic zircon age of 2.5 Ga and the believable youngest detrital zircon age of 3.4 Ga of the metasedimentary rocks (Fig. 10), we can now only limit deposition to between 3.4 and 2.5 Ga. Mainly based on zircon dating, Han et al. (2014a) concluded that the supracrustal rocks in Xingshan, ~1.5 km northwest of

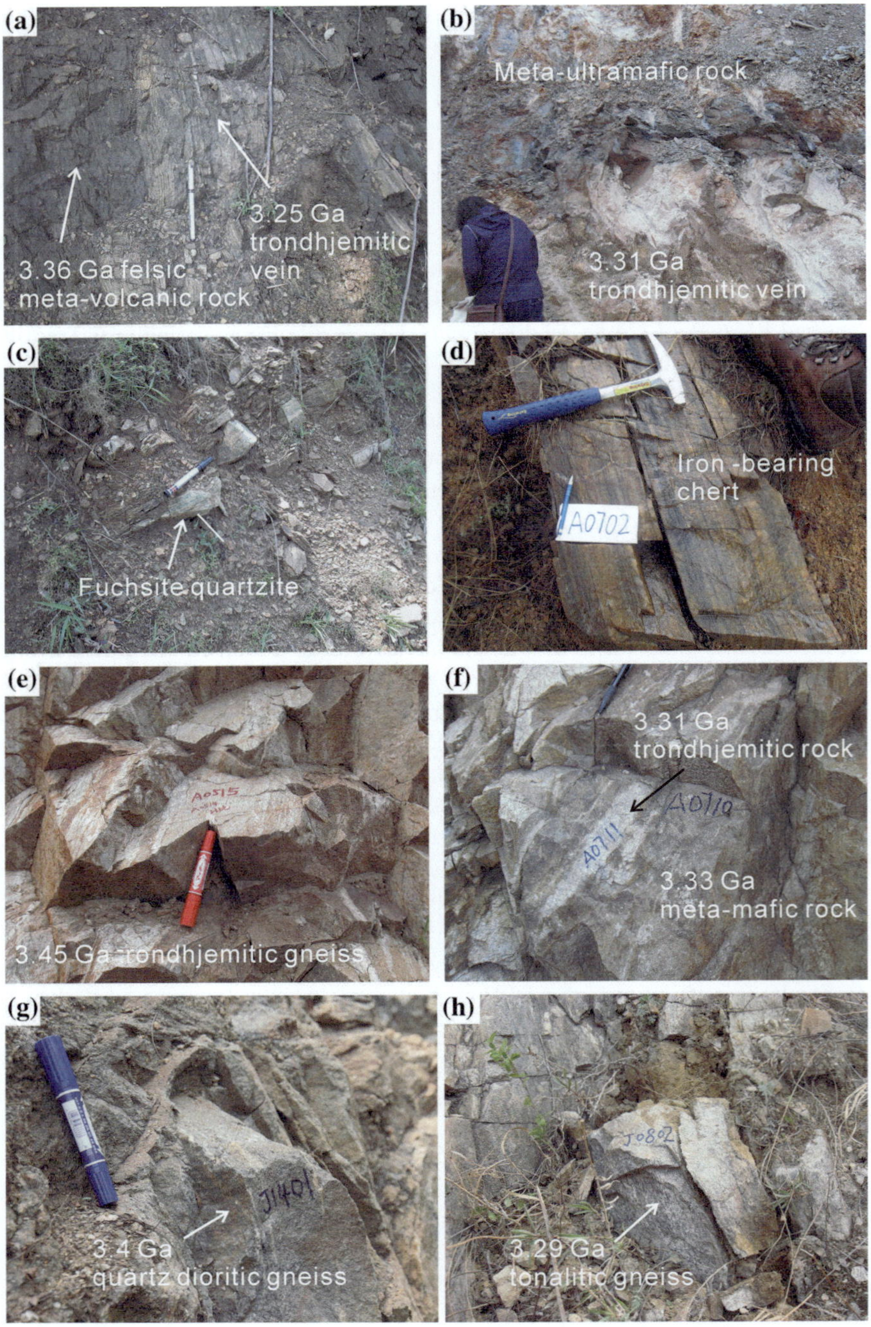
(a)
3.25 Ga
trondhjemitic
vein
3.36 Ga felsic
meta-volcanic rock
(b)
Meta-ultramafic rock
3.31 Ga
trondhjemitic vein
(c)
Fuchsite quartzite
(d)
Iron -bearing
chert
A0702
(e)
A0515
3.45 Ga trondhjemitic gneiss
(f)
3.31 Ga
trondhjemitic rock
A0710
A0711
3.33 Ga
meta-mafic rock
(g)
J1401
3.4 Ga
quartz dioritic gneiss
(h)
J0802
3.29 Ga
tonalitic gneiss

◀ **Fig. 13** Field photographs of Paleoarchean rocks in the North China Craton. **a** 3.36 Ga felsic meta-volcanic rock (the Chentaigou supracrustal rocks) intruded by 3.25 Ga trondhjemitic vein, Anshan; **b** meta-ultramafic rock (the Chentaigou supracrustal rocks) intruded by 3.31 Ga trondhjemite vein, Anshan; **c** fuchsite quartzite in the Chentaigou supracrustal rocks, Anshan; **d** iron-bearing chert in the Chentaigou supracrustal rocks, Anshan; **e** 3.45 Ga trondhjemitic gneiss in the Shengousi Complex, Anshan; **f** 3.31 Ga gneissic trondhjemite and 3.33 Ga meta-mafic rock in the Shengousi Complex, Anshan; **g** 3.4 Ga gneissic quartz diorite, Huangbaiyu, eastern Hebei; **h** 3.29 Ga gneissic trondhjemite, Huangbaiyu, eastern Hebei

Huangbaiyu, that belong to the Caozhuang Complex, were deposited at 3.39 Ga. However, the dated rock contains garnet and many 3.76–3.43 Ga zircons, similar in features to the metasedimentary rocks in Huangbaiyu. Therefore, it is possible that the 3.39 Ga zircon is detrital in origin, and this age therefore only constrains the upper depositional age of the Caozhuang supracrustal rocks.

The age distribution of >3.3 Ga detrital zircons from the Caozhuang Complex is similar to that of magmatic zircons in the Anshan area (Figs. 10 and 12). However, it appears that the detrital material was derived from the surrounding area, rather than Anshan, which is too far from Caozhuang.

Paleoarchean detrital and xenocrystic zircons were identified in many areas of the NCC such as Alax, Guyang, Wutai, Jiaozue, Dengfeng, Zhongtiao, eastern Shandong, and Bengbu (Fig. 1) (Diwu et al. 2008; Gao et al. 2006; Ji 1993; Jian et al. 2012; Jin et al. 2003; Liu et al. 2012a; Shen et al. 2005; Wan et al. 2006, 2009a, c, 2010a; Wang et al. 1998; Xie et al. 2014a; Zhang et al. 2014a, b).

2.3 *Mesoarchean (3.2–2.8 Ga)*

Mesoarchean rocks are more widespread in the NCC than Paleoarchean rocks. In Anshan, large Mesoarchean intrusive bodies include the 3.14 Ga Lishan trondhjemite, the 3.0 Ga Dong'anshan granite, the 3.0–2.9 Ga Tiejiashan granite, and part of the 3.3–3.1 Ga Chentaigou granite (Fig. 3). The Tiejiashan granite (Fig. 14a) is the oldest and largest K-rich granite pluton in the NCC (Wan et al. 2007; Wu et al. 1998) and occupies a total area of >150 km^2 (Fig. 2). The rocks vary in their REE fractionation patterns but have high ΣREE contents and strong negative Eu anomalies (Fig. 8b). The Tiejiashan granite contains supracrustal xenoliths such as BIF and quartzite (Fig. 14b, c). There are also some mica–quartz schists (Fig. 14d) that were considered to be supracrustal enclaves (Yin et al. 2006). However, they show similar geochemical features as the Tiejiashan granite, and the so-called detrital zircons in the rocks are mainly ~3.0 Ga in age. In some thin sections, K-feldspar grains occur at different sizes in the fine-grained sericite + quartz groundmass. These features suggest that sericite resulted from strong deformation and alteration of K-feldspar during shearing. Therefore, we suggest that at least some schists are due to strong deformation and alteration of granite. Besides the large Mesoarchean granitoid bodies, many small Mesoarchean granitoid bodies or

(a)
A0432
2.91 Ga K-rich granite
(b)
BIF
(c)
(d)
K-rich granite
Mica-quartz schist
(e)
3.13 Ga augen-like granite
BC
(f)
3.14 Ga monzogranite
3.14 Ga trondhjemite
(g)
J0604
~3.2 Ga gneissic tonalite
(h)
2.94 Ga anatectic granitoid

◀ **Fig. 14** Field photographs of Mesoarchean rocks in the North China Craton. **a** Mesoarchean Tiejiashan K-rich granite, Anshan; **b** BIF in Mesoarchean Tiejiashan K-rich granite, Anshan; **c** quartzite in Mesoarchean Tiejiashan K-rich granite, Anshan; **d** mica–quartz schist in Mesoarchean Tiejiashan K-rich granite, Anshan; **e** 3.13 Ga augen-like granitic gneiss in the Baijiafen Complex, Anshan; **f** 3.14 Ga trondhjemite and 3.14 Ga monzogranite, Dongshan Complex, Anshan; **g** ca. 3.2 Ga gneissic tonalite, Huangbaiyu, eastern Hebei; **h** 2.94 Ga anatectic granitoid, Huangbaiyu, eastern Hebei

veins also occur in the Baijiafen, Dongshan, and Shengousi complexes (Figs. 4, 6 and 14e, f).

In Caozhuang, eastern Hebei Province, Mesoarchean igneous rocks have also been discovered (Fig. 14g, h) (Nutman et al. 2011), but they only occur on a very small scale. However, Mesoarchean xenocrystic zircons are more widespread.

In eastern Shandong, 2.9 Ga rocks are widely distributed, extending from Mazhuanghe in the west to Hexikuang in the east, with some occurring in the southern portion of the area (Fig. 15). Most of the so-called Mesoarchean Tangjiazhuang "Group" consists of intrusive rather than supracrustal rocks. Only a few 2.9 Ga supracrustal rocks, named Huangyadi supracrustal rocks, occur on a small scale within 2.9 Ga igneous rocks (Jahn et al. 2008). Mesoarchean supracrustal rocks in eastern Shandong are much smaller in scale than thought before (BGMRSP 1991). At least three types of 2.9 Ga igneous rocks, namely quartz diorite, tonalite, and high-Si trondhjemite, have been identified (Jahn et al. 2008; Liu et al. 2011, 2013a; Wang et al. 2014a; Wu et al. 2014; Xie et al. 2014b). Of these, tonalite seems to be the most common, with a total area up to several tens of square km. 2.9 Ga gneissic tonalite occurs in the lower part along a ~200 m-long road cut northeast of Zhoujiagou. It is in contact with 2.9 Ga gneissic high-Si trondhjemite in the upper part; both show strong deformation with their foliations parallel to each other (Fig. 16a–c). At the Hexikuang reservoir dam, 2.9 Ga gneissic quartz diorite occurs as enclaves within 2.9 Ga gneissic tonalite, and these show a differently oriented foliation (Fig. 16d). Felsic veins in the quartz diorite are parallel to the foliation and are considered to be anatectic products. Both the gneissic quartz diorite and tonalite contain ca. 2.5 Ga metamorphic or anatectic zircon, and the gneissic tonalite around the enclave shows strong deformation and anatexis. It seems likely that the quartz diorite enclaves were rotated after leucosome formation during the ca. 2.5 Ga tectono-thermal event (Xie et al. 2014b). 2.9 Ga gneissic tonalites exhibit strong anatexis with trondhjemitic leucosomes and biotite-rich melanosomes in local outcrops (Fig. 16e). Some leucosome forms layers and lenses at different scales and is strongly or weakly deformed. Anatexis occurred syntectonically in a regime locally changing from compression to extension at the end of the Neoarchean because 2.5 Ga metamorphic or anatectic zircons have been identified.

2.8 Ga supracrustal rocks and TTGs have been identified in the Lushan area on the southern margin of the NCC (Fig. 17). These rocks are mainly composed of (garnet-bearing) amphibolite, hornblende–plagioclase gneiss, and biotite–plagioclase gneiss. Some amphibolites show a layered structure (Fig. 16f), mainly due to

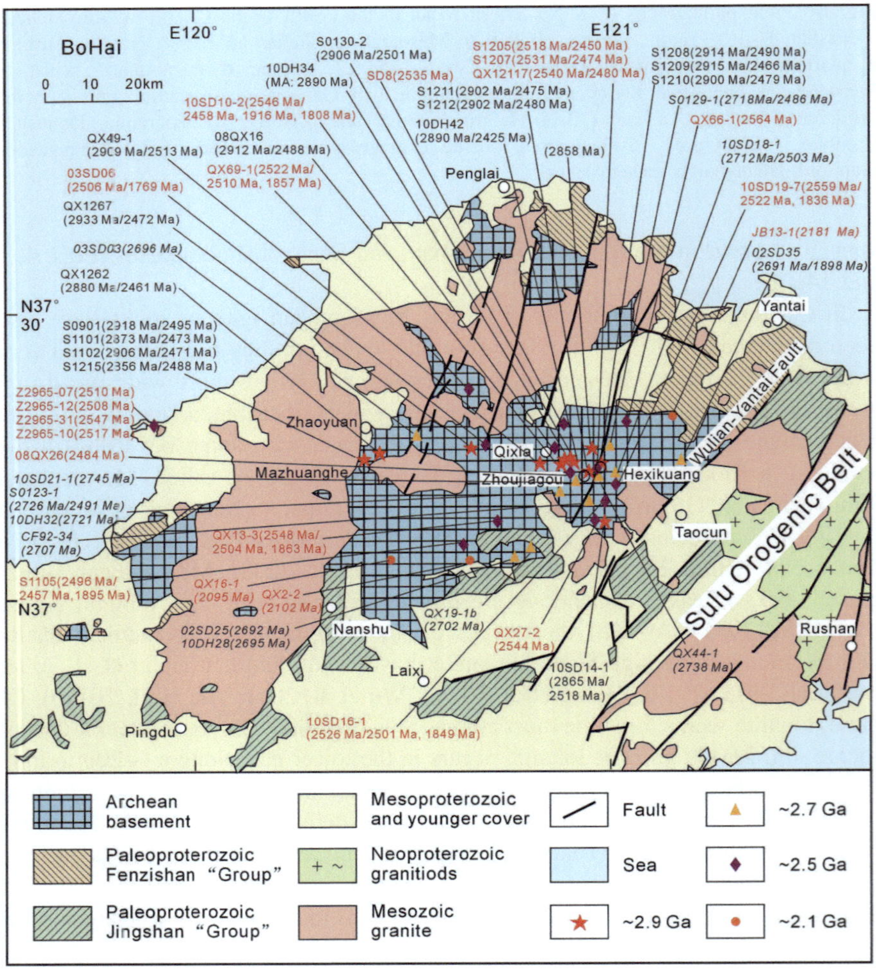

Fig. 15 Geological map of eastern Shandong. Shown also are locations of zircon dating samples, and ages in parenthesis are magmatic zircon age or magmatic zircon age/metamorphic zircon age. Data are from Jahn et al. (2008), Liu et al. (2011), Liu et al. (2011, 2013a, 2014), Liu et al. (2013b), Liu et al. (2014), Shan et al. (2015), Tang et al. (2007), Wang and Yan (1992), Wang et al. (2014a), Wu et al. (2014), Xie (2012), Xie et al. (2013, 2014b), and Zhou et al. (2008). See Fig. 1 for location of the figure in the NCC

variations in plagioclase and hornblende contents. They occur as enclaves within TTG rocks with strong deformation and local anatexis (Fig. 16g). Zircon dating of the supracrustal rocks and TTGs yielded 2.83 Ga magmatic ages and two sets of metamorphic ages at 2.79–2.77 and 2.67–2.64 Ga (Liu et al. 2009a). Huang et al. (2010) performed zircon dating on TTG-like and TTG rocks in the same area and interpreted the ages of 2.77 and 2.72 Ga as the time of crystallization of the host igneous rocks. Similar conclusions were arrived at by Diwu et al. (2010b).

(a)
2.9 Ga gneissic
high-Si trondjhemite
Fig. b
Fig. c
2.9 Ga gneissic tonalite
(b)
S1211
2.9 Ga gneissic tonalite
(c)
S1212
2.9 Ga gneissic
high-Si trondjhemite
(d)
2.91 Ga gneissic
quartz tonalite
2.91 Ga
gneissic tonalite
(e)
QX1267
2.91 Ga tonalite
showing strong anatexis (2.5 Ga)
(f)
Banded amphibolite
(g)
Anatectic veins

◀ **Fig. 16** Field photographs of Mesoarchean rocks in the North China Craton. **a** 2.9 Ga gneissic high-Si trondhjemite (*top*) in contact with gneissic tonalite (*bottom*), Qixia, eastern Shandong; **b** 2.9 Ga gneissic tonalite, local enlargement in figure a; **c** 2.9 Ga gneissic high-Si trondhjemite, local enlargement in figure a; **d** relationship between 2.91 Ga gneissic quartz diorite and 2.91 Ga gneissic tonalite, Qixia, eastern Shandong; **e** 2.91 Ga gneissic trondhjemite, Qixia, eastern Shandong; **f** 2.84 Ga amphibolites showing banded structures, Lushan, Henan; **g** anatectic veins in 2.83 Ga gneissic tonalite, Lushan, Henan

However, the dated zircons exhibit strong recrystallization and overgrowth, and it is therefore likely that these are ~2.8 Ga TTG rocks that underwent strong metamorphism in the early Neoarchean. More work is required to identify whether or not there are ~2.7 Ga rocks in the area.

Mesoarchean detrital and xenocrystic zircons were discovered all over the NCC such as Alax, Guyang, Wutai, Fuping, Dengfeng, eastern Hebei, Huoqiu, western Shandong, and eastern Shandong (Fig. 1) (Jian et al. 2012; Kröner et al. 1988; Shen et al. 2004, 2005; Wan et al. 2006, 2010a, b; Wang et al. 1998, 2014a, b, c, d, e; Zhang et al. 2014a, b).

2.4 *Neoarchean (2.8–2.5 Ga)*

2.4.1 Early Neoarchean (2.8–2.6 Ga)

Two magmatic events can be recognized in the Neoarchean crust formation and evolution, namely an early Neoarchean event (2.8–2.6 Ga) and a late Neoarchean event (2.6–2.5 Ga). 2.80–2.76 Ga rocks are rare, and the most important tectono-thermal event is 2.79–2.77 Ga metamorphism recorded in the Lushan area near the southern margin of the craton (Liu et al. 2009a). The identification of widespread ~2.7 Ga rocks is one of the most important discoveries made in recent years. These rocks are widespread, including western Shandong, eastern Shandong, Guyang, Fuping, Hengshan, Zanhuang, Zhongtiao, and Huoqiu (Fig. 1) (Cao et al. 1996; Dong et al. 2012a; Du et al. 2003, 2005, 2010; Guan et al. 2002; Han et al. 2012; Jahn et al. 2008; Jiang et al. 2010; Kröner et al. 2005a; Lu et al. 2008; Wan et al. 2010b, 2011b; Wang et al. 2014a, b, c, d, e; Wang et al. 2009a, b; Yang et al. 2013a, b; Zhu et al. 2013).

In eastern Shandong, the spatial distribution of 2.7 and 2.9 Ga rocks is uncertain. However, they show close spatial relationships with 2.7 Ga rocks occurring extensively in a southwest to northeast direction with a total area of >100 km^2 (Fig. 15). These Archean rocks underwent strong upper amphibolite- to granulite-facies metamorphism at the end of the Neoarchean (Jahn et al. 2008; Liu et al. 2011; Wu et al. 2014; Xie et al. 2014b). Similar to 2.9 Ga rocks, the 2.7 Ga assemblages exhibit strong deformation and anatexis with neosome material occurring in local outcrops (Fig. 18a). This makes it difficult to distinguish between

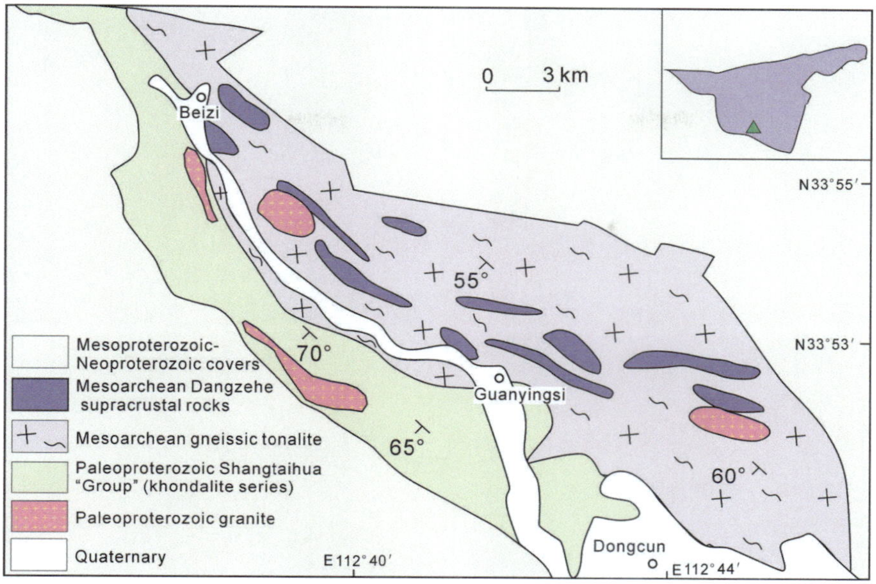

Fig. 17 Simplified geological sketch map of the Lushan area, southern margin of the North China Craton. Modified after Liu et al. (2009c). Triangle in inset map shows location of Fig. 17 in the NCC

2.9 and 2.7 Ga assemblages. At present, the only way to identify these different rocks is by zircon dating.

In western Shandong, Wan et al. (2010c, 2011b) divided the Archean basement into three belts (Fig. 19), namely a late Neoarchean belt of crustally derived granitoids in the northeast that predominantly consists of 2.53–2.49 Ga monzogranite and syenogranite (Belt A), an early Neoarchean belt in the center that is mainly composed of 2.75–2.60 Ga TTGs and supracrustal rocks (Belt B), and a late Neoarchean belt of juvenile rocks in the southwest that is dominated by granodiorite, gabbro, quartz diorite, and tonalite, with minor monzogranite and syenogranite (Belt C). Western Shandong is an area where 2.75–2.7 Ga rocks are most widely distributed in the NCC. Different types of ~2.7 Ga TTGs can be observed in contact with each other in Belt B, showing similar or different styles of deformation (Fig. 18b, c). Furthermore, western Shandong is the only area where early Neoarchean supracrustal rocks have so far been identified in the NCC. In an earlier study, the Taishan "Group," including the Yanlingguan, Liuhang, and Shancaoyu "Formations," were considered to be early Neoarchean in age (Cao et al. 1996). However, more recent studies established that only the Yanlingguan "Formation" and the lower part of the Liuhang "Formation" formed during the early Neoarchean (named the Yanlingguan-Liuhang succession). They are mainly composed of amphibolite and metamorphosed ultramafic rocks. Some of these rocks contain fine spinifex textures (Fig. 18d). Amphibolites with abundant pillow structures are abundant (Fig. 18e), and some show strong deformation (Fig. 18f, g).

(a)
~2.5 Ga leucosome
2.72 Ga gneissic tonalite
(b)
2.71 Ga granodiorite gneiss
2.71 Ga tonalite gneiss
(c)
2.71 Ga gneissic tonalite
~2.6 Ga leucosome
(d)
(e)
Carbonate+siliceous materials
(f)
(g)
Carbonate+siliceous materials
(h)

◀ **Fig. 18** Field photographs of early Neoarchean rocks in the North China Craton. **a** 2.7 Ga gneissic tonalite showing anatexis, Qixia, eastern Shandong; **b** cutting relationship between 2.71 Ga gneissic tonalite and 2.71 Ga gneissic granodiorite, Taishan, western Shandong; **c** 2.71 Ga gneissic tonalite showing anatexis and deformation, Taishan, western Shandong; **d** meta-komatiite with fine spinifex structures, Sujiagou, western Shandong; **e** amphibolite with pillow lava structures, Qixingtai, western Shandong; **f** deformed pillow lava in amphibolite, Qixingtai, western Shandong; **g** carbonate + siliceous material occurring at the end of deformed pillow lava in amphibolite, Qixingtai, western Shandong; **h** 2.69 Ga gneissic tonalite showing anatexis, Xi Ulanbulang, Yinshan

Xi Wulanbulan is the only area where 2.7 Ga tonalite was identified in the Western Block of the NCC. It extends in a north–south direction over an area of >10 km^2 (Fig. 20) and shows strong metamorphism, deformation, and anatexis (Fig. 18h) with metamorphic zircons recording an age of ~2.5 Ga. These tonalities occur together with the late Neoarchean Xinghe "Group" and other TTG rocks (Dong et al. 2012a; Ma et al. 2013a, b, c).

Wan et al. (2014a) summarized the spatial distribution, rock types, geochemistry, and Nd–Hf isotopic compositions of ~2.7 Ga granitoids in the NCC. They are mainly tonalitic in composition and show large variations in SiO_2, ΣFeO, MgO, and CaO. They can be subdivided into two types in terms of REE contents (Fig. 8c). Whole-rock Nd and Hf-in-zircon isotopic compositions indicate that the strong 2.7 Ga tectono-thermal event mainly involved juvenile additions to the continental crust with recycling of older crust only in local areas.

Ca. 2.6 Ga rocks were also reported in several areas of the NCC, including western Shandong, Hebi, Zhongtiao, Guyang, and Bayan Obe (Fan et al. 2010; Jian et al. 2012; Wan et al. 2015b; Zhang et al. 2012a, b, c, d, e, f; Zheng et al. 2012; Ma unpublished data). In most of these areas, ~2.7 Ga rocks also occur. Western Shandong is the area where both 2.7 and 2.6 Ga rocks are widely distributed (Wan et al. 2015b). They mainly occur together in Belt B (Fig. 21) and include hornblendite (meta-pyroxenite), gneissic tonalite, gneissic trondhjemite, and gneissic granite (Fig. 22a–d). Geological records are almost continuous from 2.75 to 2.59 Ga (Fig. 23). Ca. 2.6 Ga metamorphism and anatexis have also been identified (Du et al. 2003, 2005; Lu et al. 2008; Ren unpublished data). This strong tectono-thermal event resulted in the formation of migmatites in Belt B (Fig. 22e–f). These are important for considering 2.6 Ga as the boundary between the middle to early Neoarchean and late Neoarchean (2.6–2.5 Ga) in western Shandong. No ~2.7 Ga metamorphism and ~2.6 Ga supracrustal rocks have been identified there, and it appears that there was a "quiet period" between 2.60 and 2.56 Ga. We speculate that this time subdivision is applicable to the entire NCC.

2.4.2 Late Neoarchean (2.6–2.5 Ga)

2.59–2.57 Ga rocks and metamorphism are rare. However, a 2.56–2.5 Ga tectono-thermal event is very strong, resulting in extensive distribution of supracrustal and intrusive rocks and metamorphism of this age all over the NCC

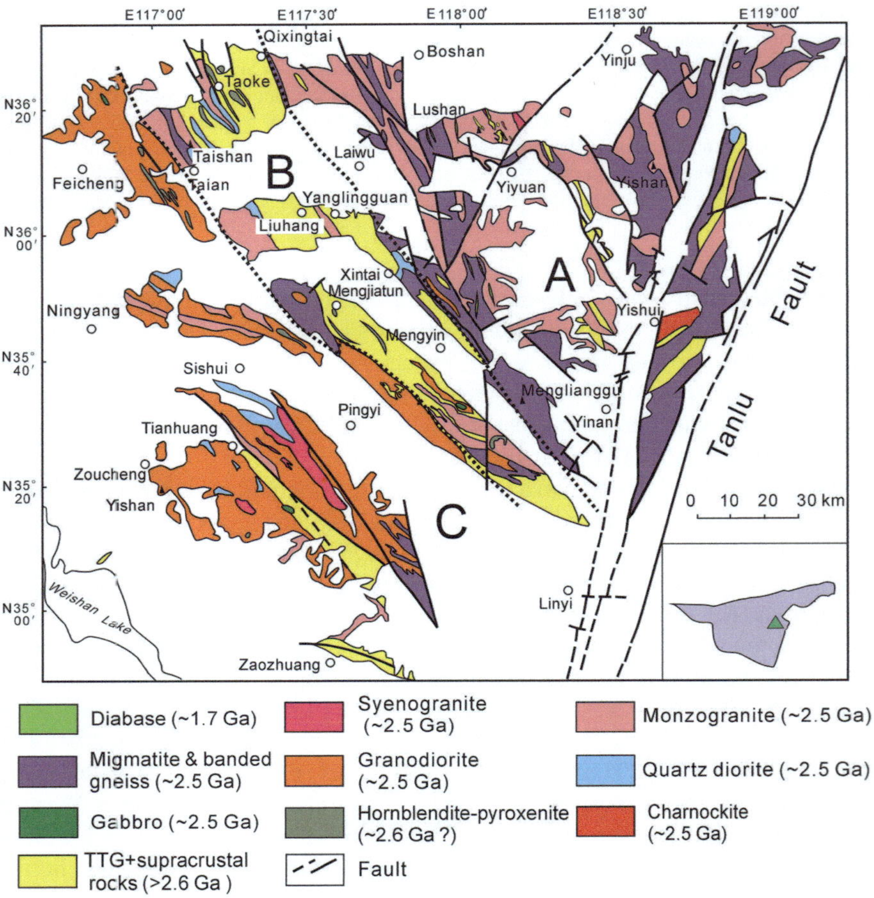

Fig. 19 Geological map of western Shandong. Modified after Wan et al. (2010c). *Belt A* A late Neoarchean crustally derived granite belt in the northeast that consists predominantly of 2.525–2.49 Ga monzogranite and syenogranite; *Belt B* An early Neoarchean belt in the center which is mainly composed of 2.75–2.60 Ga TTG and supracrustal rocks; *Belt C* A late Neoarchean belt of juvenile rocks in the southwest that is dominated by granodiorite, gabbro, quartz diorite, and tonalite, with some monzogranite and syenogranite. *Triangle* in *inset* map shows location of Fig. 19 in the NCC

(Chen 2007; Cheng et al. 2004; Cui et al. 2013, 2014; Dai et al. 2012, 2013; Dong et al. 2012b; Geng et al. 2002, 2006, 2010; Grant et al. 2009; Guo et al. 2005, 2008; Han et al. 2014a, b; He et al. 2005; Jahn et al. 1988; Jian et al. 2012; Kröner et al. 1998, 2005a, b; Li et al. 2010a, b, 2012; Liu et al. 2002, 2004a, 2007, 2009b, 2011, 2012a, 2012b; 2012c, 2014, 2011a; Lu et al. 2008; Lü et al. 2012; Ma et al. 2012, 2013a, b, c; Ma et al. 2013a, b, 2014a, b; Peng et al. 2012; Peng 2013; Peng et al. 2013; Ren et al. 2011; Shen et al. 2004, 2005, 2007; Shi et al. 2012; Song et al. 2009; Sun et al. 1991, 2010, 2014; Sun and Guan 2001; Tian et al. 2005; Wan et al.

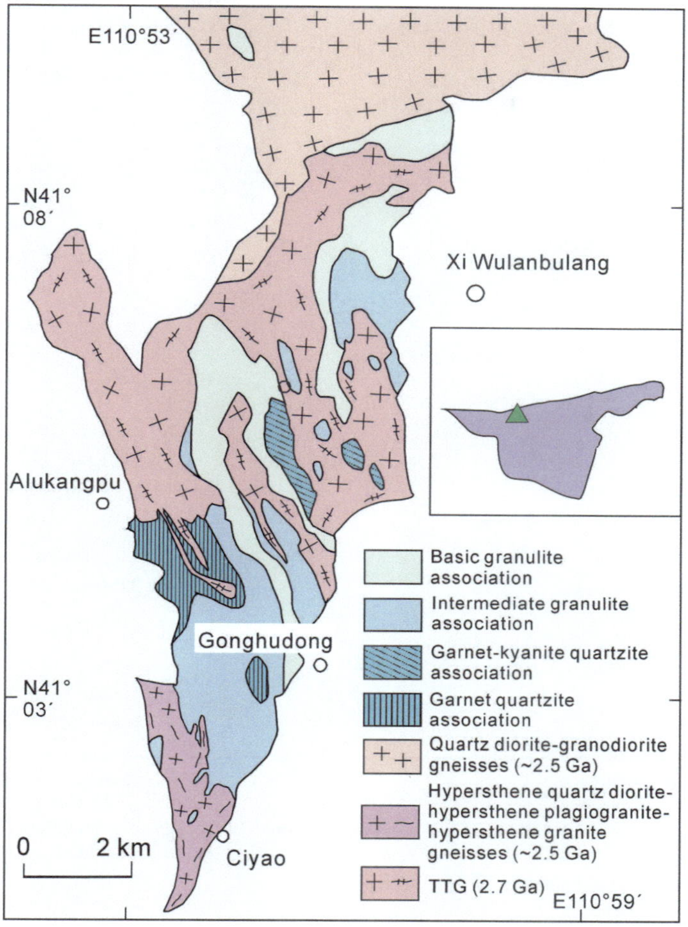

Fig. 20 Geological map of the Xi Ulanbulang area, Yinshan. Modified after Dong et al. (2012a) and Ma et al. (2013). *Triangle* in *inset* map shows location of Fig. 20 in the NCC

2005b, 2009b, c, 2010b, c, 2011a, c, 2012b, c, d, 2015a; Wang et al. 2004, 2011; Wang et al. 2014; Wang et al. 2014; Wang et al. 2010; Wang et al. 2009; Wilde et al. 2004a, 2005; Wu et al. 1998; Xiang et al. 2012; Yang et al. 2008, 2009, 2011; Yang 2013; Zhai et al. 2000, 2005; Zhang et al. 2013; Zhang et al. 2012a; Zhang et al. 2011; Zhang et al. 2014; Zhao et al. 2002, 2009; Zhao et al. 2008; Zhao et al. 2008; Zhou et al. 2009, 2011, 2014).

There are abundant Neoarchean (mainly 2.55–2.5 Ga) detrital and xenocrystic zircons in younger rocks and river sands (e.g., Diwu et al. 2012; Wan et al. 2011a). Late Neoarchean supracrustal rocks are mainly composed of mafic to felsic granulite, amphibolite, fine-grained biotite gneiss (leptinite), banded iron formation (BIF), and (fuchsite) quartzite, with the metamorphic grade ranging from

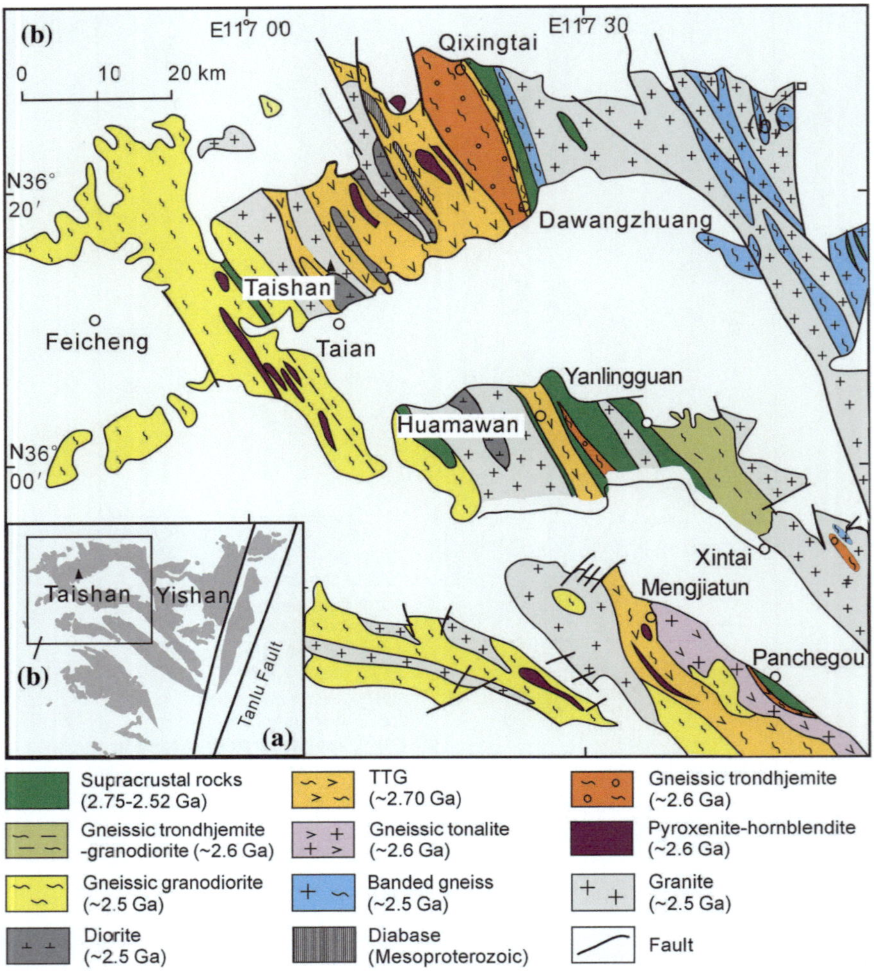

Fig. 21 Geological map of the Taishan-Mengjiatun area, western Shandong (Wan et al. 2015b)

predominantly amphibolite to granulite facies. Mafic granulite and amphibolite show light REE enrichment or flat REE patterns and are commonly rich in LILE and depleted in Nb and Ta (Wan et al. 1997b). Fine-grained biotite gneisses generally show compositional features of dacitic rocks. Late Neoarchean intrusive rocks are variable in composition, ranging from ultramafic to felsic, but TTGs and crustally derived granites constitute the main components. It is not until the late Neoarchean that granodiorites are widely distributed in the NCC. Syenogranites predominantly formed between 2.53 and 2.49 Ga and can be further subdivided into two phases with most showing massive structures and emplacement during the second phase (2.52–2.48 Ga), whereas the first phase (2.53–2.52 Ga) syenogranites show metamorphism and deformation (Wan et al. 2012b). All syenogranites share

Fig. 22 Field photographs of early Neoarchean rocks in western Shandong. **a** 2.60 Ga hornblendite, southwest of Qixingtai; **b** 2.66 Ga gneissic tonalite, Yishan; **c** gneissic trondhjemite cutting amphibolite, Taishan; **d** relationship between 2.60 Ga gneissic granodiorite and 2.60 Ga mylonitized monzogranite, Xintai; **e** 2.7 Ga gneissic tonalite showing anatexis at ~2.6 Ga, Taishan; **f** 2.7 Ga gneissic tonalite showing anatexis at ~2.6 Ga, Taishan

the same major element compositions, being high in SiO_2 and low in CaO, ΣFeO, MgO, TiO_2, and P_2O_5. However, they have variable REE compositions and can be subdivided into three types (Fig. 8d). In order to better understand the late Neoarchean geological features in various parts of the NCC, several areas are described below.

(1) Western Shandong

Late Neoarchean magmatism is widespread with peak ages of 2.53–2.52 Ga (Fig. 24). A recent study indicated that the Shancaoyu "Formation" and the upper part of the Liuhang "Formation" formed during the late Neoarchean rather than in the early Neoarchean (Wan et al. 2012c). Western Shandong is the only area where

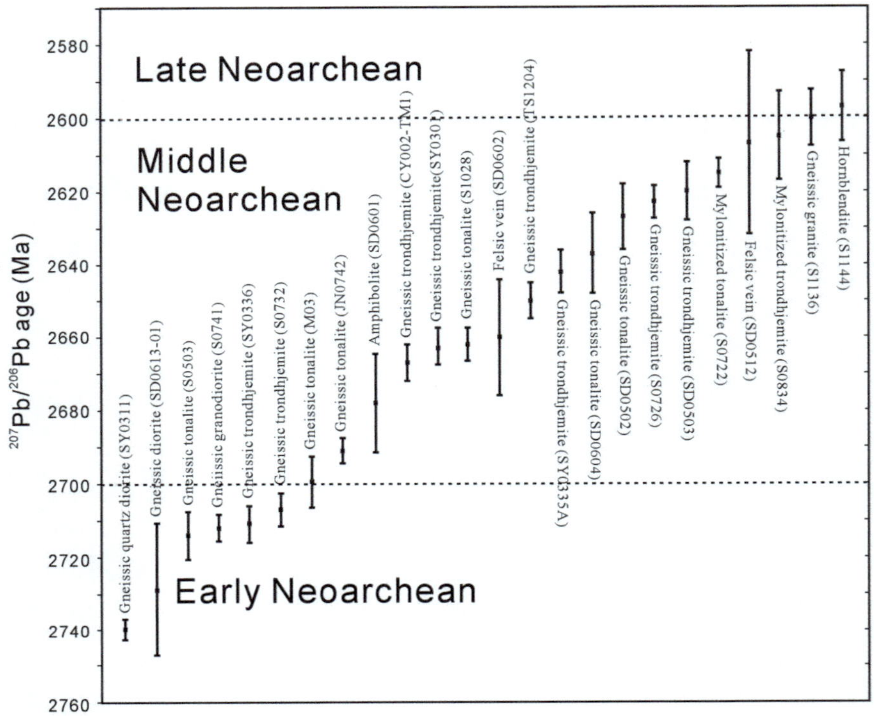

Fig. 23 Zircon age variation diagram (with *error bars*) for early to *middle Neoarchean* magmatic rocks in western Shandong. Data are from Du et al. (2003), Jiang et al. (2010), Lu et al. (2008), and Wan et al. (2011b, 2015b)

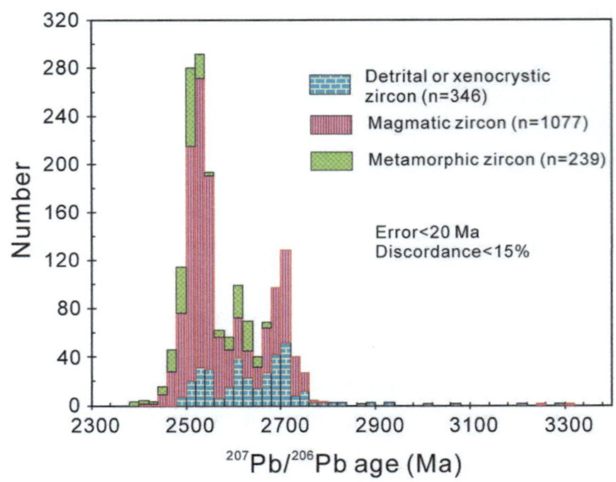

Fig. 24 Age histogram of zircons from Neoarchean rocks in western Shandong. Data are from Du et al. (2003), Jiang et al. (2010), Lu et al. (2008), Wan et al. (2010c, 2011b, 2015b), and Wan (unpublished data)

both early and late Neoarchean supracrustal rocks were identified. The late Neoarchean supracrustal rocks occur in different belts and mainly consist of meta-conglomerate, fine-grained biotite gneiss, fine-grained two-mica gneiss, mica schist, BIF, and felsic metavolcanic rocks. The metaconglomerates are interlayered with fine-grained biotite gneiss (Fig. 25a), and the conglomerate pebbles vary in size and mainly consist of TTG, monzogranite, and fine-grained felsic volcanic rocks. Some fine-grained biotite gneisses show sedimentary structures such as bedding (Fig. 25b), suggesting a clastic origin. However, in many cases, their protoliths are difficult to determine (volcanic or sedimentary) because of strong metamorphism and deformation (Fig. 25c). Mantle-derived igneous rocks in Belt C predominantly consist of gabbro, diorite, tonalite, and granodiorite. Magma mixing can be observed between different rock types (Fig. 25d, e). Different migmatite types are associated with crustally derived granites in Belt A as a result of anatexis of early supracrustal and granitoid rocks. TTGs of early and late Neoarchean ages were identified as protoliths of the anatectic rocks (Figs. 22b and 25f).

(2) Eastern Hebei

Late Neoarchean supracrustal and intrusive rocks are widespread (Fig. 26), apart from ancient crustal components documented by 3.88–3.40 Ga detrital zircons and 3.40–2.95 Ga orthogneisses in Huangbaiyu, as mentioned before. 2.55–2.53 Ga tonalites and quartz diorites are most important in the west, and 2.53–2.51 Ga granites are most important in the east, where magma mingling between granite and diorite has been also identified (Fig. 27a) (Nutman et al. 2011; Yang et al. 2008).

The Qian'an "Group" was considered to be Mesoarchean in age, but recent zircon dating indicates that most of the supracrustal rocks formed at the end of the Neoarchean (Han et al. 2014a; Zhang et al. 2011; Wan, unpublished data). Late Neoarchean supracrustal rocks commonly occur as enclaves of variable sizes in granitoids and show spatial variations in rock association, metamorphism, and deformation, with metamorphic grades varying from granulite facies in the west to lower amphibolite facies (epidote amphibolite facies) in the east and greenschist facies in local areas. These rocks have different names in different areas (Fig. 26). The Zunhua and Qian'an "Groups" and associated TTG rocks in the Zunhua-Santunying area underwent granulite-facies metamorphism and were commonly affected by amphibolite-facies retrogression. The supracrustal rocks extend in a north–south direction and are mainly composed of biotite–plagioclase gneiss, hornblende–plagioclase gneiss, two-pyroxene granulite, amphibolite, meta-ultramafic rock, and BIF. Strong metamorphism resulted in anatexis of the supracrustal rocks (Fig. 27b).

The Luanxian "Group" occurs in the Luanxian-Lulong area and is composed of fine-grained biotite gneiss, amphibolite, and BIF, whereas at Huangbaiyu, the Qian'an "Group" and Luanxian "Group" are separated by a $\sim$2.5 Ga granite. The Dantazi and Zhuzhangzi "Groups" at Qinlong exhibit lower amphibolite-facies metamorphism. The Dantazi "Group" (or Shuangshanzi "Group") is dominated by fine-grained two-mica gneiss, garnet–mica schist, two-mica quartz schist, mafic and felsic metavolcanic rocks, and BIF. It is covered unconformably by the Zhuzhangzi "Group" (or Qinglong "Group"), commonly with >100 m of conglomerate at the

Fig. 25 Field photographs of late Neoarchean rocks in western Shandong. **a** Metaconglomerates interlayered with metasediments; pebbles are mainly TTG and felsic volcanic rocks, upper part of Liuhang "Formation," Huamawan; **b** metasedimentary rock showing bedding, Shancaoyu "Formation," Qixingtai; **c** mylonitized fine-grained biotite gneiss (S0801) of Shancaoyu "Formation," Panchegou; **d** magma mingling between ~2.5 Ga granodiorite and quartz diorite, showing large variations in proportion, southeast of Tianhuang; **e** magma mingling between ~2.5 Ga syenogranite and quartz diorite, southeast of Tianhuang; **f** 2.52 Ga gneissic tonalite showing anatexis, Yinshan

base (Fig. 27c). Furthermore, the Zhuzhangzi "Group" contains fine-grained two-mica gneiss, fine-grained biotite gneiss, mica schist, quartz schist, biotite–hornblende schist, and BIF. Ca. 2.5 Ga BIFs are well developed in eastern Hebei and contain important iron deposits. The supracrustal rocks in the Qinlong area

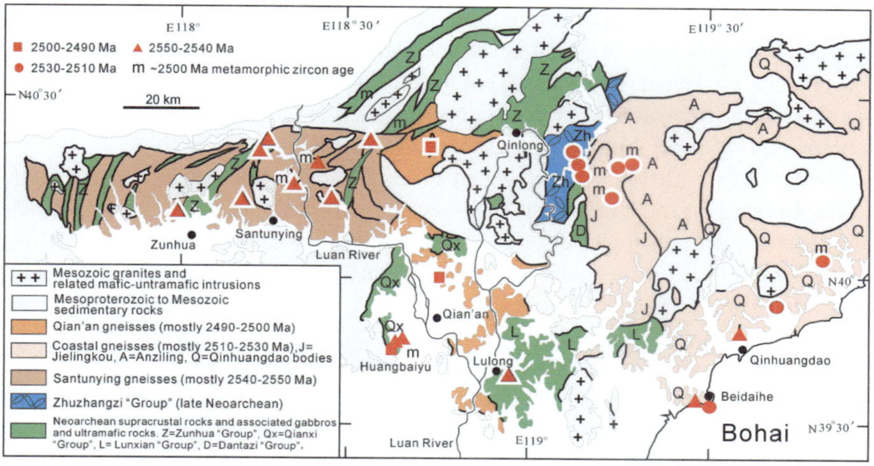

Fig. 26 Geological map of eastern Hebei, modified after Nutman et al. (2011). For location in the NCC, see Fig. 1

Fig. 27 Field photographs of late Neoarchean rocks in eastern Hebei. **a** Magma mingling between ~2.5 Ga granite and diorite, Beidaihe; **b** anatexis of granulite-facies gneiss (the Zunhua "Group"), north of Santunying; **c** metaconglomerate at the bottom of the Zhuzhangzi "Group," southeast of Qinlong; **d** anatexis of TTG rock, west of Anziling

commonly underwent lower amphibolite-facies metamorphism, but the TTG rocks east of Qinlong show strong anatexis and deformation (Fig. 27d). In the eastern coastal area of Hebei Province, weakly deformed 2.53–2.51 Ga granites occur together with subordinate granodiorites, diorites, and magnesian gabbros and contain slightly older Neoarchean granitoid inclusions.

Nutman et al. (2011) suggested that Neoarchean magmatism in eastern Hebei was not a single protracted event but was marked by temporally, geographically, and geochemically distinct pulses of igneous activity at 2.55–2.535, 2.53–2.52 Ga, and 2.50–2.49 Ga, respectively, with the latter accompanied by granulite-facies metamorphism.

(3) Fuping-Wutai-Hengshan

This is an important and famous exposure of Neoarchean rocks in the NCC with a long history of research. Based mainly on the work of Cheng et al. (2004), Guan et al. (2002), Han et al. (2012), Kröner et al. (2005a, 2005b, 2006), Li et al. (2010a), Liu et al. (2002), Liu et al. (2004a), Miao (2003), Qian et al. (2013), Trap et al. (2007, 2008), Wan et al. (2010a), Wang et al. (2010), Wei et al. (2014), Wilde et al. (2004a, 2004b, 2005), Wu et al. (1989, 1998), Zhang et al. (2006); Zhang et al. (2006); Zhang et al. (2007, 2009), and Zhao et al. (2011), some common features of the basement are summarized as follows (Fig. 28).

(1) The early Precambrian basement is composed of the Fuping Complex in a southeastern belt, the Wutai Complex in a central belt, and the Hengshan Complex in a northwestern belt. All of these are composed of intrusive and supracrustal rocks of different ages.
(2) Early Precambrian intrusive rocks mainly formed during the late Neoarchean (2.55–2.5 Ga), including TTGs (mainly tonalite and granodiorite) with some monzogranite and K-rich granite. 2.2–2.0 Ga intrusive rocks occur on a relatively small scale and are predominantly crustally derived granites with a few mantle-derived mafic–ultramafic rocks. Early Neoarchean (~2.7 Ga) TTGs and early Neoarchean and older detrital and xenocrystic zircons were also identified in the Fuping and Hengshan areas.
(3) No early Neoarchean or older supracrustal rocks have so far been found, but this does not mean that they do not occur in these areas. Late Neoarchean supracrustal rocks in the Wutai area are composed of mafic to felsic metavolcanic and clastic metasedimentary rocks, BIF, and minor limestone (the Wutai "Group"). In Fuping, late Neoarchean supracrustal rocks, named the Fuping "Group," are mainly composed of mafic to intermediate metavolcanic and clastic metasedimentary rocks and minor BIF and marble. Some or many biotite–plagioclase gneisses and hornblende–plagioclase gneisses of the Suojiazhuang-Yuanfang unit I of the Fuping "Group" may be strongly metamorphosed and deformed intrusive rocks (Yang, unpublished data). Late Neoarchean supracrustal rocks rarely occur as enclaves in the Hengshan Complex, including garnet-bearing biotite–plagioclase gneiss, BIF, and amphibolite. High-pressure garnet two-pyroxene granulites are derived from gabbroic dikes.

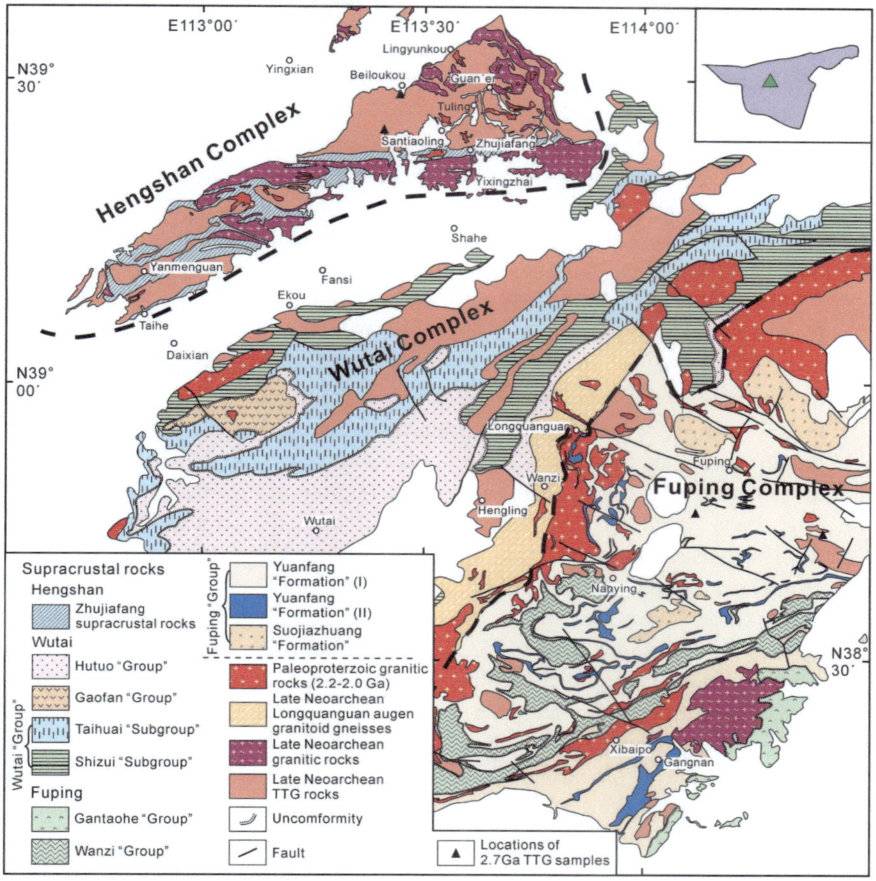

Fig. 28 Geological map of Hengshan-Wutai-Fuping, modified after Miao (2003), Cheng et al. (2004), Liu et al. (2004b), Li et al. (2010a, b), Zhao et al. (2011), and Qian et al. (2013). *Triangle* in *inset* map shows location of Fig. 28 in the NCC. The Yuanfang "Formation"(I) of the Fuping "Group" is mainly composed of anatectic biotite–plagioclase gneiss, hornblende–biotite plagioclase gneiss, and fine-grained biotite–plagioclase (two feldspar) gneiss (leptinite), and some of them may be metamorphic igneous rocks; Yuanfang "Formation"(II) of the Fuping "Group" is composed mainly of amphibolite, fine-grained garnet–biotite gneiss, fine-grained feldspar gneiss, fine-grained hornblende gneiss, banded iron formation, calc-silicate rock, and marble

(4) In Wutai, the Gaofan "Group," originally named as the Gaofan "Subgroup" of the Wutai "Group," unconformably overlies the Wutai "Group" and was deposited between 2.47 and 2.14 Ga. It should therefore be excluded from the Wutai "Group." The Hutou "Group" unconformably overlies the Gaofan "Group" and is considered to be $\sim$2.14 Ga or younger in age. The Wanzi "Group" in the Fuping Complex was originally considered to be late Neoarchean in age, but some authors considered it to be early Paleoproterozoic (Guan et al. 2002; Li et al. 2005). It is composed of

metasedimentary rocks including marble and is similar to the khondalite sequences of the northwestern NCC in rock association and metamorphism.

(5) Whole-rock Nd and Hf-in-zircon isotopic studies indicate that the formation of all early and most late Neoarchean TTGs involved juvenile magmatic additions to the continental crust; most late Neoarchean and early Paleoproterozoic (2.2–2.0 Ga) granitoids formed by recycling of early Neoarchean or older crustal material.

(6) The Fuping Complex and the northern portion of the Hengshan Complex underwent upper amphibolite- to granulite-facies metamorphism, whereas the Shizui "Subgroup" of the Wutai "Group" and the southern portion of the Hengshan Complex underwent lower amphibolite-facies metamorphism, and the Taihuai "Subgroup" of the Wutai "Group" underwent greenschist-facies metamorphism. Metamorphic grades vary from granulite facies to greenschist facies from both sides to the center. Consequently, the late Neoarchean granitoid bodies in Wutai commonly show weaker metamorphism and deformation than their equivalents in the Fuping and Hengshan complexes.

(7) Fold axes, foliations, and lineations mainly extend in a west–east or southwest–northeast direction. Three episodes of folding and two phases of ductile thrusting and shearing have been identified (Li et al. 2010a). There are two major shear zones: One is the west–east-striking Zhujiafang shear zone separating the granulite-facies Hengshan Complex in the north from the amphibolite-facies Hengshan Complex in the south. Another is the southwest–northeast-striking Longquanguan shear zone separating the Wutai "Group" in the northeast from the Fuping Complex in the southwest.

(8) A strong late Paleoproterozoic tectono-thermal event is recorded by well-developed 1.95–1.83 Ga metamorphic zircons. This event resulted in strong metamorphism and deformation in the Fuping and Hengshan complexes and the lower part of the Wutai Complex and formation of the above two shear zones. ~2.5 Ga metamorphic zircons have also been identified in Fuping (Yang, unpublished data), but their significance is uncertain.

(4) Yinshan

Little was known about the early Precambrian basement of the Yinshan Block, but new data were acquired in recent years (Chen 2007; Dong et al. 2012a, b; Fan et al. 2010; Jian et al. 2012; Li et al. 1987; Liu et al. 2012d, 2014; Ma et al. 2013; Ma 2013a, b, 2014a, b; Mei 1991; Zhang et al. 2003; Zhang and Liu 2004; Zhang et al. 2014). The main features of the basement are summarized as follows (Figs. 29 and 30).

(1) There are 2.7 Ga tonalites and 2.6 Ga granites and older xenocrystic zircons in ~2.5 Ga granitoid rocks. The 2.7 Ga tonalite exposure is >10 km^2 in size. Zhang et al. (2014) obtained a zircon age of 2.51 Ga for a charnockite exposed near a 2.7 Ga tonalite dated by Ma et al. (2013) and interpreted this to reflect emplacement of the charnockite precursor. However, considering strong recrystallization of the zircon cores, the significance of the 2.51 Ga age is

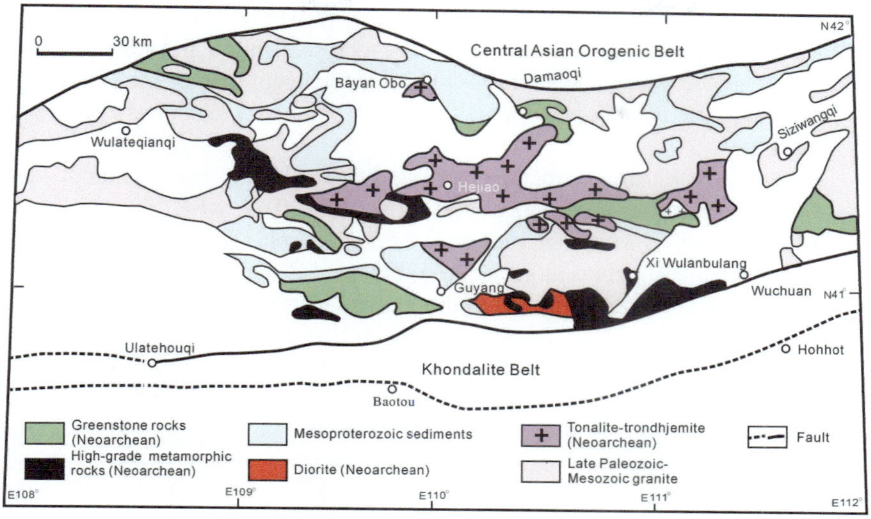

Fig. 29 Geological map of Yinshan, modified after Jian et al. (2012). For location in the NCC, see Fig. 1

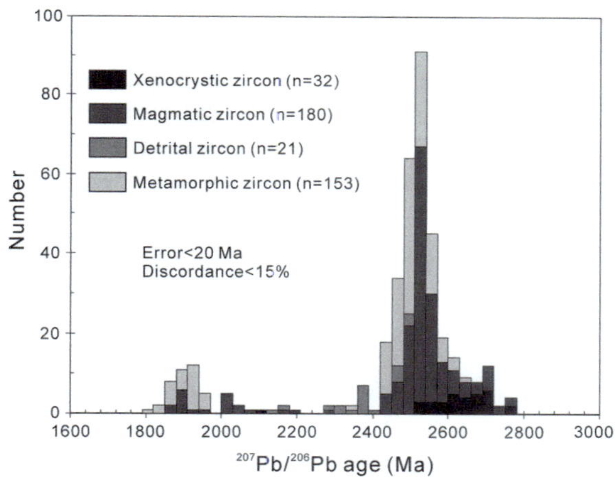

Fig. 30 Age histogram of zircons from early Precambrian rocks in Yinshan Block. Data are from Dong et al. (2012a, b), Fan et al. (2010), Jian et al. (2012), Liu et al. (2012d, 2014), Ma et al. (2013); Ma et al. (2013a, b, 2014a, b), and Zhang et al. (2014)

uncertain. We suggest that more >2.6 Ga rocks and zircons will be discovered in Yinshan as more work is carried out.

(2) 2.55–2.5 Ga magmatism was widespread and resulted in the formation of intrusive rocks composed of tonalite–trondhjemite and subordinate gabbro, diorite, granite, and syenogranite, whereas granodiorite is rare. It appears that

most tonalite–trondhjemite formed during an early phase in this event. It also seems likely that, for some rocks, strong recrystallization and Pb loss caused the magmatic zircon apparent ages to become younger than their original ages. Middle to late Paleoproterozoic (2.2–1.9 Ga) granitoid rocks were identified in local areas.

(3) Late Neoarchean supracrustal rocks include metavolcanic and volcano-sedimentary sequences including BIFs. Two types of supracrustal sequence can be recognized in terms of metamorphic grade. The high-grade metamorphic supracrustal rocks (the Xinghe "Group") consist of mafic to felsic granulite with some containing garnet, and the low-grade metamorphic supracrustal rocks (the Sheerteng "Group" or greenstone belt) consist of greenschist- to lower amphibolite-facies mafic to felsic metavolcanic rocks and minor metamorphosed ultramafic rocks, clastic metasediments, and BIF, with some mafic lavas showing relict pillows. The supracrustal and associated intrusive rocks broadly extend in an east–west direction. Some metasedimentary rocks of the high- and low-grade supracrustal sequences are Paleoproterozoic in age (Z.Y. Xu, unpublished data).

(4) Whole-rock Nd and Hf-in-zircon isotopic analyses of 2.7, 2.6, and 2.55–2.50 Ga intrusive as well as 2.55–2.50 Ga supracrustal rocks indicate that addition of significant juvenile material from the mantle occurred at 2.8–2.7 Ga, and both juvenile magmatism and crustal recycling played significant roles during magmatic events at the end of the Neoarchean.

(5) A late Neoarchean tectono-thermal event was widespread in the Yinshan Block as indicated by abundant ~2.5 Ga metamorphic zircons. Late Paleoproterozoic (1.94–1.86 Ga) metamorphic zircons were also found at several localities. These metamorphic zircons are rare in the high-grade metamorphic rocks, and this was interpreted to be due to late Paleoproterozoic metamorphism having occurred in a dominantly "dry" system caused by previous late Neoarchean high-grade metamorphism (Ma et al. 2012; Wan et al. 2011a, b, c)

3 Distribution of Zircon Ages and Isotope Geochemistry

3.1 Zircon Age Distribution

Wan et al. (2011a) compiled a total of 7586 early Precambrian zircon age data for the entire NCC and arrived at the conclusion that the most significant zircon-forming tectono-thermal events occurred in the late Neoarchean to the earliest Paleoproterozoic (2.55–2.48 Ga) and in the late Paleoproterozoic (1.95–1.80 Ga), with age peaks at ~2.52 and ~1.85 Ga, respectively. In the present study, we compiled additional zircon age data bringing the total to 15060. The geographic distribution of these data is similar to those compiled by Wan et al. (2011a).

The main differences include the following: (1) More data were obtained from the Western Block, including the basement of the Ordos basin. (2) Only data are considered where the 1σ error and discordance are less than 20 Ma and 15 %, thus excluding many results obtained by LA-ICP-MS. (3) In areas where rocks underwent strong metamorphism during the late Neoarchean and late Paleoproterozoic, such as eastern Shandong and Daqingshan, some published ages of 2.45–2.0 Ga are not included in our compilation. Although these data satisfy the above analytical conditions, they may still be geologically meaningless because of likely partial resetting of the U–Pb isotopic systems in the zircons and/or partial Pb loss during high-grade metamorphism.

Our new compilation of zircon data supports the earlier conclusions of Wan et al. (2011a) that, considering the NCC as an entity, there is a continuous age record from 3.8 to 1.8 Ga. Two tectono-thermal events are very significant in the late Neoarchean to the earliest Paleoproterozoic and late Paleoproterozoic history of the craton, reflected by age peaks at ~2.52 and ~1.85 Ga, respectively (Fig. 31a). However, the age valleys are slightly different in our new compilation at ~3.6, ~3.2, ~2.85, ~2.65, ~2.25, and 2.0 Ga. Although there is an age valley at ~2.3 Ga worldwide for zircon ages (Condie et al. 2009), rocks of this age were widely identified along the southern margin of the NCC (Diwu et al. 2007, 2014; Huang et al. 2012a b, 2013; Jiang et al. 2011; Wang et al. 2012).

As indicated by Wan et al. (2011a), the Eastern Block shows an important difference to the Western Block and the Trans-North China Orogen (TNCO) in showing abundant ages >3.0 Ga (Fig. 31b–d) because of the well-studied ancient rocks in Anshan and eastern Hebei. The discovery of many >3.3 Ga detrital zircons in the Zhongtiao, Dengfeng, and Jiaozuo areas (Diwu et al. 2008; Gao et al. 2006; Liu et al. 2012a; Yin et al. 2015) reduces the previous age difference between the TNCO and Eastern Block. Significant progress was recently made in dating of early Precambrian basement rocks in the Western Block, including the discovery of 2.7 Ga TTGs and identification of strong magmatism and metamorphism at ~2.5 Ga in the Yinshan Block (Dong et al. 2012a, b; Jian et al. 2012; Liu et al. 2014; Ma et al. 2012, 2013a, b, 2014a, b; Zhang et al. 2014). More data were also obtained from the TNCO and Eastern Block. However, these new results do not change the earlier conclusion that the Eastern Block, Western Block, and TNCO have almost identical late Neoarchean age distribution patterns with zircon ages varying from 2.55 to 2.48 Ga and a prominent age peak at ~2.52 Ga. The prominent age peak in the Western Block seems slightly shifted toward younger ages (Fig. 31d), probably due to the significant early and late Paleoproterozoic metamorphic events.

3.2 *Whole-Rock Nd Isotopic Composition*

Wu et al. (2005b) compiled Nd isotopic data from NCC rocks and constructed depleted mantle Nd model age histograms that indicated an important period of

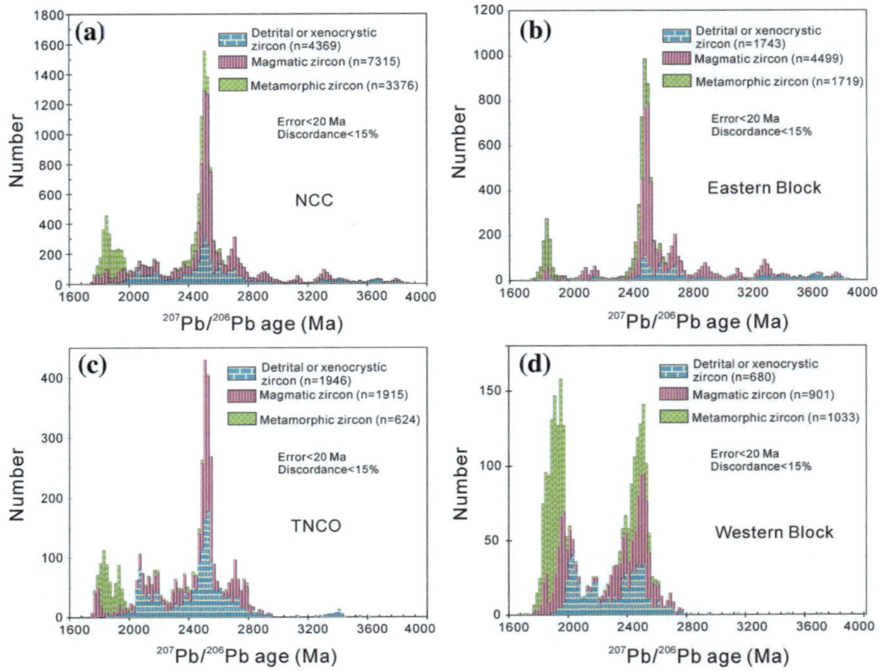

Fig. 31 Age histogram for zircons from the early Precambrian basement of the North China Craton. **a** All data for the North China Craton; **b** data for the Eastern Block; **c** data for the Trans-North China Orogen (TNCO); **d** data for the Western Block. See text for interpretation and references

crustal growth during 3.0–2.6 Ga and minor events at 3.6–3.2 and ~2.2 Ga. Many samples in their compilation were Paleoproterozoic or even younger in age. In order to better understand the Archean geological evolution, we only used data of Archean rocks in our compilation, including some 2.5–2.45 Ga samples. The total number of samples for our $\varepsilon_{Nd}(t)$ versus formation age diagram and the Nd model age histogram are 1103 and 871, respectively. The data are mainly from references published within the last twenty years (more than 70 papers, most are listed in the "References") together with our unpublished data. Similar to the zircon ages, the whole-rock Nd isotopic data show an uneven geographic distribution. Most results come from the eastern Block, and rock samples older than 2.9 Ga are mainly from Anshan, eastern Hebei, and eastern Shandong. We divided the samples into four types, namely ultramafic to intermediate rocks (including gabbro, diorite, quartz diorite, metamorphic ultramafic rocks, mafic granulite, amphibolite, greenschist, fine-grained hornblende gneiss), TTG and related rocks (e.g., metadacite, fine-grained biotite gneiss), crustally derived granites, and metasedimentary rocks. All parameters and equations used for calculation of depleted mantle Nd model ages are those of Jahn et al. (1990) and Wu et al. (2005b). The crystallization ages of most geological bodies in the NCC were determined from zircon dating.

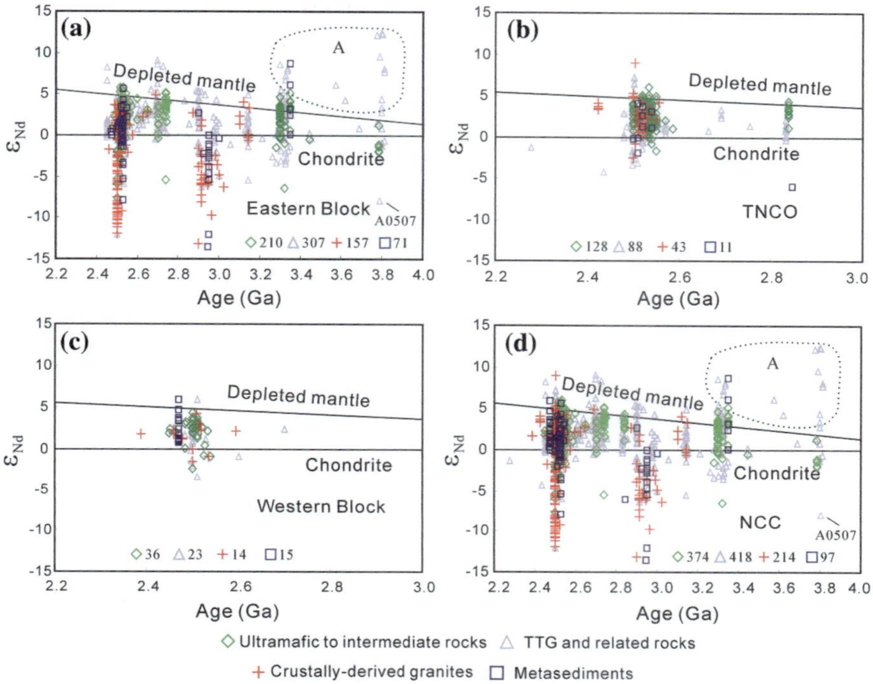

Fig. 32 $\varepsilon_{Nd}(t)$ versus formation age diagram for Archean rocks of the North China Craton. **a** Data for the Eastern Block; **b** data for the Trans-North China Orogen (TNCO); **c** data for the Western Block; **d** all data for the North China Craton. See text for interpretation and references

This made it possible to construct $\varepsilon_{Nd}(t)$ versus crystallization age diagrams for the Archean rocks (Fig. 32a–d). Some important features are as follows.

(1) Juvenile, mantle-derived material was added to the crust during several major periods, namely at 3.8, 3.35–3.3, 3.1, 2.9, 2.8, and 2.75–2.5 Ga. Crustal recycling occurred during almost every period of juvenile crustal additions. Some 3.8 Ga rocks from Anshan have negative $\varepsilon_{Nd}(t)$ values, probably suggesting that crustal recycling was already an important process in the Eoarchean. This is similar to the most ancient rocks in Canada (Acasta gneiss, Iizuka et al. 2009) and the Ancient Gneiss Complex of Swaziland, southern Africa (Kröner et al. 2014). Crustal recycling became more dominant with the evolution of continental crust as indicated by $\varepsilon_{Nd}(t)$ values becoming more negative with time.

(2) Neoarchean (2.75–2.5 Ga) rocks from the Eastern and Western blocks and TNCO show similar Nd isotopic features. However, the Eastern Block shows higher maturity in crustal evolution than the Western Block and TNCO at the end of Archean.

(3) Young, crustally derived rocks inherited the Nd isotopic features from their source material in different areas to variable degrees. In Anshan, for example, some 2.5 Ga crustally derived granites have very negative $\varepsilon_{Nd}(t)$ values due to long geological history up to 3.8 Ga. In contrast, in western Shandong, where the majority of the oldest rocks formed during 2.75–2.7 Ga as addition of juvenile material from depleted mantle source, the 2.5 Ga crustally derived granites commonly have depleted mantle Nd model ages of 2.8 2.7 Ga.

(4) Some rocks older than 3.3 Ga show very high $\varepsilon_{Nd}(t)$ values (Fig. 32a, d). Many of these are from the Anshan area and have very low Sm and Nd contents and are strongly deformed and metamorphosed. Therefore, the above positive anomalies are not considered to be original features of the rocks but are likely due to analytical uncertainty and/or more probably to post-crystallization Nd mobility during metamorphism and other late, fluid-induced geological processes. Some rock samples reported in the earlier literature from the Baijiafen Complex may not be ~3.8 Ga in age, and thus, their $\varepsilon_{Nd}(t)$ values were incorrectly calculated.

The $f_{Sm/Nd}$ values for Archean rock samples selected for the Nd model age histogram are limited to between –0.2 and –0.6 in order to reduce the uncertainties in the model age calculations caused by strong Sm/Nd fractionation during geological processes (Jahn et al. 1990; Wu et al. 2005b). In the single-stage Nd model age (depleted mantle Nd model age) histogram, the data are concentrated between 2.8 and 2.6 Ga with a model age peak at ~2.75 Ga (Fig. 33a). Different rock types show similar single-stage Nd model ages between 3.1 and 2.5 Ga. Compared with TTGs, however, ultramafic to intermediate rocks commonly have young Nd model ages. A model age valley is seen at ~3.1 Ga, and the model ages show a plateau distribution between 3.9 and 3.1 Ga. Two-stage model ages (crustal model ages) show a similar distribution but are commonly shifted toward older ages by 100–50 Ma (Fig. 33a, b).

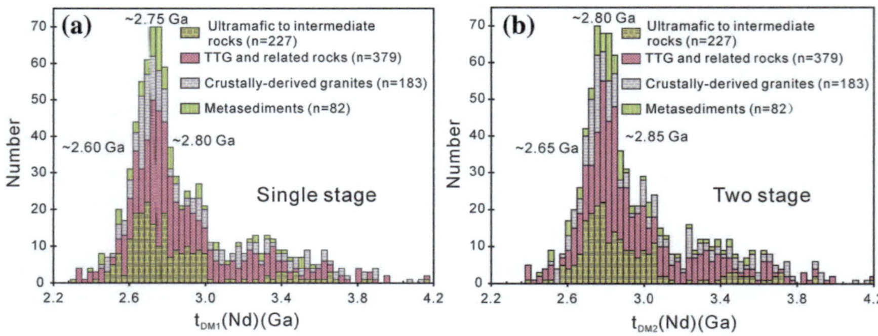

Fig. 33 Nd model age histograms for Archean rocks of the North China Craton. **a** *Single-stage* model age (depleted mantle model age); **b** *two-stage* model age (crustal model age). See text for further interpretation and references

3.3 Hf-in-Zircon Isotopic Composition

Geng et al. (2012) and Wang and Liu (2012) summarized the Hf isotopic features of zircons from early Precambrian basement rocks of the NCC and arrived at the conclusion that crust formation and growth of the craton mainly occurred during the early Neoarchean (2.8–2.7 Ga). In this paper, we used 8564 and 8736 data to compile $\varepsilon_{Hf}(t)$ versus formation age diagrams and Hf model age histograms, respectively. These data are mainly from Archean rocks of the NCC (some are 2.49–2.45 Ga in age, and some detrital zircon data are from Paleoproterozoic metasedimentary rocks). The data sources are mainly from references published within the last ten years (more than 80 papers, many listed in the "References"), together with our unpublished data, and the geographic distribution of the data is similar to those of zircon U–Pb ages with most data for >2.9 Ga rocks being from Anshan, eastern Hebei, and eastern Shandong. Zircon subdivisions are the same in Sect. 3.1. All parameters and equations used for calculation and diagram are those used by Bouvier et al. (2008), Griffin et al. (2000), and Söderlund et al. (2004).

In $\varepsilon_{Hf}(t)$ versus formation age diagrams, data points far below the depleted mantle evolution line have two possible interpretations: (1) the rock formed by melting of older crustal material; (2) zircons with apparently depleted mantle Hf isotopic compositions underwent lead loss, and therefore, their age assessment is wrong. In order to avoid such possibly erroneous data, we only used weighted mean $^{207}Pb/^{206}Pb$ ages and upper concordia intercept ages for magmatic and metamorphic zircons. We also excluded data where zircon analyses have a discordance of >40 %. For detrital zircons, individual data were only used when the 1σ error is less than 20 Ma and the discordance is less than 15 %. Considering the entire NCC, the Hf-in-zircon isotopic data show a similar distribution as the whole-rock Nd isotopic compositions, namely addition of juvenile material to the crust at 3.8–3.55, 3.45, 3.35–3.3, 2.9, and 2.8–2.5 Ga (Fig. 34a–d). Crustal recycling began as early as 3.8 Ga, lasted from 3.8 to 3.25 Ga and, between 3.25 and 2.90 Ga, played a more important role than addition of juvenile material. Ca. 2.90 Ga granitoids in Anben and eastern Shandong show different zircon Hf isotopic features, reflecting differences in their early geological histories. It seems that the original features of Hf-in-zircon isotopic compositions are better preserved than those of whole-rock Nd isotopic compositions. Compared to the whole-rock Nd isotopic data, Hf-in-zircon data reveal significant additions of juvenile material and stronger crustal recycling. This is probably due to a much larger data set and may thus more objectively reflect the Archean geological evolution of the NCC. For the Eastern Block and TNCO, whole-rock Nd isotopic data show obvious differences in the $\varepsilon_{Nd}(t)$ versus formation age diagram (Fig. 32a–b), whereas the Hf-in-zircon isotopic data are similar in the $\varepsilon_{Hf}(t)$ versus formation age diagram (Fig. 34a–b). For example, many detrital zircons from the Zhongtiao area of the TNCO have distinctly negative $\varepsilon_{Hf}(t)$ values, suggesting that >2.9 Ga rocks may occur in this area, and therefore, the TNCO shows similar isotopic features in early crustal evolution as the Eastern Block. In many cases, metamorphic zircons inherited the Hf isotopic

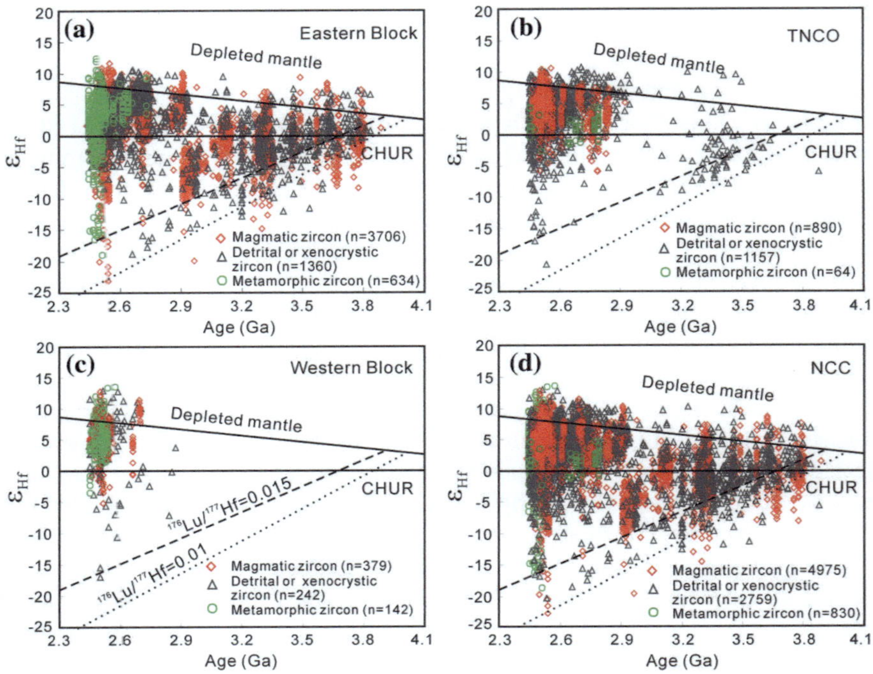

Fig. 34 $\varepsilon_{Hf}(t)$ versus formation age diagrams for zircons of Archean rocks from the North China Craton. **a** Data for the Eastern Block; **b** data for the Trans-North China Orogen (TNCO); **c** data for the Western Block; **d** all data for the North China Craton. *Dotted* and *dashed lines* represent felsic crust with $^{176}Lu/^{177}Hf$ being 0.01 and 0.015, respectively. See text for interpretation and references

features of their magmatic zircons in the same rocks unless metamorphic garnet occurs in the metamorphic rocks. It may therefore be incorrect to assign a metamorphic age to the $^{176}Hf/^{177}Hf$ ratio, but it should be the igneous age of the host rock. However, if using the igneous ages, metamorphic zircon analyses will occur in the same area as the magmatic zircons, making it more difficult to distinguish them. Considering this and presence of garnet in only a few metamorphic rocks, we still use metamorphic zircon ages in this paper. It is easily observed that metamorphic zircons are similar in their Hf isotopic features to magmatic zircons.

Zircons do not generally crystallize from magmas derived directly from mantle sources. Therefore, their single-stage Hf model ages (depleted mantle Hf model ages) are geologically meaningless, and corresponding histograms are not shown here. Two-stage Hf model ages (crustal Hf model ages) are commonly shifted by about 100–50 Ma toward old ages compared with depleted mantle Hf model ages. In the two-stage Hf model age histogram, the data are concentrated between 2.85 and 2.6 Ga with an age peak at ~2.8 Ga (Fig. 35). The age valley occurs at ~3.3 Ga. Compared with the whole-rock Nd model ages, the zircon Hf model ages show an obvious plateau distribution but with a larger age variation between 4.1 and 3.2 Ga. We point out that the continuous whole-rock Nd and zircon Hf model

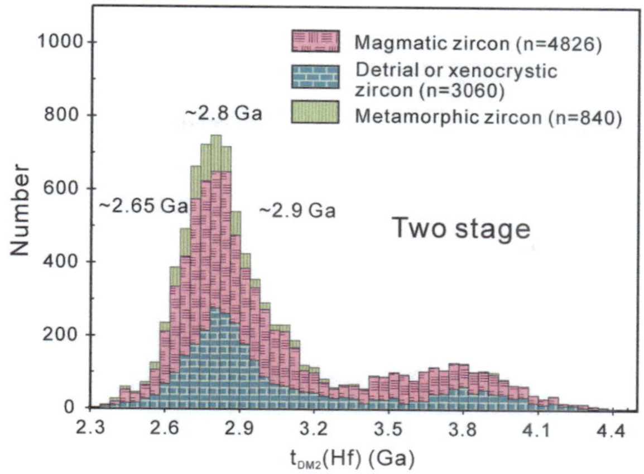

Fig. 35 *Two-stage* Hf model age (crustal Hf model age) histogram for zircons of Archean rocks from the North China Craton. Some are 2.49–2.45 Ga in age, and some detrital zircons are from Paleoproterozoic metasedimentary rocks. See text for interpretation and references

age distributions do not mean that juvenile material was continuously added to the continental crust. We rather suggest that the patterns in whole-rock $\varepsilon_{Nd}(t)$ and zircon $\varepsilon_{Hf}(t)$ versus formation age diagrams are partly a result of mixing of juvenile and recycled material.

4 Formation and Evolution of the Archean Basement of the NCC

4.1 Temporal Evolution

Rocks older than 2.8 Ga only occur in local areas such as Anshan, eastern Hebei, eastern Shandong, Lushan, and Xinyang, amounting to less than 5 % of the area of the NCC Archean basement. However, detrital and xenocrystic zircons older than 2.8 Ga occur over a wider area. Rocks >3.4 Ga consist mainly of granitoids, but ultramafic to mafic rocks of these ages are also present, but most were not dated due to lack of magmatic zircon. Almost all rocks and zircons were involved in a late Neoarchean tectono-thermal event. Rocks >2.8 Ga do not show an increasing trend in distribution with time, but this may be due to uneven reworking of old rocks during later geological processes. 2.75–2.6 Ga magmatic rocks have recently been identified in many areas, and 2.55–2.5 Ga rocks occur more widely than thought before.

Only a few >2.6 Ga supracrustal rocks have been identified, including the 3.36 Ga Chentaigou supracrustals and the 3.0–2.9 Ga Tiejiashan supracrustals in Anshan, the 2.9 Ga Huangyadi supracrustals in eastern Shandong, the 2.82 Ga

Xiataihua supracrustals in Lushan, and the 2.75–2.71 Ga Yanlingguan-Liuhang succession in western Shandong. However, 2.56–2.51 Ga supracrustal rocks occur in almost every late Neoarchean terrane, although their proportions are small (commonly <10 %). Basaltic, intermediate, and felsic volcano-sedimentary rocks are the predominant protoliths of the supracrustal sequences.

Intrusive rocks show variations in rock types with time. The 3.8–3.1 and 2.9–2.7 Ga intrusive rocks are mainly trondhjemitic and tonalitic in composition, respectively, but some gabbroic and dioritic rocks also occur. The oldest large-scale K-rich granite body is the crustally derived 3.0–2.9 Ga Tiejiashan pluton. Granodiorites only began to occur in the late Neoarchean on a large scale, together with tonalites and trondhjemites, whereas crustally derived late Neoarchean granites (mainly monzogranites and syenogranites) are widely distributed.

TTG rocks with ages between 3.45 and 3.3 Ga exhibit variable REE contents, resulting in weakly to strongly fractionated REE patterns. This is considered to be a result of cooling of the Earth (Wan et al. 2005a). However, Moyen (2011) indicated that variable REE patterns in TTGs do not have any relationships with the cooling Earth. Some late Neoarchean rock types such as syenogranite show large variations in REE and trace element compositions, suggesting that the source regions and conditions of magma formation became more complex with time.

The NCC exhibits a continuous evolution from 3.8 to 2.5 Ga, but there are age valleys at ~3.5, ~3.2, 2.85, and ~2.65 Ga. Whole-rock Nd isotopes and Hf-in-zircon isotopes indicate that both juvenile additions and crustal recycling played important roles during almost every tectono-thermal event. The Neoarchean was the most important period for the formation of continental crust in the NCC, and this will be discussed below in more detail.

Metamorphic zircons older than 2.8 Ga were rarely found, and their formation may be related to local events. The oldest (2.76 Ga) well-developed metamorphic zircons are from 2.82 Ga rocks in the Lushan area. Widespread 2.6 Ga metamorphic zircon ages were obtained from rocks in western Shandong. The most important Archean metamorphic event occurred at the end of the Neoarchean, as indicated by widespread ~2.5 Ga metamorphic zircon ages all over the craton. These may suggest that significant crustal thickening only occurred after the Mesoarchean and reached a climax at the end of the Neoarchean due to subduction/collision and/or underplating. Ca. 2.5 Ga metamorphic zircons have not been identified in some areas such as Wutai, but this may be due to the low metamorphic grade in these rocks. In Anshan, there are 3.8–2.5 Ga rocks and zircons, and in eastern Hebei, 3.88–3.4 Ga detrital zircons and 3.4–3.0 Ga and 2.5 Ga rocks have been identified. Many old zircons show overgrowth and recrystallization, suggesting that they became involved in later crustal recycling and metamorphism.

The most important Archean ore deposits in the NCC are BIFs and some massive Cu–Zn sulfide deposits. Although BIFs show large variations in formation age from the Paleoarchean to the early Paleoproterozoic, they are predominantly late Neoarchean in age (2.55–2.51 Ga) (Wan et al. 2012d; Zhang et al. 2012b). It seems that a stable depositional environment was necessary for the formation of large-scale BIF deposits. Gold deposits mainly formed in the Mesozoic, but it is generally considered that Archean mafic–ultramafic supracrustal rocks (greenstone belts) were the main sources (Zhai 2010).

4.2 Ancient Material Records Beneath the NCC

Understanding the geological, geochemical, and geochronological processes during the early Precambrian evolution of the NCC is mainly based on studying the exposed basement as shown above. However, some progress has been made to recognize ancient material in the deep crust of the NCC. There are two ways to obtain information from this cratonic region. One is from rock samples recovered from drill holes that penetrated the basement. One such study revealed late Paleoproterozoic magmatism and metamorphism in basement rocks beneath the Songliao basin in the northeastern NCC (Pei et al. 2007), whereas another study indicated that the basement beneath the Ordos basin in the western NCC was involved in a late Paleoproterozoic tectono-thermal event (Hu et al. 2012; Wan et al. 2013).

Wan et al. (2014b) recently carried out zircon dating, Hf-in-zircon isotopic analyses, and a whole-rock geochemical study of igneous and metasedimentary rocks from basement covered by Mesoproterozoic and younger sedimentary rocks in the Central Hebei Basin (CHB). The CHB extends in a NE–SW direction and covers an area of >35,000 km^2 (Fig. 36). Based on drill core data and geophysical investigations (NCOCP 2012), the basement is composed of greenschist- to upper amphibolite-facies magmatic and supracrustal rocks that locally experienced anatexis. The bottom of the basin shows irregular elevations with the greatest depth being >5000 m. This study identified late Neoarchean magmatic and Paleoproterozoic metasedimentary rocks that were subjected to late Neoarchean to early Paleoproterozoic and late Paleoproterozoic tectono-thermal events, similar to those identified in the early Precambrian basement around the basin. Wan et al. (2014b) concluded that the basement beneath the CHB is part of the NCC. On the basis of this study, the authors suggested that early Precambrian rocks are extensive beneath the Mesoproterozoic and younger sedimentary cover all over the NCC.

Another way to reveal ancient material in the deep crust is to study rock enclaves (xenoliths) and xenocrystic zircons brought to the surface by eruption of volcanoes and/or intrusion of magmas from a deep source. Zheng et al. (2012), Zhang et al. (2012a, b), and Zhang (2014) summarized progresses made in the investigation of such rocks and zircons. Young igneous rocks containing old xenoliths and xenocrystic zircons include Paleozoic kimberlite and lamproite, Mesozoic volcanic or intrusive rocks, and Cenozoic basalts (Fig. 37). The xenoliths are considered to be derived from the lower crust and upper mantle. The lower crustal xenoliths consist of high-grade metamorphic rocks generally less than 10 cm in diameter. Many igneous rocks also contain xenocrystic zircons that were derived from lower crustal sources or were captured during magma ascent. These samples provide important information on the age and composition of the deep NCC crust.

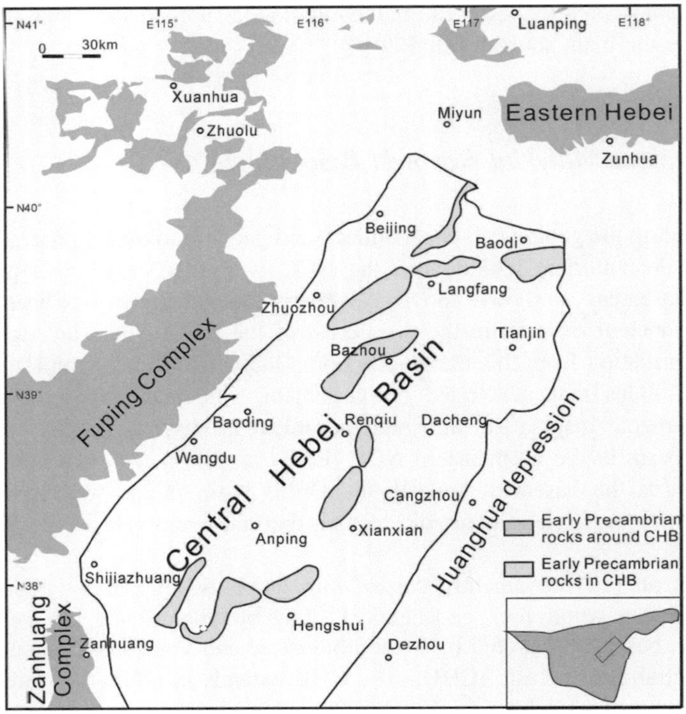

Fig. 36 Geological map of the Central Hebei Basin and surrounding areas, modified after NCOCP (2012) and Wan et al. (2014b). *Inset* shows location in the NCC

The concordia diagram with zircon age data from xenoliths shows several clusters around 2.5, 1.9–1.8, and 0.1 Ga (Fig. 38a). Some data for old zircons from xenoliths of the Xinyang volcanic rocks plot along a discordia line with an upper concordia intercept age of ~3.6 Ga. Note that there is significant lead loss in some of these zircons. Thus, a cumulative histogram compiling all data may lead to an overestimate of zircon crystallization events. Several lead loss trends can be recognized: (1) ancient lead loss in Paleoarchean zircon grains at ca. 2.5 and 1.8 Ga; (2) ancient lead loss in Neoarchean zircon grains at 1.8 Ga; and (3) lead loss in Neoarchean and Paleoproterozoic zircons during the Phanerozoic. The concordia diagram of zircon xenocrysts shows a similar age pattern, including lead loss (Fig. 38b). These features were also observed in the zircon age distribution of the exposed early Precambrian basement. However, there are no zircon grains with Paleoarchean ages, consistent with the limited occurrence of such rocks in the NCC.

All early Paleoproterozoic zircon data, including those with variable lead loss, are shown in Fig. 38c and d. The Hf-in-zircon isotopic data from the xenoliths show that the oldest age group has even more negative $\varepsilon_{Hf}(t)$ (Fig. 38c) than the Paleoarchean zircons in the Anben and Caozhuang areas, indicating that they represent recycled crust. The ~2.5 Ga zircon grains have both negative and

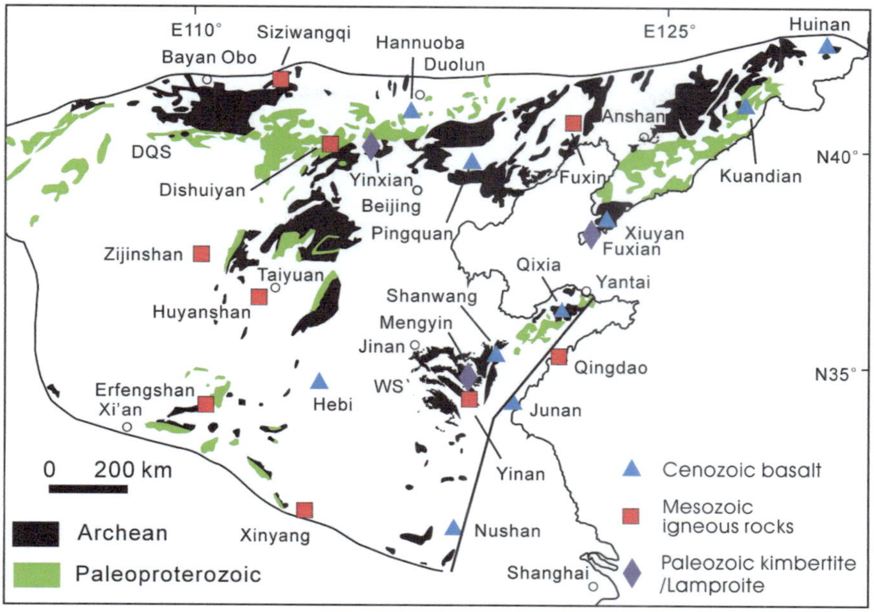

Fig. 37 Geological map showing locations of rocks containing xenoliths and xenocrystic zircons in the North China Craton and adjacent area. Paleozoic kimbertite/lamproite: Mengyin, Fuxin, and Yingxian; Mesozoic volcanic/intrusive rocks: Xinyang, Erfengshan, Yinan, Qingdao, Huyanshan, Zijinshan, Dishuiyan, Siziwangqi, and Fuxian; Cenozoic basalt: Nushan, Hebi, Junan, Shanwang, Qixia, Xiuyan, Pingquan, Hannuoba, Huinan, and Kuandian. Locations of rocks containing xenoliths and xenocrystic zircons are from HF Zhang et al. (2012a, b) and Zheng et al. (2012)

positive $\varepsilon_{Hf}(t)$, implying that juvenile material was added to the crust and recycling of older material also occurred. A similar pattern was observed in the 1.9–1.8 Ga zircons. Specifically, Phanerozoic zircon grains mainly show negative $\varepsilon_{Hf}(t)$ values, and some grains have extremely low $\varepsilon_{Hf}(t)$ of –40 and even lower. There is no doubt that some Archean material was recycled during Phanerozoic tectono-magmatic events. The Hf isotopic data of the zircon xenocryst show a similar pattern (Fig. 38d).

Both the data from xenoliths and zircon xenocrysts show similar Hf model age distributions (Fig. 38e, f) as those from basement rocks exposed on the surface. The major peak is at about 2.8 Ga, and very old model ages near the beginning of the formation of continental crust on the Earth suggest that there may be some very old material in the deep crust of the NCC.

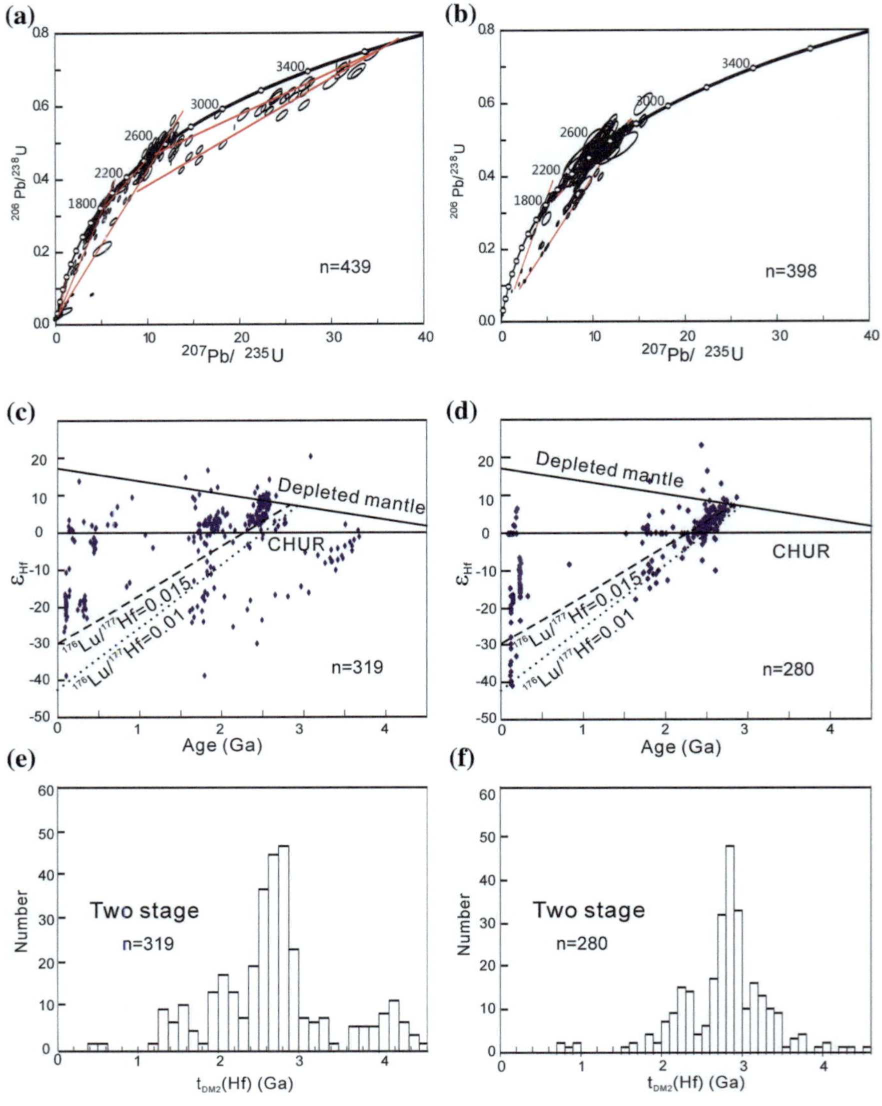

Fig. 38 Ages and Hf isotopes of zircons from deep crust in the North China Craton. **a** and **b** U–Pb concordia diagrams; **c** and **d** age versus $\varepsilon_{Hf}(t)$ diagrams; **e** and **f** Hf model age histograms. **a**, **c**, and **e** zircons from xenoliths; **b**, **d**, and **f** xenocrystic zircons. Data are from Gao et al. (2004), YS Liu et al. (2004b), Ying et al. (2011), Zhang (2012), Zhang et al. (2012a, b, c), and Zheng et al. (2004a, b, c, 2008)

4.3 Major Periods of Continent Formation

Formation of continental crust in the NCC can be traced back to the Eoarchean. Geological records earlier than Mesoarchean were most likely partly destroyed during later tectono-metamorphic processes, and it seems likely that the Neoarchean was a major period of crust formation in the NCC. This means that some crucial processes occurred in the Earth during this time, probably indicating a transformation in the global tectonic regime due to changes in the thermal state of the Earth from hot to cool. Zircon age histograms indicate that the most important tectono-thermal event occurred at ~2.5 Ga. Evidence for addition of juvenile material at ~2.5 Ga includes the following: (1) Ca. 2.5 Ga supracrustal rocks (greenstone belts), commonly with a high proportion of metabasalt, occur in almost every terrane; (2) Ca. 2.5 Ga gabbroic to dioritic rocks are widely distributed all over the NCC (Li et al. 2010b; Ma et al. 2012; Wan et al. 2010c; Wan, unpublished data); and (3) some ~2.5 Ga TTGs exhibit whole-rock Nd and Hf-in-zircon isotopic compositions similar to the depleted mantle (Diwu et al. 2012; Geng et al. 2012; Liu et al. 2009b; Wang and Liu 2012; this study). However, mafic to intermediate assemblages (including volcanic and intrusive rocks) constitute only a small portion of the granitoid–greenstone belts, and 2.5 Ga TTGs with depleted mantle model ages are relatively rare (Figs. 32d and 34d).

In western Shandong, >2.8 Ga rocks have not been identified, but voluminous ~2.5 Ga crustally derived granites show the same whole-rock Nd and Hf-in-zircon isotopic features as the 2.75–2.7 Ga rocks. Therefore, the ~2.5 Ga granitoids are considered to have been derived from 2.75–2.7 Ga sources (Wan et al. 2010c, 2011b). In areas where >2.8 Ga rocks are exposed, ~2.5 Ga granites with Nd and Hf isotopic compositions of 2.7 Ga juvenile rocks could have formed as a result of mixtures of depleted mantle-derived magmas and older crustal material. However, >2.8 Ga rocks are rare in the NCC, so this may not have been a significant process for the formation of ~2.5 Ga granitoids. On the other hand, some magmatic zircons from ~2.5 Ga gabbros, diorites, and granodiorites show Hf isotopic enrichment, similar to the Nd isotopic composition of some ~2.5 Ga amphibolites. This may be a feature of a mantle source or a result of crustal contamination of magmas derived from a depleted mantle. There are three potential sources for the ~2.5 Ga magmatic rocks with similar Nd and Hf isotopic compositions as the 2.7 Ga juvenile rocks (Wan et al. 2014a): (1) Ca. 2.7 Ga granitoids that constituted precursors for the ~2.5 Ga crustally derived granitoids; (2) Ca. 2.7 Ga mafic rocks that also constituted precursors for the ~2.5 Ga intermediate and more felsic rocks; and (3) a mantle source with ~2.7 Ga depleted mantle model ages from which some mafic rocks could have been derived. The possibility cannot be excluded that ~2.5 Ga granitoids with ~2.7 Ga depleted mantle model ages resulted from crustal recycling of ~2.5 Ga mantle-derived rocks with the same isotopic features; however, rocks formed in this way must be limited in volume. Therefore, the conclusion can be drawn that the most important period of addition of juvenile material occured in the early Neoarchean rather than in the late Neoarchean, consistent with rocks and detrital and xenocrystic zircons of these ages occurring in many areas of the NCC. Fundamentally, therefore, the NCC is similar to many other cratons elsewhere in that tectono-thermal and crust-forming events at ~2.7 Ga are globally well developed. The main difference between the NCC and many other cratons is that a superimposed ~2.5 Ga tectono-thermal event was particularly strong in the former (Wan et al. 2010c, 2011b, 2014a, 2015b; Zhai and Santosh 2011).

4.4 Tectonic Subdivision of the NCC

The early Precambrian subdivision of the craton is mainly based on the distribution of late Neoarchean to Paleoproterozoic rocks because of insufficient data for the earlier geological evolution. The subdivision is debated in the literature, and several schemes have been proposed as summarized by Zhai and Santosh (2011) and Zhao and Zhai (2013).

Wu et al. (1998) subdivided the NCC into five blocks, namely the Jiaoliao, Qianhuai, Jinji, Yuwan, and Mongshan blocks (Fig. 39a). The Jiaoliao and Qianhuai blocks were considered to have assembled along the Jiao-Liao-Ji Belt, resulting from an east-dipping subduction zone to form a larger block at ~2.5 Ga which then collided with other blocks to result in final amalgamation of the NCC during the late Paleoproterozoic.

Zhao et al. (2001) suggested a 3-fold subdivision of the craton, namely the Eastern Block, Western Block, and Trans-North China Orogen (TNCO). Late Archean anticlockwise P-T paths for the two blocks were interpreted to have resulted from several mantle plumes, resulting in magmatic underplating and high-grade metamorphism at ~2.5 Ga, whereas the clockwise path in the TNCO was interpreted to reflect continental collision between the Eastern and Western blocks, leading to Paleoproterozoic assembly of the NCC at ~1.85 Ga. Zhao et al. (2005) later modified this model into the currently favored 6-fold subdivision of the craton (Fig. 39b), resulting from the speculative interpretation of major collisional zones within both the Eastern and Western blocks.

Li et al. (2002), Kusky and Li (2003), and Kusky et al. (2007) suggested that collision between the Eastern and Western blocks occurred in the late Neoarchean (~2.5 Ga) along the Central Orogenic Belt (COB) to form a unified NCC. The COB is similar in spatial distribution to the TNCO of Zhao et al. (2001, 2005) with differences being that the COB extends farther northeast into southern Jilin (Fig. 39c) and that the subduction polarity was west-dipping, rather than east-dipping, as suggested by Zhao et al. (2001, 2005).

Zhai and Santosh (2011) subdivided the NCC into seven microblocks, mainly based on the spatial distribution of ancient rocks and tectonic boundaries revealed by several granite–greenstone belts. These were named the Alashan (ALS), Xuhuai (XH), Xuchang (XCH), Jining (JN), Ordos (OR), Qianhuai (QH), and Jiaoliao (JL) blocks, with very different boundaries from those of other authors. They

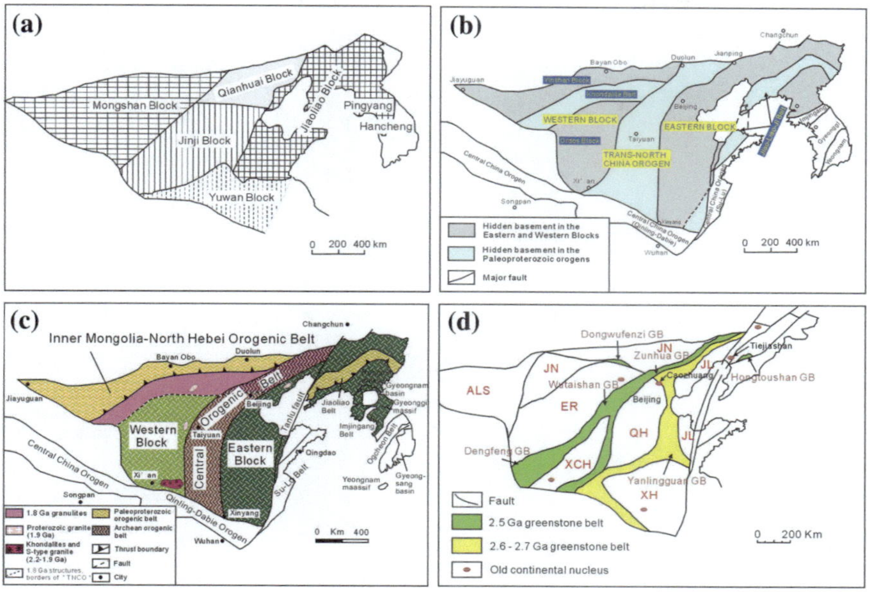

Fig. 39 Different tectonic subdivisions of the North China Craton. **a** Wu et al. (1998); **b** Zhao et al. (2005); **c** Li et al. (2002); **d** Zhai and Santosh (2011)

suggested that these microblocks were welded together along Neoarchean granite–greenstone belts at ~2.5 Ga (Fig. 39d).

In order to understand the tectonic setting of the NCC during the late Neoarchean, it is necessary to determine the spatial distribution of ancient (≥2.6 Ga) continental domains in the craton. A strong tectono-thermal event at ~2.6 Ga in western Shandong is an important reason for considering 2.6 Ga as a chronological boundary between the early and late Neoarchean. With regard to the spatial distribution of these ancient crustal domains, rocks and zircons of different origins and ages have different geodynamic interpretations. Rocks ≥2.6 Ga themselves represent ancient crust, whereas ≥2.6 Ga detrital zircons in late Neoarchean and Paleoproterozoic metasedimentary rocks suggest the existence of ancient crust in nearby areas, and ≥2.6 Ga xenocrystic zircons in young (<1.8 Ga) intrusive rocks may indicate the existence of old material in the deep crust. However, old detrital zircons in young sedimentary rocks have no significance because they may have undergone multirecycling. Figure 1 shows ≥2.6 Ga detrital and xenocrystic zircons from Paleoproterozoic or older rocks. Based on the spatial distribution of ancient rocks and zircons, three ancient terranes can be delineated, namely the Eastern Ancient Terrane, Southern Ancient Terrane, and Central Ancient Terrane (Fig. 40).

(1) Eastern Ancient Terrane

This is the best understood ancient terrane and occurs along the eastern margin of the NCC in a NE–SW direction, including Anben, eastern Hebei, eastern Shandong, and western Shandong. The oldest rocks and zircons mainly occur in

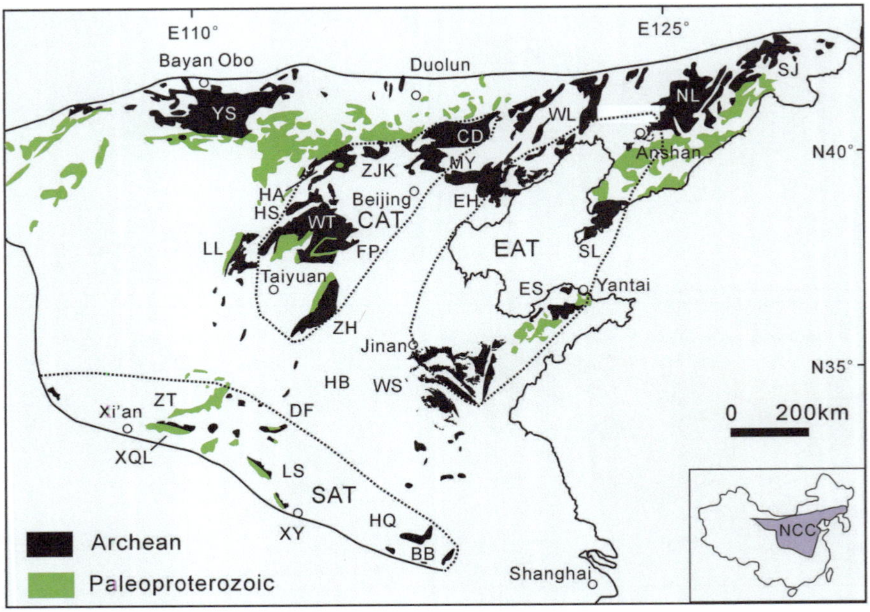

Fig. 40 Distribution of ancient (>2.6 Ga) terranes in the North China Craton. *EAT* Eastern Ancient Terrane; *SAT* Southern Ancient Terrane; *CAT* Central Ancient Terrane. Abbreviations are as in Fig. 1

this terrane. Besides the 3.8–2.9 Ga rocks in Anshan and 3.4–3.0 Ga rocks and 3.8–3.3 Ga detrital zircons in eastern Hebei, 2.9–2.6 Ga rocks were identified in eastern Shandong and western Shandong. The spatial relationships between the old rocks in the terrane may not have changed significantly since the early Neoarchean (2.6 Ga). In western Shandong, the margin of the ancient terrane is the boundary between belts B and C (Fig. 19). Although granitoids in Belt A are mainly ~2.5 Ga in age, they are derived from melting of older basement. Syenogranites and monzogranites are also widely distributed in eastern Hebei, Jinzhou, and Anben, and they constitute the largest crustally derived late Neoarchean granite belt in the NCC. It is notable that these ~2.5 Ga granitoids inherited their compositional features from older basement. As mentioned above, in Anshan, recycling of >3.0 Ga crust played a profound role in the formation of the ~2.5 Ga syenogranites, as evidenced by the presence of older xenocrystic zircons as well as whole-rock Nd and Hf-in-zircon isotopic compositions (Wan et al. 2015a). In western Shandong, on the other hand, the ~2.5 Ga crustally derived granites in Belt A have similar whole-rock Nd and Hf-in-zircon isotopic compositions as the adjacent early Neoarchean rocks (Wan et al. 2010c, 2011b).

(2) Southern Ancient Terrane

This terrane occurs along the southern margin of the NCC in a nearly E–W direction. 3.65, 2.83–2.82, 2.7, and 2.7–2.6 Ga rocks have been identified in

Xinyang, Lushan, Huoqiu, and Zhongtiao, respectively, and 3.6–2.6 Ga detrital and xenocrystic zircons were discovered in additional areas (Liu et al. 2012b). Importantly, as pointed out before, the captured or detrital 4.1–3.5 Ga zircons in Paleozoic volcano-sedimentary rocks of the northern Qinling Orogenic Belt may be derived from the southern margin of the NCC. This suggests that the Southern Ancient Terrane may have a long geological history back to 4.1 Ga ago. Widespread identification of >2.6 Ga rocks and zircons in this terrane is an important progress made in recent years.

(3) Central Ancient Terrane

This includes the Zanhuang, Fuping, Hengshan, Zhangjaikou, and Chengde areas where 2.7 Ga TTG rocks and zircons were identified, although the TTG rocks commonly occur at small scales. It is likely that more rocks and zircons older than 2.6 Ga will be discovered in this terrane, and the widespread presence of crustally derived granites in the Central Ancient Terrane is also consistent with the existence of older material in the deep crust. However, there are also many ~2.5 Ga TTG rocks in the terrane, and this makes it uncertain whether or not the Central Ancient Terrane existed prior to 2.6 Ga.

Besides the above three ancient terranes, >2.6 Ga rocks and zircons were identified in other areas of the NCC. For example, in Xi Wulanbulang, 2.7 Ga TTG rocks with a relatively wide distribution as well as older detrital and xenocrystic zircons were discovered (Jian et al. 2012; Dong et al. 2012a, 2012b; Ma et al. 2013). Ca. 2.5 Ga zircons in late Neoarchean supracrustal and intrusive rocks have the same Hf isotopic compositions as zircons from the 2.7 Ga TTG rocks, providing evidence for crustal recycling or contamination of early Neoarchean crust. Therefore, there may also be an ancient terrane in the Yinshan Block.

4.5 Tectonic Regime

Mantle plume activity resulting in magmatic underplating and plate tectonics are considered to have been the most important processes involving the formation and evolution of early continental crust (Van Kranendonk et al. 2014), but the timing of initiation of plate tectonics is debatable. Some authors suggested that present-type plate tectonics only began in the late Paleoproterozoic or later, whereas others considered that tectonic regimes similar to those of today occurred as early as the Eoarchean (e.g., Nutman et al. 2013). Therefore, there are different opinions on whether mantle plume activity (or underplating) and/or arc magmatism or both played key roles in the Neoarchean, an important period when much of the global continental crust formed (Bédard 2013; Condie and Kröner 2013; Dostal and Mueller 2013; Halla et al. 2009; Manikyamba and Kerrich 2012; Mohan et al. 2013; Wyman 2013a, b). Magmatic underplating is considered to be related to mantle plume activity or, more probably, to mantle overturn (Davies 1995; Rey et al. 2003) and may have been an important mechanism in continental growth and reworking (Frost et al. 2001; Warren and Ellis 1996). Mantle plumes generally last for only

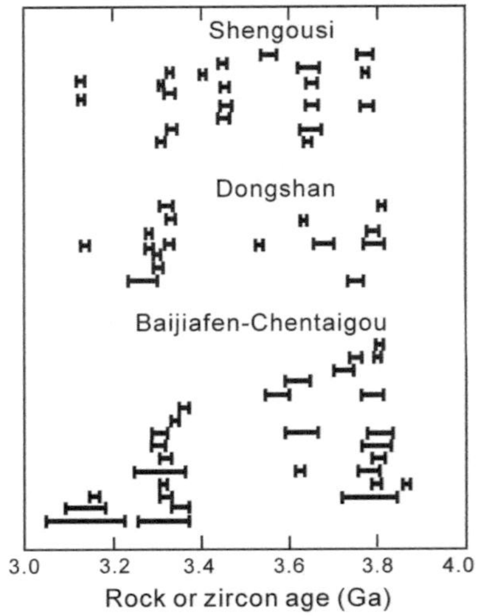

Fig. 41 Comparison of the Archean magmatic records of the Baijiafen, Dongshan, and Shengousi complexes in the Anshan area. Data are from Liu et al. (1992, 2008), Song et al. (1996), Wan et al. (2005a, b, c, 2012a), Wu et al. (2008), and Wan (unpublished data)

5–10 Ma (Abbott and Isley 2002), a much shorter time span than the late Neoarchean (2.55–2.50 Ga) igneous activity in the NCC. Mantle overturn may have led to longer magmatism.

We now have a better understanding of the early (>2.8 Ga) evolution of the NCC through detailed studies in some areas. In Anshan, magmatism almost continuously lasted for a long time from 3.8 to 2.9 Ga (Fig. 12), where multiple and complex phases of igneous activity were recorded in all three complexes (Fig. 41). The rock types include both mantle-derived and crustally derived rocks (Dong et al. 2013; Liu et al. 1992, 2008; Song et al. 1996; Wan et al. 2005a, 2007, 2012a, 2015a; Zhou et al. 2007; Zhou et al. 2009). These data suggest that long-term magmatism related to mantle activity widely occurred in Anshan and adjacent area. In eastern Hebei, detrital zircons record almost continuous ages ranging from 3.88 to 3.4 Ga, although only 3.4–3.0 Ga Paleoarchean–Mesoarchean rocks were discovered until now. We suggest that widespread magmatism due to mantle overturn activity may have been the main mechanism of continental growth and reworking before the Mesoarchean, consistent with the underplating model.

Western Shandong and eastern Shandong are areas where ~2.7 Ga rocks are most widely distributed in the NCC. Metabasalts with REE depletion and enrichment have been identified in western Shandong, and komatiites with well-preserved spinifex textures (Fig. 18d) were considered to be ~2.7 Ga in age (Polat et al. 2006a). In eastern Shandong, ~2.7 Ga TTGs formed about 200 Ma later than ~2.9 Ga TTGs, and this suggests that subduction may not have been responsible for the formation of the ~2.7 Ga rocks because a 200 Ma time span is too short for a full Wilson cycle (Jahn et al. 2008). Furthermore, the majority of TTG rocks in both western Shandong and eastern Shandong show strongly fractionated REE patterns and Nb–Ti–P depletion (Jahn et al. 2008; Wan et al. 2014a). Based on rock associations, TTG compositions, and the geological evolution, mafic magma underplating was considered as a viable process for the formation of the supracrustal and intrusive rocks (Jahn et al. 2008; Polat et al. 2006a). The tectonic setting of ~2.7 Ga rocks in other areas is not so clear because of their small size and poor data. However, Yang et al. (2013a) favored a subduction model for the formation of ~2.7 Ga TTG rocks in the Zanhuang area.

Underplating and arc magmatism have been proposed for crustal growth in the late Neoarchean. The former model was mainly suggested from studies of the eastern NCC (Geng et al. 2006; Yang et al. 2008; Zhao et al. 1998, 1999, 2001). Zhao and Zhai (2013) summarized the main features supporting this view as follows: (1) An exceptionally large exposure of granitoid intrusions formed over a short time period (2.55–2.50 Ga) and shows no systematic age progression across a ~800-km-wide block; (2) generation of komatiitic magmas with eruption temperatures as high as ~1650 °C; (3) dominant domal structures; (4) bimodal volcanic assemblages in the greenstone sequences; (5) affinities of mafic rocks to continental tholeiitic basalts; and (6) metamorphism with anticlockwise P-T paths involving isobaric cooling. Jian et al. (2012) attributed the strong late Neoarchean (2.55–2.50 Ga) magmatism and metamorphism in the Yinshan Block to episodic mantle upwelling/melting as a result of spontaneous delamination of the lower crust and periodic delamination of melt residues during granitoid production. They considered this to have occurred in a continental environment because of the presence of >2.6 Ga rocks and zircons.

Some authors suggested a scenario of arc magmatism followed by collisional orogeny, although there are different opinions on the timing and spatial distribution of collisional belts (Kröner et al. 2005a, b; Kusky and Li 2003; Kusky et al. 2007; Li et al. 2002; Nutman et al. 2011; Polat et al. 2006b; Wan et al. 2005b, c, 2010c, 2012c; Wilde et al. 2005; Wu et al. 1998; Zhang et al. 2007, 2009; Zhao et al. 2001, 2002, 2005). The main points include the following: (1) Igneous rocks of different ages and compositions occur in different zones and spatially show an asymmetrical distribution in some areas such as western Shandong and eastern Hebei; (2) compared with TTGs, syenogranites and monzogranites commonly formed during a slightly late phase; (3) older rocks (2.56–2.525 Ga) commonly underwent stronger deformation and metamorphism than younger rocks (2.525–2.48 Ga); (4) intrusive and volcanic activities occurred during almost a same period, and the intrusive rocks were uplifted quickly to the surface to form the source for sediments in the late Neoarchean basins, suggesting an active tectonic environment; (5) there are metabasaltic rocks with depletion and enrichment features in many supracrustal belts; and (6) intrusive rocks of different compositions are commonly depleted in Nb and Ta and thus display a subduction-related chemical signature. Nutman et al. (2011) indicated that the complexity of the late Neoarchean eastern Hebei arc magmatism matches that found in long-lived arc systems that involved older continental crust, such as the Andean margin of South America (Hildreth and Moorbath 1988), the Paleoproterozoic of South Greenland (Garde et al. 2002), and the Neoarchean crustal development in southern India (Chadwick et al. 2007).

In fact, some of the above features can be intercepted by both the underplating and subduction models. More data, including geophysical surveys, are required in order to better understand the tectonic regime in the NCC during the late Neoarchean. These include (1) the difference in the spatial distribution of Archean rocks at present and in the late Neoarchean and (2) the compositions and ages of the unexposed basement. It is difficult to explain that the above two different tectonic settings worked at the same time with underplating having occurred beneath the Eastern and Western blocks when both these blocks moved toward each other resulting in subduction/collision. On the basis of similarities in rock association, ages of formation, and geological evolution in different areas, we suggest that only one tectonic setting played an important role in the NCC during the late Neoarchean.

It is notable that globally, apart from the NCC, ~2.5 Ga rocks only occur in a few areas such as southern India (Chadwick et al. 2007; Clark et al. 2009; Dey et al. 2012; Jayananda et al. 2000; Moyen et al. 2003), Antarctica (Corvino and Henjes-Kunst 2007; Clark et al. 2012; Duclaux et al. 2008; Tsunogae et al. 2014), and northern Australia (Drüppel et al. 2009; McCready et al. 2004). The NCC underwent a much stronger tectono-thermal event at ~2.5 Ga than many other cratons characterized by ~2.7 Ga events (Condie 2000; Condie et al. 2009). It may be possible that all these continental terranes once belonged to a single block (Clark et al. 2012; Wan et al. 2011a; Zhao et al. 2003), but more work is required to identify this. If underplating was active at ~2.5 Ga, it may have modified an early Neoarchean supracontinent previously formed by a widespread and strong ~2.7 Ga tectono-thermal event. On the other hand, if arc magmatism was active at this time, the areas where ~2.5 Ga TTG rocks are well developed may represent an ancient subduction/collision belt between two large blocks of the early Neoarchean. In both cases, the different cratons may be dispersed remnants of what was once a single continent at the end of the Neoarchean, consistent with many cratons showing fragmented features (Bleeker 2003).

All late Neoarchean tectono-magmatic belts so far proposed in the NCC, such as the Liao-Ji-Lu magmatic belt (Wu et al. 1998), the Central Orogenic Belt (Li et al. 2002), and the greenstone belts (Zhai and Santosh 2011), only contain a relatively small portion of the late Neoarchean areas. Therefore, these are not consistent with the arc magmatism model. In order to overcome this problem, we propose a multi-island arc model in which the three ancient terranes, and possibly other old terranes, occurred in an oceanic domain, and amalgamation of these terranes due to subduction/collision resulted in the formation of supracrustal and intrusive rocks as well as juvenile additions and crustal recycling and finally assembled the NCC at the end of the Neoarchean. This is consistent with rocks of continental and oceanic affinities in many areas of the NCC. In the western Liaoning, Dengfeng and Huai'an areas, for example, there are supracrustal and intrusive rocks apparently derived from depleted mantle sources, but crustally derived granites and sediments have also been identified (Diwu et al. 2011; Liu et al. 2009b; Wang et al. 2011). The late Neoarchean may have been a period when plate tectonics began to play an important role in crust formation, although subduction/collision would have been different in pattern and scale from present-day processes because the Earth was hotter than now and the oceanic lithosphere was thicker, softer, and more buoyant, and continent blocks were smaller.

4.6 Craton Stabilization

Although there are different opinions on the late Neoarchean tectono-thermal event in the NCC, it is accepted by many authors, including those who favored the arc magmatism model (Nutman et al. 2011; Wan et al. 2010c, 2011b), that the latest Neoarchean event was related to an extensional tectonic setting, probably due to magmatic underplating. Most latest Neoarchean crustally derived granites have formation ages of 2.52–2.49 Ga and either are undeformed and massive or show weak deformation. They intruded into earlier, deformed rocks and are associated with undeformed, mantle-derived gabbroic and dioritic rocks of the same age. In western Shandong, for example, intrusive rocks older than 2.525 Ga commonly show strong deformation, whereas intrusive rocks younger than 2.525 Ga commonly are undeformed or only weakly deformed (Fig. 42). This suggests that the tectonic regime probably changed from compressional to extensional between 2.53 and 2.52 Ga (Wan et al. 2010c). Similar scenarios were proposed for other areas of the NCC such as Wutai and Zanhuang (Wilde et al. 2005; Yang et al. 2013a).

It is also evident that migmatites occur on large scales and are associated with crustally derived granites in many areas such as western Shandong (Fig. 19). Migmatization occurred at the same time or slightly earlier than the formation of granites. Metamorphism and associated crustally derived granites were identified all

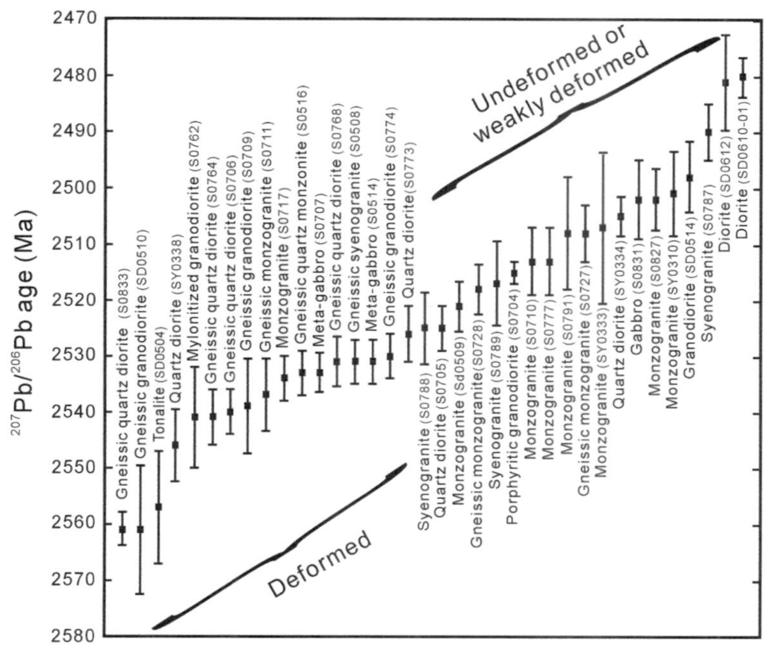

Fig. 42 Zircon age variation diagram (with *error bars*) for different types of intrusive rocks in western Shandong. Data are from Lu et al. (2008) and Wan et al. (2010c)

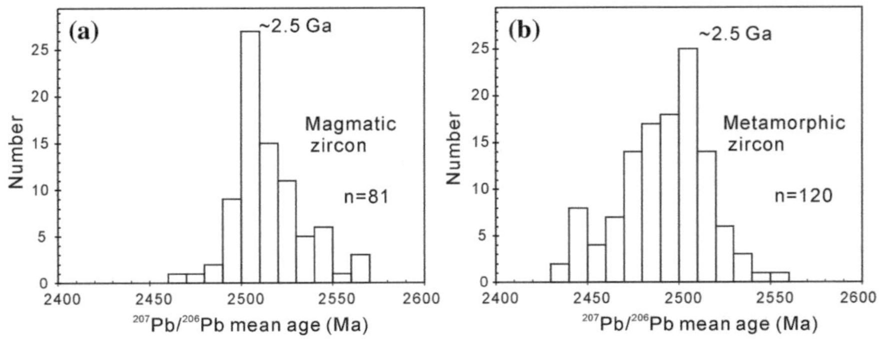

Fig. 43 Weighted mean ^{207}Pb/^{206}Pb age histograms for magmatic zircons from crustally derived granites (**a**) and metamorphic zircons from high-grade metamorphic rocks (**b**) in the North China Craton. See text for interpretation and references

over the NCC, including eastern Shandong, western Shandong, northern Liaoning, western Liaoning, eastern Hebei, Dengfeng, Yinshan, and Daqingshan, with metamorphic ages identical to crystallization ages of many crustally derived granites, mainly ranging from 2.53 to 2.49 Ga and 2.53 to 2.45 Ga, respectively (Fig. 43) (Bai et al. 2014; Chen et al. 2006; Cui et al. 2013; Dai et al. 2013; Deng et al. 2014; Dong et al. 2012a, b; Geng et al. 2006; Grant et al. 2009; Guan et al. 2002; Guo et al. 2013; Han et al. 2014a; Jahn et al. 2008; Jian et al. 2012; Kröner et al. 1998, 2005a; Li et al. 2009, 2010b, 2011a, b 2013a, b; Lu et al. 2008; Lü et al. 2012; Ma et al. 2013, 2013a; Nutman et al. 2011; Peng et al. 2012, 2013a, b; Ren 2010; Ren et al. 2013; Shen et al. 2004; Shi et al. 2012; Sun et al. 2010; Wan et al. 2005b, 2009c, 2010c, 2012b, d, 2015a, unpublished data; Wang et al. 2000, 2011, 2012, 2013, 2014a, b; Wilde et al. 1997, 2005; Wu et al. 2013, 2014; Xie et al. 2013, 2014b; Yang et al. 2008, 2011; Zhang et al. 2011, 2012a, 2014; Zhao et al. 2002, 2008b2009, 2011; Zhu et al. 2015). The association of undeformed, mantle-derived rocks (gabbro, diorite) of the same age (mainly 2.52–2.49 Ga) suggests that underplating may have played an important role in causing metamorphism and anatexis in the lower crust to produce crustally derived granites during extension in the NCC basement at the end of the Neoarchean. These magmas moved to higher crustal levels to form intrusive bodies at different scales, including syenogranites. The formation of large-scale granite batholiths and widespread high-grade metamorphism and anatexis are considered to have been important processes to cause cratonic stabilization of the NCC at the end of the Neoarchean, a convenient time to set 2.5 Ga as the Archean–Proterozoic boundary (Wan et al. 2012b, 2015a; Yang et al. 2011).

Metamorphic zircon ages are younger than magmatic zircon ages of crustally derived granites in some areas, down to 2.45 Ga or even later, although some may be the result of partial resetting of the U–Pb isotopic system in the zircons due to the strong late Paleoproterozoic tectono-thermal event. It is uncertain whether the metamorphism lasted from the late Neoarchean to the earliest Paleoproterozoic (~2.45 Ga) or whether the earliest Paleoproterozoic metamorphism was a different event. In western Liaoning, metamorphism was considered to be one event and lasted from the latest Archean to early Paleoproterozoic (Kröner et al. 1998; Liu et al. 2011a). In Daqingshan, however, the earliest Paleoproterozoic Daqingshan supracrustal sequence (mainly metasediments) contains 2.45 2.40 Ga metamorphic zircons, suggesting that here the earliest Paleoproterozoic metamorphism was a separate tectono-thermal event.

Cratonic stabilization implies that the NCC was already a single tectonic unit at the end of the Neoarchean as suggested by Wan et al. (2011a), Zhai and Peng (2007), and Zhai and Santosh (2011). The Eastern and Western blocks and the TNCO share many common features, and some have been discussed before: (1) These tectonic units display almost identical late Neoarchean zircon age spectra with age peaks at ~2.52 Ga; (2) there is evidence for >2.8 Ga material, although this is more evident in the Eastern Block; (3) there are ~2.7 Ga TTGs in all three blocks; (4) there are ~2.5 Ga supracrustal rocks with similar rock associations with BIFs being typical components; (5) there was strong juvenile magma addition from mantle sources to the crust between 2.8 and 2.7 Ga; (6) there are abundant ~2.5 Ga TTGs, associated with minor gabbroic and dioritic rocks; (7) the old rocks underwent metamorphism and anatexis between 2.52 and 2.48 Ga; (8) crustally derived late Neoarchean granites occur in all three blocks and are commonly younger than the TTGs; (9) crustal recycling of ~2.7 Ga rocks at ~2.5 Ga played an more important role than juvenile addition; and (10) young sediments in different areas commonly contain late Neoarchean and late Paleoproterozoic detrital zircons.

In general, it is difficult to determine the original relationships between the different blocks just by comparing rock types and ages. Considering the limited distribution of the ~2.5 Ga tectono-thermal event worldwide, however, the obvious similarity of the Eastern and Western blocks may suggest that they formed in a similar tectonic setting, supporting the conclusion that the NNC had already been a single unit at the end of the Neoarchean.

5 Summary and Conclusions

(1) The NCC underwent a long and complex evolution from 3.8 to 2.5 Ga. Rocks older than 2.8 Ga occur only locally, with Anshan and eastern Hebei being the most important areas where >3.8 Ga rocks and crustal components have been identified.

(2) The most important tectono-thermal event in the NCC occurred at ~2.5 Ga, and this is different from many other cratons worldwide. However, whole-rock Nd and Hf-in-zircon isotopic compositions indicate that the late Mesoarchean to early Neoarchean is the most important period for rapid production of continental crust. This is similar to several other cratons.

(3) The deep crust of the NCC shows similar evidence for the presence of ancient material as the surface exposures, and the available data suggest that the most important tectono-thermal events occurred in the late Neoarchean and late Paleoproterozoic, and juvenile material was added to the crust during the late Mesoarchean to early Neoarchean.
(4) We suggest that three ancient terranes older than 2.6 Ga can be delineated in the NCC, namely the Eastern, Southern, and Central Ancient Terranes.
(5) Vertical crustal growth is considered to have been the main mechanism of continental evolution prior to the Mesoarchean, and this is mainly based on geological features in the Anshan area. It is still uncertain whether magmatic underplating or arc magmatism played a decisive role in the evolution of the NCC during the late Neoarchean. We favor a multi-island arc model related to amalgamation through subduction/collision of different ancient terranes.
(6) The NCC may have been a single tectonic unit at the end of the late Neoarchean, due to cratonic stabilization as indicated by the formation of widespread and voluminous granites and extensive high-grade metamorphism and anatexis all over the craton, probably as a result of mantle underplating.

Acknowledgements We are grateful to the members of the Beijing SHRIMP Center for zircon separation, mount making, CL imaging, dating, and other help. Chunli Guo, Kejun Hou, Liewen Xie, Yueheng Yang, Honglin Yuan, Chunrong Diwu, and Hongying Zhou assisted in Hf-in-zircon isotopic analysis, and Suohan Tang, Jinhui Wang, Chaofeng Li, Jinghui Guo, Fukun Cheng, and Hongying Zhou assisted in whole-rock Nd isotopic analysis. We also thank Vickie Bennett, Michael Brown, Bin Chen, Yuelong Chen, Kent Condie, Xiaozhong Ding, Chunrong Diwu, Lilin Du, Clark R.L. Friend, Yuansheng Geng, Yongchang Gu, Jinghui Guo, Baofu Han, Jianmin Hu, Ping Jian, Wei Jin, Huaikun Li, Huimin Li, Houmin Li, Jianghai Li, Jinyi Li, Sanzhong Li, Tingdong Li, Xianhua Li, Xuping Li, Yanhe Li, Shoufa Lin, Fulai Liu, Junlai Liu, Shuwen Liu, Xiaochun Liu, Yongshun Liu, Zhihong Liu, Songnian Lu, Changqian Ma, Peisen Miao, Zhiyao Ni, Allen Nutman, Dominic Papineau, Peng Peng, Jishun Ren, M. Santosh, Qihan Shen, Min Sun, Yong Sun, Jeff D. Vervoort, Huichu Wang, Renmin Wang, Shijin Wang, Shiyan Wang, Tao Wang, Zejiu Wang, Chunjing Wei, Ian Williams, Simon A. Wilde, Chunming Wu, Fuyan Wu, Yuanbao Wu, Qinghui Xiao, Wenliang Xu, Zhongyuan Xu, Chonghui Yang, Jinhui Yang, Zhensheng Yang, Hongfu Zhang, Lianchang Zhang, Fengqing Zhao, Minggou Zhai, Guochun Zhao, Taiping Zhao, Yue Zhao, Jianping Zheng, Jianbo Zhou, Xiaodong Zhou, Xiangkun Zhu, and others for discussions and help in this study. We particularly thank Academician Minggou Zhai for originating this book. This paper is the result of long-term work in the NCC, supported financially by the Major State Basic Research Program of the People's Republic of China (2012CB416600, 2015FY310100), the National Natural Science Foundation of China (40172044, 40672127, 41002062, 41472167, 41472169, 49473172, 49872024), and the Key Program of the Ministry of Land and Resources of China (1212010711815, 1212011120151, 12120114021301, 12120113013700, 1212011220465, 1212011120142, 1212010811033, DKD2001020-3, DKD9904011, DKD20010209, J1.3.1). We express our sincere appreciation to the distinguished geologist Academician Yuqi Cheng who is one of the most important founders of metamorphic petrology, metamorphic geology, and geochronology in China and who has made significant contributions to the geology of China.

References

Abbott DH, Isley AE (2002) The intensity, occurrence, and duration of superplume events and eras over geological time. J Geodyn 34:265–307

Bureau of Geology Mineral Resources of Shandong Province (BGMRSP) (1991) Regional geology of Shandong Province. Geological Publishing House, Beijing, pp 1–595 (in Chinese with English abstract)

Bai J et al (1986) The early Precambrian geology of Wutaishan. Tianjing Publishing press, Tianjin, pp 1–46 (in Chinese with English abstract)

Bai J, Huang X-G, Wang H-C, Guo J-J (1996) The Precambrian evolution of China, 2nd edn. Geological Publishing House, Beijing, pp 1–259 (in Chinese)

Bai X, Liu S-W, Guo R-R, Zhang L-F, Wang W (2014) Zircon U-Pb-Hf isotopes and geochemistry of Neoarchean dioritic-trondhjemitic gneisses, Eastern Hebei, North China Craton: constraints on petrogenesis and tectonic implications. Precambr Res 251:1–20

Bédard J (2013) How many arcs can dance on the head of a plume? A 'comment' on: a critical assessment of Neoarchean 'plume only' geodynamics: evidence from the Superior Province, by Derek Wyman. Precambr Res 229:189–197

Black LP, Williams IS, Compston W (1986) Four zircon ages from one rock: the history of a 3930 Ma-old granulite from Mount Sones, Enderby Land, Antarctica. Contrib Miner Petrol 94:427–437

Bleeker W (2003) The late Archean record: a puzzle in ca. 35 pieces. Lithos 71:99–134

Bowring SA, Williams IS (1999) Priscoan (4.00-4.03 Ga) orthogneisses from northwestern Canada. Contrib Miner Petrol 134:3–16

Bouvier A, Vervoort JD, Patchett PJ (2008) The Lu-Hf and Sm-Nd isotopic composition of CHUR: constraints from unequilibrated chondrites and implications for the bulk composition of terrestrial planets. Earth Planet Sci Lett 273:48–57

Cao G-Q et al (1996) Early Precambrian geology of western Shandong. Geological Publishing House, Beijing, pp 1–193 (in Chinese with English abstract)

Chadwick B, Vasudev VN, Hegde GV, Nutman AP (2007) Structure and SHRIMP U/Pb zircon ages of granites adjacent to the Chitradurga Schist Belt: implications for Neoarchean convergence in the Dharwar Craton, southern India. J Geol Soc India 69:5–24

Chen L (2007) Geochemistry and Geochronology of the Guyang Greenstone Belt. Postdoctoral Report. Institute of Geology and Geophysics, Chinese Academy of Sciences, Beijing, pp 1–40 (in Chinese)

Chen T (1988) The geology and petrological characteristics of granitic rocks in the Caozhuang-Bailonggang area, Qian'an, Hebei. Bull Inst Geol, Chin Acad Geol Sci 18:82–97 (in Chinese with English abstract)

Chen B, Liu S-W, Geng Y-S, Liu C-Q (2006) Zircon U-Pb ages, Hf isotopes and significance of the late Archean-Paleoproterozoic granitoids from the Wutai-Lüliang terrain, North China. Acta Petrologica Sinica 22:296–304

Cheng Y-Q, Sheng Q-H, Wang Z-J (1982) Preliminary study of the metamorphosed basic Volcano-Sedimentary Yanlingguan formation of the Taishan Group of Xintai, Shandong. Geological Publishing House, Beijing, pp 1–72 (in Chinese)

Cheng Y-Q, Yang C-H, Wan Y-S, Liu Z-X, Zhang X-P, Du L-L, Zhang S-G, Wu J-S, Gao J-F (2004) Early Precambrain geological characters and anatectic reconstruction of crust in the north part of middle Taihang Mountain. Geological Publishing House, Beijing, pp 1–191 (in Chinese)

Clark C, Collins AS, Timms NE, Kinny PD, Chetty TRK, Santosh M (2009) SHRIMP U-Pb age constraints on magmatism and high-grade metamorphism in the Salem Block, southern India. Gondwana Res 16:27–36

Clark C, Kinny PD, Harley SL (2012) Sedimentary provenance and age of metamorphism of the Vestfold Hills, East Antarctica: evidence for a piece of Chinese Antarctica? Precambr Res 196–197:23–45

Condie KC (2000) Episodic continental growth models: afterthoughts and extensions. Tectonophysics 322:153–162
Condie KC, Belousova E, Griffin WL, Sircombe KN (2009) Granitoid events in space and time: constraints from igneous and detrital zircon age spectra. Gondwana Res 15:228–242
Condie KC, Kröner A (2013) The building blocks of continental crust: evidence for a major change in the tectonic setting of continental growth at the end of the Archean. Gondwana Res 23:394–402
Corvino AF, Henjes-Kunst F (2007) A record of 2.5 and 1.1 billion year old crust in the Lawrence Hills, Antarctic southern Prince Charles Mountains. Terra Antartica 14:13–30
Cui M-L, Zhang L-C, Wu H-Y, Xu Y-X, Li W-J (2014) Timing and tectonic setting of the Sijiaying banded iron deposit in the eastern Hebei province, North China Craton: constraints from geochemistry and SIMS zircon U-Pb dating. J Asian Earth Sci 94:240–251
Cui P-L, Sun J-G, Sha D-M, Wang X-J, Zhang P, Gu A-L, Wang Z-Y (2013) Oldest zircon xenocryst (4.17 Ga) from the North China Craton. Int Geol Rev 55:1902–1908
Dai Y-P, Zhang L-C, Wang C-L, Liu L, Cui M-L, Zhu M-T, Xiang P (2012) Genetic type, formation age and tectonic setting of the Waitoushan banded iron formation, Benxi, Liaoning Province. Acta Petrologica Sinica 28:3574–3594 (in Chinese with English abstract)
Dai Y-P, Zhang L-C, Zhu M-T, Wang C-L, Liu L (2013) Chentaigou BIF-type iron deposit, Anshan area associated with Archean crustal growth: constraints from zircon U-Pb dating and Hf isotope. Acta Petrologica Sinica 29:2537–2550 (in Chinese with English abstract)
Davies GF (1995) Punctuated tectonic evolution of the earth. EarthPlanet Sci Lett 136:363–379
Deng H, Kusky T, Polat A, Wang J-P, Wang L, Fu J-M, Wang Z-S, Yuan Y (2014) Geochronology, mantle source composition and geodynamic constraints on the origin of Neoarchean mafic dikes in the Zanhuang Complex, Central Orogenic Belt, North China Craton. Lithos 205:359–378
Dey S, Pandey UK, Rai AK, Chaki A (2012) Geochemical and Nd isotope constraints on petrogenesis of granitoids from NW part of the eastern Dharwar craton: possible implications for late Archaean crustal accretion. J Asian Earth Sci 45:40–56
Diwu C-R, Sun Y, Lin C-L, Liu X-M, Wang H-L (2007) Zircon U-Pb ages and Hf isotopes and their geological significance of Yiyang TTG gneisses from Henan Province, China. Acta Petrologica Sinica 23:253–262 (in Chinese with English abstract)
Diwu C-R, Sun Y, Yuan H-L, Wang H-L, Zhong X-P, Liu X-M (2008) U-Pb ages and Hf isotopes for detrital zircons from quartzite in the Paleoproterozoic Songshan Group on the southwestern margin of the North China Craton. Chin Sci Bull 53:2828–2839
Diwu C-R, Sun Y, Dong Z-C, Wang H-L, Chen D-L, Chen L, Zhang H (2010a) In situ geochronology of Hadean zircon xenocryst (4.1 ~ 3.9 Ga) west of the Northern Qinling Orogenic Belt. Acta Petrologica Sinica 26:1171–1174 (in Chinese with English abstract)
Diwu C-R, Sun Y, Lin C-L, Wang H-L (2010b) LA-(MC)-ICP-MS U-Pb zircon geochronology and Lu-Hf isotope compositions of the Taihua complex on the southern margin of the North China Craton. Chin Sci Bull 55:2557–2571
Diwu C-R, Sun Y, Guo A-L, Wang H-L, Liu X-M (2011) Crustal growth in the North China Craton at ~2.5 Ga: evidence from in situ zircon U-Pb ages, Hf isotopes and whole-rock geochemistry of the Dengfeng Complex. Gondwana Res 20:149–170
Diwu C-R, Sun Y, Zhang H, Wang Q, Guo A-L, Fan L-G (2012) Episodic tectonothermal events of the western North China Craton and North Qinling Orogenic Belt in central China: constraints from detrital zircon U-Pb ages. J Asian Earth Sci 47:107–122
Diwu C-R, Sun Y, Wilde SA, Wang H-L, Dong Z-C, Zhang H, Wang Q (2013) New evidence for ~4.45 Ga terrestrial crust from zircon xenocrysts in Ordovician ignimbrite in the North Qinling Orogenic Belt, China. Gondwana Research 23:1484–1490
Diwu C-R, Sun Y, Zhao Y, Lai S-C (2014) Early Paleoproterozoic (2.45-2.20 Ga) magmatic activity during the period of global magmatic shutdown: Implications for the crustal evolution of the southern North China Craton. Precambr Res (in press)

Dong X-J, Xu Z-Y, Liu Z-H, Sha Q (2012a) 2.7 Ga granitic gneiss in the Northern Foot of Daqingshan Mountain, Central Inner Mongolia, and its geological implications. Earth Sci 37(Suppl.):45–52 (in Chinese with English abstract)

Dong X-J, Xu Z-Y, Liu Z-H, Sha Q (2012b) Zircon U-Pb geochronology of Archean high-grade metamorphic rocks from Xi Ulanbulang area, central Inner Mongolia. Sci China Ser D-Earth Sci 55:204–212

Dong C-Y, Wan Y-S, Zhang Y-H, Yang Z-Q, Liu D-Y (2013) 3.3-3.1 Ga magmatism recorded in an outcrop of the Dongshan Complex in the Anshan area, North China Craton: evidence from geochemistry and SHRIMP zircon dating. Acta Petrologica Sinica 29:414–420 (in Chinese with English abstract)

Dostal J, Mueller WU (2013) Deciphering an Archean mantle plume: Abitibi greenstone belt, Canada. Gondwana Res 23:493–505

Drüppel K, McCready AJ, Stumpfl EF (2009) High-K granites of the Rum Jungle Complex, N-Australia: insights into the Late Archean crustal evolution of the North Australian Craton. Lithos 111:203–219

Du L-L, Zhuang YX, Yang CH, Wan Y-S, Wang X-S, Wang S-J, Zhang L-F (2003) Characters of zircons in the Mengjiatun formation in Xintai of Shandong and their chronological significance. Acta Geol Sinica 77:359–366 (in English with Chinese abstract)

Du L-L, Zhuang Y-X, Yang C-H, Wan Y-S, Wang X-S (2005) SHRIMP U-Pb zircon chronology of fine-grained amphibolite in the Mengjiatun area, western Shandong. Acta Geoscientica Sinica 26:429–434 (in English with Chinese abstract)

Du L-L, Yang C-H, Zhuang Y-X, Wei R-Z, Wan Y-S, Ren L-D, Hou K-J (2010) Hf isotopic compositions of zircons from 2.7 Ga metasedimentary rocks and biotite-plagioclase gneiss in the Mengjiatun Formation Complex, western Shandong Province. Acta Geol Sinica 84:991–1001 (in Chinese with English abstract)

Duclaux G, Rolland Y, Ruffet G, Ménot RP, Guillot S, Peucat JJ, Fanning M, Rey P, Pêcher A (2008) Superimposed Neoarchaean and Paleoproterozoic tectonics in the Terre Adélie Craton (East Antarctica): Evidence from Th-U-Pb ages on monazite and $^{40}Ar/^{39}Ar$ ages. Precambr Res 167:316–338

Gao L-Z, Zhao T, Wan Y-S, Zhao X, Ma Y-S, Yang S-Z (2006) Report on 3.4 Ga SHRIMP zircon age from the Yuntaishan Geopark in Jiaozuo, Henan Province. Acta Geologica Sinica 80:52–57

Gao S, Rudnick RL, Yuan H-L, Liu X-M, Liu Y-S, Xu W-L, Ling W-L, Ayers J, Wang X-C, Wang Q-H (2004) Recycling lower continental crust in the North China craton. Nature 432:892–897

Garde AA, Hamilton MA, Chadwick B, Grocott J, McCaffrey KJW (2002) The Ketilidian orogen of South Greenland: geochronology, tectonics, magmatism, and fore-arc accretion during Paleoproterozoic oblique convergence. Can J Earth Sci 39:765–793

Geng Y-S, Wan Y-S, Shen Q-H (2002) Early Precambrian basic volcanism and crustal growth in the North China Craton. Acta Geol Sinica 76:199–202 (in Chinese with English abstract)

Geng Y-S, Liu F-L, Yang C-H (2006) Magmatic event at the end of the Archean in eastern Hebei Province and its geological implication. Acta Geol Sinica 80:819–833

Geng Y-S, Shen Q-H, Ren L-D (2010) Late Neoarchean to Paleoproterozoic magmatic events and tectonothermal systems in the North China Craton. Acta Petrologica Sinica 26:1945–1966 (in Chinese with English abstract)

Geng Y-S, Du L-L, Ren L-D (2012) Growth and reworking of the early Precambrian continental crust in the North China Craton: constraints from zircon Hf isotopes. Gondwana Res 21:517–529

Grant ML, Wilde SA, Wu F-Y, Yang J-H (2009) The application of zircon cathodoluminescence imaging, Th-U-Pb chemistry and U-Pb ages in interpreting discrete magmatic and high-grade metamorphic events in the North China Craton at the Archean/Proterozoic boundary. Chem Geol 261:155–171

Griffin WL, Pearson NJ, Belousova E, Jackson SE, Achterbergh EV, O'Reilly SY, Shee SR (2000) The Hf isotope composition of cratonic mantle: LAM-MC-ICPMS analysis of zircon megacrysts in kimberlites. Geochimica Cosmochimica Acta 64:133–147

Guan H, Sun M, Wilde SA, Zhou X-H, Zhai M-G (2002) SHRIMP U-Pb zircon geochronology of the Fuping Complex: implications for formation and assembly of the North China Craton. Precambr Res 113:1–18

Guo J-H, Sun M, Chen F-K, Zhai M-G (2005) Sm-Nd and SHRIMP zircon geochronology of high-pressure granulites in the Sanggan area, North China Craton: timing of Paleoproterozoic continental collision. J Asian Earth Sci 24:629–642

Guo L-S, Liu S-W, Liu Y-L, Tian W, Yu S-Q, Li Q-G, Lü Y-J (2008) Zircon Hf isotopic features of TTG gneisses and formation environment of Precambrian Sushui complex in Zhongtiao mountains. Acta Petrologica Sinica 24:139–148 (in Chinese with English abstract)

Guo R-R, Liu S-W, Santosh M, Li Q-G, Bai X, Wang W (2013) Geochemistry, zircon U-Pb geochronology and Lu-Hf isotopes of metavolcanics from eastern Hebei reveal Neoarchean subduction tectonics in the North China Craton. Gondwana Res 24:664–686

Fan H-R, Yang K-F, Hu F-F, Wang K-Y, Zhai M-G (2010) Zircon geochronology of basement rocks from the Bayan Obo area, Inner Mongolia, and tectonic implications. Acta Petrologica Sinica 26:1342–1350 (in Chinese with English abstract)

Frost CD, Bell JM, Frost BR, Chamberlain KR (2001) Crustal growth by magmatic underplating: isotopic evidence from the northern Sherman batholith. Geology 29:515–518

Halla J, Hunen JV, Heilimo E, Hölttäd P (2009) Geochemical and numerical constraints on Neoarchean plate tectonics. Precambr Res 174:155–162

Han B-F, Xu Z, Ren R, Li L-L, Yang J-H, Yang Y-H (2012) Crustal growth and intracrustal recycling in the middle segment of the Trans-North China Orogen, North China Craton: a case study of the Fuping Complex. Geol Mag 149:729–742

Han C-M, Xiao W-J, Su B-X, Chen Z-L, Zhang X-H, Ao S-J, Zhang J-E, Zhang Z-Y, Wan B, Song D-F, Wang Z-M (2014a) Neoarchean Algoma-type banded iron formations from Eastern Hebei, North China Craton: SHRIMP U-Pb age, origin and tectonic setting. Precambr Res 251:212–231

Han C-M, Xiao W-J, Su B-X, Sakyi PA, Chen Z-L, Zhang X-H, Ao S-J, Zhang J, Wan B, Zhang Z-Y, Wang Z-M, Ding J (2014b) Formation age and genesis of the Gongchangling Neoarchean banded iron deposit in eastern Liaoning Province: constraints from geochemistry and SHRIMP zircon U-Pb dating. Precambr Res 254:306–322

He Y-H, Sun Y, Chen L, Li H-P, Yuan H-L, Liu X-M (2005) Zircon U-Pb chronology of Longshan complex by LA-ICP-MS and its geological significance. Acta Petrologica Sinica 21:125–134 (in Chinese with English abstract)

Hildreth W, Moorbath S (1988) Crustal contributions to arc magmatism in the Andes of Central Chile. Contrib Miner Petrol 98:455–489

Hu J-M, Liu X-S, Li Z-H, Zhao Y, Zhang S-H, Liu X-C, Qu H-J, Chen H (2012) SHRIMP U-Pb zircon dating of the Ordos Basin basement and its tectonic significance. Chin Sci Bull 58:118–127

Huang D-M, Zhang D-H, Wang S-Y, Zhang Y-X, Dong C-Y, Liu D-Y, Wan Y-S (2012a) 2.3 Gamagmatism and 1.94 Ga metamorphism in the Xiatang Area, southern margin of the North China Craton. Geol Rev 58:565–573 (in Chinese with English abstract)

Huang X-L, Niu Y-L, Xu Y-G, Yang Q-J, Zhong J-W (2010) Geochemistry of TTG and TTG-like gneisses from Lushan-Taihua complex in the southern North China Craton: implications for late Archean crustal accretion. Precambr Res 182:43–56

Huang X-L, Wilde SA, Yang QJ, Zhong JW (2012b) Geochronology and petrogenesis of gray gneisses from the Taihua Complex at Xiong'er in the southern segment of the Trans-North China Orogen: implications for tectonic transformation in the Early Paleoproterozoic. Lithos 134–135:236–252

Huang X-L, Wilde SA, Zhong J-W (2013) Episodic crustal growth in the southern segment of the Trans-North China Orogen across the Archean-Proterozoicboundary. Precambr Res 233:337–357

Iizuka T, Komiya T, Johnson SP, Kon Y, Maruyama S, Hirata T (2009) Reworking of Hadean crust in the Acasta gneisses, northwestern Canada: evidence from in-situ Lu-Hf isotope analysis of zircon. Chem Geol 259:230–239

Jahn B-M, Auvray B, Cornichet J, Bai Y-L, Shen Q-H, Liu D-Y (1987) 3.5 Ga Old amphibolites from eastern Hebei province, China: field occurrence, petrography, Sm-Nd isochron age and REE chemistry. Precambr Res 34:311–346

Jahn B-M, Auvray B, Shen Q-H, Liu D-Y, Zhang Z-Q, Dong Y-J, Ye X-J, Zhang Q-Z, Cornichet J, Mace J (1988) Archean crustal evolution in China: the Taishan Complex, and evidence for juvenile crustal addition from long-term depleted mantle. Precambr Res 38:381–403

Jahn B-M, Zhou X-H, Li JL (1990) Formation and tectonic evolution of Southeastern China and Taiwan: isotopic and geochemical constraints. Tectonophysics 183:145–160

Jahn B-M, Liu D-Y, Wan Y-S, Song B, Wu J-S (2008) Archean crustal evolution of the Jiaodong peninsula, China, as revealed by zircon SHRIMP geochronology, elemental and Nd-isotope geochemistry. Am J Sci 308:232–269

Jayananda M, Moyen FF, Martin H, Peucat JJ, Auvray B, Mahabaleswar B (2000) Late Archaean (2550-2520 Ma) juvenile magmatism in the Eastern Dharwar Craton, Southern India: constraints from geochronology Nd-Sr isotopes and whole rock geochemistry. Precambr Res 99:225–254

Ji Z-Y (1993) New data on isotope age of the Proterozoic metamorphic rocks from northern Jiaodong and its geological significance.Shandong. Geology 9:40–51 (in Chinese with English abstract)

Jian P, Kröner A, Windley BF, Zhang Q, Zhang W, Zhang L-Q (2012) Episodic mantle melting-crustal reworking in the late Neoarchean of the northwestern North China Craton: Zircon ages of magmatic and metamorphic rocks from the Yinshan Block. Precambr Res 222–223:230–254

Jiang N, Guo J-H, Zhai M-G, Zhang S-Q (2010) ~2.7 Ga crust growth in the North China craton. Precambr Res 179:37–49

Jiang Z-S, Wang G-D, Xiao L-L, Diwu C-R, Lu J-S, Wu C-M (2011) Paleoproterozoic metamorphic P-T-t path and tectonic significance of the Luoning metamorphic complex at the southern terminal of the Trans-North China Orogen, Henan Province. Acta Petrologica Sinica 27:3701–3717 (in Chinese with English abstract)

Jin K, Xu W-L, Wang Q-H, Gao S, Liu X-C (2003) Formation time and sources of the Huaiguang "Migmatitic Granodiorite" in Bengbu, Anhui Provinece: evidence from SHRIMP zircon U-Pb geochronology. Acta Geoscientia Sinica 24:331–335 (in Chinese with English abstract)

Kinny PD (1986) 3820 Ma zircons from a tonalitic Amitsoq gneiss in the Godthab district of southern West Greenland. Earth Planet Sci Lett 79:337–347

Kröner A, Compston W, Zhang G-W, Guo A-L, Todt W (1988) Age and tectonic setting of Late Archean greenstone-gneiss terrain in Henan Province, China, as revealed by single grain zircon dating. Geology 16:211–215

Kröner A, Cui W-Y, Wang S-Q (1998) Single zircon ages from high-grade rocks of the Jianping Complex, Liaoning Province, NE China. J Asian Earth Sci 16:519–532

Kröner A, Wilde SA, O'Brien PJ, Li J-H, Passchier CW, Walte NP, Liu D-Y (2005a) Field relationships, geochemistry, zircon ages and evolution of a late Archaean to Palaeoproterozoic lower crustal section in the Hengshan terrain of northern China. Acta Geologica Sinica 79: 605–629

Kröner A, Wilde SA, Li J-H, Wang K-Y (2005b) Age and evolution of a late Archean to Paleoproterozoic upper to lower crustal section in the Wutaishan/Hengshan/Fuping terrain of northern China. J Asian Earth Sci 24:577–595

Kröner A, Wilde SA, Zhao G-C, O'Brien PJ, Sun M, Liu D-Y, Wan Y-S, Liu S-W, Guo J-H (2006) Zircon geochronology and metamorphic evolution of mafic dykes in the Hengshan Complex of northern China: evidence for late Palaeoprotero-zoic extension and subsequent high-pressure metamorphism in the North ChinaCraton. Precambr Res 146:45–67

Kröner A, Hoffmann JE, Xie H, Münker C, Hegner E, Wan Y, Hofmann A, Liu D, Yang J (2014) Generation of early Archaean grey gneisses through crustal melting of older crust in the eastern Kaapvaal craton, southern Africa. Precambr Res 255:833–846

Kusky TM, Li J-H (2003) Paleoproterozoic tectonic evolution of the North China Craton. J Asian Earth Sci 22:383–397

Kusky TM, Li J-H, Santosh M (2007) The Paleoproterozoic North Hebei Orogen: North China Craton's collisional suture with the Columbia supercontinent. Gondwana Res 12:4–28
Li B, Jin W, Zhang J-H, Wang Y-F, Cai L-B, Wang Q-L (2013) Composition and structural characteristics of Archean granite in Chentaigou area of Anshan. Global Geol 32:191–199 (in Chinese with English abstract)
Li C-D, Zhang F-Q, Miao L-C, Du Y-L, Hua Y-Q, Xu Y-W, Kang S-M (2009) Zircon SHRIMP U-Pb geochronology of the Zhangsangou Formation Complex in the northeastern margin of North China Craton and its geological significance. Acta Geol Sinica 83:642–650 (in Chinese with English abstract)
Li C-J, Bao Z-W, Zhao Z-H, Qiao Y-L (2012) Zircon U-Pb age and Hf isotopic compositions of the granitic gneisses from the Sanggan complex in the Zhangjiakouk area: constraints on the early evolution of North China Craton. Acta Petrologica Sinica 28:1057–1072 (in Chinese with English abstract)
Li J-H, Qian X-L, Hou G-T, Liu S-W, Chen J (2000) Late Paleoproterozoic to early Mesoproterozoic tectonic framework and major tectono-thermal episodes of north China: new interpretation of "Luliang orogeny". Earth Sci 25:15–20 (in Chinese with English abstract)
Li J-H, Kusky TM, Huang X-N (2002) Archean podiform chromitites and mantle tectonites in ophiolitic mélange, North China Craton: a record of early oceanic mantle processes. GSA Today 12:4–11
Li J-H, Yang C-H, Du L-L, Wan Y-S, Liu Z-X (2005) SHRIMP U-Pb geochronology evidence for the formation time of the Wanzi Group at Pingshan County, Hebei Province. Geol Rev 51:201–207 (in Chinese with English abstract)
Li S-X, Ji S-K, Ma Z-H, Huo G-P, Tian Y-Q, Yang W-K (1986) Geology of metasedimentary iron deposit in Wutaishan. Jilin Science and Technology Press, Changchun, pp 1–299 (in Chinese)
Li S-X, Liu X-S, Zhang L-Q (1987) Granite-greenstone belt in Sheerteng area, Inner Mongolia, China. J Changchun Univ Geol 17:81–102 (in Chinese with English abstract)
Li S-Z, Zhao G-C, Wilde SA, Zhang J, Sun M, Zhang G-W, Dai L-M (2010a) Deformation history of the Hengshan-Wutai-Fuping Complexes: implications for the evolution of the Trans-North China Orogen. Gondwana Res 18:611–631
Li T-S, Zhai M-G, Peng P, Chen L, Guo J-H (2010b) Ca. 2.5 billion year old coeval ultramafic-mafic and syenitic dykes in Eastern Hebei: implicaitons for cratonization of the North China Craton. Precambr Res 180:143–155
Liaoning Bureau of Geology and Mineral Resources Exploration (LBGMRE) (1975a) Geological map of Liaoyang sheet (1:200000)
Liaoning Bureau of Geology and Mineral Resources Exploration (LBGMRE) (1975b) Geological map of Yingkou sheet (1:200000)
Liaoning Bureau of Geology and Mineral Resources Exploration (LBGMRE) (1976) Geological map of Shenyang sheet (1:200000)
Liu C-H, Zhao G-C, Sun M, Zhang J, Yin C-Q (2012a) U-Pb geochronology and Hf isotope geochemistry of detrital zircons from the Zhongtiao Complex: constraints on the tectonic evolution of the Trans-North China Orogen. Precambr Res 222–223:159–172
Liu D-Y, Nutman AP, Compston W, Wu J-S, Shen Q-H (1992) Remnants of 3800 Ma crust in the Chinese part of the Sino-Korean Craton. Geology 20:339–342
Liu D-Y, Wilde SA, Wan Y-S, Wu J-S, Zhou H-Y, Dong C-Y, Yin X-Y (2008) New U-Pb and Hf isotopic data confirm Anshan as the oldest preserved segment of the North China Craton. Am J Sci 308:200–231
Liu D-Y, Wilde SA, Wan Y-S, Wang S-Y, Valley JW, Kita N, Dong C-Y, Xie H-Q, Yang C-X, Zhang Y-X, Gao L-Z (2009a) Late Mesoarchean-early Neoarchean tectonothermal events at the southern margin of the North China Craton: evidence of multiple events from SHRIMP U-Pb dating and hafnium isotope analysis of zircons from metamorphosed supracrustal rocks and tonalites. Chem Geol 261:140–154
Liu F, Guo J-H, Lu X-P, Diwu C-Y (2009b) Crustal growth at ~2.5 Ga in the North China Craton: evidence from whole-rock Nd and zircon Hf isotopes in the Huai'an gneiss terrane. Chin Sci Bull 54:4704–4713

Liu J-H, Liu F-L, Liu P-H, Wang F, Ding Z-J (2011a) Polyphase magmatic and metamorphic events from early Precambrian metamorphic basement in Jiaobei area: evidence from the Zircon U-Pb dating of TTG and granitic gneisses. Acta Petrologica Sinica 27:943–960 (in Chinese with English abstract)

Liu J-H, Liu F-L, Ding Z-J, Liu P-H, Wang F (2012b) The zircon Hf isotope characteristics of ~2.5 Ga magmatic event and implication for the crustal evolution in the Jiaobei terrane, China. Acta Petrologica Sinica 28:2697–2704

Liu J-H, Liu F-L, Ding Z-J, Liu C-H, Yang H, Liu P-H, Wang F, Meng E (2013a) The growth, reworking and metamorphism of early Precambrian crust in the Jiaobei terrane, the North China Craton: constraints from U-Th-Pb and Lu-Hf isotopics ystematics, and REE concentrations of zircon from Archean granitoid gneisses. Precambr Res 224:287–303

Liu L, Zhang L-C, Dai Y-P, Wang C-L, Li Z-Q (2012) Formation age, geochemical signatures and geological significance of the Sanheming BIF-type iron deposit in the Guyang greenstone belt, Inner Mongolia: Acta Petrologica Sinica 28:3623–3637 (in Chinese with English abstract)

Liu L, Zhang L-C, Dai Y-P (2014) Formation age and genesis of the banded iron formations from the Guyang Greenstone Belt, Western North China Craton. Ore Geol Rev 63:388–404

Liu S-J, Wan Y-S, Sun H-Y, Nutman A, Xie H-Q, Dong C-Y, Ma M-Z, Du L-L, Liu D-Y, Jhan B-M (2013b) Paleo- to Eoarchean crustal materials in eastern Hebei, North China Craton: new evidence from SHRIMP U-Pb dating and in-situ Hf isotopic studies in detrital zircons of supracrustal rocks. J Asian Earth Sci 78:4–17

Liu S-W, Pan Y, Li J-H, Li G-G, Zhang J (2002) Geological and isotopic geochemical constraints on the evolution of the Fuping Complex, North China Craton. Precambr Res 117:41–56

Liu S-W, Pan Y-M, Xie Q-L, Zhang J, Li Q-G (2004a) Archean geodynamics in the Central Zone, North China Craton: constraints from geochemistry of two contrasting series of granitoids in the Fuping and Wutai complexes. Precambr Res 130:229–249

Liu S-W, Lü Y-J, Feng Y-G, Zhang C, Tian W, Yan Q-R, Liu X-M (2007) Geology and zircon U-Pb isotopic chronology of Dantazi Complex, northern Hebei Province. Geol J China Univ 13:484–497 (in Chinese with English abstract)

Liu S-W, Santosh M, Wang W, Bai X, Yang P-T (2011b) Zircon U-Pb chronology of the Jianping Complex: implications for the Precambrian crustal evolution history of the northern margin of North China Craton. Gondwana Res 20:48–63

Liu S-W, Lü Y-J, Wang W, Yang P-T, Bai X, Feng Y-G (2011c) Petrogenesis of the Neoarchean granitoid gneisses in northern Hebei Province. Acta Petrologica Sinica 27:909–921 (in Chinese with English abstract)

Liu X-C, Zhao Y, Song B, Liu J, Cui J-J (2009c) SHRIMP U-Pb zircon geochronology of high-grade rocks and charnockites from the eastern Amery Ice Shelf and southwestern Prydz Bay, East Antarctica: constraints on Late Mesoproterozoic to Cambrian tectonothermal events related to supercontinent assembly. Gondwana Res 16:342–361

Liu Y-S, Gao S, Yuan H-L, Zhou L, Liu X-M, Wang X-C, Hu Z-C, Wang L-S (2004b) U-Pb zircon ages and Nd Sr, and Pb isotopes of lower crustal xenoliths from North China Craton: insights on evolution of lower continental crust. Chem Geol 211:87–109

Lu L-Z, Xu X-C, Liu F-L (1996) Early Precambrian Khondalite series of North China. Changchun Publishing House, Changchun, pp 1–272 (in Chinese)

Lu S-N, Chen Z-H, Xiang Z-Q (2008) The world geopark of Taishan: geochronological framework of Ancient Intrusives. Geological Publishing House, Beijng, pp 1–90 (in Chinese)

Lü B, Zhai M-G, Li T-S, Peng P (2012) Zircon U-Pb ages and geochemistry of the Qinglong volcano-sedimentary rock series in Eastern Hebei: implication for ~2500 Ma intra-continental rifting in the North China Craton. Precambr Res 208–211:145–160

Ma M-Z, Wan Y-S, Santosh M, Xu Z-Y, Xie H-Q, Dong C-Y, Liu D-Y (2012) Decoding multiple tectonothermal events in zircons from single rock samples: SHRIMP zircon U-Pb data from the late Neoarchean rocks of Daqingshan, North China Craton. Gondwana Res 22:810–827

Ma M-Z, Xu Z-Y, Zhang L-C, Dong C-Y, Liu S-J, Liu D-Y, Wan Y-S (2013a) SHRIMP dating and Hf isotope analysis of zircons from the early Precambrian basement in the Xi Ulanbulang area, Wuchuan, Inner Mongolia. Acta Petrologica Sinica 29:501–516 (in Chinese with English abstract)

Ma X-D, Fan H-R, Santosh M, Guo J-H (2013b) Geochemistry and zircon U-Pb chronology of charnockites in the Yinshan Block, North China Craton: tectonic evolution involving Neoarchaean ridge subduction. Int Geol Rev 55:1688–1704

Ma X-D, Guo J-H, Liu F, Qian Q, Fan H-R (2013c) Zircon U-Pb ages, trace elements and Nd-Hf isotopic geochemistry of Guyang sanukitoids and related rocks: implications for the Archean crustal evolution of the Yinshan Block, North China Craton. Precambr Res 230:61–78

Ma X-D, Fan H-R, Santosh M, Guo J-H (2014a) Chronology and geochemistry of Neoarchean BIF-type iron deposits in the Yinshan Block, North China Craton: implications for oceanic ridge subduction. Ore Geol Rev 63:405–417

Ma X-D, Fan H-R, Santosh M, Liu X, Guo J-H (2014b) Origin of sanukitoid and hornblendite enclaves in the Dajitu pluton from the Yinshan Block, North China Craton: product of Neoarchaean ridge subduction? Int Geol Rev. doi:10.1080/00206814.2014.929055

Manikyamba C, Kerrich R (2012) Eastern Dharwar Craton, India: Continental lithosphere growth by accretion of diverse plume and arc terranes. Geosci Front 3:225–240

McCready AJ, Stumpfl EF, Lally J, Ahmad M, Gee RD (2004) Polymetallic mineralization at the Browns deposit, Rum Jungle Mineral Field, Northern Territory, Australia. Econ Geol 99:257–277

Mei H-L (1991) P-T path of basic granulite in the Zhulagou complex, Inner Mongolia, China. Acta Petrologica et Minerlogica 10:133–142 (in Chinese with English abstract)

Miao P-S (2003) Tectonic style of the high-grade metamorphic terrain of Hengshan in the North China Craton. Tianjing Publishing press, Tianjing, pp 1–86 (in Chinese with English abstract)

Mohan MR, Piercey SJ, Kamber BS, Sarma DS (2013) Subduction related tectonic evolution of the Neoarchean eastern Dharwar Craton, southern India: new geochemical and isotopic constraints. Precambr Res 227:204–226

Moyen JF (2011) The composite Archaean grey gneisses: petrological significance, and evidence for a non-unique tectonic setting for Archaean crustal growth. Lithos 123:21–36

Moyen JF, Martin H, Jayananda M, Auvray B (2003) Late Archaean granites: a typology based on the Dharwar Craton (India). Precambr Res 127:103–123

North China Oilfield Company of PetroChina (NCOCP) (2012) Introduction on Drilling Cores Relating to Precambrian Basement in the Central Hebei Basin. Unpublished Ms, pp 1–18

Nutman AP, McGregor VR, Friend CRL, Bennett VC, Kinny PD (1996) The Itsaq Gneiss Complex of southern West Greenland: the world's most extensive record of early crustal evolution (3900-3600 Ma). Precambr Res 78:1–39

Nutman AP, Wan Y-S, Liu D-Y (2009) Integrated field geological and zircon morphology evidence for ca. 3.8 Ga rocks at Anshan: Comment on "Zircon U-Pb and Hf isotopic constraints on the Early Archean crustal evolution in Anshan of the North China Craton" by Wu et al. [Precambr Res 167 (2008) 339–362]. Precambr Res 172:357–360

Nutman AP, Wan Y-S, Du L-L, Friend CRL, Dong C-Y, Xie H-Q, Wang W, Sun H-Y, Liu D-Y (2011) Multistage late Neoarchaean crustal evolution of the North China Craton, eastern Hebei. Precambr Res 189:43–65

Nutman AP, Bennett VC, Friend CRL, Hidaka H, Yi K, Lee SR, Kamiichi T (2013) The Itsaq gneiss complex of Greenland: episodic 3900 to 3600 Ma juvenile crust formation and recycling in the 3600 to 3600 Ma Isukasian orogeny. Am J Sci 313:877–911

Nutman AP, Maciejowski R, Wan Y-S (2014) Protoliths of enigmatic Archaean gneisses established from zircon inclusion studies: case study of the Caozhuang quartzite, E. Hebei, China. Geosci Front 5:445–455

O'Neil J, Maurice C, Stevenson RK, Larocque J, Cloquet C, David J, Francis D (2007) The geology of the 3.8 Ga Nuvvuagittuq (Porpoise Cove) greenstone belt, northeastern Superior Province, Canada. In: van Kranendonk MJ, Smithies RH, Bennett VC (eds) Earth's oldest rocks, developments in Precambrian Geology, vol 15. pp 219–250

Pei F-P, Xu W-L, Yang D-B, Zhao Q-G, Liu X-M, Hu Z-C (2007) Zircon U-Pb geochronology of basement metamorphic rocks in the Songliao Basin. Chin Sci Bull 52:942–948

Peng M-S, Zhang P, Yang H-Z, Wang X-J, Qiao S-Y (2013a) Geochemical characteristics and geological significance of iron deposit in Nanfen, Liaoning. Sci Technol Rep 31:33–37 (in Chinese with English abstract)

Peng P, Li Y, Liu F, Wang F (2012) Geological relation of late Archean lithologic units in northwest Hebei, North China Craton: implication for building of early continental crust. Acta Petrologica Sinica 28:3531–3544 (in Chinese with English abstract)

Peng T-P, Wilde SA, Fan W-M, Peng B-X (2013b) Late Neoarchean potassic high Ba-Sr granites in the Taishan granite-greenstone terrane: petrogenesis and implications for continental crustal evolution. Chem Geol 344:23–41

Polat A, Li J, Fryer B, Kusky T, Gagnon J, Zhang S (2006a) Geochemical characteristics of the Neoarchean (2800-2700 Ma) Taishan greenstone belt, North China Craton: Evidence for plume-craton interaction. Chem Geol 230:60–87

Polat A, Herzberg C, Münker C, Rodgers R, Kusky T, Li J, Fryer B, Delaney J (2006b) Geochemical and petrological evidence for a suprasubduction zone origin of Neoarchean (ca. 2.5 Ga) peridotites, central osorgenic belt, North China Craton. Bull Geol Soc Am 118:771–784

Qian J-H, Wei C-J, Zhou X-W, Zhang Y-H (2013) Metamorphic P-T paths and new zircon U-Pb age data for garnet-mica schist from the Wutai Group, North China Craton. Precambr Res 233:282–296

Qian X-L, Cui W-Y, Wang S-Q, Wang G-Y (1985) Geology of Precambrian iron ores in eastern Hebei Province, China. Hebei Science and Tecnology Press, Shijiazhuang, pp 1–273 (in Chinese)

Qian X-L, Wang R-M et al. (1994) Geological evolution of the granulite terrain in north part of the North China Craton. Seismological Press, Beijing, pp 1–234 (in Chinese)

Ren K-X, Yan G-H, Cai J-H, Zou T-R, Mu B-L, Li F-T (2013) SHRIMP U-Pb zircon age and its implications of quartz syenite in Hadamengou, Inner Mongolia. Adv Geosci 3:42–47 (in Chinese with English abstract)

Ren R, Han B-F, Zhang Z-C, Li J-F, Yang Y-H, Zhang Y-B (2011) Zircon U-Pb and Hf isotopic studies of basement gneiss and overlying Meso-Neoproterozoic sedimentary rocks from the Changping area, Beijing, and their geological implications. Acta Petrologica Sinica 27:1721–1745 (in Chinese with English abstract)

Ren Y-W (2010) The study of granite-greenstone belt in Xihongshan Area, Inner Mongolia. Ms. Thesis, Jilin University, pp 1–64 (in Chinese with English abstract)

Rey PF, Philippot P, Thebaud N (2003) Contribution of mantle plumes, crustal thickening and greenstone blanketing to the 2.75-2.65 Ga global crisis. Precambr Res 127:43–60

Shan H-X, Zhai M-G, Wang F, Zhou Y-Y, Santosh M, Zhu X-Y, Zhang H-F, Wang W (2015) Zircon U-Pb ages, geochemistry, and Nd-Hf isotopes of the TTG gneisses from the Jiaobei terrane: Implications for Neoarchean crustal evolution in the North China Craton. J Asian Earth Sci (in press)

Shen B-F, Luo H, Li S-B, Li J-J, Peng X-L, Hu X-D, Mao D-B, Liang R-X (1994a) Geology and metallization of Archean Greenstone Belt in North China Plateform. Geological Publishing House, Bejing, pp 1–202 (in Chinese with English abstract)

Shen B-F, Luo H, Han G-G, Dai X-Y, Jin W-S, Hu X-D, Li S-B, Bi S-Y (1994b) Archean geolgy and metallization in northern Liaoning Province and southern Jilin Province. Geological Publishing House, Beijing, pp 1–255 (in Chinese with English abstract)

Shen Q-H, Zhang Y-F, Gao J-F, Wang P (1990) Study on Archaean metamorphic rocks in midsouthern Nei Mongol of China. In: Bulletin of Institute of Geology of Chinese Academy of Geological Scienes. Geological Publishing House, Beijing, pp 1–192 (in Chinese)

Shen Q-H, Xu H-F, Zhang Z-Q, Gao J-F, Wu J-S, Ji C-L (1992) Early Precambrain granulites of China. Geological Publishing House, Beijing, pp 1–237 (in Chinese)

Shen Q-H, Shen K, Geng Y-S, Xu H-F (2000) Compositions and geological evolution of Yishui Complex, Shandong. Geological Publishing House, Bejing, pp 1–179 (in Chinese)

Shen Q-H, Song B, Xu H-F, Geng Y-S, Shen K (2004) Emplacement and metamorphism ages of the Caiyu and Dashan igneous bodies, Yishui County, Shandong Province: zircon SHRIMP chronology. Geol Rev 50:275–284 (in Chinese with English abstract)

Shen Q-H, Geng Y-S, Song B, Wan Y-S (2005) New information from the surface outcrops and deep crust of Archean rocks of the North China and Yangtze Blocks, and Qinling-Dabie Orogenic Belt. Acta Geol Sinica 79:616–627 (in Chinese with English abstract)

Shen Q-H, Zhao Z-R, Song B, Song H-X (2007) Geology, petrochemistry and SHRIMP zircon U-Pb dating of the Mashan and Xueshan granitoids in Yishui County, Shandong Province. Geol Rev 53:180–186 (in Chinese with English abstract)

Shi Y-R, Wilde SA, Zhao X-T, Ma Y-S, Du L-L, Liu D-Y (2012) Late Neoarchean magmatic and subsequent metamorphic events in the northern North China Craton: SHRIMP zircon dating and Hf isotopes of Archean rocks from Yunmengshan Geopark, Miyun, Beijing. Gondwana Res 21:785–800

Song B, Nutman AP, Liu D-Y, Wu J-S (1996) 3800 to 2500 Ma crustal evolution in the Anshan area of Liaoning Province, northeastern China. Precambr Res 78:79–94

Song H-X, Zhao Z-R, Shen Q-H, Song B (2009) Study on petrochemistry and hafnium isotope in Yishui Complex, Shandong Province. Acta Petrologica Sinica 25:1872–1882 (in Chinese with English abstract)

Söderlund U, Patchett PJ, Vervoort JD, Isachsen CE (2004) The ^{176}Lu decay constant determined by Lu-Hf and U-Pb isotope systematics of Precambrian mafic intrusions. Earth Planet Sci Lett 219:311–324

Sun D-Z et al (1984) The early Precambrian geology of eastern Hebei. Tiangjing Science and Technology Press, Tianjing, pp 1–273 (in Chinese with English abstract)

Sun D-Z, Li H-M, Lin Y-X, Zhou H-F, Zhao F-Q, Tang M (1991) Precambrian geochronology, chronotectonic framework and model of chronocrustal structure of the Zhongtiao mountains. Acta Geol Sinica 65:216–231 (in Chinese with English abstract)

Sun D-Z, Hu H-F (1993) The tectonic framework of Precambrian in Zhongtiaoshan. Geological Publishing House, Beijing, pp 1–180 (in Chinese with English abstract)

Sun H-Y, Dong C-Y, Xie H-Q, Wang W, Ma M-Z, Liu D-Y, Nutman A, Wan Y-S (2010) The formation age of the Neoarchean Zhuzhangzi and Dantazi groups in the Qinglong area, eastern Hebei Province: evidence from SHRIMP U-Pb zircon dating. Geol Rev 56:888–898 (in Chinese with English abstract)

Sun M, Guan H (2001) Zircon U-Pb ages of the Fuping Complex and their implications: some comments on the geochronological study of the Precambrian high-grade metamorphic terranes. Acta Petrologica Sinica 17:145–156 (in Chinese with English abstract)

Sun X-H, Zhu X-Q, Tang H-S, Zhang Q, Luo T-Y (2014) Reprint of "The Gongchangling BIFs from the Anshan-Benxi area, NE China: petrological-geochemical characteristics and genesis of high-grade iron ores". Ore Geol Rev 63:374–387

Tang J, Zheng Y-F, Wu Y-B, Gong B, Liu X-M (2007) Geochronology and geochemistry of metamorphic rocks in the Jiaobei terrane: constraints on its tectonic affinity in the Sulu orogen. Precambrian Res 152:48–82

Tian W, Liu S-W, Liu C-Q, Yu S-Q, Li Q-G, Wang Y-R (2005) Zircon SHRIMP geochronology and geochemistry of TTG rocks of the Sushui Complex in the Zhongtiao area, and its geological significance. Nat Sci Prog 15:1476–1484 (in Chinese)

Trap P, Faure M, Lin W, Monié P (2007) Late Palaeoproterozoic (1900–1800 Ma) nappe stacking and polyphase deformation in the Hengshan-Wutaishan area: implication for the understanding of the Trans-North China Belt, North China Craton. Precambr Res 156:85–106

Trap P, Faure M, Lin W, Monié P, Bruguier O (2008) Contrasted tectonic styles for the Paleoproterozoic evolution of the North China Craton. Evidence for a $\sim$2.1 Ga thermal and tectonic event in the Fuping Massif. J Struct Geol 30:1109–1125

Tsunogae T, Dunkley DJ, Horie K, Endo T, Miyamoto T, Kato M (2014) Petrology and SHRIMP zircon geochronology of granulites from Vesleknausen, Lützow-Holm Complex, East Antarctica: Neoarchean magmatism and Neoproterozoic high-grade metamorphism. Geosci Front 5:167–182

Van Kranendonk MJ, Smithies RH, Griffin WL, Huston DL, Hickman AH, Champion DC, Anhaeusser CR, Pirajno F (2014) Making it thick: a volcanic plateau origin of Palaeoarchean continental lithosphere of the Pilbara and Kaapvaal cratons. In: Roberts NMW, Van

Kranendonk M, Parman S, Shirey S, Clift PD (eds) Continent formation through time, vol 389. Geological Society, London, Special Publications, pages unknown
Wan Y-S (1993) The formation and evolution of the iron-bearing rock series of the Gongchangling area, Liaoning Province. Beijing Science and Technology Publishing House, Beijing, pp 1–108 (in Chinese with English abstract)
Wan Y-S, Zhang Z-Q, Wu J-S, Song B, Liu D-Y (1997a) Geochemical and Nd isotopic characteristics of some rocks from the Paleoarchaean Chentaigou supracrustal, Anshan area, NE China. Cont Dyn 2:39–46
Wan Y-S, Geng Y-S, Wu J-S (1997b) Geochemical compositions of early Precambrian meta-basaltic rocks in the North China Craton and its adjacent area. In: Cheng Y-Q (ed) Essay collection of early Precambrian geology of the North China Platform 39–59 (in Chinese)
Wan Y-S, Song B, Liu D-Y, Li H-M, Yang C, Zhang Q-D, Yang C-H, Geng Y-S, Shen Q-H (2001) Geochronology and geochemistry of 3.8-2.5 Ga ancient rock belt in the Dongshan scenic park, Anshan area. Acta Geologica Sinica 75:363–370 (in Chinese with English abstract)
Wan Y-S, Liu D-Y, Song B, Wu J-S, Yang C-H, Zhang Z-Q, Geng Y-S (2005a) Geochemical and Nd isotopic compositions of 3.8 Ga meta-quartz dioritic and trondhjemitic rocks from the Anshan area and their geological significance. J Asian Earth Sci 24:563–575
Wan Y-S, Song B, Yang C, Wu J-S (2005b) Zircon SHRIMP U-Pb geochronology of Archaean rocks from the Fushun-Qingyuan area, Liaoning Province and its geological significance. Acta Geol Sinica 79:78–87 (in Chinese with English abstract)
Wan Y-S, Song B, Geng Y-S, Liu D-Y (2005c) Geochemical characteristics of Archaean basement in the Fushun-Qingyuan area, northern Liaoning Province and its geological significance. Geol Rev 51:128–137 (in Chinese with English abstract)
Wan Y-S, Song B, Liu D-Y, Wilde SA, Wu J-S, Shi Y-R, Yin X-Y, Zhou H-Y (2006) SHRIMP U-Pb zircon geochronology of Paleoproterozoic metasedimentary rocks in the North China Craton: evidence for a major Late Paleoproterozoic tectonothermal event. Precambr Res 149:249–271
Wan Y-S, Liu D-Y, Yin X-Y, Wu J-S, Wilde SA (2007) SHRIMP geochronology and Hf isotope composition of zircons from the Tiejiashan granite and supracrustal rocks in the Anshan area, Liaoning province. Acta Petrologica Sinica 23:241–252 (in Chinese with English abstract)
Wan Y-S, Liu D-Y, Dong C-Y, Nutman A, Wilde SA, Wang W, Xie H-Q, Yin X-Y, Zhou H-Y (2009a) The oldest rocks and zircons in China. Acta Petrologica Sinica 25:1793–1807 (in Chinese with English abstract)
Wan Y-S, Liu D-Y, Dong C-Y, Xu Z-Y, Wang Z-J, Wilde SA, Yang Y-H, Liu Z-H, Zhou H-Y (2009b) The Precambrian Khondalite Belt in the Daqingshan area, North China Craton: evidence for multiple metamorphic events in the Palaeoproterozoic. In: Reddy SM, Mazumder R, Evans DAD, Collins AS (eds) Palaeoproterozoic supercontinents and global evolution: Journal of the London Geological Society, vol 323. Special Publications, pp 73–97
Wan Y-S, Liu D-Y, Wang S-Y, Zhao X, Dong C-Y, Zhou H-Y, Yin X-Y, Yang C-X, Gao L-Z (2009c) Early Precambrian crustal evolution in the Dengfeng area: constraints from geochemistry and SHRIMP U-Pb zircon dating. Acta Geol Sinica 83:982–999 (in Chinese with English abstract)
Wan Y-S, Miao P-S, Liu D-Y, Yang C-H, Wang W, Wang H-C, Wang Z-J, Dong C-Y, Du L-L, Zhou H-Y (2010a) Formation ages and source regions of the Palaeoproterozoic Gaofan, Hutuo and Dongjiao groups in the Wutai and Dongjiao areas of the North China Craton from SHRIMP U-Pb dating of detrital zircons: resolution of debates over their stratigraphic relationships. Chin Sci Bull 55:1278–1284
Wan Y-S, Dong C-Y, Wang W, Xie H-Q, Liu D-Y (2010b) Archaean basement and a Palaeoproterozoic collision orogen in the Huoqiu area at the southeastern margin of the North China Craton: evidence from SHRIMP U-Pb zircon geochronology. Acta Geol Sinica 84:91–104
Wan Y-S, Liu D-Y, Wang S-J, Dong C-Y, Yang E-X, Wang W, Zhou H-Y, Ning Z-G, Du L-L, Yin X-Y, Xie H-Q, Ma M-Z (2010c) Juvenile magmatism and crustalrecycling at the end of the

Neoarchean in Western Shandong Province, NorthChina Craton: evidence from SHRIMP zircon dating. Am J Sci 310:1503–1552

Wan Y-S, Liu D-Y, Wang W, Song T-R, Kröner A, Dong C-Y, Zhou H-Y, Yin X-Y (2011a) Provenance of Meso- to Neoproterozoic cover sediments at the Ming Tombs, Beijing, North China Craton: An integrated study of U-Pb dating and Hf isotopic measurement of detrital zircons and whole-rock geochemistry. Gandwana Res 20:219–242

Wan Y-S, Liu D-Y, Wang S-J, Yang E-X, Wang W, Dong C-Y, Zhou H-Y, Du L-L, Yang Y-H, Diwu C-R (2011b) ~2.7 Ga juvenile crust formation in the North China Craton (Taishan-Xintai area, western Shandong Province): further evidence of an understated event from U-Pb dating and Hf isotopic composition of zircon. Precambr Res 186:169–180

Wan Y-S, Liu D-Y, Dong C-Y, Liu S-J, Wang S-J, Yang E-X (2011c) U-Th-Pb behaviorof zircons under high-grade metamorphic conditions: a case study of zircon dating of meta-diorite near Qixia, eastern Shandong. Geosci Front 2:137–146

Wan Y-S, Liu D-Y, Nutman A, Zhou H-Y, Dong C-Y, Yin X-Y, Ma M-Z (2012a) Multiple 3.8–3.1 Ga tectono-magmatic events in a newly discovered area of ancient rocks (the Shengousi Complex), Anshan, North China Craton. J Asian Earth Sci 54–55:18–30

Wan Y-S, Dong C-Y, Liu D-Y, Kröner A, Yang C-H, Wang W, Du L-L, Xie H-Q, Ma M-Z (2012b) Zircon ages and geochemistry of late Neoarchean syenogranites in the North China Craton: a review. Precambr Res 222–223:265–289

Wan Y-S, Wang S-J, Liu D-Y, Wang W, Kröner A, Dong C-Y, Yang E-X, Zhou H-Y, Xie H-Q, Ma M-Z (2012c) Redefinition of depositional ages of Neoarchean supracrustal rocks in western Shandong Province, China: SHRIMP U-Pb zircon dating. Gondwana Res 21:768–784

Wan Y-S, Dong C-Y, Xie H-Q, Wang S-J, Song M-C, Xu Z-Y, Wang S-Y, Zhou H-Y, Ma M-Z, Liu D-Y (2012d) Formation ages of early Precambrian BIFs in North China Craton: SHRIMP zircon U-Pb dating. Acta Geol Sinica 86:1447–1478 (in Chinese with English abstract)

Wan Y-S, Xie H-Q, Yang H, Wang Z-J, Liu D-Y, Kröner A, Wilde SA, Geng Y-S, Sun L-Y, Ma M-Z, Liu S-J, Dong C-Y, Du L-L (2013) Is the Ordos Block Archean or Paleoproterozoic in age? Implications for the Precambrian evolution of the North China Craton. Am J Sci 313:683–711

Wan Y-S, Xie S-W, Yang C-H, Kröner K, Ma M-Z, Dong C-Y, Du L-L, Xie H-Q, Liu D-Y (2014a) Early Neoarchean (~2.7 Ga) tectono-thermal events in the North China Craton: a synthesis. Precambr Res 247:45–63

Wan Y-S, Zhao X-Z, Wang Z-J, Liu D-Y, Kröner A, Dong C-Y, Xie H-Q, Geng Y-S, Zhang Y-H, Fan R-L, Sun H-Y (2014b) SHRIMP zircon dating and LA-ICPMS Hf analysis of early Precambrian rocks from drill holes into the basement beneath the Central Hebei Basin, North China Craton. Geosci Front 5:471–484

Wan Y-S, Ma M-Z, Dong C-Y, Xie H-Q, Xie S-W, Ren P, Liu D-Y (2015a) Widespread late Neoarchean reworking of Meso- to Paleoarchean continental crust in the Anshan-Benxi area, North China Craton, as documented by U-Pb-Nd-Hf-O isotopes. Am J Sci (in press)

Wan Y-S, Dong C-Y, Wang S-J, Kröner A, Xie H-Q, Ma M-Z, Zhou H-Y., Xie S-W, Liu D-Y (2015b) Middle Neoarchean magmatism in western Shandong, North China Craton: SHRIMP zircon dating and LA-ICP-MS Hf isotope analysis. Precambr Res (in press)

Wang A-D, Liu Y-C (2012) Neoarchean (2.5-2.8 Ga) crustal growth of the North China Craton revealed by zircon Hf isotope: a synthesis. Geosci Front 3:147–173

Wang A-D, Liu Y-C, Gu X-F, Hou Z-H, Song B (2012a) Late-Neoarchean magmatism and metamorphism at the southeastern margin of the North China Craton and their tectonic implications. Precambr Res 220–221:65–79

Wang C-L, Zhang L-C, Lan C-Y, Dai Y-P (2014a) Petrology and geochemistry of the Wangjiazhuang banded iron formation and associated supracrustal rocks from the Wutai greenstone belt in the North China Craton: implications for their origin and tectonic setting. Precambr Res 255:603–626

Wang C-Q, Cui W-Y, Kröener A, Nemchin AA (2000) Zircon isotopic ages from magnetite quartzites of the Jianping metamorphic complex, western Liaoning Province. Chin Sci Bull 45:547–551

Wang E-D, Xia J-M, Fu J-F, Jia S-S, Men Y-K (2014b) Formation mechanism of Gongchangling high-grade magnetite deposit hosted in Archean BIF, Anshan-Benxi area, Northeastern China. Ore Geol Rev 57:308–321

Wang G-D, Wang H, Chen H-X, Lu J-S, Xiao L-L, Wu C-M (2012b) U-Pb dating of zircons from metamorphic rocks of the Taihua metamorphic complex, Mt. Huashan, southern margin of the Tans-North China Orogen. Acta Geol Sinica 86:1541–1551 (in Chinese with English abstract)

Wang H-L, Chen L, Sun Y, Liu X-M, Xu X-Y, Chen J-L, Zhang H, Diwu C-R (2007) 4.1 Ga xenocrystal zircon from Ordovician volcanic rocks in western part of North Qinling Orogenic Belt. Chin Sci Bull 52:3002–3010

Wang J, Kusky T, Polat A, Wang L, Deng H, Wang S (2013a) A late Archean tectonic mélange in the Central Orogenic Belt, North China Craton. Tectonophysics 608:929–946

Wang L-G, Qiu Y-M, McNaughton NJ, Groves DI, Luo Z-K, Huang J-Z, Miao L-C, Liu Y-K (1998) Constraints on crustal evolution and gold metallogeny in the Northwestern Jiaodong Peninsula, China, from SHRIMP U-Pb zircon studies of granitoids. Ore Geol Rev 13:275–291

Wang L-M, Yan T-M (1992) The Archaean tonalites in Qixia area, Shandong. Shandong Geol 8:80–87 (in Chinese with English abstract)

Wang Q-Y, Zheng J-P, Pan Y-M, Dong Y-J, Liao F-X, Zhang Y, Zhang L, Zhao G, Tu Z-B (2014c) Archean crustal evolution in the southeastern North China Craton: new data from the Huoqiu Complex. Precambr Res 255:294–315

Wang R-M, Wan Y-S, Cheng S-H, Feng Y-G (2009a) Modern-style subduction processes in the Archean: evidence from the Shangyi Complex in North China Craton. Acta Geol Sinica 83:535–543

Wang W, Yang E-X, Wang S-J, Du L-L, Xie H-Q, Dong C-Y, Wan Y-S (2009b) Petrography of Archean metamorphic pillow basalt in Taishan Rock Group, western Shandong, and SHRIMP U-Pb dating of zircons from intruding trondhjemite. Geol Rev 5:737–744 (in Chinese with English abstract)

Wang W, Wang S-J, Liu D-Y, Li P-Y, Dong C-Y, Xie H-Q, Ma M-Z, Wan Y-S (2010a) Formation age of the Neoarchaean Jining Group (banded iron formation) in the western Shandong Province: constraints from SHRIMP zircon U-Pb dating. Acta Petrologica Sinica 26:1171–1174 (in Chinese with English abstract)

Wang W, Liu S-W, Bai X, Yang P-T, Li Q-G, Zhang L-F (2011) Geochemistry and zircon U-Pb-Hf isotopic systematics of the Neoarchean Yixian-Fuxin greenstone belt, northern margin of the North China Craton: implications for petrogenesis and tectonic setting. Gondwana Res 20:64–81

Wang W, Liu S-W, Santosh M, Bai X, Li Q-G, Yang P-T, Guo R-R (2013b) Zircon U-Pb-Hf isotopes and whole-rock geochemistry of granitoid gneisses in the Jianping gneissic terrane, Western Liaoning Province: constraints on the Neoarchean crustal evolution of the North China Craton. Precambr Res 224:184–221

Wang W, Zhai M-G, Li T-S, Santosh M, Zhao L, Wang H-Z (2014d) Archean-Paleoproterozoic crustal evolution in the eastern North China Craton: zircon U-Th-Pb and Lu-Hf evidence from the Jiaobei terrane. Precambr Res 241:146–160

Wang W, Zhai M-G, Tao Y-B, Santosh M, Lü B, Zhao L, Wang S-J (2014e) Late Neoarchean crustal evolution of the eastern North China Craton: a study on the provenance and metamorphism of paragneiss from the Western Shandong Province. Precambr Res 255:583–602

Wang Z-H, Wilde SA, Wan J-L (2010b) Tectonic setting and significance of 2.3-2.1 Ga magmatic events in the Trans-North China Orogen: new constraints from the Yanmenguan mafic-ultramafic intrusion in the Hengshan-Wutai-Fuping area. Precambr Res 178:27–42

Wang Z-J, Shen Q-H, Wan Y-S (2004) SHRIMP U-Pb zircon geochronology of the Shipaihe "metadiorite mass" from Dengfeng County, Henan Province. Acta Geoscientica Sinica 25:295–298 (in Chinese with English abstract)

Warren RG, Ellis DJ (1996) Mantle underplating, granite tectonics, and metamorphic P-T-t paths. Geology 24:663–666

Wei C-J, Qian J-H, Zhou X-W (2014) Paleoproterozoic crustal evolution of the Hengshan-Wutai-Fuping region, North China Craton. Geosci Front 5:485–497

Wilde SA, Dorsett-Bain HL, Jinglan L (1997) The identication of a Late Pan-African granulite facies event in northeastern China: SHRIMP U-Pb zircon dating of the Mashan Group at Liu Mao, Heilongjiang Province, China. In: Proceedings of the 30th international geological congress. VSP International Science Publishers, Amsterdam, pp 59–74
Wilde SA, Cawood PA, Wang K-Y, Nemchin AA, Zhao G-C (2004a) Determining Precambrian crustal evolution in China: a case-study from Wutaishan, Shanxi Province, demonstrating the application of precise SHRIMP U-Pb geochronology. J Lond Geol Soc 226:5–25
Wilde SA, Zhao G-C, Wang K-Y, Sun M (2004b) First SHRIMP U-Pb ages for Hutuo Group in Wutaishan: further evidence for Palaeoproterozoic amalgamation of North China Craton. Chin Sci Bull 49:83–90
Wilde SA, Cawood PA, Wang K-Y, Nemchin AA (2005) Granitoid evolution in the Late Archean Wutai Complex, North China Craton. J Asian Earth Sci 24:597–613
Wilde SA, Valley JW, Kita NT, Cavosie AJ, Liu D-Y (2008) SHRIMP U-Pb and CAMECA 1280 Oxygen isotope results from ancient detrital zircons in the Caozhuang Quartzite, Eastern Hebei, North China Craton: evidence for crustal reworking 3.8 Ga ago. Am J Sci 308:185–199
Wu F-Y, Yang J-H, Liu X-M, Li T-S, Xie L-W, Yang Y-H (2005a) Hf isotopes of the 3.8 Ga zircons in eastern Hebei Province, China: implications for early crustal evolution of the North China Craton. Chin Sci Bull 50:2473–2480
Wu F-Y, Zhao G-C, Wilde S-A, Sun D (2005b) Nd isotopic constraints on crustal formation in the North China Craton. J Asian Earth Sci 24:523–545
Wu F-Y, Zhang Y-B, Yang J-H, Xie L-W, Yang Y-H (2008) Zircon U-Pb and Hf isotopic constraints on the Early Archean crustal evolution in Anshan of the North China Craton. Precambr Res 167:339–362
Wu J-S, Geng Y-S, Xu H-F, Jin L-G, He S-Y, Sun S-W (1989) Metamorphic geology of the Fuping Group. Bull Inst Geol Chin Acad Geol Sci 19:1–213 (in Chinese with English abstract)
Wu J-S, Geng Y-S, Shen Q-H, Wan Y-S, Liu D-Y, Song B (1998) Archaean geology characteristics and tectonic evolution of China-Korea Paleo-continent. Geological Publishing House, Beijing., pp 1–212 (in Chinese)
Wu M-L, Zhao G-C, Sun M, Li S-Z, He Y-H, Bao Z-A (2013) Zircon U-Pb geochronology and Hf isotopes of major lithologies from the Yishui Terrane: Implications for the crustal evolution of the Eastern Block, North China Craton. Lithos 170–171:164–178
Wu M-L, Zhao G-C, Sun M, Li S-Z, Bao Z-A, Yuk TP, Eizenhöefer PR, He Y-H (2014) Zircon U-Pb geochronology and Hf isotopes of major lithologies from the Jiaodong Terrane: implications for the crustal evolution of the Eastern Block of the North China Craton. Lithos 190–191:71–84
Wyman DA (2013a) A critical assessment of Neoarchean "plume only" geodynamics: evidence from the Superior Province. Precambr Res 229:3–19
Wyman DA (2013b) A reply to "How many arcs can dance on the head of a plume?" by Jean Bédard, Precambrian Research, 2013. Precambr Res 229:198–202
Xiang P, Cui M-L, Wu H-Y, Zhang X-J, Zhang L-C (2012) Geological characteristics, ages of host rocks and its geological significance of the Zhoutaizi iron deposit in Luanping, Hebei Province. Acta Petrologica Sinica 28:3655–3669 (in Chinese with English abstract)
Xie S-W (2012) Zircon chronology and geochemical characteristics of Jiaolai basin and its basement. China University of Geosciences (Wuhan), Wuhan, pp 1–142
Xie H-Q, Wan Y-S, Wang S-J, Liu D-Y, Xie S-W, Liu S-J, Dong C-Y, Ma M-Z (2013) Geology and zircon dating of trondhjemitic gneiss and amphibolite in the Tangezhuang area, eastern Shandong. Acta Petrologica Sinica 29:619–629 (in Chinese with English abstract)
Xie S-W, Wang S-J, Xie H-Q, Liu S-J, Dong C-Y, Ma M-Z, Liu D-Y, Wan Y-S (2014a) SHRIMP U-Pb dating of detrital zircons from the Fenzishan Group in eastern Shandong, North China craton. Acta Petrologica Sinica 30:2989–2998 (in Chinese with English abstract)
Xie S-W, Xie H-Q, Wang S-J, Kröner A, Liu S-J, Zhou H-Y, Ma M-Z, Dong C-Y, Liu D-Y, Wan Y-S (2014b) Ca. 2.9 Ga granitoid magmatism in eastern Shandong, North China Craton: zircon dating, Hf-in-zircon isotopic analysis and whole-rock geochemistry. Precambr Res 255:538–562

Yang C-H, Du L-L, Ren L-D, Wan Y-S, Song H-X, Yuan Z-L, Wang S-Y (2009) SHRIMP U-Pb ages and stratigraphic comparison of Angou group, on the southern margin of North China Craton. Acta Petrologica Sinica 25:1853–1862 (in Chinese with English abstract)

Yang C-H, Du L-L, Ren L-D, Song H-X, Wan Y-S, Xie H-Q, Liu Z-X (2011) Petrogenesis and geodynamic setting of Jiandeng potassic granite at the end of the Neoarchean in Zanhuang Complex, North China Craton. Earth Sci Front 18:62–78 (in Chinese with English abstract)

Yang C-H, Du L-L, Ren L-D, Song H-X, Wan Y-S, Xie H-Q, Geng Y-S (2013a) Delineation of the ca. 2.7 Ga TTG gneisses in the Zanhuang Complex, North China Craton and its geological implications. J Asian Earth Sci 72:178–189

Yang J-H, Wu F-Y, Wilde SA, Zhao G-C (2008) Petrogenesis and geodynamics of Late Archean magmatism in eastern Hebei, eastern North China Craton: geochronological, geochemical and Nd-Hf isotopic evidence. Precambr Res 167:125–149

Yang X-Q, Li H-M, Xue C-J, Li L-X, Liu M-J, Chen J (2013b) Geochemical characteristics of two iron ores from the Waitoushan Iron deposit, Liaoning Province: constraints on ore-forming mechanism. Acta Geol Sinica 87:1580–1592 (in Chinese with English abstract)

Yin X-Y, Wan Y-S, Liu D-Y, Wilde SA, Zhou H-Y, Wu J-S (2006) Formation time of supracrustal rocks in the Tiejiashan granite in the Anshan area: evidence from detrital zircon SHRIMP dating. Acta Petrologica et Mineralogica 25:282–286

Yin X-Y, Zhou H-Y, Liu D-Y, Gao L-Z, Dong C-Y, Wan Y-S (2015) Ancient Material Records in the North China Craton: SHRIMP U-Pb Dating and LA-MC-ICPMS Hf analysis of zircons from Archean Metamorphic Rocks in the Jiaozou Area, Henan. Geol Rev 61:183–194 (in Chinese with English abstract)

Ying J-F, Zhang H-F, Tang Y-J (2011) Crust-mantle interaction in the central North China Craton during the Mesozoic: evidence from zircon U-Pb chronology, Hf isotope and geochemistry of syenitic-monzonitic intrusions from Shanxi province. Lithos 125:449–462

Zhai M-G, Bian A-G, Zhao T-P (2000) The amalgamation of the supercontinent of North China Craton at the end of Neo-Archaean and its breakup during late Palaeoproterozoic and Meso-Proterozoic. Sci China Ser D-Earth Sci 43:219–232

Zhai M-G, Guo J-H, Liu W-J (2005) Neoarchean to Paleoproterozoic continental evolution and tectonic history of the North China Craton: a review. J Asian Earth Sci 24:547–561

Zhai M-G, Peng P (2007) Paleoproterozoic events in the North China Craton. Acta Petrologica Sinica 23:2665–2682 (in Chinese with English abstract)

Zhai M-G (2010) Tectonic evolution and metallogenesis of North China Craton. Miner Deposits 29:24–36 (in Chinese with English abstract)

Zhai M-G, Santosh M (2011) The early Precambrian odyssey of the North China Craton: a synoptic overview. Gondwana Res 20:6–25

Zhang H-F (2012) Destruction of ancient lower crust through magma underplating beneath Jiaodong Peninsula, North China Craton: U-Pb and Hf isotopic evidence from granulite xenoliths. Gondwana Res 21:281–291

Zhang H-F (2014) Neoarchean recycling of ^{18}O-enriched supracrustal materials into the lower crust: zircon record from the North China Craton. Precambr Res 248:60–71

Zhang H-F, Zhai M-G, Santosh M, Diwu C-R, Li S-R (2011a) Geochronology and petrogenesis of Neoarchean potassic meta-granites from Huai'an Complex: Implications for the evolution of the North China Craton. Gondwana Res 20:82–105

Zhang H-F, Ying J-F, Santosh M, Zhao G-C (2012a) Episodic growth of Precambrian lower crust beneath the North China Craton: a synthesis. Precambr Res 222–223:255–264

Zhang H-F, Yang Y-H, Santosh M, Zhao X-M, Ying J-F, Xiao Y (2012b) Evolution of the Archean and Paleoproterozoic lower crust beneath the Trans-North China Orogen and the Western Block of the North China Craton. Gondwana Res 22:73–85

Zhang H-F, Ying J-F, Tang Y-J, Li X-H, Feng C, Santosh M (2012c) Phanerozoic reactivation of the Archean North China Craton through episodic magmatism: evidence from zircon U-Pb geochronology and Hf isotopes from the Liaodong Peninsula. Gondwana Res 19:446–459

Zhang H-F, Wang J-L, Zhou D-W, Yang Y-H, Zhang G-W, Santosh M, Yu H, Zhang J (2014a) Hadean to Neoarchean episodic crustal growth: Detrital zircon records in Paleoproterozoic quartzites from the southern North China Craton. Precambr Res 254:245–257

Zhang J, Zhao G-C, Sun M, Wilde SA, Li S-Z, Liu S-W (2006a) High-pressure mafic granulites in the Trans-North China Orogen: tectonic significance and age. Gondwana Res 9:349–362

Zhang J, Zhao G-C, Li S-Z, Sun M, Liu S-W, Wilde SA, Kröner A, Yin C-Q (2007) Deformation history of the Hengshan Complex: implications for the tectonic evolution of the Trans-North China Orogen. J Struct Geol 29:933–949

Zhang J, Zhao G-C, Li S-Z, Sun M, Wilde SA, Liu S-W, Yin C-Q (2009) Deformational history of the Fuping Complex and new U-Th-Pb geochronological constraints: implications for the tectonic evolution of the Trans-North China Orogen. J Struct Geol 31:177–193

Zhang J, Zhang H-F, Lu X-X (2013) Zircon U-Pb age and Lu-Hf isotope constraints on Precambrian evolution of continental crust in the Songshan area, the south-central North China Craton. Precambr Res 226:1–20

Zhang J-J, Zhao L, Liu S-W (2006b) Structures of syn-deformational granites in the Longquanguan shear zone and their monazite electronic microprobe dating. Acta Geol Sinica 80:864–874 (in Chinese with English abstract)

Zhang L-C, Zhai M-G, Zhang X-J, Xiang P, Dai Y-P, Wang C-L, Franco P (2012d) Formation age and tectonic setting of the Shirengou Neoarchean Banded iron deposit in eastern Hebei Province: constrains from geochemistry and SIMS zircon U-Pb dating. Precambr Res 222–223:325–388

Zhang L-C, Zhai M-G, Wan Y-S, Guo J-H, Dai Y-P, Wang C-L, Liu L (2012e) Study of the Precambrian BIF-iron deposits in the North China Craton: progreese and problems. Acta Petrologica Sinica 28:3431–3445 (in Chinese with English abstract)

Zhang Q-S, Yang Z-S, Wang Y-J (1988) Early crust and mineral deposits of Liaodong Peninsula. China Geological Publishing House, Beijing, pp 1–451 (in Chinese with English abstract)

Zhang R-Y, Zhang C-L, Diwu C-R, Sun Y (2012f) Zircon U-Pb geochronology, geochemistry and its geological implications for the Precambrian granitoids in Zhongtiao Mountain, Shanxi Province. Acta Petrologica Sinica 28:3559–3573 (in Chinese with English abstract)

Zhang X-J, Zhang L-C, Xiang P, Wan B, Francon P (2011) Zircon U-Pb age, Hf isotopes and geochemistry of Shuichang Algoma-type banded iron-formation, North China Craton on the ore-forming age and techonic setting. Gondwana Res 20:137–148

Zhang X-H, Yuan L-L, Xue F-H, Zhai M-G (2014b) Neoarchean metagabbro and charnockite in the Yinshan block, western North China Craton: petrogenesis and tectonic implications. Precambr Res 255:563–582

Zhang Y-Q, Wang T, Jia H-Y, Zhang Z-X (2003) U-Pb ages of zircons from the Xiulanbulang hypersthene-plagioclase granulite in the north Daqing Mountains, Central Inner Mongolia. Geol China 30:394–399 (in Chinese with English abstract)

Zhang Y-Q, Liu Y-J (2004) Zircon U-Pb age of the quartz diorite in North Daqing Mountains, central Inner Mongolia. Geol Miner Resour South China 4:22–27 (in Chinese with English abstract)

Zhao G-C (2014) Precambrian evolution of the North China Craton. Elsevier, Amsterdam, pp 1–194

Zhao G-C, Wilde SA, Cawood PA, Lu L-Z (1998) Thermal evolution of Archean basement rocks from the eastern part of the North China carton and its bearing on tectonic setting. Int Geol Rev 40:706–721

Zhao G-C, Wilde SA, Cawood PA, Lu L-Z (1999) Thermal evolution of two types of mafic granulites from the North China craton: implications for both mantle plume and collisional tectonics. Geol Mag 136:223–240

Zhao G-C, Wilde SA, Cawood PA, Sun M (2001) Archean blocks and their boundaries in the North China Craton: lithological, geochemical, structural and P-T path constrains and tectonic evolution. Precambr Res 107:45–73

Zhao G-C, Wilde SA, Cawood PA, Sun M (2002) SHRIMP U-Pb zircon ages of the Fuping complex: implications for Late Archean to Paleoproterozoic accretion and assembly of the North China Craton. Am J Sci 302:191–226

Zhao G-C, Sun M, Wilde SA (2003) Correlations between the Eastern Block of the North China Craton and the South Indian Block of the Indian Shield: an Archaean to Palaeoproterozoic link. Precambr Res 122:201–233

Zhao G-C, Sun M, Wilde SA, Li S-Z (2005) Late Archean to Paleoproterozoic evolution of the North China Craton: key issues revisited. Precambr Res 136:177–202

Zhao G-C, Wilde SA, Sun M, Guo J-H, Kröner A, Li S-Z, Li X-P, Zhang J (2008a) SHRIMP U-Pb zircon geochronology of the Huai'an Complex: constraints on late Archean to Paleoproterozoic magmatic and metamorphic events in the Trans-North China Orogen. Am J Sci 308:270–303

Zhao G-C, Zhai M-G (2013) Lithotectonic elements of Precambrian basement in the North China Craton: review and tectonic implications. Gondwana Res 23:1207–1240

Zhao R-F, Guo J-H, Peng P, Liu F (2011) 2.1 Ga crustal remelting event in Hengshan Complex: evidence fom U-Pb dating and Hf-Nd isotopic study on potassic granites. Acta Petrologica Sinica 27:1607–1623 (in Chinese with English abstract)

Zhao, Z.-P., et al., 1993. Precambrian crustal evolution of the Sino-Korean Paraplatform. Beijing: Science Press: 1–444 (in Chinese)

Zhao Z-R, Song H-X, Shen Q-H, Song B (2008b) Geological and geochemical char acter istics and SHRIMP U-Pb zircon dating of the Yinglingshan granite and its xenoliths in Yishui County, Shandong, China. Geol Bull China 27:1551–1558 (in Chinese with English abstract)

Zhao Z-R, Song H-X, Shen Q-H, Song B (2009) The petrogeochemiscal characters and SHRIMP U-Pb zircon ages of the meta-mafic rocks from the Yushui Complex, in Yishui County, Shandong Province. Geol Rev 55:286–299 (in Chinese with English abstract)

Zheng J-P, Griffin WL, O'Reilly SY, Lu F-X, Wang C-Y, Zhang M, Wang F-Z, Li H-M (2004a) 3.6 Ga lower crust in central China: new evidence on the assembly of the NCC. Geology 32:229–232

Zheng J-P, Griffin WL, O'Reilly SY, Lu F-X, Yu C-M, Zhang M, Li H-M (2004b) U-Pb and Hf-isotope analysis of zircons in mafic xenoliths from Fuxian kimberlites: evolution of the lower crust beneath the North China craton. Contrib Miner Petrol 148:79–103

Zheng J-P, Lu F-X, Yu S-M, Tang H-Y (2004c) An in situ Zircon Hf isotopic U-Pb age and trace element study of banded granulite xenolith from Hannuoba basalt: tracking the early evolution of the lower crust in the North China Craton. Chin Sci Bull 49:3277–3285

Zheng J-P, Griffin WL, O'Reilly SY, Hu B-Q, Zhang M, Tang H-Y, Su Y-P, Zhang Z-H, Pearson N, Wang F-Z, Lu F-X (2008) Continental collision and accretion recorded in the deep lithosphere of central China. Earth Planet Sci Lett 269:496–506

Zheng J-P, Griffin WL, Ma Q, O'Reilly SY, Xiong Q, Tang H-Y, Zhao J-H, Yu C-M, Su Y-P (2012) Accretion and reworking beneath the North China Craton. Lithos 149:61–78

Zhou H-Y, Liu D-Y, Wan Y-S, Wilde SA, Wu J-S (2007) 3.3 Ga magmatic events in the Anshan area: new SHRIMP age and geochemical constraints. Acta Petrologica et Mineralogica 26:123–129 (in Chinese with English abstract)

Zhou H-Y, Liu D-Y, Wan Y-S, Dong C-Y (2009a) 3.3-3.1 Ga magmatism in the Dongshan Complex, Anshan: evidence from SHRIMP U-Pb zircon dating. Geol Bull China 27:2122–2126 (in Chinese with English abstract)

Zhou J-B, Wilde SA, Zhao G-C, Zheng C-Q, Jin W, Zhang X-Z, Cheng H (2008) SHRIMP U-Pb zircon dating of the Neoproterozoic Penglai Group and Archean gneisses from the Jiaobei Terrane, North China, and their tectonic implications. Precambr Res 160:323–340

Zhou Y-Y, Zhao T-P, Xue L-W, Wang S-Y, Gao J-F (2009b) Petrological, geochemical and chronological constraints for the origin and geological significance of Neoarchean TTG gneiss in the Songshan area, North China Craton. Acta Petrologica Sinica 25:331–347 (in Chinese with English abstract)

Zhou Y-Y, Zhao T-P, Wang C-Y, Hu G-H (2011) Geochronology and geochemistry of 2.5 to 2.4 Ga granitic plutons from the southern margin of the North China Craton: implications for a tectonic transition from arc to post-collisional setting. Gondwana Res 20:171–183

Zhou Y-Y, Zhao T-P, Zhai M-G, Gao J-F, Sun Q-Y (2014) Petrogenesis of the Archean tonalite-trondhjemite-granodiorite (TTG) and granites in the Lushan area, southern margin of the North China Craton: implication for crustal accretion and transformation. Precambr Res 255:514–537

Zhu M-T, Zhang L-C, Dai Y-P, Wang C-L (2015) In situ zircon U-Pb dating and O isotopes of the Neoarchean Hongtoushan VMS Cu-Zn deposit in the North China Craton: implication for the ore genesis. Ore Geol Rev 67:354–367

Zhu X-Y, Zhai M-G, Chen F-K, Lü B, Wang W, Peng P, Hu B (2013) ~2. 7 Gacrustal growth in the North China Craton: evidence from zircon U-Pb ages and Hf isotopes of the Sushui Complex in the Zhongtiao terrane. J Geol 121:239–254

Paleoproterozoic Granulites in the North China Craton and Their Geological Implications

Jinghui Guo, Mingguo Zhai, Peng Peng, Shujuan Jiao, Lei Zhao and Haozheng Wang

Abstract It is well known that there are widespread Paleoproterozoic high-grade rocks, mainly garnet amphibolites and granulites, in the North China Craton. The granulites could be classified into two types, the high-pressure granulites and (ultra-) high-temperature granulites. The high-pressure granulites mainly distribute in the Huai'an-Xuanhua, Hengshan, Fuping, Zanhuang, Qianlishan–Helanshan, and Jiaobei regions (the central and east of the craton), while the (ultra-)high-temperature granulites distribute mainly in the Liangcheng–Jining–Zhuozi and the Daqingshan regions (west of the northern margin of the craton). Nevertheless, (ultra-) high-temperature granulites have been also discovered in northeast Hebei, north Henan, Shandong and Liaoning provinces, and the Korean peninsula, and some drilling cores from under the Ordos basin, though some of them might be formed in the Late Archean. The Paleoproterozoic high-grade metamorphism has partly overprinted on the Late Archean low- to high-grade rocks. Both types of granulites have recorded clockwise *P-T* paths, which might indicate a process similar to those happened in the Phanerozoic orogenic belts. In the past 20–30 years, several different tectonic models have been introduced to interpret the environments of the Paleoproterozoic granulites in the North China Craton. However, there are several things need to be clarified before any conclusion, e.g., (1) the distribution of the different types of high-grade rocks; (2) the relationship between the high-pressure granulites and the (ultra-)high-temperature granulites; and (3) the exhumation rate and genetic mechanism of these rocks. These Paleoproterozoic high-grade rocks in the North China Craton are more complex than any high-grade rocks in the Phanerozoic orogens, and await further study.

Keywords North China Craton · Paleoproterozoic · High-pressure granulite · Ultra-high-temperature granulite

---

J. Guo (✉) · M. Zhai (✉) · P. Peng · S. Jiao · L. Zhao · H. Wang
Institute of Geology and Geophysics, Chinese Academy of Sciences, 100029 Beijing, China
e-mail: jhguo@mail.iggcas.ac.cn

M. Zhai
e-mail: mgzhai@mail.iggcas.ac.cn

M. Zhai (ed.), *Precambrian Geology of China*,
Springer Geology, DOI 10.1007/978-3-662-47885-1_3

1 Distribution of the Paleoproterozoic Granulites in the North China Craton

It is well known that there are widespread Paleoproterozoic high-grade rocks in the North China Craton (simplified as NCC, Fig. 1), and among them, the granulites could be classified into two types, the high-pressure granulites and (ultra-) high-temperature granulites (e.g., Zhai et al. 2005; Zhai and Santosh, 2011; Zhao et al. 2005, 2012; Lu et al. 2008; Zhao and Zhai, 2013; Peng et al. 2014; Fig. 2). This Paleoproterozoic metamorphism has partly overprinted on the Late Archean metamorphism, and thus, their tectonic environments are of great interesting to the Precambrian geologists. This chapter will summarize the occurrences, petrological series, mineral assemblages, metamorphic conditions, and history of the granulites.

The high-pressure granulites were first discovered in the Huai'an-Xuanhua region, and then, similar rocks have been illustrated in the Hengshan, Fuping, Zanhuang, Qianlishan–Helanshan, and Jiaobei regions (Figs. 1 and 2; Zhai et al. 1992, 1996; Guo et al. 1993, 2001, 2005; Li and Qian, 1994; Li et al. 1998; Liu et al. 1998, 2006, 2013; Zhao et al. 2000, 2005, 2014; O'Brien et al. 2005; Yin et al. 2009, 2014; Zhou et al. 2010; Xiao et al. 2010, 2013; Tam et al. 2012a, b). The existence of high-pressure granulite in the craton has led to an extensive discussion on its tectonic implications, and one debate is whether the genesis of these rocks was similar to those exposure in the Phanerozoic orogens.

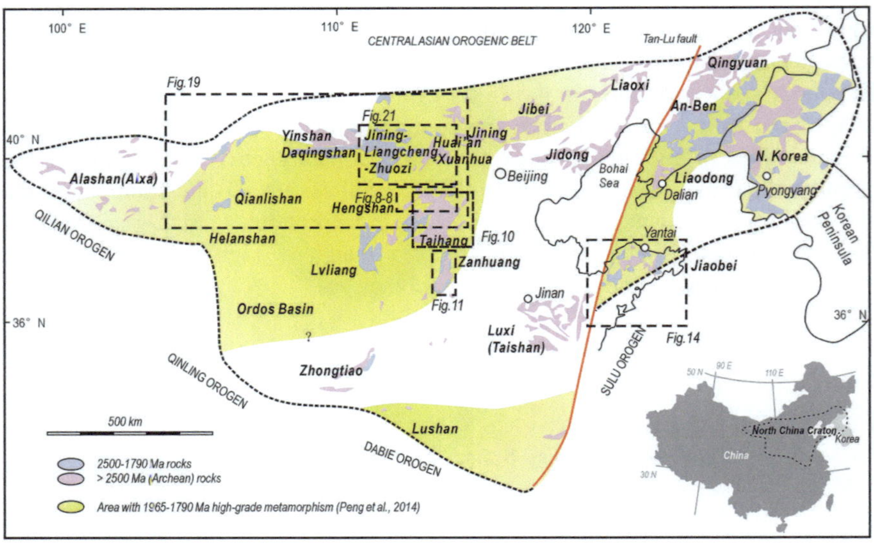

Fig. 1 Distribution of the Paleoproterozoic high-grade metamorphic rocks (granulites) in the North China Craton

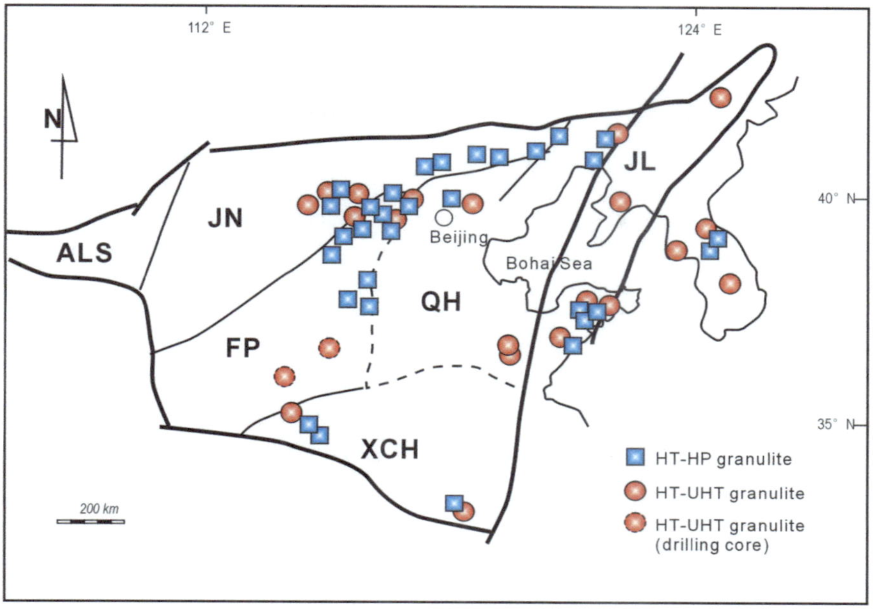

Fig. 2 A simplified map showing the distribution of the high-temperature–high-pressure granulite and high-temperature/ultra-high-temperature granulite in the North China Craton. The background map shows the subdivision of palaeo-terrains in the Late Archean after Zhai et al. (2005)

The discovery of ultra-high-temperature granulites in the NCC is another significant progresses in the NCC (e.g., Guo et al. 2006, 2012; Santosh et al. 2007a, 2009; Jiao and Guo 2011; Jiao et al. 2011; Liu et al. 2012), following the discovery of the high-pressure granulites. Ultra-high-temperature metamorphism takes place in the deep crust at extreme temperature conditions of 900–1100 °C and medium pressure (7–13 kbar: Harley 1998). Understanding the conditions and processes of ultra-high-temperature metamorphism provides important constraints on the nature of coupling between the crust, subcrustal lithosphere, and asthenospheric mantle, as well as the tectonic setting of regional metamorphic belts (Harley 1998, 2008; Kelsey 2008). The (ultra-)high-temperature granulites were reported from the Liangcheng–Jining–Zhuozi area and the Daqingshan area (Figs. 1 and 2; Guo et al. 2006, 2012; Santosh et al. 2007a, 2009; Jiao and Guo 2011; Jiao et al. 2011; Liu et al. 2012). Nevertheless, high-temperature granulites have been also discovered in northeast Hebei, northern Henan, Shandong and Liaoning provinces, and the Korean peninsula, and some drilling cores from under the Ordos basin, though some of them might form in the Late Archean (Fig. 2; Zhai et al. 2005; Zhai and Santosh 2011).

2 High-Pressure Granulite in the North China Craton

2.1 *Huai'an-Xuanhua Region*

The Huai'an-Xuanhua region is located in the central North China Craton (Fig. 1). Gray gneisses, felsic to mafic granulites, and minor supracrustal rocks are the dominant rock types in this region, most of which underwent granulite facies metamorphism. Zhai et al. (1992a, b) first reported high-pressure mafic granulite in this region, and they occur as blocks and boudins with varying sizes within a 2 km wide and several kilometer long area, which could be parts of the lowermost crust. The most significantly known high-pressure mafic granulite block exposes near Manjinggou with an outcropping area of ~ 20 km^2, bounded by granulitic gray gneisses to the north and shear zone accompanying some anorthositic and gabbroic dykes and veins to the south (Zhai et al. 1993: Fig. 3).

There are two kinds of high-pressure granulites, represented by the Manjinggou outcrop and the Xiwangshan outcrop (Guo et al. 1993, 2002). High-pressure granulites in the Manjinggou outcrop (Fig. 3) occur as small sheetlike bodies or lenticular enclaves within two-pyroxene granulites of intercalated tonalitic and gabbroic to dioritic compositions (Guo et al. 1993), and they could be originally mafic dykes (Peng et al. 2005). High-pressure basic granulites in the Xiwangshan outcrop (Fig. 4) are characterized by parallel, narrow, and elongate lenses of garnetiferous basic

Fig. 3 Geological sketch map of distribution of HP basic granulites in Manjinggou (from Guo et al. 1993)

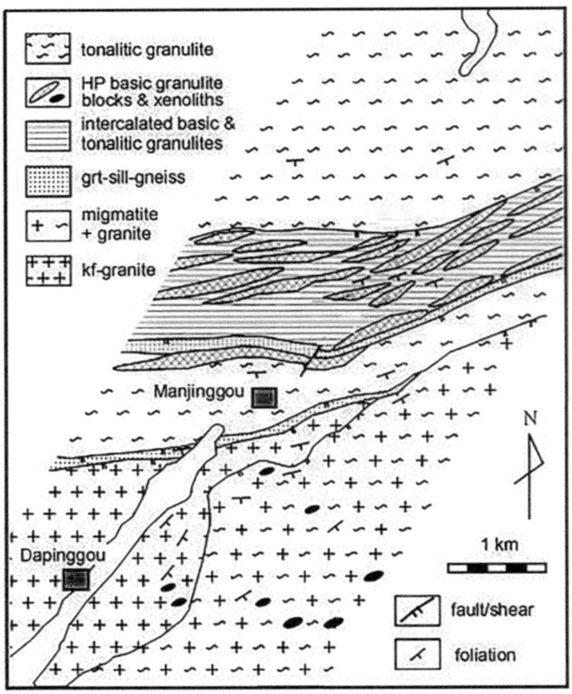

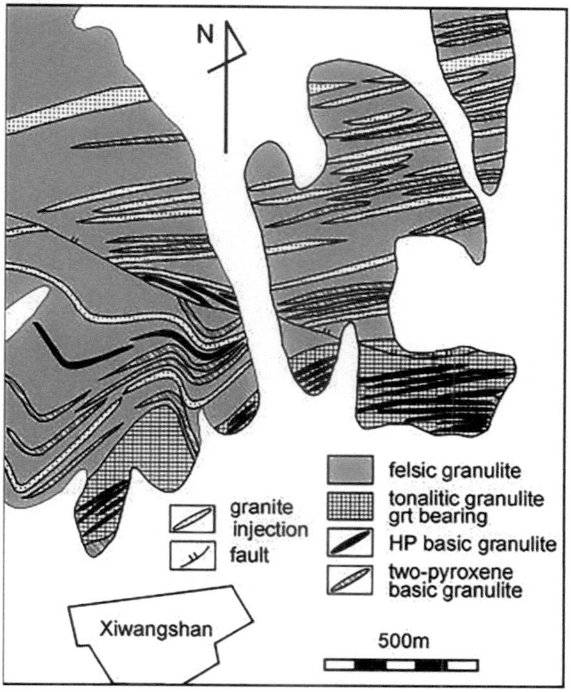

Fig. 4 Geological sketch map of distribution of HP basic granulites in Xiwangshan (Li et al. 1998)

high-pressure granulites associated with garnet-bearing tonalitic granulites or garnet-bearing two-pyroxene granulites (Li et al. 1998). High-pressure basic granulites can be also classified as two types according to their country rocks, being associated with tonalitic granulites and metapelites (Li et al. 1998), or being associated with gray gneisses (Guo et al. 1993), and Zhang et al. (2014) suggest that both occurrences have experienced similar tectonothermal evolution with their country rocks.

Typical mineral assemblage of the high-pressure granulites in the Huai'an-Xuanhua region are garnet + pyroxene + plagioclase, and the garnet grains occur as porphyroblasts ranging from 2 to 20 mm in diameter. Pyroxene and plagioclase grains are the main matrix minerals (Fig. 5). Four metamorphic stages including prograde, peak, near-isothermal decompression, and amphibolite facies retrogression have been proposed. The initial stage of prograde metamorphism is defined by the inclusions of abundant quartz and minor clinopyroxene, plagioclase, titanite, apatite, rutile, and ilmenite in the low X_{Ca} ($X_{Ca} = Ca/(Ca + Mg + Fe + Mn)$) core of compositionally zoned porphyroblastic garnets. The mineral assemblage of peak metamorphic stage is dominated by the sparse inclusions of clinopyroxene + quartz or clinopyroxene + plagioclase + quartz + accessory minerals (titanite, apatite, rutile) in garnet and high-X_{Ca} region of garnet with flat interior compositional profiles (Zhai et al. 1993; Guo et al. 2002). The near-isothermal decompression metamorphic stage is characterized by the mineral assemblage of kelyphitic and coronitic Opx + Pl ± Mt, or hornblende surrounding garnet.

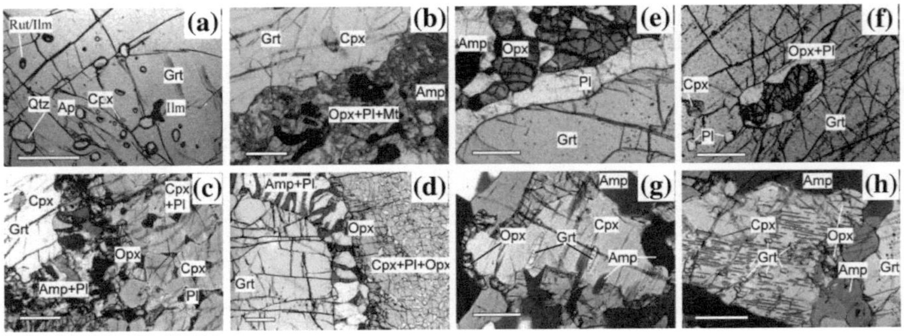

Fig. 5 Photomicrographs showing typical textures of the HP granulites. All *scale bars* are 0.5 mm. **a** Inclusions in core of garnet with growth zoning; **b** kelyphite of Opx + Pl ± Mt around garnet and Cpx inclusion in garnet; **c** garnet with Cpx inclusion, kelyphite of Amp + Pl around garnet itself surrounded by Opx aggregates, and Cpx aggregates with minor Pl at triple junctions; **d** kelyphite of Amp + Pl around garnet itself surrounded by Opx aggregates and a symplectite-like Cpx + Pl + Opx domain; **e** Pl + Opx corona around garnet; **f** coronitic Opx + Pl in garnet, and Cpx, Pl inclusions in garnet; **g** garnet exsolution lamellae in clinopyroxene which is itself rimmed by Opx and Amp whereby amphibole crosses the Cpx-Opx boundary; **h** garnet and amphibole lamellae in clinopyroxene (from Guo et al. 2002)

There are also cooling phenomena such as oriented garnet exsolution lamellae. The final amphibole facies retrogressive stage is represented by the replacing of symplectite on pyroxenes or garnet with hornblende and plagioclase. Zhai et al. (1993), and Guo et al. (2002) defined clockwise *P-T* paths comprising prograde, near-isothermal decompression (ITD), and near-isobaric cooling (IBC) stages (Fig. 6). It reveals that the prograde stage (M1) is at 670–730 °C and 9–12 kbar; peak stage (M2) is at 800–900 °C and 12–14 kbar; M3 stage has similar temperature with M2 but its pressure is 2–4 kbar lower than M2; and M4 stage is 200–300 °C and 0.5–3 kbar lower than M3. Metamorphic ages of 1850–1800 Ma have been reported (e.g., Guo et al. 2005; Peng et al. 2005; Zhao et al. 2008; Wang et al. 2010).

2.2 *Hengshan Region*

Hengshan region is located in northern Shanxi Province. This region mainly consists of gray gneisses, mafic granulites, pegmatites, and some supracrustal rocks (Wang et al. 1991; Li and Qian 1994; Tian et al. 1996) (Fig. 7). High-pressure granulites occur in gray gneisses as boudins and sheets with varied sizes. Retrograde eclogites, first reported by Zhai et al. (1996), are enveloped in mylonitized granitic gneiss as lenses or coexist with gray tonalitic gneiss. The eclogite facies metamorphism is determined by reintegrated mineral composition from the residue of breakdown omphacite, the omphacite inclusions in garnet (Fig. 8), and the presence of rutile surrounded by titanite and ilmenite (Fig. 8).

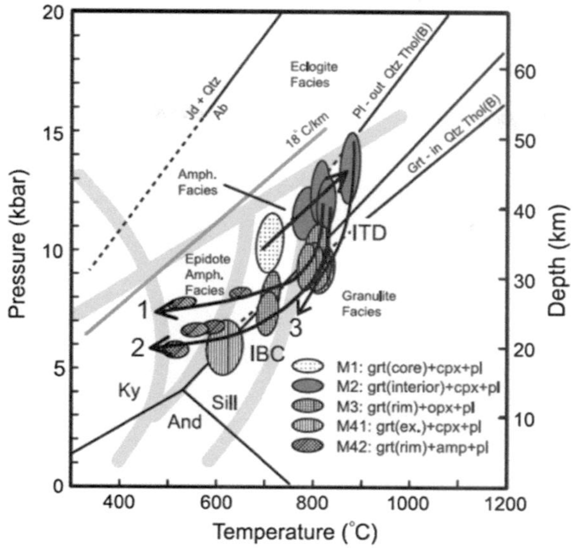

Fig. 6 *P-T* path results of high-pressure granulites in Huai'an region. Path lines 1,2 from Guo et al. (2002), path line 3 from Zhai et al. (1993). Ab = Jd + Qtz (Holland 1980); Grt-in and Pl-out for quartz tholeiite B metabasite composition (Green and Ringwood 1967); aluminosilicate stability fields (Salje 1986); fields of metamorphic facies and subfacies (Spear 1993)

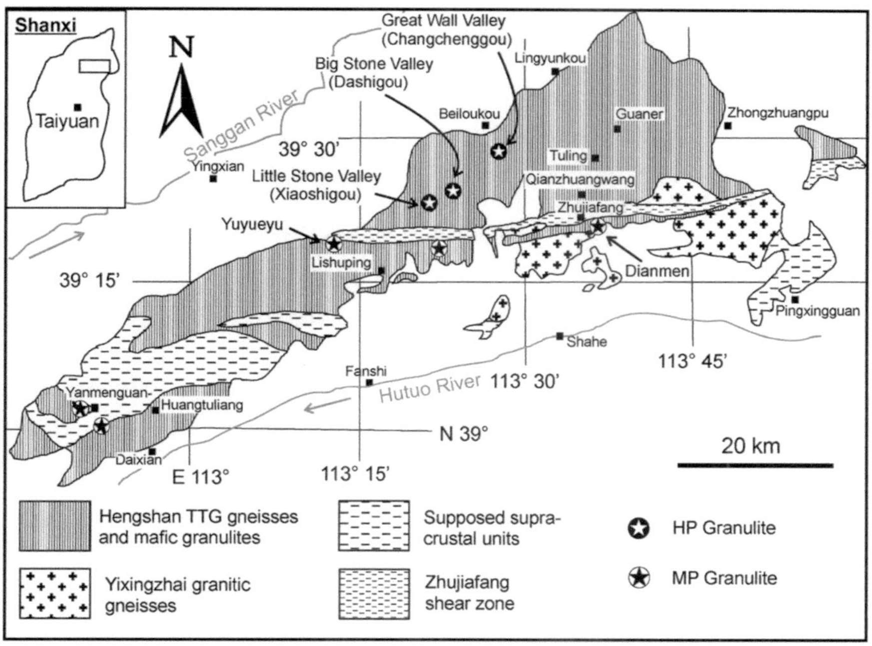

Fig. 7 Simplified geological map of Hengshan complex (after Li and Qian 1994) with main mafic granulites distribution (O'Brien et al. 2005)

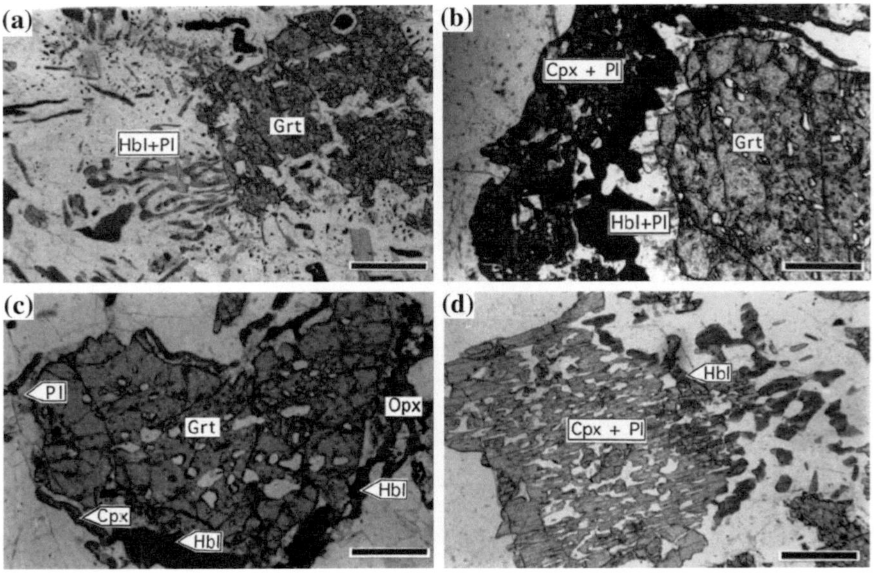

Fig. 8 Petrographic photographs of retrograde eclogite (high-pressure granulite) in Hengshan region (Zhao et al. 2001a). *Scale bars* are 0.1 mm

Four stages of metamorphic mineral assemblages have been described (Zhai 1997; Zhao et al. 2001a; O'Brien et al. 2005): M1, prograde assemblage of quartz and rutile inclusions within the cores of porphyroblastic garnet and omphacite pseudomorphs; M2, peak assemblage consisting of clinopyroxene, garnet, sodic plagioclase, quartz, and hornblende; M3, near-isothermal decompression leading to the formation of plagioclase + orthopyroxene + clinopyroxene symplectite surrounding garnet; and M4, decompression-cooling stage of hornblende + plagioclase symplectites rimming garnet. Symplectite of plagioclase and hornblende on rounded to subhedral porphyroblastic garnet, which is called 'white eye socket', is visible on the specimen. Lathlike symplecitite of diopside and plagioclase characteristically suggests the breakdown of jadeite-rich pyroxene. It yields peak (M2) *P-T* of 770–840 °C/13.5–15.5 kbar, M3 of 6.5–8.0 kbar/750–830 °C, and M4 of 4.5–6 kbar/680–790 °C (Fig. 9). Most metamorphic ages are 1900–1830 Ma (e.g., Chang et al. 1999; Kröner et al. 2005a, b; Jia 2007; Pang et al. 2010; Shi et al. 2012). Qian et al. (2015) have recently obtained a metamorphic age of 1964 ± 25 Ma from the garnet amphibolite.

2.3 Wutai–Fuping Region

Wutai–Fuping region mainly consists of TTG gneisses, metamorphosed felsic volcanic rocks, mafic granulites, amphibolites, metapelites, and banded iron

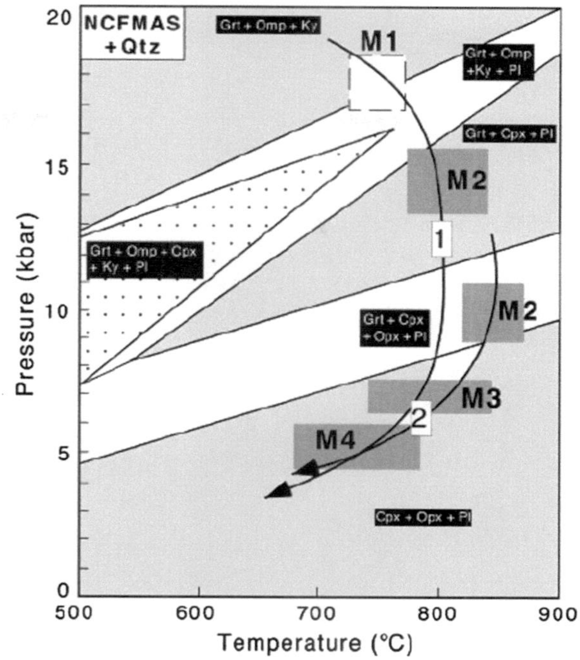

Fig. 9 *P-T* path of high- and medium-pressure granulites from the Hengshan region (results estimated from *P-T* pseudosection of NCFMAS system). Path line 1 represents high-pressure granulite evolution history, and path line 2 represents medium-pressure granulite evolution history (Zhao et al. 2001a)

formation (Figs. 1 and 10; Zhao et al. 2000). Mafic granulites are generally presented at Daliushu and Fangli districts, which occur as different scale lenses or layers within TTG gneisses. There are also some other amphibolite facies volcanic or supracrustal rocks occurring as lenticular or boudin-like or layered bodies within TTG gneisses (Liu 1996). The Fuping mafic granulites record a four-stage metamorphic evolutionary history, including the prograde process (M1) represented by inclusions of amphibole, plagioclase, clinopyroxene within garnet; the peak stage (M2) represented by coarse and paragenetic minerals of amphibole, clinopyroxene, garnet, plagioclase, and unfrequent hypersthene, quartz; M3 stage represented by vermicular orthopyroxene + clinopyroxene + plagioclase + magnetite symplectites surrounding garnet; and M4 represented by coronitic hornblende and plagioclase surrounding garnet (Liu 1996; Zhao et al. 1999, 2000). *P-T* estimation of peak stage is 8.5–9.5 kbar/870–930 °C (Liu 1996). Their *P-T* paths comprise a near-isothermal decompression, following the clockwise *P-T* path (Liu and Liang 1997; Zhao et al. 1999; Qian et al. 2013). The metamorphic ages are 1965–1800 Ma (Liu et al. 2006; Trap et al. 2007, 2012; Qian et al. 2013).

2.4 *Zanhuang Region*

Zanhuang region is south to the Wutai–Fuping area (Figs. 1 and 11). Regional structural history and mineral geochronology shows that the Zanhuang massif was

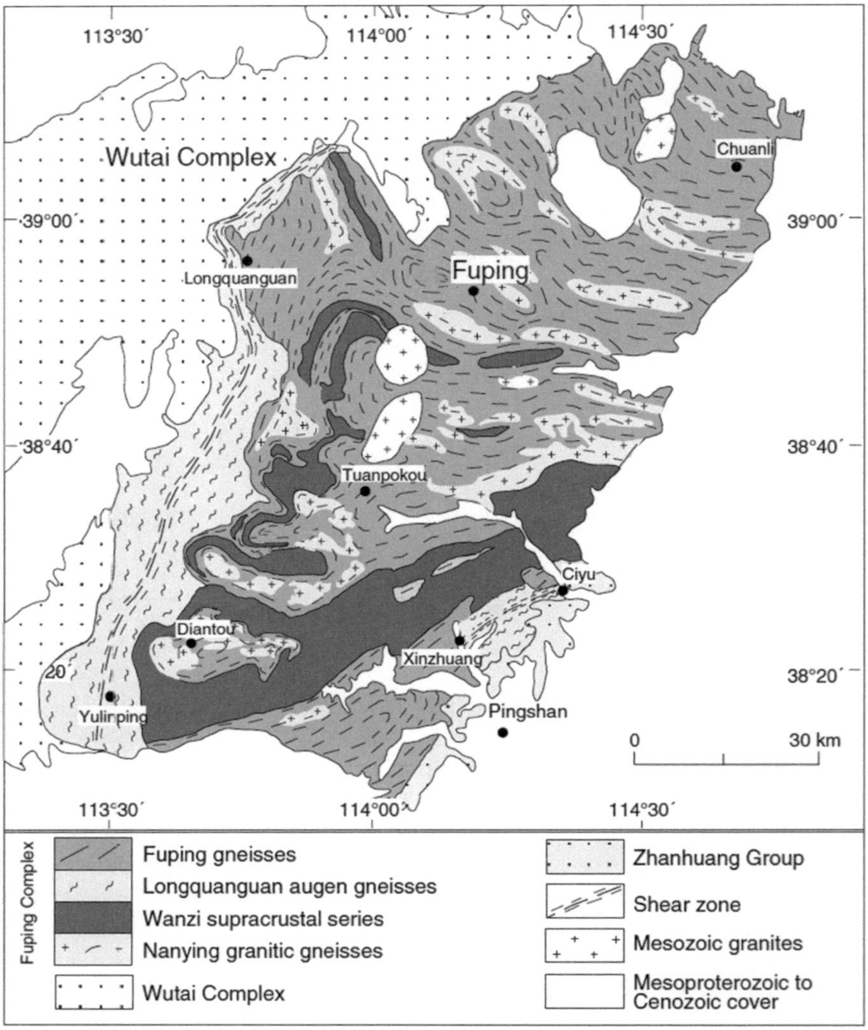

Fig. 10 Simplified geological map of the Fuping Complex (Zhao et al. 2000, 2002)

subjected to a number of deformation events during Paleoproterozoic (Wang 2003; Trap et al. 2009b). The high-grade rocks occurring in the Archean metamorphic complexes (Fig. 11), have experienced high-pressure amphibolite facies metamorphism. The amphibolites comprise of quartz + plagioclase + hornblende + garnet + clinopyroxene ± biotite, while the metapelites comprise of quartz + biotite + plagioclase + garnet ± kyanite ± K-feldspar (Figs. 12 and 13). Both amphibolite and metapelitic rocks developed symplectite on garnet grains (Figs. 12 and 13). Three-stage metamorphism has been deciphered from both the amphibolite and metapelite rocks, and these comprise a clockwise *P-T* path (Xiao et al. 2010).

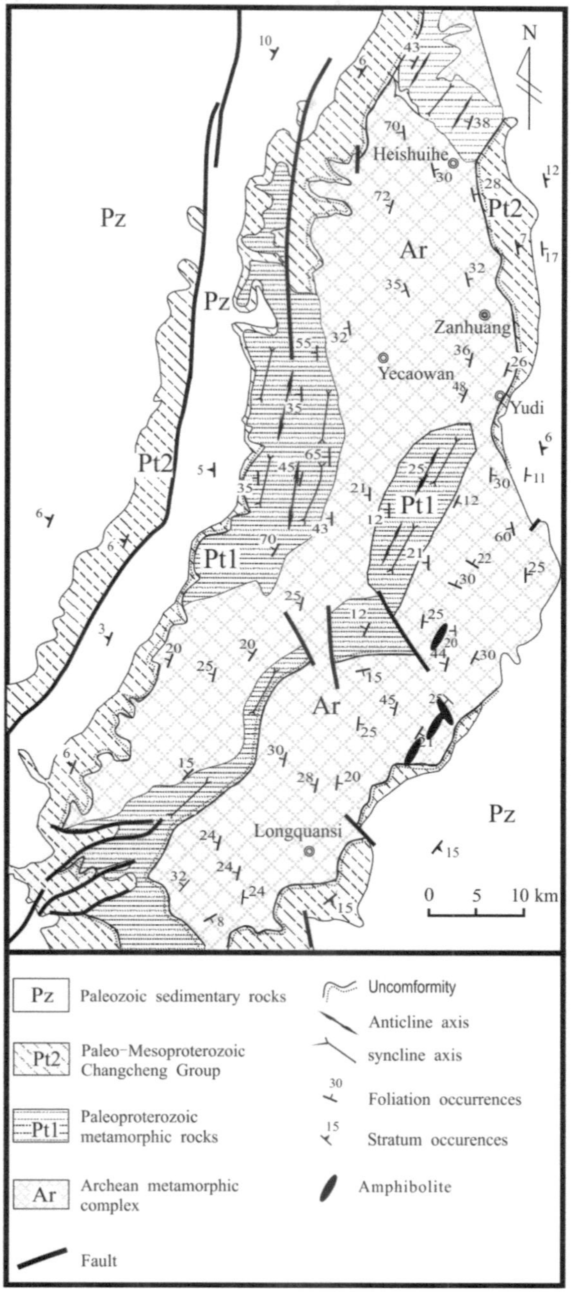

Fig. 11 Simplified geological map of the Zanhuang region (Xiao et al. 2010)

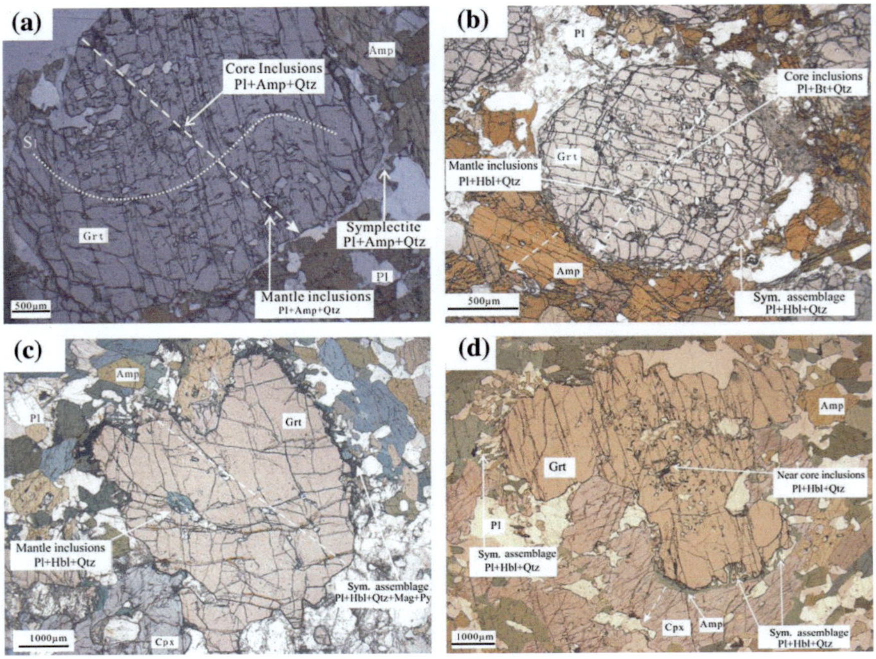

Fig. 12 Representative photomicrographs of amphibolites in the Zanhuang region (Xiao et al. 2010)

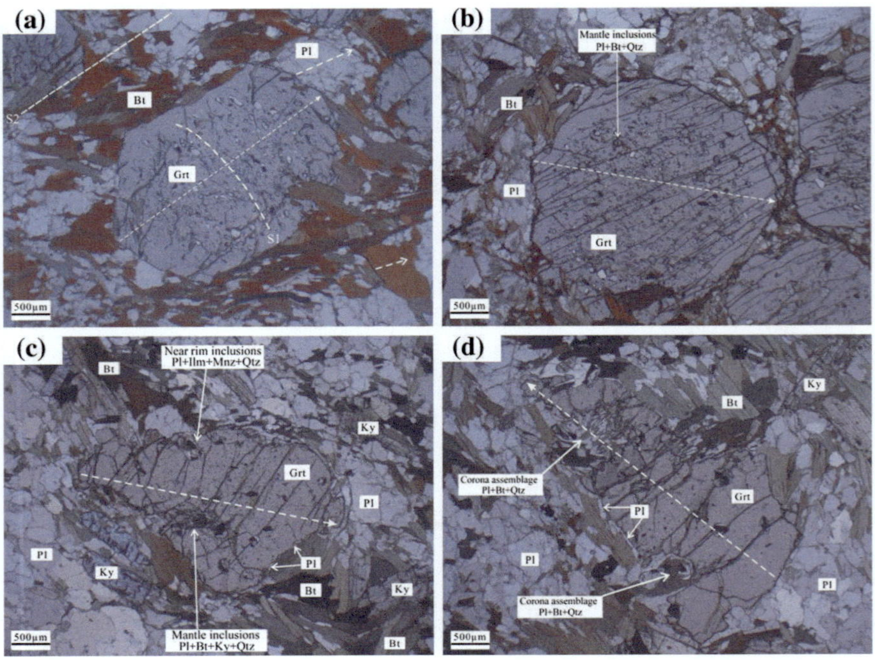

Fig. 13 Representative photomicrographs of metapelites in the Zanhuang region (Xiao et al. 2010)

Metamorphic conditions are 650–710 °C/8.2–9.2 kbar for prograde metamorphic stage, >810 °C and >12.5 kbar for peak metamorphic stage, and 660–680 °C/4.4–4.5 kbar for retrograde metamorphic stage (Xiao et al. 2010). The metamorphism occurred at 1860–1840 Ma (Xiao et al. 2013).

2.5 *Jiaobei Region*

The Jiaobei region is located in the eastern part of the NCC (Figs. 1 and 14). High-pressure mafic granulites in the Jiaobei terrane were first reported by Li et al. (1997) and Liu et al. (1998). These high-grade rocks are composed of mafic garnet granulites, garnet–hypersthene granulites, and garnet amphibolite. There are some ultramafic rocks which have also experienced granulite facies metamorphism. In addition, although low- to medium-pressure pelitic granulites are dominated in the Jingshan Group, minor high-pressure pelitic granulites have also been discovered (Zhou et al. 2004, 2008).

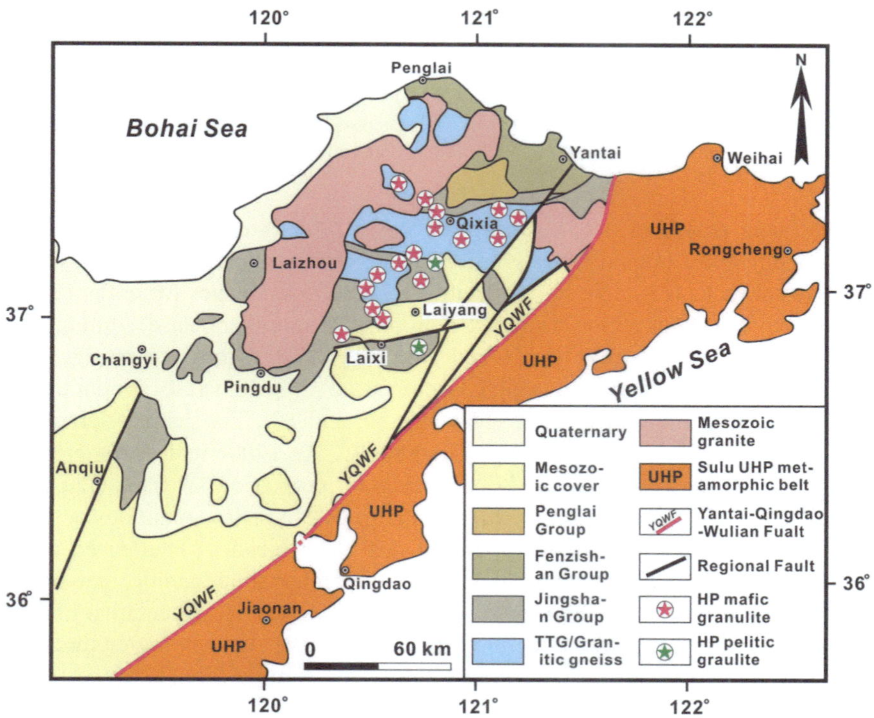

Fig. 14 Simplified geological map showing the geological setting of the high-pressure granulites in the Jiaobei region (Liu et al. 2013)

Fig. 15 Representative field photographs from the Jiaobei region showing the relationships between HP granulites and their country rocks. **a** and **b** Mafic granulites occur as lenses in TTG gneisses. **c** Mafic granulites occur as dykes in TTG gneisses. **d** Retrograde textures preserved by mafic granulites usually referred to as 'white eye.' After Liu et al. (2013)

The mafic granulites occur as irregular lenses or deformed dykes within tonalitic–trondhjemitic–granodioritic gneisses and granitic gneisses, and sometimes supracrustal rocks (Fig. 15). Representative mineral assemblages for high-pressure mafic granulites in this region are garnet + clinopyroxene + plagioclase ± orthopyroxene ± amphibole ± quartz (Figs. 16 and 17). They all show obvious retrogressive symplectite on garnet grains, which is composed of plagioclase ± clinopyroxene ± orthopyroxene ± amphibole (Figs. 15, 16, and 17). The peak metamorphic stage conditions are ~1.5 GPa and 850–880 °C. An isothermal decompression followed by isobaric cooling can be seen from all proposed *P-T* paths (Liu et al. 1998; Tam et al. 2012a; Liu et al. 2013). High-pressure pelitic granulites occur as thin layers ranging from several to tens of meters in thickness interleaved in the Jingshan Group meta-sedimentary rocks. Felsic and migmatitic granitic gneisses are the direct country rocks of high-pressure pelitic granulites. Representative mineral assemblages of high-pressure pelitic granulites in the Jiaobei region are garnet + kyanite + perthite + anti-perthite + muscovite + rutile (Fig. 18) (Zhou et al. 2004). Based on thermodynamic modeling, P and T conditions for peak metamorphic stage were estimated at 1.0–1.25 GPa and 800–840 °C (Zhou et al. 2004) and 1.5–1.6 GPa and 860–890 °C (Tam et al. 2012b). Similar *P-T* paths for the

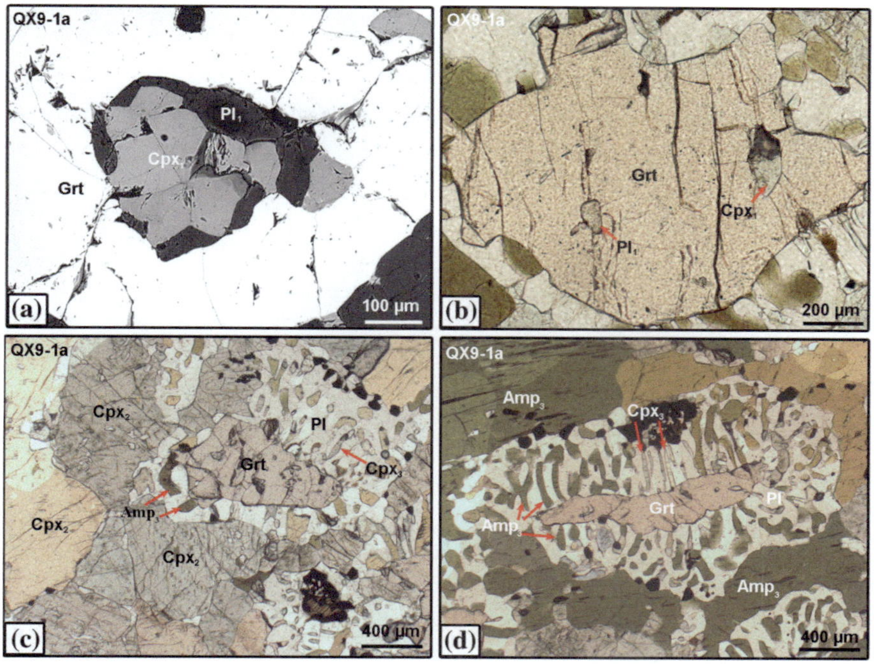

Fig. 16 Representative photomicrographs showing typical textures and mineral assemblages of HP mafic granulites. **a** BSE image and **b**, **c**, and **d** PPL (plane polarized light) image. After Liu et al. (2013)

mafic and politic granulites have been reconstructed; they are clockwise with an isobaric cooling stage following earlier isothermal decompression stage (Liu et al. 1998, 2013; Zhou et al. 2004; Tam et al. 2012a, b). Metamorphic ages range from 1960 to 1790 Ma have been reported in the region (Zhou et al. 2008; Tam et al. 2011; Liu et al. 2013; Zhao et al. 2014), though it is unclear whether these ages refer to a single metamorphic event or not.

2.6 *Qianlishan–Helanshan Region*

The Qianlishan–Helanshan region is located in the western part of the NCC (Figs. 1 and 19). In this region, Paleoproterozoic metamorphic rocks are dominated by high-grade supracrustal rocks accompanied by some S-type granites. These supracrustal rocks, which are usually termed as khondalite series, consist of graphite-bearing sillimanite–garnet gneiss, garnet quartzite, felsic paragneiss, calc-silicate rocks, and some marbles. High-pressure pelitic granulites were described by Zhou et al. (2010) and Yin et al. (2014). Their mineral assemblages are garnet + kyanite + perthite + anti-perthite + plagioclase + biotite + sillimanite + quartz (Fig. 20).

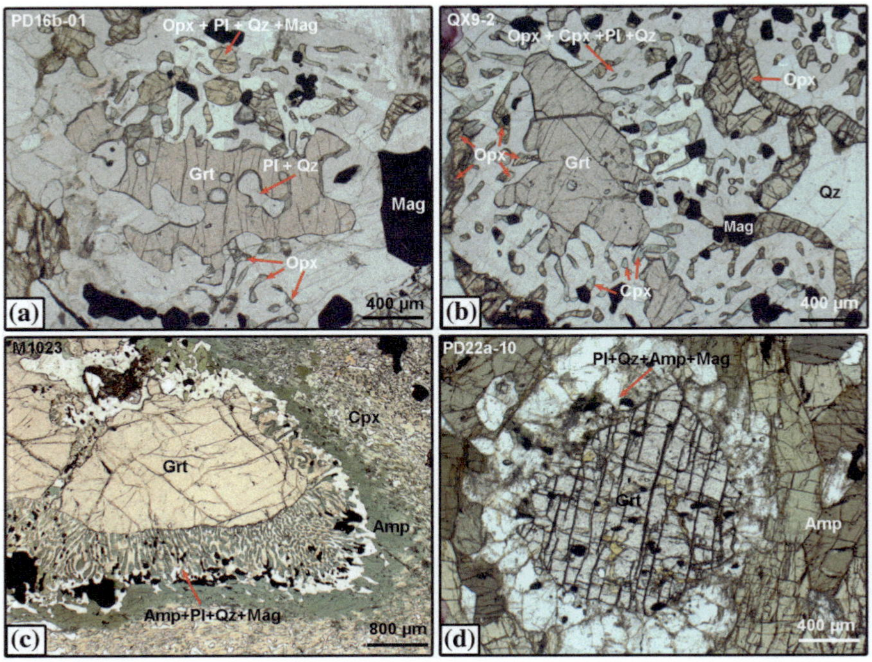

Fig. 17 Representative photomicrographs showing typical textures and mineral assemblages of HP mafic granulites. After Liu et al. (2013)

Peak metamorphic p and t are estimated at 1.1–1.2 GPa and 834–845 °C (Yin et al. 2014) and 1.4–1.5 GPa and 850–870 °C (Zhou et al. 2010) for the high-pressure pelitic granulites in the Qianlishan and Helanshan areas, respectively. The *P-T* paths are both clockwise, characterized by isothermal decompression and then isobaric cooling (Yin et al. 2014; Zhou et al. 2010). Apparently, two metamorphic age groups at ~1950 and ~1920–1870 Ma have been reported (Yin et al. 2009, 2011).

2.7 Liaohe, Lvliang, and Other Regions

The Liaohe Group is an important Paleoproterozoic stratigraphic unit in the northeastern part of the NCC and is traditionally subdivided into the North and South Liaohe Groups. The Liaohe Group consists of an assemblage that is transitional from a lower arkose- and volcanic-rich sequence, through a middle carbonate-rich sequence to an upper argillaceous sequence. The large volumes of granitoids that are tectonically associated with the Liaohe Group were emplaced mainly at 2100–2200 Ma. Li et al. (2006) and Luo et al. (2008) suggest that the

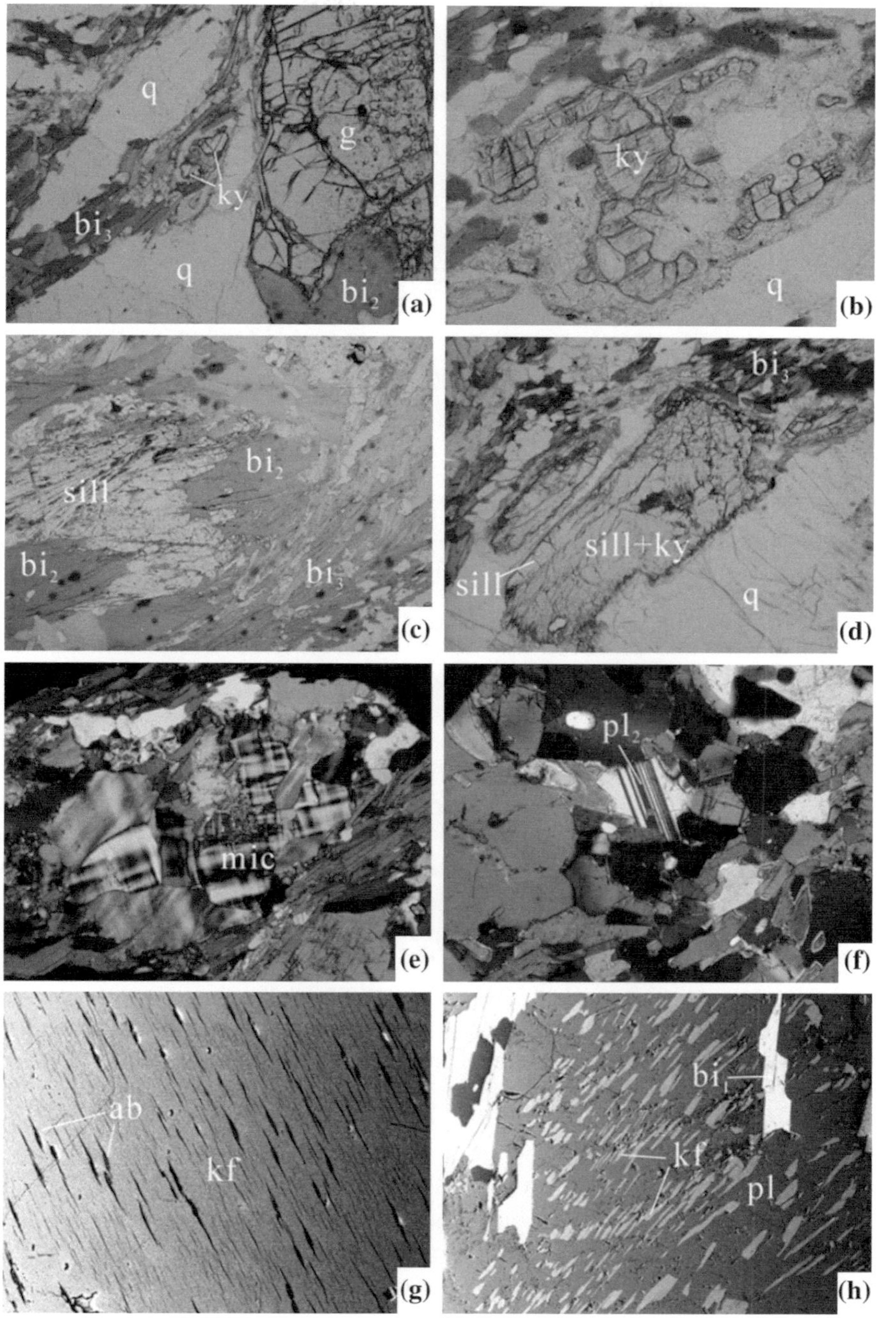

Fig. 18 Representative photomicrographs of HP pelitic granulites of the Jiaobei region show mineral assemblages. After Zhou et al. (2004)

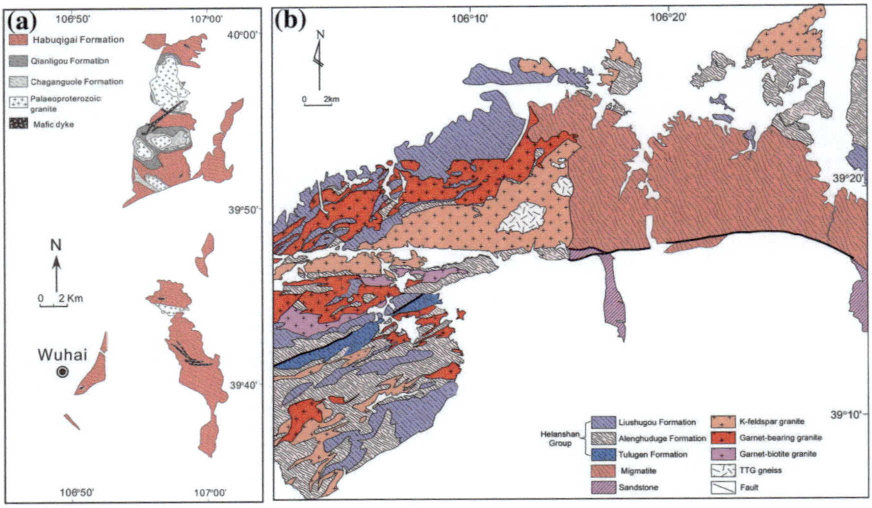

Fig. 19 Simplified geological map of the Qianlishan **a** and Helanshan **b** regions in the NCC (Yin et al. 2014)

North and South Liaohe Groups were developed on the same Archean basement (Li et al. 2006; Luo et al. 2008). However, the metamorphic rocks of the North Liaohe Group show typical metamorphic minerals of garnet + staurolite + kyanite and clockwise *P-T* path, whereas those in the South Liaohe Group usually show typical metamorphic minerals of garnet + staurolite + andalusite + cordierite + sillimanite and anticlockwise *P-T* path (He and Ye 1998; Lu et al. 1996). And the metamorphic grade in both the North and South Liaohe Groups is amphibolite facies (He and Ye 1998; Li et al. 2001). The reported metamorphic age ranges from 1960 to 1850 Ma (Li and Zhao 2007; Luo et al. 2004, 2008).

The Lvliang region is west to the Zanhuang. Based on detailed structural studies, Trap et al. (2009a) divided the associated Early Precambrian rocks into six main units, which are as follows: (1) TTG gneiss and Al-rich metasedimentary rocks; (2) a mafic, turbiditic, and volcanic unit; (3) an orthogneiss and meta-volcanic unit; (4) a lower gneissic TTG and migmatitic unit; (5) an unconformable late-orogenic weakly metamorphosed sedimentary series; and (6) several generations of post-tectonic granites. Garnet-bearing metamorphic rocks, i.e., the metapelites and amphibolites, show metamorphic grade of amphibolite facies. Representative mineral assemblages are garnet + sillimanite + kyanite + biotite + plagioclase + quartz for the metapelites but garnet + amphibole + plagioclase + quartz for the amphibolites (Santosh et al. 2015; Trap et al. 2009a). Metamorphic *P-T* paths recorded by these rocks are all clockwise (Zhao et al. 2000). The metamorphic age is at 1890–1870 Ma (Trap et al. 2009a).

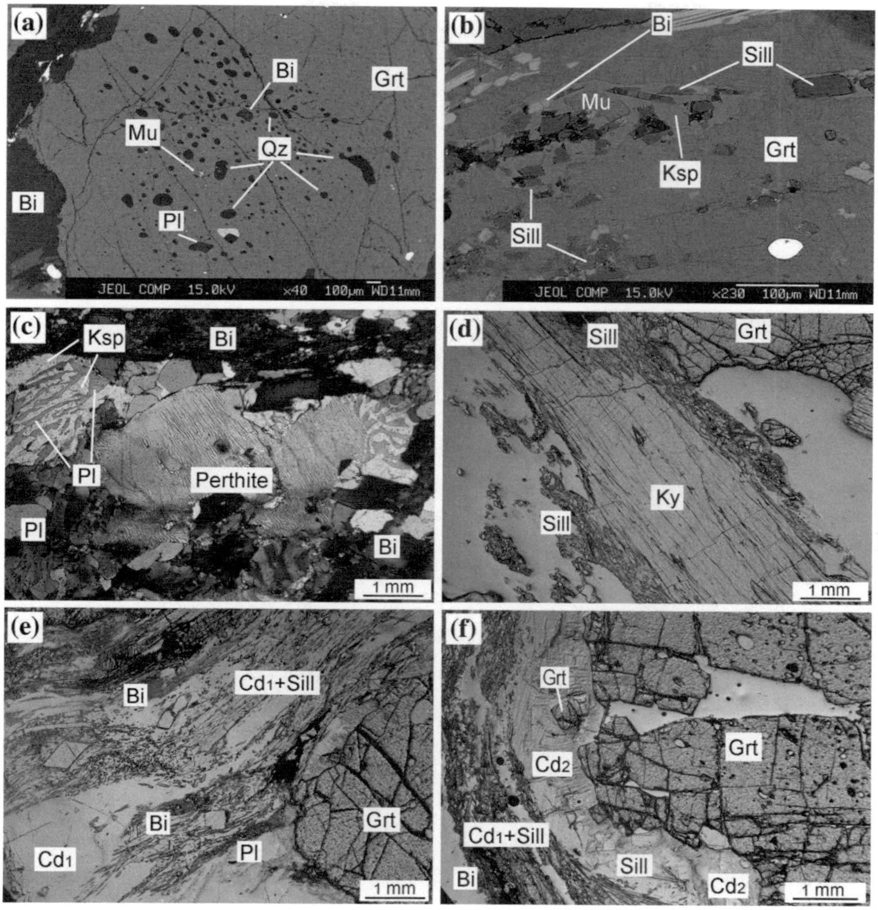

Fig. 20 Representative photomicrographs of high-pressure granulites in the Qianlishan–Helanshan region (Yin et al. 2014)

Some other places in the NCC, such as the Taihua and Northern Hebei regions, also experienced amphibolite to granulite facies metamorphism, and the metamorphic *P-T* paths recorded by metamorphic rocks in these regions are all clockwise (Lu et al. 2013; Zhao et al. 2000). For example, garnetiferous amphibolites in the Taihua region have a mineral assemblage of garnet + amphibole + clinopyroxene + orthopyroxene + plagioclase, and geothermo-barometric computation results show that these amphibolites recorded peak P and T conditions of 0.9–1.1 Gpa, 740–810 °C (Lu et al. 2013). The metamorphic ages are range from 1950 to 1750 Ma (Lu et al. 2013).

3 (Ultra-)High-Temperature Granulite in the North China Craton

3.1 Jining–Liangcheng–Zhuozi Region

The (ultra-)high-temperature granulites in this region are first reported and illustrated by Santosh et al. (2006, 2007a) and their following works at the Tianpishan Hill in Tuguiwula Town (Fig. 21). This locality is mainly composed of inter-layered aluminous metasediments in association with graphite-bearing garnet biotite gneiss, garnet-bearing felsic gneiss, and S-type granite. The rocks show prominent compositional layering with garnet–sillimanite ± cordierite ± spinel-rich melanosomes and quartzo-feldspathic leucosomes indicating partial melting (Fig. 22). These rocks are further intruded by late pink granite, granitic pegmatite, and mafic dyke. An early NW trending and a later SW trending lineations are identified (Santosh et al. 2007a). The garnet gneiss belts are strongly folded, and the enclosing S-type granites also suffered strong ductile deformation with sinistral shear sense. At least four generations of deformation can be recognized within the metapelites involving D1 early thrust, D2 extension, D3 strike-slip shear deformation (mainly sinistral), and D4 intrusion of mafic dyke swarms (Santosh et al. 2007a).

The sapphirine (Spr)-bearing Mg–Al-rich granulites at Tianpishan locality occur as several centimeter- to decimeter-sized concordant bands and layers within metapelites. According to different proportions of the characteristic minerals, four

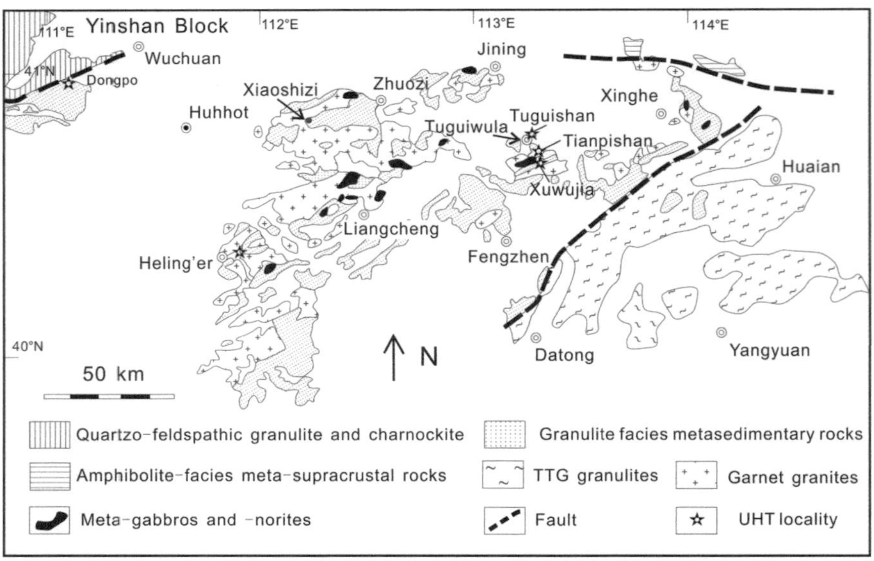

Fig. 21 Distribution of the high-grade metamorphic rocks in the Jining–Liangcheng–Zhuozi area (modified after Guo et al. 2001)

Fig. 22 Field photographs of the Spr-bearing UHT granulites at the Tianpishan locality

types of lithologies are divided here: the garnet–sillimanite–sapphirine granulite, orthopyroxene–sillimanite granulite, quartzo-feldspathic garnet–spinel granulite, and the quartzo-feldspathic garnet–sillimanite granulite. Spr is usually medium-grained (0.2–0.5 mm) and surrounded by quartz with or without reaction textures (Fig. 23). Orthopyroxene and sillimanite are usually in direct contact, but sometimes isolated by a thin domain of cordierite (Santosh et al. 2007a). Almost all K-feldspars have experienced exsolution to transform into perthites or mesoperthites, indicating high temperature and late slow cooling (Jiao and Guo 2011). Low-Zn and Cr Spinel is also identified in association with quartz (Santosh et al. 2007a). Both sapphirine and quartz occur as inclusions in garnet and sillimanite and also in the matrix. Orthopyroxene normally occurs in association with sillimanite in the matrix. It suggests the following reaction in the FMAS system: Spr + Qz + Grt = Opx + Sil (Santosh et al. 2007a). This reaction indicates cooling from the Spr-bearing field to the Opx + Sil-bearing field (Kelsey 2008). Late decompression micro-textures are representative by extensive cordierite corona and Crd + Opx (containing low alumina) symplectite around Grt or large-sized high-alumina Opx (Liu et al. 2011). The inferred reactions are Opx + Sil + Qz = Crd, Grt + Sil + Qz = Crd, and Grt + Qz = Crd + Opx. Therefore, an anticlockwise *P-T* path characterized by isobaric cooling after peak metamorphism, followed by isothermal decompression and late cooling, is retrieved (Fig. 24; Santosh et al. 2007a, 2012). Thermometers, especially the Ti-in-zircon and Zr-in-rutile, and two-feldspar thermometers yield robust *P-T* conditions of ~900–990 °C and ~8–10 kbar (Santosh et al. 2007a; Liu et al. 2010; Jiao and Guo 2011; Jiao et al. 2011). SHRIMP zircon U-Pb dating and monazite chemical dating confine the timing of the peak metamorphism to be ~1920 Ma (Santosh et al. 2007a, b).

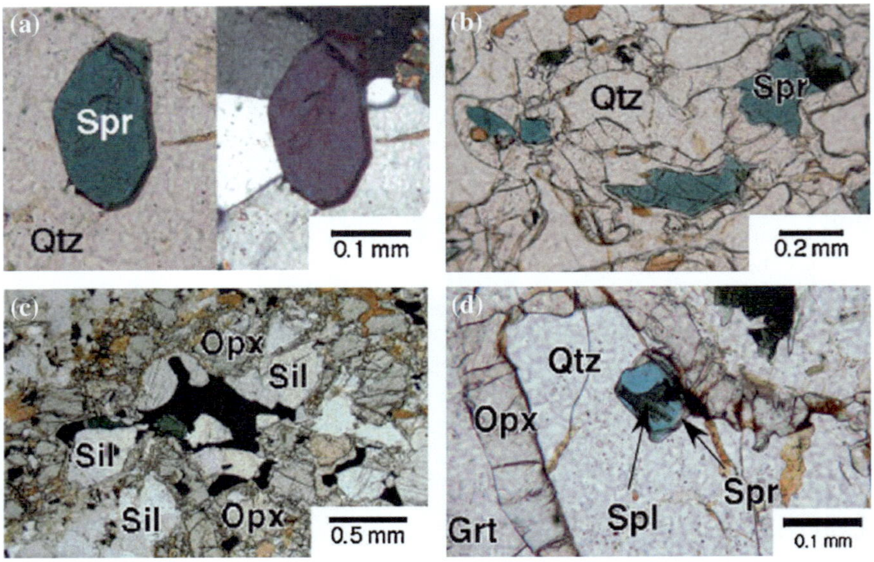

Fig. 23 Photomicrographs of mineral assemblages in the Spr-bearing UHT granulites at the Tianpishan locality (From Santosh et al. 2012)

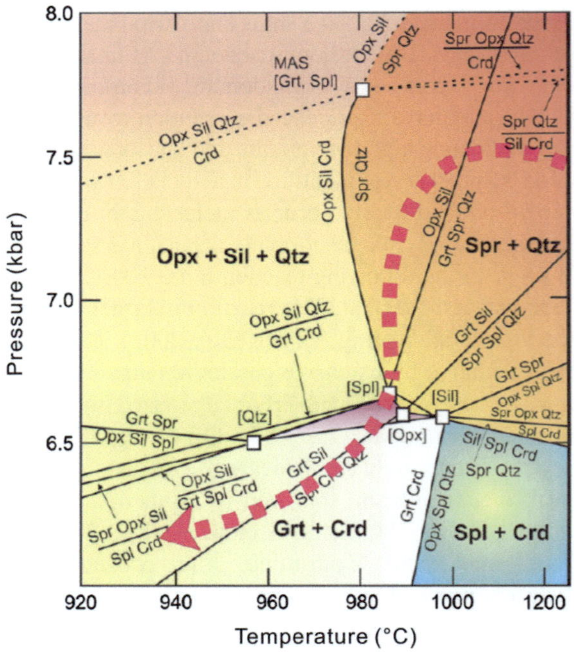

Fig. 24 *P-T* path retrieved for the Spr-bearing UHT granulites at the Tianpishan locality (From Santosh et al. 2012)

In addition, based on two-feldspar thermometer and single mineral trace elements thermometers (i.e., Zr-in-zircon and Zr-in-rutile), three other ultra-high-temperature granulites localities are identified in this region, such as the Tuguishan, Xuwujia, and Helinger (Fig. 21; Jiao and Guo 2011; Jiao et al. 2011; Liu et al. 2012). However, the diagnostic ultra-high-temperature mineral assemblages are absent, as the bulk rock composition is not typically Mg–Al rich. Associated with the ~1920 Ma UHT metamorphism, there were extensive coeval gabbroic and noritic to dioritic magmatism in the region (Fig. 21; Peng et al. 2010). This contemporary mafic magma is possibly cogenetic with the UHT metamorphism. Moreover, this mafic magma has also possibly caused the extensive lower crustal anatexis and generation of S-type and I-type granites in the region (Peng et al. 2012). The garnetite occurring extensively within the metasedimentary rocks and granites might be the melting residual. One of the largest-sized garnetites localized in the Xiaoshizi area has recorded the isothermal decompression metamorphism during the orogenic extension stage at ~1890 Ma (Jiao et al. 2013a, b).

3.2 *Daqingshan Region*

(Ultra-)high-temperature granulites in the Daqingshan region are reported by Jin (1989), and Liu et al. (1993a, b). Guo et al. (2006, 2012) identified the outcrop and defined its (ultra-)high-temperature metamorphism. The sapphirine granulites from the Daqingshan region are typical silica-undersaturated residual granulites and contain up to 30 % sapphirine, garnet (30–50 %), spinel (5–15 %), sillimanite (5–15 %), biotite (10–20 %), and plagioclase (10–20 %) with minor cordierite, rutile, and ilmenite, but without quartz and orthopyroxene. The sapphirine granulite is always in contact with a gabbronorite sill.

The silica-undersaturated Spr-bearing granulites in the Daqingshan region are located in the easternmost part of the Daqingshan–Ulashan Complex, which is separated from the Archean Wuchuan Complex by the Jiuguan–Xiashihao Fault (Fig. 21). The most typical outcrop is exposed in Dongpo village (Fig. 25). The dominant rocks are quartzo-feldspathic gneisses, in which granulite facies metapelites and associated marbles, and calc-silicate rocks occur as thick layers, some of which extend up to 20 km (Fig. 25; Guo et al. 2012). The sapphirine granulites occur as a 3–5-m-thick layer within the hosting quartzo-feldspathic gneiss (Fig. 26). There is a ~10-m-wide meta-gabbroic sill between the sapphirine granulite layer and the quartzo-feldspathic gneiss, which is composed of Pl + Cpx + Opx + Hb (Fig. 26). This sill even preserves its igneous texture, as indicated by tabular, subhedral plagioclase, and clinopyroxene crystals. The Spr-bearing granulite layer is separated from the quartzo-feldspathic gneiss by a 1-m-thick quartz +plagioclase vein (Fig. 26). Smaller rootless quartz +plagioclase veins are also found within the Spr-bearing granulite layer. Both types of quartz +plagioclase veins show a slightly foliation defined by elongated quartz and plagioclase; they may have been derived from the partial melting of the pelitic granulites (Guo et al. 2012).

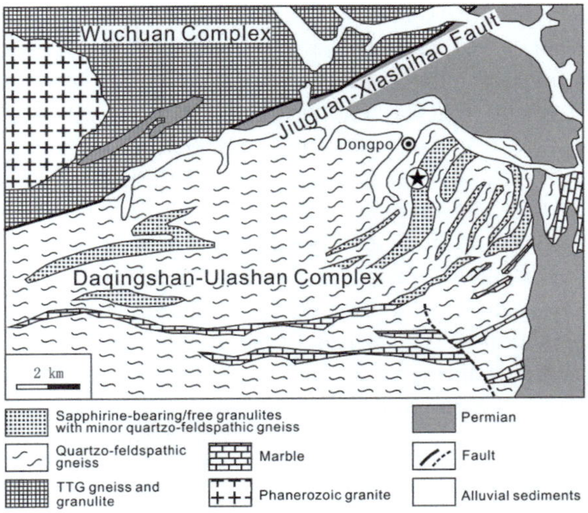

Fig. 25 Geological sketch map around Dongpo Village, Daqingshan, Inner Mongolia (from Guo et al. 2012)

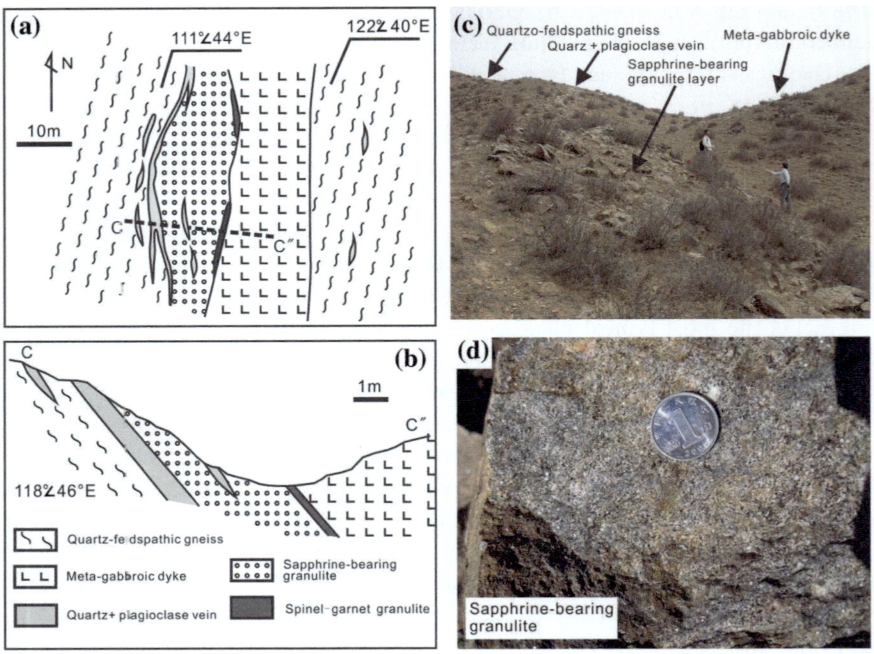

Fig. 26 A geological map and field photographs showing the occurrences of the Spr-bearing UHT granulite and associated rocks including a meta-gabbroic dyke in the Daqingshan region (Guo et al. 2012)

The Spr-bearing granulites can be divided into three types of lithologies: spinel–garnet–sillimanite–biotite–plagioclase–sapphirine gneiss, sapphirine granulite which is the dominant type, and spinel–garnet granulite that occurs closely to the meta-gabbroic dyke as a ~10-cm-wide reaction rim. According to detailed petrographic studies, several metamorphic stages (M1–M4; Fig. 28) that indicate an isothermal decompression are revealed. The most typical stage is the formation of Spr +Pl symplectite among Grt poikiloblast, Bt, and Sil, which is then replaced by late Spl +Pl symplectite and/or Crd corona (Fig. 27; Guo et al. 2012). Phase equilibria modeling in the NCKFMASH system using the bulk rock composition of Spr-bearing ultra-high-temperature granulite reveals a *P-T* path involving the peak ultra-high-temperature (910–980 °C) metamorphism (M1) followed by near-isothermal decompression (M2 and M3) and cooling (M4) (Fig. 28). It suggests the granulites underwent rapid exhumation or uplift following the peak metamorphism. Therefore, it is likely that the iso-extensional underplating or intrusion of the mantle-derived magma has heated up the preexisting melting residual in the lower crust and caused the ultra-high-temperature metamorphism (Guo et al. 2012).

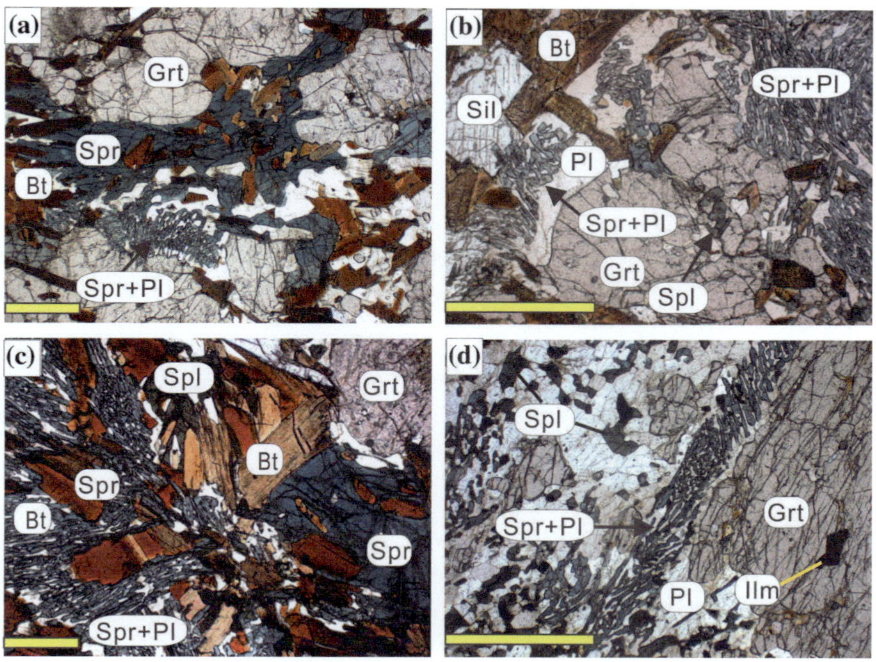

Fig. 27 Photomicrographs showing Spr +Pl symplectite between Grt and Sil from the Spr-bearing UHT granulites in the Daqingshan region (Guo et al. 2012)

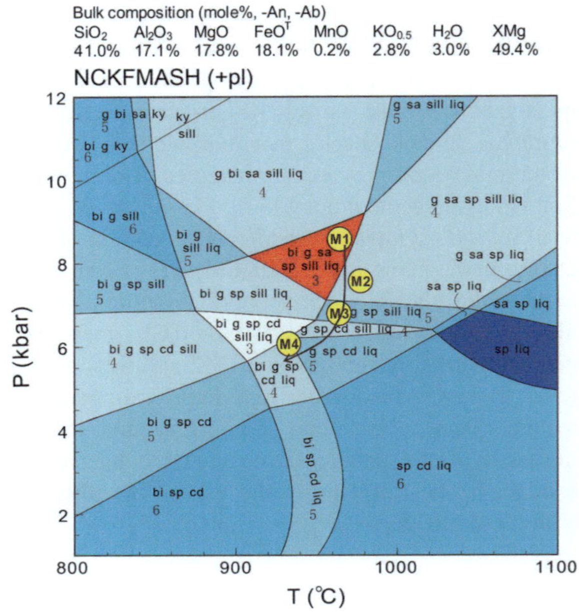

Fig. 28 *P-T* pseudosection and *P-T* path for the Spr-bearing UHT granulites in the Daqingshan region (Guo et al. 2012)

4 Geological Implications of the Paleoproterozoic Granulite

On the basis of integration of lithological, metamorphic, structural, and geochronologic data, tectonic model for the whole of the NCC has been proposed in many ways. Continental collision models have been invoked to interpret the genesis of the Paleoproterozoic mafic high-pressure granulites and retrograde eclogites in the NCC even in the earliest investigations (Zhai et al. 1993, 1996; Guo et al. 1993). That is strongly supported by the clockwise *P-T* paths with relatively high peak metamorphic pressures at 12–16 kbar, and significant isothermal decompression indicated by the widespread symplectic plagioclase and pyroxenes and coronitic plagioclase ('white-socket' texture) around the coarse-grained garnet porphyroblasts. A coherent terrain with many mafic high-pressure granulite outcrops in the Hengshan-Huai'an region has been considered to represent the lower crust of the NCC (Zhai et al. 2001). Zhai et al. (2005) and Zhai (2009) have discussed the possibility of mantle upwelling during the exhumation of the high-grade granulite.

The sapphirine-bearing ultra-high-temperature granulites with metamorphic temperatures over 900 °C have been identified in the metapelites in the west part of the NCC (Jin et al. 1991; Lu et al. 1992; Lu and Jin 1993; Liu et al. 1993a, b, 2000; Guo et al. 2006, 2012; Santosh et al. 2007a, 2009; Jiao and Guo 2011; Jiao et al. 2011; Liu et al. 2012). Their metamorphic temperatures are 100–200 °C higher than the surrounding granulites. Field relations and geochronological data suggest that the UHT metamorphism is cogenetic with the coeval 1.93–1.91 Ga Xuwujia gabbronorites, which represent a high-magnesium mantle-derived magmatism generated probably by ridge subduction (Peng et al. 2010, 2012; Guo et al. 2012).

Nevertheless, several linear tectonic belts have been defined, i.e., the Jinyu Mobile Belt/Central Zone/Trans-North China Orogen/Sanggan Belt, the Fengzhen Mobile Belt/Khondalite Belt/Inner Mongolia Suture Zone, and the Jiao-Liao-Ji (Liaoji) Mobile/Orogenic Belt (e.g., Zhai et al. 1993, 2005; Guo et al. 2005, 2012; Zhao et al. 2000, 2001a, b, 2005; Li et al. 1998, 2000; Kusky et al. 2001, 2007; Kusky and Li 2003; Zhai and Liu 2003; Zhai and Peng 2007; Kröner et al. 2005a, b; Liu et al. 2005; Santosh et al. 2007a, b, 2012, 2015; Trap et al. 2007, 2009a, b, 2012; Zhai and Santosh 2011; Zhao and Zhai 2013; Peng et al. 2014). Each belt contains both mafic and pelitic high-pressure granulites with clockwise isothermal decompressional *P-T* paths indicating a collisional-like process in the final stage. However, there are several things need to be clarified before any interpretation, and these include (1) the spatial–temporal distribution of the two types of high-grade rocks; (2) the relationship between the high-pressure granulites and the (ultra-) high-temperature granulite; and (3) the exhumation rate and mechanism of the rocks (Zhai et al. 2005; Zhao et al. 2005; Kusky et al. 2007; Zhai 2009; Zhai and Santosh 2011; Zhao and Zhai 2013; Peng et al. 2014). These Paleoproterozoic high-grade rocks in the North China Craton are more complex than any high-grade rocks in the Phanerozoic orogens, and await further study.

References

Chang XY, Chen YD, Zhu BQ (1999) U-Pb zircon isotope age of metabasites from Hengshan grey gneiss. Acta Min Sin 19:263–266 (in Chinese with English abstract)

Green DH, Ringwood AE (1967) An experimental investigation of the gabbro to eclogite transformation and its petrological applications. Geochimica Cosmochimica Acta 31:767–833

Guo J, Peng P, Chen Y, Jiao S, Windley BF (2012) UHT sapphirine granulite metamorphism at 1.93–1.92 Ga caused by gabbronorite intrusions: Implications for tectonic evolution of the northern margin of the North China Craton. Precambr Res 222–223:124–142

Guo JH, O'Brien PJ, Zhai MG (2002) High pressure granulites in the Sanggan area, North China craton: metamorphic evolution, *P-T* paths and geotectonic significance. J Metamorph Geol 20:741–756

Guo JH, Sun M, Chen FK, Zhai MG (2005) Sm-Nd and SHRIMP U-Pb zircon geochronology of high pressure granulites in the Sanggan area, North China Craton: timing of Paleoproterozoic continental collision. J Asian Earth Sci 24:629–642

Guo JH, Wang SS, Sang HQ, Zhai MG (2001) ^{40}Ar-^{39}Ar age spectra of garnet porphyroblast: Implications for metamorphic age of high pressure granulite in the North China Craton. Acta Petrologica –Sinica 17:436–442 (in Chinese with English abstract)

Guo JH, Zhai MG, Zhang YG, Li YG, Yan YH, Zhang WH (1993) Early Precambrian Manjinggou high pressure granulite melange belt on the south edge of the Huaian complex, North China craton: geological features, petrology and isotopic geochronology. Acta Petrologica Sinica 9:329–341 (in Chinese with English abstract)

Guo JH, Chen Y, Peng P, Liu F, Chen L, Zhang LQ (2006) Sapphirine granulite from Daqingshan area, Inner Mongolia –1.85 Ga ultrahigh temperature (UHT) metamorphism. In: Proceedings of national conference on petrology and geodynamics in China (Nanjing), pp 215–218

Harley SL (1998) On the occurrence and characterization of ultrahigh temperature crustal metamorphism. In: Treloar PJ, O'Brien PJ (Eds) What Drives Metamorphism and Metamorphic Reactions? Special Publication Geological Society of London, pp 81–107
Harley SL (2008) Refining the *P-T* records of UHT crustal metamorphism. J Metamorph Geol 26:125–154
He GP, Ye HW (1998) Two types of early Proterozoic Metamorphism and its tectonic signigicance in eastern Liaining and southern Jilin areas. Acta Petrologica Sinica 14:25–35 (In Chinese)
Holland TJB (1980) Reaction Albite = Jadeite + Quartz Determined Experimentally in the Range 600-1200-Degrees-C. Am Mineral 65(1–2):129–134
Jia XM (2007) Study on the Mineralogy and Chronology of theNianzigou Rutile (PhD thesis)
Jiao S, Guo J, Harley SL, Windley BF (2013a) New constraints from Garnetite on the P–T Path of the Khondalite Belt: implications for the tectonic evolution of the North China Craton. J Petrol 54:1725–1758
Jiao S, Guo J, Mao Q, Zhao R (2011) Application of Zr-in-rutile thermometry: a case study from ultrahigh temperature granulites of the Khondalite belt, North China Craton. Contrib Miner Petrol 162:379–393
Jiao SJ, Guo JH (2011) Application of the two-feldspar geothermometer to ultrahigh temperature (UHT) rocks in the Khondalite belt, North China craton and its implications. Am Mineral 96:250–260
Jiao SJ, Guo JH, Harley SL, Peng P (2013b) Geochronology and trace element geochemistry of zircon, monazite and garnet from the garnetite and/or associated other high-grade rocks: Implications for Paleoproterozoic tectonothermal evolution of the Khondalite Belt, North China Craton. Precambr Res 237:78–100
Jin W (1989) Early Precambrian geological evolution and metamorphic study on the basement of the north margin the North China platform. PhD thesis of the Changchun College of Geology, China (in Chinese)
Jin W, Li SX, Liu XS (1991) The metamorphic dynamics of early Precambrian high-grade metamorphic rocks series in Daqing-Ulashan area, Inner Mongolia. Acta Petrologica Sinica 4:27–35 (in Chinese with English abstract)
Kelsey DE (2008) On ultrahigh temperature crustal metamorphism. Gondwana Res 13:1–29
Kröner A, Wilde SA, Li JH, Wang KY (2005a) Age and evolution of a late-Archean to early Palaeozoic upper to lower crustal section in the Wutais-han/Hengshan/Fuping terrain of northern China. J Asian Earth Sci 24:577–595
Kröner A, Wilde SA, O'Brien PJ, Li JH, Passchier CW, Walte NP, Liu DY (2005b) Field relationships, geochemistry, zircon ages and evolution of a late Archean to Paleoproterozoic lower crustal section in the Hengshan Terrain of Northern China. Acta Geol Sin (Engl Ed) 79:605–629
Kusky TM, Li J-H, Tucker RT (2001) The Archean Dongwanzi ophiolite complex, North China Craton: 2.505 billion year old oceanic crust and mantle. Science 292:1142–1145
Kusky TM, Li J-H (2003) Paleoproterozoic tectonic evolution of the North China Craton. J Asian Earth Sci 22:383–397
Kusky TM, Li J-H, Santosh M (2007) The Paleoproterozoic North Hebei Orogen: North China Craton's collisional suture with the Columbia supercontinent. Gondwana Res 12:4–28
Li SZ, Han ZZ, Liu YJ, Yang ZS, Ma R (2001) Continental dynamics and regional metamorphism of the Liaohe Group. Geological Rev 47:9–18 (in Chinese)
Li YG, Zhai MG, Liu WJ, Guo JH (1997) Sm-Nd geochronology of the high pressure basic granulite, in Laixi, eastern Shandong. Scientia Geologica Sinica 32:283–290 (In Chinese)
Li JH, Qian XL (1994) The Early Precambrian Crustal Evolution of Hengshan Metamorphic Terrain, North China Craton. Shanxi Science and Technology Press, China, pp 1–116 (in Chinese with English abstract)
Li J-H, Qian X-L, Huang X-N, Liu S-W (2000) Tectonic framework of North China block and its cratonization in the Early Precambrian. Acta Petrologica Sinica 16:1–10 (in Chinese with English abstract)

Li JH, Zhai MG, Qian XL, Guo JH (1998) The geological occurrence, regional tectonic setting and exhumation of late Archean high pressure granulite with the highgrade metamorphic terranes, north to central portion of North China craton. Acta Petrologica Sinica 14:176–189 (in Chinese with English abstract)

Li S, Zhao G (2007) SHRIMP U-Pb zircon geochronology of the Liaoji granitoids: Constraints on the evolution of the Paleoproterozoic Jiao-Liao-Ji belt in the Eastern Block of the North China Craton. Precambr Res 158:1–16

Li S, Zhao G, Sun M, Han Z, Zhao G, Hao D (2006) Are the South and North Liaohe Groups of North China Craton different exotic terranes? Nd isotope constraints. Gondwana Res 9:198–208

Liu WJ, Zhai MG, Li YG (1998) Metamorphism of the high pressure basic granulites in Laixi, eastern Shandong, China. Acta Petrologica Sinica 14:449–459 (In Chinese)

Liu JZ, Qiang XK, Liu XS, Ouyang ZY (2000) Dynamics and genetic grids of sapphirine-bearing spinel gneiss in Daqing Mountain orogen zone, Inner Mongolia. Acta Petrologica Sinica 16:245–255 (in Chinese with English abstract)

Liu P, Liu F, Liu C, Wang F, Liu J, Yang H, Cai J, Shi J (2013) Petrogenesis, P-T–t path, and tectonic significance of high pressure mafic granulites from the Jiaobei terrane, North China Craton. Precambr Res 233:237–258

Liu SW, Liang HH (1997) Metamorphism of Al-rich gneisses from the Fuping Complex, Taihang Mountain, China. Acta Petrologica Sinica 13:303–312

Liu S, Bai X, Li J, Santosh M (2011) Retrograde metamorphism of ultrahigh temperature granulites from the khondalite belt in Inner Mongolia, North China Craton: evidence from aluminous orthopyroxenes. Geol J 46:263–275

Liu SJ, Li JH, Santosh M (2010) First application of the revised Ti-in-zircon geothermometer to Paleoproterozoic ultrahigh temperature granulites of Tuguiwula, Inner Mongolia, North China Craton. Contrib Miner Petrol 159:225–235

Liu SJ, Tsunogae T, Li WS, Shimizu H, Santosh M, Wan YS, Li JH (2012) Paleoproterozoic granulites from Heling'er: Implications for regional ultrahigh temperature metamorphism in the North China Craton. Lithos 148:54–70

Liu SW (1996) Study on the *P-T* path of granulites in Fuping area, Hebei province. Geol J Univ 2:75–84

Liu S-W, Pan Y-M, Xie Q-L, Zhang J, Li Q-G, Yang B (2005) Geochemistry of the Paleoproterozoic Nanying granitic gneisses in the Fuping Complex: implications for the tectonic evolution of the Central Zone, North China Craton. J Asian Earth Sci 24:643–658

Liu SW, Zhao GC, Wilde SA, Shu GM, Sun M, Li QG, Tian W, Zhang J (2006) Th–U–Pb monazite geochronology of the Lüliang and Wutai Complexes: constrains on the tectonothermal evolution of the Trans-North China Orogen. Precambr Res 148:205–225

Liu XS, Jin W, Li SX, Xu XC (1993a) Two types of Precambrian high-grade metamorphism, Inner Mongolia. China. J Metamorph Geol 11:499–510

Liu XS, Jin W, Li SX, Xu XC (1993b) Two types of Precambrian high-grade metamorphism, Inner Mongolia. China. J Metamorph Geol 11:499–510

Lu J-S, Wang G-D, Wang H, Chen H-X, Wu C-M (2013) Metamorphic *P-T*-t paths retrieved from the amphibolites, Lushan terrane, Henan Province and reappraisal of the Paleoproterozoic tectonic evolution of the Trans-North China Orogen. Precambr Res 238:61–77

Lu L, Xu X, Liu F (1996) Early Precambrian Khondalite series in North China. Changchun Publishing House, Changchun, pp 1–277 (In Chinese)

Lu LZ, Jin SQ (1993) *P-T*-t paths and tectonic history of an early Precambrian granulite facies terrane, Jining district, southeastern Inner Mongolia. China. J Metamorph Geol 11:483–498

Lu LZ, Jin SQ, Xu XT, Liu FL (1992) Petrogenesis and Mineralization of Khondalite Series in Southeastern Inner Mongolia. Jilin Science and Technology Press, Changchun, China. 156p (in Chinese with English abstract)

Lu SN, Zhao GC, Wang HC, Hao GJ (2008) Precambrian metamorphic basement and sedimentary cover of the North China Craton: A review. Precambr Res 160:77–93

Luo Y, Sun M, Zhao G, Li S, Ayers JC, Xia X, Zhang J (2008) A comparison of U-Pb and Hf isotopic compositions of detrital zircons from the North and South Liaohe Groups: Constraints on the evolution of the Jiao-Liao-Ji Belt, North China Craton. Precambr Res 163:279–306

Luo Y, Sun M, Zhao G, Li S, Xu P, Ye K, Xia X (2004) LA-ICP-MS U-Pb zircon ages of the Liaohe Group in the Eastern Block of the North China Craton: constraints on the evolution of the Jiao-Liao-Ji Belt. Precambr Res 134:349–371

O'Brien PJ, Walte N, Li JH (2005) The petrology of two distinct granulite types in the Hengshan Mts, China, and tectonic implications. J Asian Earth Sci 24:615–627

Pang EC, Xu YJ, Shi GH, Jia M, Zhang ZX (2010) Geochemistry and chronologyof Hongtang ore-bearing rocks in the Daixian rutile deposit, Shanxi province. Acta Petrol Mineral 29:497–506 (in Chinese with English abstract)

Peng P, Guo JH, Windley BF, Liu F, Chu Z, Zhai MG (2012) Petrogenesis of Late Paleoproterozoic Liangcheng charnockites and S-type granites in the central-northern margin of the North China Craton: Implications for ridge subduction. Precambr Res 222–223:107–123

Peng P, Guo JH, Zhai MG, Bleeker W (2010) Paleoproterozoic gabbronoritic and granitic magmatism in the northern margin of the North China craton: Evidence of crust-mantle interaction. Precambr Res 183:635–659

Peng P, Wang X, Windley BF, Guo J, Zhai M, Li Y (2014) Spatial distribution of ~1950–1800 Ma metamorphic events in the North China Craton: Implications for tectonic subdivision of the craton. Lithos 202–203:250–266

Peng P, Zhai MG, Zhang HF, Guo JH (2005) Geochronological constraints on the Paleoproterozoic evolution of the North China Craton: SHRIMP zircon ages of different types of mafic dikes. Int Geol Rev 47:492–508

Qian JH, Wei CJ, Clarke GL, Zhou XW (2015) Metamorphic evolution and Zircon ages of garnet-orthoamphibole rocks in southern Hengshan, North China Craton: insights into the regional Paleoproterozoic P-T-t history. Precambr Res 256:223–240

Qian JH, Wei CJ, Zhou XW, Zhang YH (2013) Metamorphic P-T paths and new zircon U–Pb age data for garnet-mica schist from the Wutai Group, North China Craton. Precambr Res 233:282–296

Salje E (1986) Heat-Capacities and Entropies of Andalusite and Sillimanite - the Influence of Fibrolitization on the Phase-Diagram of the Al_2SiO_5 Polymorphs. Am Mineral 71(11–12): 1366–1371

Santosh M, Liu SJ, Tsunogae T, Li JH (2012) Paleoproterozoic ultrahigh temperature granulites in the North China Craton: Implications for tectonic models on extreme crustal metamorphism. Precambr Res 222–223:77–106

Santosh M, Sajeev K, Li JH (2006) Extreme crustal metamorphism during Columbia supercontinent assembly: Evidence from North China Craton. Gondwana Res 10:256–266

Santosh M, Sajeev K, Li JH, Liu SJ, Itaya T (2009) Counterclockwise exhumation of a hot orogen: The Paleoproterozoic ultrahigh temperature granulites in the North China Craton. Lithos 110:140–152

Santosh M, Tsunogae T, Li JH, Liu SJ (2007a) Discovery of sapphirine-bearing Mg-Al granulites in the North China Craton: Implications for Paleoproterozoic ultrahigh temperature metamorphism. Gondwana Res 11:263–285

Santosh M, Wilde SA, Li JH (2007b) Timing of Paleoproterozoic ultrahigh temperature metamorphism in the North China Craton: Evidence from SHRIMP U-Pb zircon geochronology. Precambr Res 159:178–196

Santosh M, Yang Q-Y, Teng X, Tang L (2015) Paleoproterozoic crustal growth in the North China Craton: Evidence from the Lüliang Complex. Precambr Res 263:197–231

Shi GH, Li XH, Li QL, Chen ZY, Deng J, Liu YX, Kang ZJ, Pang EC, Xu YJ, Jia XM (2012) Ion Microprobe U-Pb Age and Zr-in-Rutile Thermometry of Rutilefrom the Daixian Rutile Deposit in the Hengshan Moutains, Shanxi Province. China Econ Geol 107:525–535

Spear FS (1993) Metamorphic Phase Equilibria and Pressure-Temperature-Time Paths. Mineralogical Society of America, Washington DC, Monograph Series

Tam PY, Zhao G, Liu F, Zhou X, Sun M, Li S (2011) Timing of metamorphism in the Paleoproterozoic Jiao-Liao-Ji Belt: new SHRIMP U-Pb zircon dating of granulites, gneisses and marbles of the Jiaobei massif in the North China Craton. Gondwana Res 19:150–162
Tam PY, Zhao G, Sun M, Li S, Wu M, Yin C (2012a) Petrology and metamorphic P-T path of high pressure mafic granulites from the Jiaobei massif in the Jiao-Liao-Ji Belt, North China Craton. Lithos 155:94–109
Tam PY, Zhao G, Zhou X, Sun M, Guo J, Li S, Yin C, Wu M, He Y (2012b) Metamorphic P-T path and implications of high pressure pelitic granulites from the Jiaobei massif in the Jiao-Liao-Ji Belt, North China Craton. Gondwana Res 22:104–117
Tian YQ, Ma ZH, Yu KR, LiuZ.H., Peng, Q.M., 1996. The early Precambrian geology of Wutai-Hengshan Mts., Shanxi, China. Field Trip Guide T315, 30th International Geological Congress, Beijing, China, Geological Publishing House, Beijing: 52
Trap P, Faure M, Lin W, Breton NL, Monié P (2012) Paleoproterozoic tectonic evolution of the Trans-North China Orogen: toward a comprehensive model. Precambr Res 222:191–211
Trap P, Faure M, Lin W, Meffre S (2009a) The Luliang Massif: a key area for the understanding of the Paleoproterozoic Trans-North China Belt, North China Craton. Geological Society, London, Special Publications 323:99–125
Trap P, Faure M, Lin W, Monié P (2007) Late Paleoproterozoic (1900–1800 Ma) nappe-stacking and polyphase deformation in the Hengshan-Wutaishan area: implications for the understanding of the Trans-North China Belt, North China Craton. Precambr Res 156:85–106
Trap P, Faure M, Lin W, Monié P, Meffre S, Melleton J (2009b) The Zanhuang Massif, the second and eastern suture zone of the Paleoproterozoic Trans-North China Orogen. Precambr Res 172:80–98
Wang J, Wu YB, Gao S, Peng M, Liu XC, Zhao LS, Zhou L, Hu ZC, Gong HJ, Liu YS (2010) Zircon U-Pb and trace element data from rocks of the Huai'an Complex: New insights into the late Paleoproterozoic collision between the Eastern and Western Blocks of the North China Craton. Precambr Res 178:59–71
Wang RM, Chen ZZ, Chen F (1991) Grey gneisses and high pressure granulite enclaves in Hengshan area, Shanxi Province, and their geological significance. Acta Petrologica Sinica 7:36–45 (in Chinese)
Wang Y (2003) Structural evolution and 40Ar/39Ar dating of the Zanhuang metamorphic domain in the North China Craton: constraints on Paleoproterozoic tectonothermal overprinting. Precambr Res 122:159–182
Xiao L-L, Wang G-D, Wang HAO, Jiang Z-S, Diwu C-R, Wu C-M (2013) Zircon U-Pb geochronology of the Zanhuang metamorphic complex: reappraisal of the Paleoproterozoic amalgamation of the Trans-North China Orogen. Geol Mag 150:756–764
Xiao L-L, Wu C-M, Zhao G-C, Guo J-H, Ren L-D (2010) Metamorphic P-T paths of the Zanhuang amphibolites and metapelites: constraints on the tectonic evolution of the Paleoproterozoic Trans-North China Orogen. Int J Earth Sci 100:717–739
Yin C, Zhao G, Guo J, Sun M, Xia X, Zhou X, Liu C (2011) U-Pb and Hf isotopic study of zircons of the Helanshan Complex: Constrains on the evolution of the Khondalite Belt in the Western Block of the North China Craton. Lithos 122:25–38
Yin C, Zhao G, Sun M, Xia X, Wei C, Zhou X, Leung W (2009) LA-ICP-MS U-Pb zircon ages of the Qianlishan Complex: Constrains on the evolution of the Khondalite Belt in the Western Block of the North China Craton. Precambr Res 174:78–94
Yin C, Zhao G, Wei C, Sun M, Guo J, Zhou X (2014) Metamorphism and partial melting of high pressure pelitic granulites from the Qianlishan complex: constraints on the tectonic evolution of the Khondalite Belt in the North China Craton. Precambr Res 242:172–186
Zhai MG, Guo JH, Yan YH, Li YG, Han XL (1993) Discovery of high pressure basic granulite terrain in North China Archean craton and preliminary study. Sci China (Ser B) 36:1402–1408
Zhai MG (1997) Recent advances in the study of granulites from the North China craton. Int Geol Rev 39:325–341
Zhai MG, Guo JH, Li JH, Li YG, Yan YH, Zhang WH (1996) Retrograded eclogites in the Archean North China craton and their geological implication. Chinese Sci Bull 41:315–326

Zhai MG, Guo JH, Liu WJ (2001) An exposed cross-section of Early Precambrian continental lower crust in North China craton. Physics and Chemistry of the Earth, Part A: Solid Earth and Geodesy 26:781–792

Zhai MG, Guo JH, Liu WJ (2005) Neoarchean to Paleoproterozoic continental evolution and tectonic history of the North China Craton: a review. J Asian Earth Sci 24:547–561

Zhai M-G, Guo J-H, Yan Y-H, Li Y-G (1992a) The discovery of high pressure basic granulite in the Archean North China Craton and preliminary study. Sci China (Ser B) 12:1325–1300 (in Chinese)

Zhai MG, Guo JH, Yan YH, Li YG, Han XL (1992) Discovery of high pressure basic granulite terrain in North China craton and preliminary study. Sci China(B) 12:1325–1330

Zhai M-G, Liu W-J (2003) Paleoproterozoic tectonic history of the North China Craton: a review. Precambr Res 122:183–199

Zhai M-G, Peng P (2007) Paleoproterozoic events in the North China Craton. Acta Petrologica Sinica 23:2665–2682 (in Chinese with English abstract)

Zhai MG, Santosh M (2011) The early Precambrian odyssey of North China Craton: a synoptic overview. Gondwana Res 20:6–25

Zhai MG (2009) Two kinds of granulites (HT-HP and HT-UHT) in North China Craton: their genetic relation and geotectonic implications. Acta Petrologica Sinica 25(8):1553–1571 (in Chinese with English abstract)

Zhang HF, Zhai MG, Santosh M, Wang HZ, Zhao L, Ni ZY (2014) Paleoproterozoic granulites from the Xinghe graphite mine, North China Craton: Geology, zircon U-Pb geochronology and implications for the timing of deformation, mineralization and metamorphism. Ore Geol Rev 63:478–497

Zhao GC, Cawood PA, Wilde SA, Lu LZ (2001a) High pressure granulites (retrograded eclogites) from the Hengshan Complex, North China Craton: Petrology and tectonic implications. J Petrol 42:1141–1170

Zhao GC, Wilde SA, Cawood PA, Sun M (2002) SHRIMP U-Pb zircon ages of the Fuping Complex: Implications for late Archean to Paleoproterozoic accretion and assembly of the North China Craton. Am J Sci 302:191–226

Zhao GC, Cawood PA, Lu LZ (1999) Petrology and *P-T* history of the Wutai amphibolites: implications for tectonic evolution of the Wutai Complex, China. Precambr Res 93:181–199

Zhao GC, Cawood PA, Wilde SA, Sun M, Lu L (2000) Metamorphism of basement rocks in the Central Zone of the North China Craton: implications for Paleoproterozoic tectonic evolution. Precambr Res 103:55–88

Zhao GC, Cawood PA, Wilde SA, Sun M, Zhang J, He YH, Yin CQ (2012) Amalgamation of the North China Craton: key issues and discussion. Precambr Res 222–223:55–76

Zhao G-C, Wilde SA, Cawood PA, Sun M (2001b) Archean blocks and their boundaries in the North China Craton: lithological, geochemical, structural and *P-T* path constraints and tectonic evolution. Precambr Res 107:45–73

Zhao GC, Sun M, Wilde SA, Li SZ (2005) Late Archean to Paleoproterozoic evolution of the North China Craton: key issues revisited. Precambr Res 136:177–202

Zhao GC, Wilde SA, Sun M, Guo JH, Kroener A, Li SZ, Li XP, Zhang J (2008) Shrimp U-Pb zircon geochronology of the Huai'an Complex: Constraints on late Archean to paleoproterozoic magmatic and metamorphic events in the Trans-North China Orogen. Am J Sci 308:270–303

Zhao GC, Zhai MG (2013) Lithotectonic elements of Precambrian basement in the North China Craton: Review and tectonic implications. Gondwana Res 23:1207–1240

Zhao L, Li T, Peng P, Guo J, Wang W, Wang H, Santosh M, Zhai M (2014) Anatomy of zircon growth in high pressure granulites: SIMS U-Pb geochronology and Lu-Hf isotopes from the Jiaobei Terrane, eastern North China Craton. Gondwana Res. doi:10.1016/j.gr.2014.1010.1009

Zhou XW, Zhao GC, Geng YS (2010) Helanshan high pressure pelitic granulite: petrologic evidence for collision event in the western block of the North China Craton. Acta Petrologica Sinica 26:2113–2121 (In Chinese)
Zhou X, Wei C, Geng Y, Zhang L (2004) Discovery and implications of the high pressure pelitic granulite from the Jiaobei massif. Chin Sci Bull 49:1942–1948
Zhou X, Zhao G, Wei C, Geng Y, Sun M (2008) EPMA U-Th-Pb monazite and SHRIMP U-Pb zircon geochronology of high pressure pelitic granulites in the Jiaobei massif of the North China Craton. Am J Sci 308:328–350

Late Paleoproterozoic–Neoproterozoic (1800–541 Ma) Mafic Dyke Swarms and Rifts in North China

Peng Peng

Abstract The Late Paleoproterozoic–Neoproterozoic (1800–541 Ma) sedimentation, volcanism and igneous events, and rifting in the North China Craton can be categorized into different stages: (1) Stage 1 (1800–1730 Ma, Early Statherian): It is represented by the Taihang (1780–1770 Ma) and Miyun ($\sim$1730 Ma) dyke swarms, the Xiong'er volcanic province ($\sim$1780 Ma), and a little clastic sedimentation related to the evolution of the Xiong'er rift. (2) Stage 2 (1730–1600 Ma, Late Statherian): It is represented by the Damiao–Shachang anorthosite–rapakivi granite–dyke complexes (1710–1680 Ma), the Laiwu ($\sim$1680 Ma) and Taishan ($\sim$1620 Ma) dyke swarms, the Dahongyu lavas ($\sim$1620 Ma), and some clastic rocks and a little limestone related to the Yan-Liao rift, the Xiong'er rift, and probably the Bayan Obo rift. (3) Stage 3 (1600–1400 Ma, Calymmian/Jixian): It is represented by the dolomite-dominated strata in the Yan-Liao rift, the Xiong'er rift, and probably the Bayan Obo rift, with little volcanism and magmatism. (4) Stage 4 (1400–1200 Ma, Ectasian): It is represented by the $\sim$1320 Ma dykes–sills–granites and $\sim$1230 Ma Licheng dyke swarm, and the sedimentation dominated by clastic rocks with a little marlstone and limestone possibly limited in the Yan-Liao rift. (5) Stage 5 (1200–1000 Ma, Stenian): It is unclear whether there were any magmatism and deposition during this period. (6) Stage 6 (1000–800 Ma, Tonian): It is represented by the Dashigou ($\sim$925 Ma) and Qianlishan ($\sim$810 Ma) dyke swarms, the Sariwon (Dalian–Chulan–Zenghekou) sills (925–890 Ma), and clastic rocks/carbonate-dominated sediments within the Xuhuai rift. (7) Stage 7 (800–541 Ma, Cryogenian–Ediacaran): It is uncertain yet if there were any igneous events or sedimentation during this period, although some strata in Henan province are candidates. Multiple stages of rifting indicate a prolonged and stepwise rifting lasting for 1000 Ma, with the center of the rifts shifted from the south (the Xiong'er rift, 1780–1730 Ma) to the north (the Yan-Liao rift, 1730–1200 Ma) and to the southeast (the Xuhuai rift, 1000–800 Ma) of the North China Craton.

P. Peng (✉)
State Key Laboratory of Lithospheric Evolution, Institute of Geology and Geophysics Chinese Academy of Sciences, 100029 Beijing, China
e-mail: pengpengwj@mail.iggcas.ac.cn

M. Zhai (ed.), *Precambrian Geology of China*,
Springer Geology, DOI 10.1007/978-3-662-47885-1_4

Keywords North China · Paleoproterozoic · Neoproterozoic · Mafic dyke swarm · Rift

It is well known that there was widespread metamorphism at 1950–1800 Ma in the North China Craton (e.g., Zhai et al. 1996; Zhao et al. 2001; Kusky and Li 2003; Guo et al. 2005, 2012; Liu et al. 2005; Kusky et al. 2007; Zhai and Peng 2007; Zhao 2014; Peng et al. 2014); and thus, ca. 1800 Ma was traditionally thought to be the timing between the Paleoproterozoic and the Mesoproterozoic in China; this is quite different from an international time framework (Fig. 1). There was a series of intra-continental rifts from 1800 Ma and intermittently continued to ca. 800 Ma (Fig. 2). In this chapter, the Late Paleoproterozoic to the Neoproterozoic (1800–541 Ma) strata, rift systems, and igneous events, mainly mafic dyke swarms, sills, and volcanic rocks in North China, will be summarized.

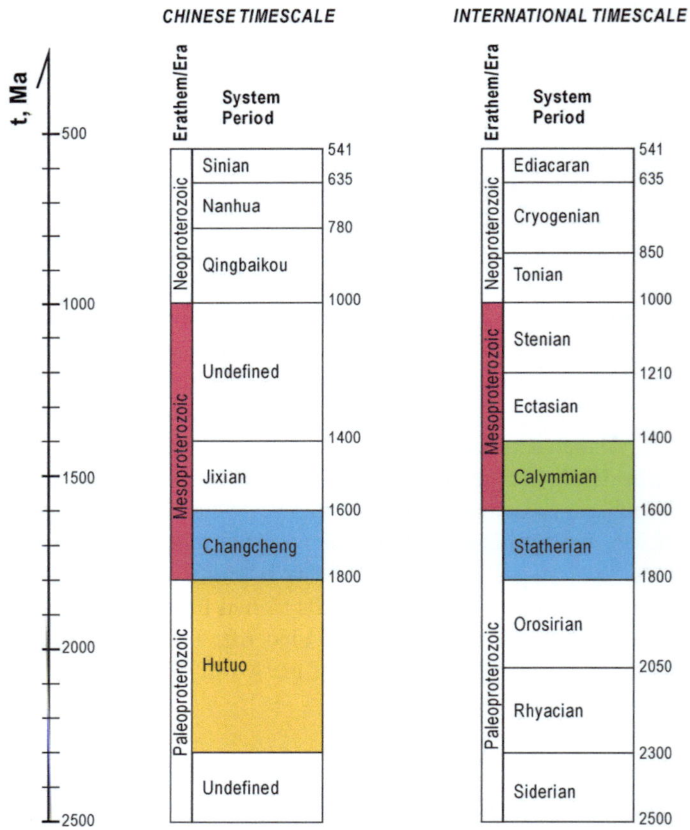

Fig. 1 Timescales of Proterozoic as released by the Chinese National Commission on Stratigraphy in 2009 and the International Commission on Stratigraphy in 2014

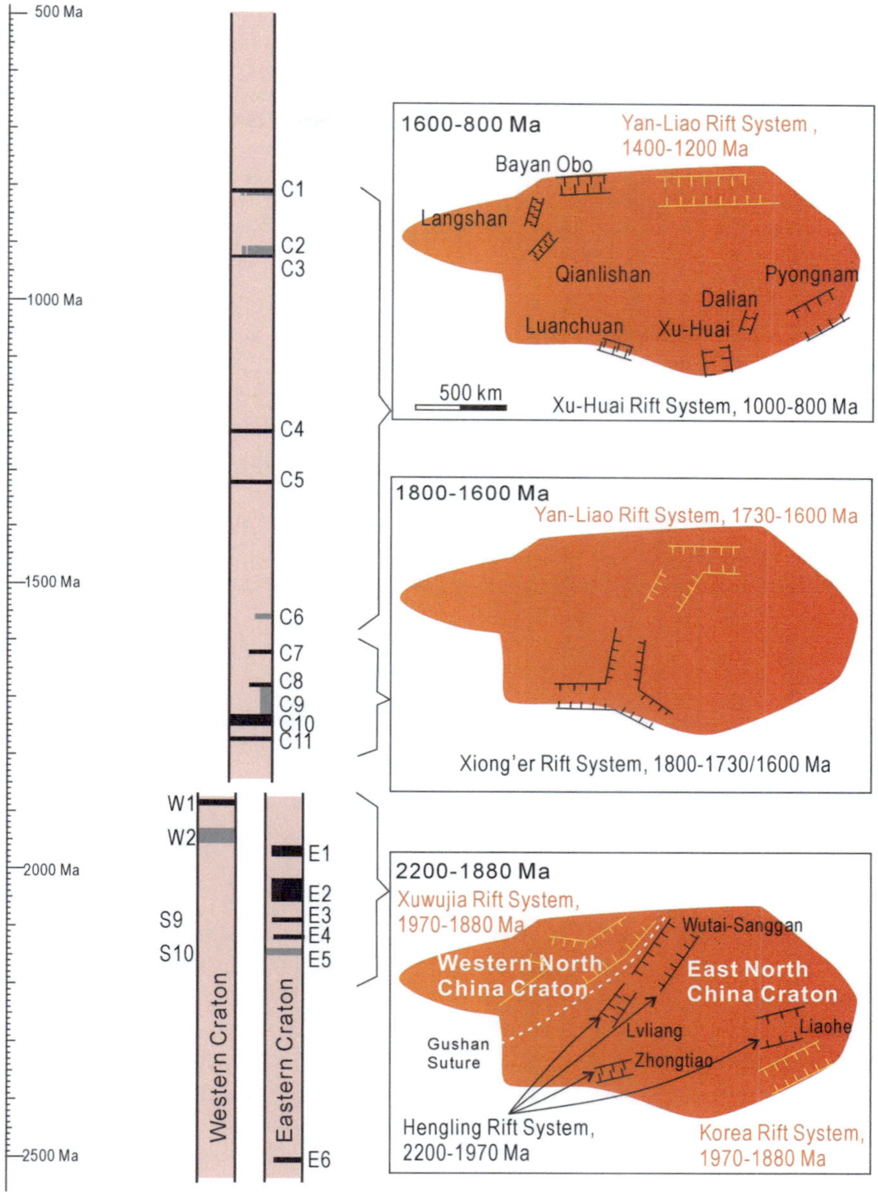
500 Ma
1000 Ma
1500 Ma
2000 Ma
2500 Ma
C1
C2
C3
C4
C5
C6
C7
C8
C9
C10
C11
W1
W2
S9
S10
Western Craton
Eastern Craton
E1
E2
E3
E4
E5
E6
1600-800 Ma
Yan-Liao Rift System , 1400-1200 Ma
Bayan Obo
Langshan
Qianlishan
Pyongnam
Dalian
Luanchuan
Xu-Huai
500 km
Xu-Huai Rift System, 1000-800 Ma
1800-1600 Ma
Yan-Liao Rift System, 1730-1600 Ma
Xiong'er Rift System, 1800-1730/1600 Ma
2200-1880 Ma
Xuwujia Rift System, 1970-1880 Ma
Wutai-Sanggan
Western North China Craton
East North China Craton
Lvliang
Liaohe
Gushan Suture
Zhongtiao
Hengling Rift System, 2200-1970 Ma
Korea Rift System, 1970-1880 Ma

◀ **Fig. 2** The 2200–800 Ma major igneous events and rift systems in North China. Note that the shape of the craton was reconstructed back to the Precambrian (the strike slip movement along the Tan-Lu fault was reconstructed). *C1* Qianlishan dykes and Langshan volcanics; *C2* Sariwon, Chulan, Dalian, and Luanchuan sills; *C3* Dashigou dykes; *C4* Licheng dykes; *C5* Datong dykes, Chaoyang, Chengde, and Liaoyang sills; *C6* Gaoyuzhuang volcanics/sills; *C7* Taishan dykes; *C8* Laiwu dykes; *C9* Damiao anorthosites, Shachang rapakivi granites; *C10* Miyun dykes; *C11* Taihang dykes, Xiong'er volcanics. (8) Western North China Craton: *W1* Halaqin dykes/volcanics; *W2* Xuwujia dykes. (9) Eastern North China Craton: *E1* Xiwangshan dykes; *E2* Yixingzhai dykes; *E3* Zanhuang sills; *E4* Haicheng sills; *E5* Hengling dykes/sills; *E6* Huangbaiyu dykes. After Peng (2015)

The Chinese Mesoproterozoic–lower Neoproterozoic stratotypes were mostly based on the Jixian Section in Tianjin City (Chen et al. 1980; Fig. 1), and this section comprises the Chinese Changcheng, Jixian, and Qingbaikou Systems with ages of 1800–1600, 1600–1000, and 1000–800 Ma, respectively. However, recent identification of some Mesoproterozoic aged sills in the Xiamaling Fm. (the Qingbaikou Group) has changed this time framework (Li et al. 2009; Zhang et al. 2009). In 2009, the newly published timescale by the Chinese National Commission on Stratigraphy had added an unnamed system between the Jixian and Qingbaikou Systems and reset the timing of the Jixian System to 1600–1400 Ma (Fig. 1).

1 Major Mafic Dyke Swarms and Other Igneous Events

1.1 The 1800–1730 Ma Igneous Events

1.1.1 The 1780–1770 Ma Taihang Dyke Swarm

The Taihang dyke swarm consists of hundreds of NNW-trended (315°–345°) dykes, as well as some E–W-oriented (250°–290°) mainly in the central parts of North China, with an area of >450 × 800 km^2 (Peng 2015). It is named after the Taihang Mts. where dykes are well outcropped (Qian and Chen 1987). The E–W-trended dykes, mainly in the Lvliang but fewer in the Wutai and Zhongtiao Mts., are sometimes grouped as the Lvliang swarm (Figs. 3 and 4). The individual dykes are up to 60 km long and 100 m wide, with typical widths of 10–40 m. These dykes were emplaced at 1780–1770 Ma (Halls et al. 2000; Peng et al. 2005; Han et al. 2007; Peng 2015). Reconstruction of some rotations among active blocks of North China during Mesozoic reveals a fanning geometry for the dykes with magma flow from the southern margin (Peng et al. 2008; Hou et al. 2008a, b). It has also been revealed that the dykes in the Yinshan area were uplifted and exhumed from crustal levels up to 20 km deep (Hou et al. 2001a; Peng et al. 2008).

The ~1780 Ma Xiong'er Group volcanic province has an areal extent of 350 × 500 km^2 and a thickness of 3–7 km and is dominated by thick and continuous lava flows, with rare volcaniclastic thin layers (~2 vol%, Zhao et al. 2002). It is

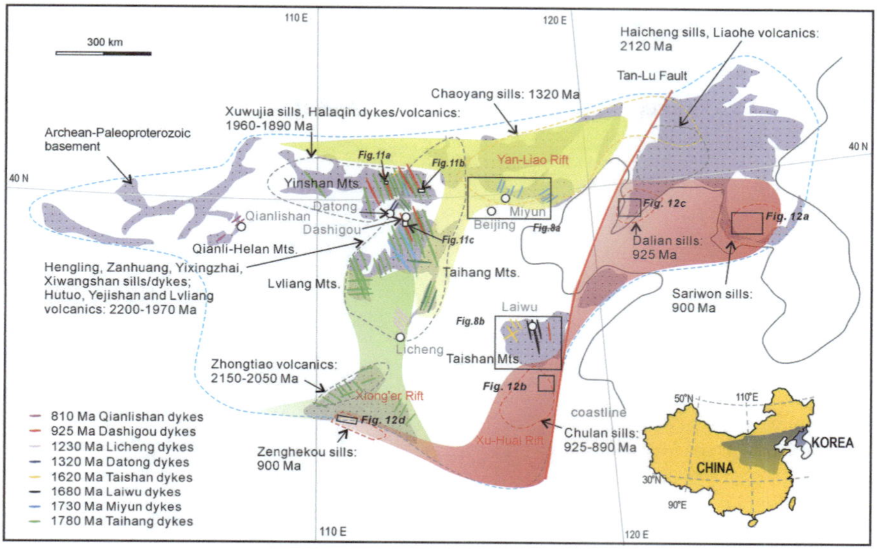

Fig. 3 Sketched map showing the distribution of Proterozoic mafic dyke swarms and related associations in the North China Craton. After Peng (2015)

natural to connect the coeval Taihang dykes and the Xiong'er volcanics as they have roughly overlapped compositions, and more importantly, they are spatially correlated (Peng et al. 2008): They possibly constitute a ~1780 Ma large igneous province by definition, and it was possibly driven by a paleoplume (Peng et al. 2008). Nevertheless, some others have demonstrated that these events were resulted from syn-orogenic/post-collisional uplifting and Andean-style collision (Wang et al. 2004, 2008; He et al. 2009; Zhao et al. 2009a) or intra-continental rift (Hou et al. 2001a, b, 2006; Fig. 5).

The recognition of the southern margin of the eastern North China Craton as the magma center of the Taihang dyke swarm and Xiong'er volcanic province is important; it means that there could have been another paleocontinent which has been connected with North China and which has been drifted away later. A group of paleogeographic reconstruction has been made for the Late Paleoproterozoic based on the 1780 Ma magmatic events (e.g., Peng et al. 2005; Hou et al. 2008a, b). Figure 6 is one hypothetic model of the paleocontinents in the Late Paleoproterozoic, which was based on the comparison of 1780–1730 and ~925 Ma dyke swarms and rift systems (Peng 2015).

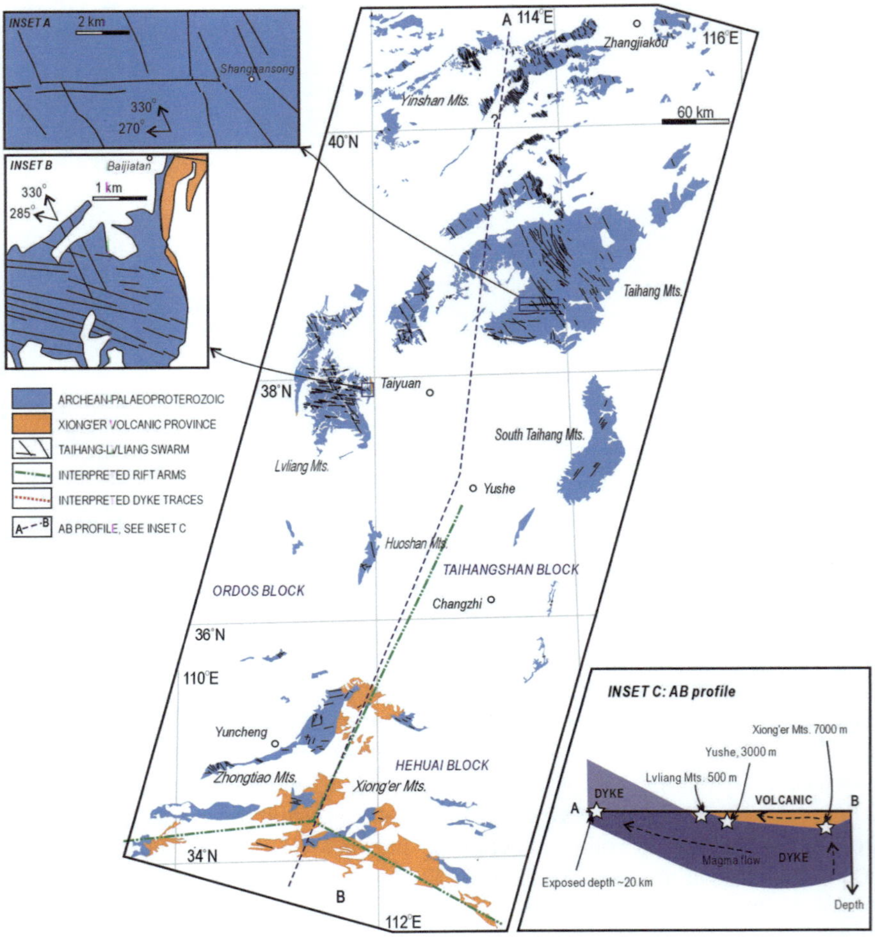

Fig. 4 Map showing the distribution of the Taihang–Lvliang swarm and Xiong'er volcanic province (after Peng 2010). *Insets A* and *B* show enlarged maps of local areas, which show the relationship between the Taihang dykes and the Xiong'er volcanic flows. *Inset C* is a hypothesis profile of the study area

1.1.2 The 1730 Ma Miyun Dyke Swarm

The ca. 1730 Ma Miyun dyke swarm has a total areal extent of 500×300 km^2 and consists of NE-trended dykes in the north and east, and NNW-trended in the central of North China (Fig. 3). The dykes are generally several kilometers long and 10–50 m wide. They are typically diabase and composed of plagioclase and clinopyroxene. This swarm is unconformably overlain by the Changcheng Group (Peng et al. 2012). Conglomerates, weathering surface, and direct contacts between the Changcheng Group and the dated dyke ($\sim$1731 Ma) were discovered in Miyun

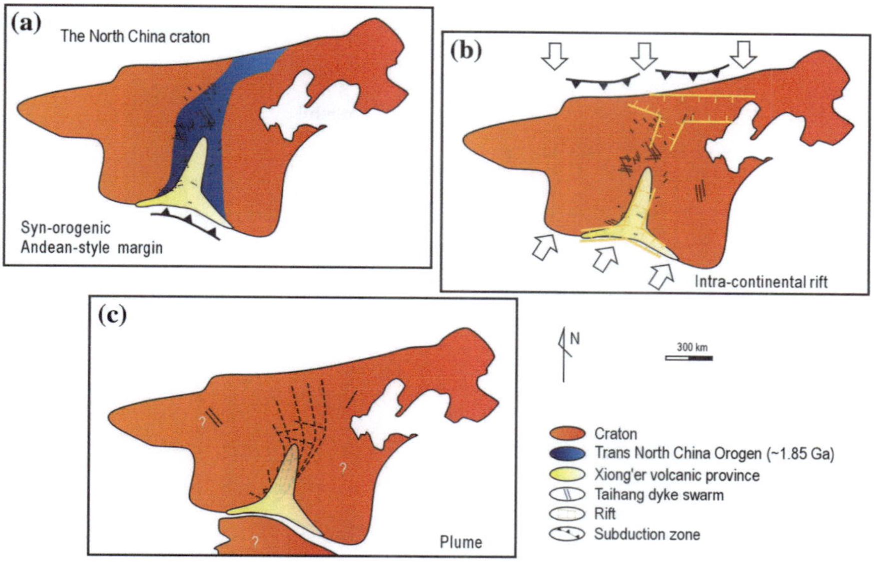

Fig. 5 Cartoons showing tectonic environments for the 1780 Ma dyke swarm and volcanic province. **a** Post-orogenic (e.g., Wang et al. 2004, 2008; He et al. 2009; Zhao et al. 2009); **b** intra-continental rifts evolving along with a paleosubduction (e.g., Hou et al. 2006); **c** plume-related resulted in continental breakup (Peng et al. 2005). After Peng (2010)

(Fig. 7). This clearly indicates that the Changcheng Group is younger than ~1731 Ma.

1.2 *The 1730–1600 Ma Igneous Events*

1.2.1 The 1730–1680 Ma Damiao–Shachang Anorthosite–Rapakivi Granite–Dyke Complexes

The 1730–1680 Ma Damiao–Shachang anorthosite–rapakivi granite–dyke complexes occur as small plutons or stocks of anorthosite and rapakivi granite and dykes in the Yan-Liao rift (Fig. 8). The distribution of these plutons was likely controlled by the E–W fault systems. These rocks were thought to be anorogenic, possibly related to a global event (Rämö et al. 1995; Yang et al. 2005; Zhang et al. 2007; Zhao et al. 2009b; Jiang et al. 2011; Wang et al. 2013).

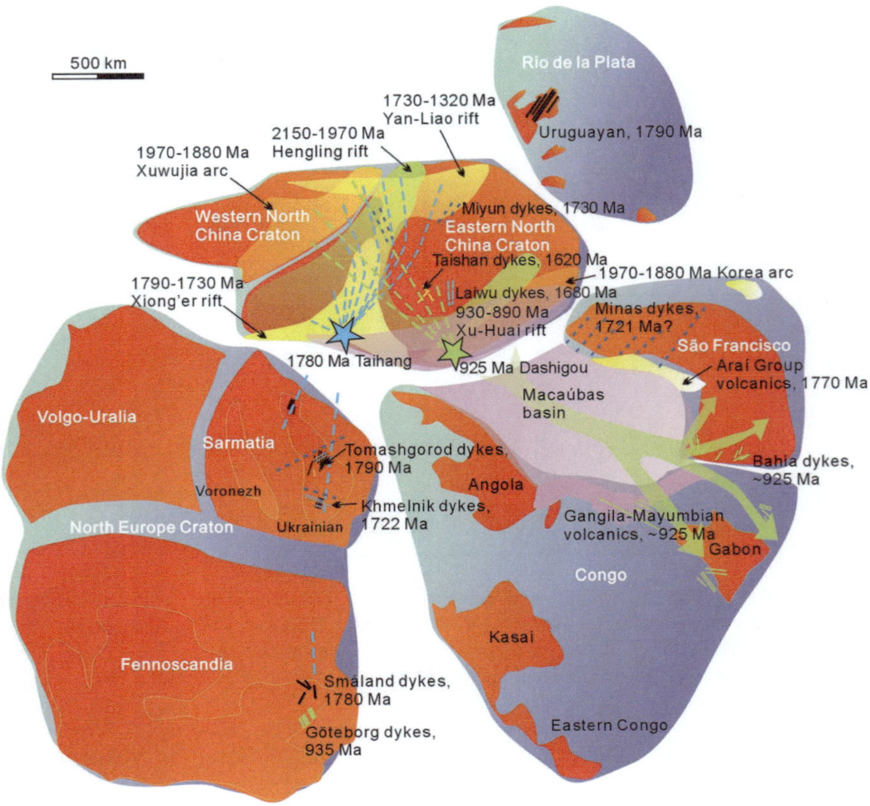

Fig. 6 Hypothetical paleogeography map of North China at ca. 1800 Ma. After Peng (2015)

1.2.2 The 1680 Ma Laiwu Dyke Swarm

The Laiwu dykes are newly discovered in Taishan Mts. (Fig. 8). Most of these dykes are 10–35 m wide, strike NNW (340°–350°), and are subvertical (65°–85° W). They generally outcrop continuously for several kilometers. They are composed of typical diabase (Li et al. 2015). Volcanic rocks with similar ages may exist in the Tuanshanzi Fm. in the Yan-Liao rift (Li et al. 1995a, b).

1.2.3 The 1620 Ma Taishan Dykes and Dahongyu High-Potassium Volcanics

The 1620 Ma Taishan dykes are distributed west to the Laiwu dykes (Fig. 8). Two biggest of them are NNW- and NE-oriented, respectively. Both of them are

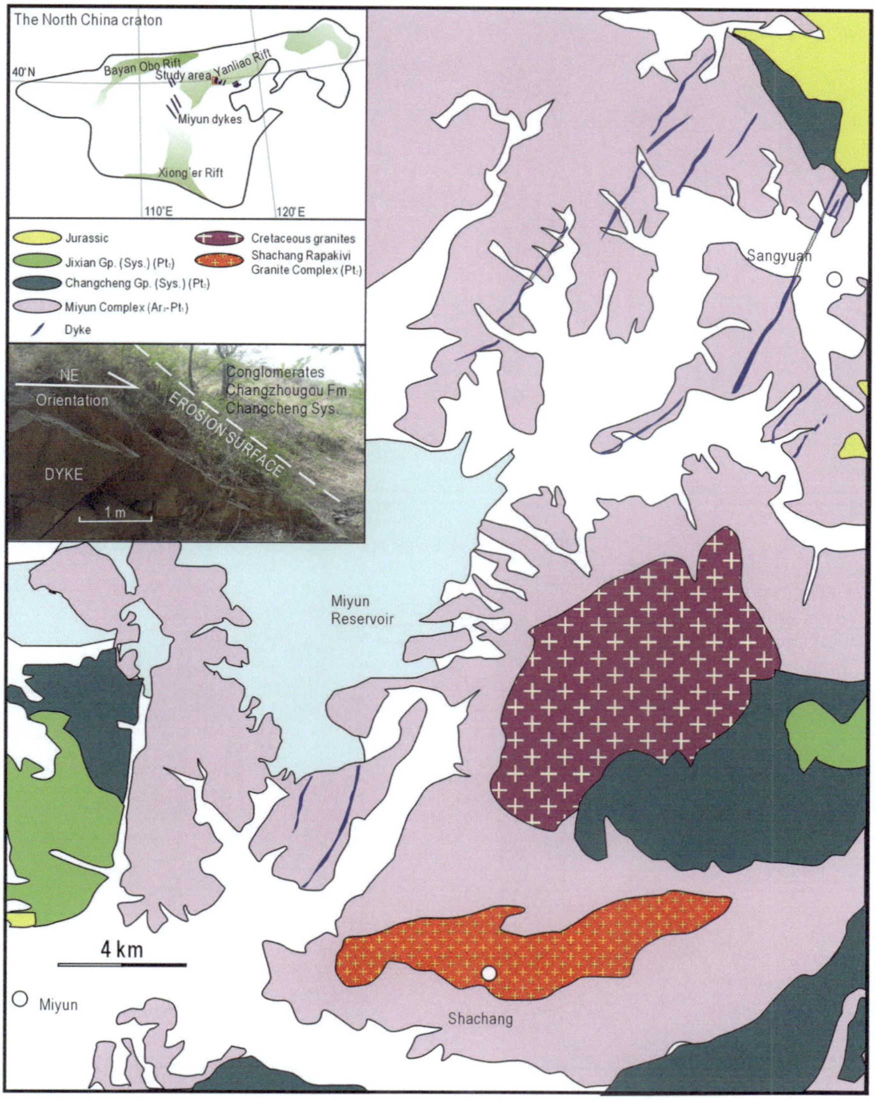

Fig. 7 The 1730 Ma Miyun dyke swarm in Miyun area. After Peng et al. (2012)

~25 km long and up to 40 m wide. The rocks of these dykes are composed of clinopyroxene (~40 vol%) and plagioclase (~55 vol%) (Li et al. 2015). There are also some coeval high-potassium volcanics and sills distributed in the Dahongyu Fm. in the Yan-Liao rift (Lu et al. 2008).

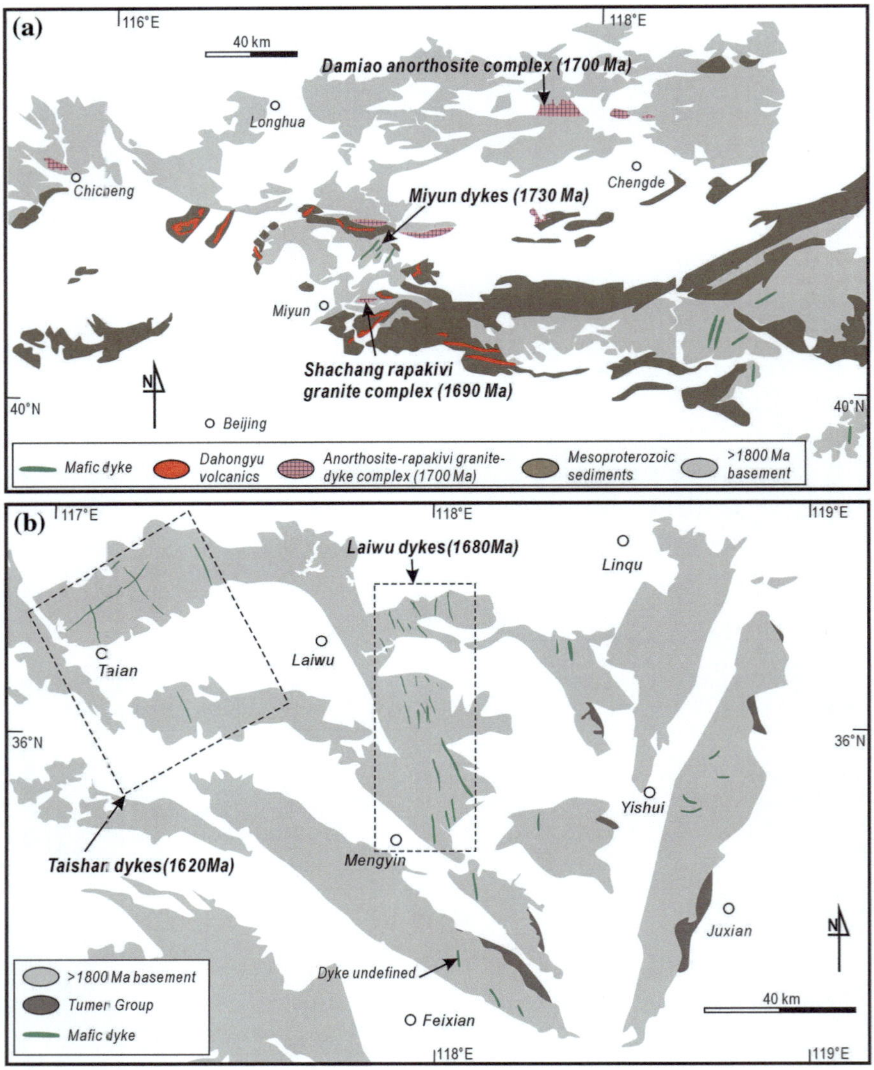

Fig. 8 Distribution of the 1710–1680 Ma anorogenic anorthosite–rapakivi granite–dyke complex and the 1620 Ma Dahongyu volcanic layers (**a**), and the 1680 Ma Laiwu and 1620 Ma Taishan dyke swarms (**b**). After Li et al. (2015)

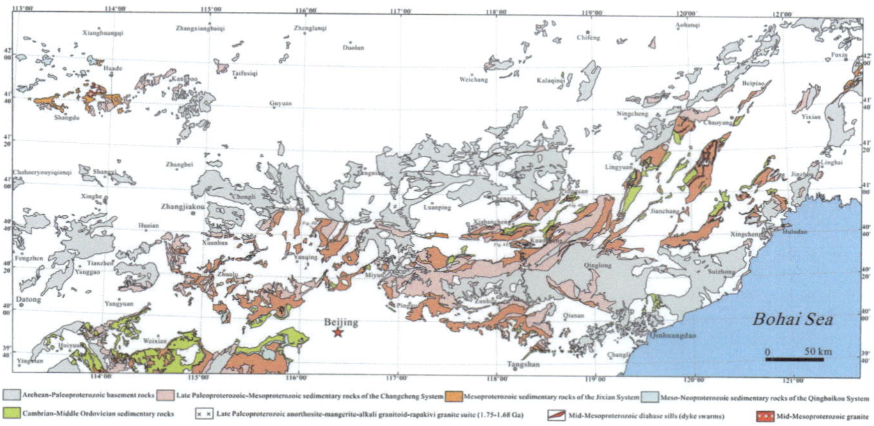

Fig. 9 Dustribution of the ~1320 Ma sills in the adjacent areas of the Yan-Liao rift. Copied from Zhang et al. (2012a)

1.3 The 1400–1200 Ma Igneous Events

1.3.1 The 1320 Ma Datong Dyke Swarm, the (Chaoyang–Kuancheng–Huailai) Sill Swarm and the Granites

The 1320 Ma dykes were discovered from Datong area (Fig. 3; Peng 2015). There are also some carbonatite dykes, possibly with ages between 1400 and 1200 Ma, in/around the Bayan Obo REE (rare earth element)-Nb–Fe-deposit (e.g., Fan et al. 2006; Le Bas et al. 2007; Yang et al. 2011).

The 1320 Ma diabase sills are widespread mainly in the Mesoproterozoic strata in the Yan-Liao rift (Fig. 9; Li et al. 2009; Zhang et al. 2009, 2012a, b). They are several meters to several hundred meters thick, and many of them continue for a few kilometers or even up to several tens of kilometers. They have typical ophitic textures with similar mineral compositions of pyroxene and plagioclase, and minor magnetite and hornblende (Zhang et al. 2009). Zhang et al. (2012a) reported ~1331–1313 Ma granites in Shangdu–Huade area (Fig. 9).

1.3.2 The 1230 Licheng Dyke Swarm

The 1230 Ma Licheng dykes are distributed in the Licheng area in the middle of Shanxi province (Fig. 3; Peng 2015). These NW-trending dykes are ~50 m wide cut the Mesoproterozoic strata and are covered by the Cambrian conglomerates. The rocks are gabbroic. There are also some 1209 Ma gabbros in Taishan Mts. (Peng et al. 2013).

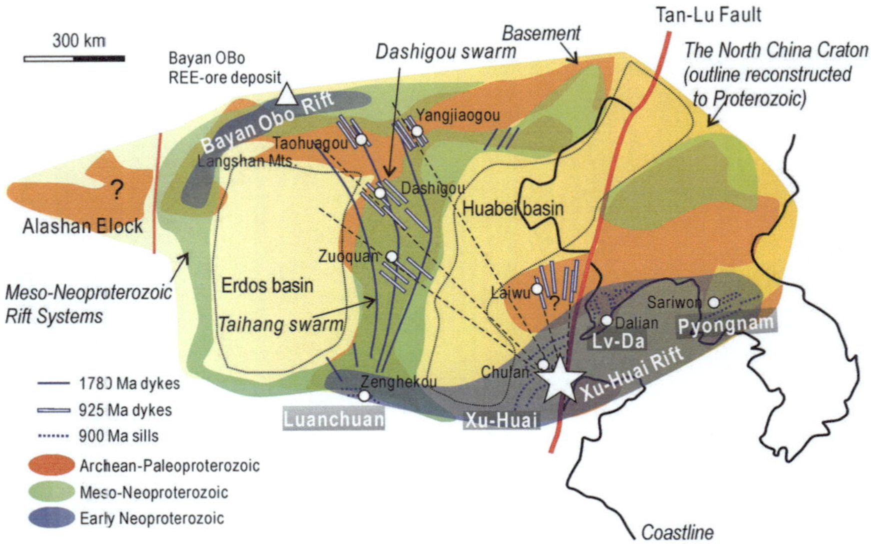

Fig. 10 The relationship of the 925 Ma Dashigou dykes, the 925–900 Ma sills, and the Xu-Huai rift system. After Peng et al. (2011a)

1.4 The 1000–800 Ma Igneous Events

1.4.1 The 925 Ma Dashigou Dyke Swarm

The 925 Ma Dashigou mafic dyke swarm is distributed in the central and a few in the east of North China (Figs. 3, 10 and 11; Peng et al. 2011a). The majority in the central North China covers an area of 200 × 450 km^2. They are typically 10–50 m wide and up to 10–20 km long. Their trends vary from 305°–340° to ~010° from west to east. The rocks comprise gabbro to diabase, unmetamorphosed with a typical assemblage of clinopyroxene and plagioclase, and minor hornblende, K–feldspar, and magnetite, with or without olivine.

1.4.2 The 925–900 Ma Sariwon (Dalian–Chulan–Zenghekou) Sill Swarm

The 925–900 Ma Sariwon sills are distributed along the southeastern margin of the eastern North China Craton, i.e., the Pyongnam Basin (Sariwon, Korea), the Dalian Basin (Liaoning), the Xu-Huai Basin (Chulan), and the Luanchuan Basin (Zenghekou) (Figs. 10 and 12; Peng et al. 2011b; Wang et al. 2011, 2012). The individuals are several meters to 150 m thick and extend for kilometers. The rocks are typical dolerite, metamorphosed up to greenschist facies, and are composed of clinopyroxene and feldspar, superimposed with a metamorphic assemblage of

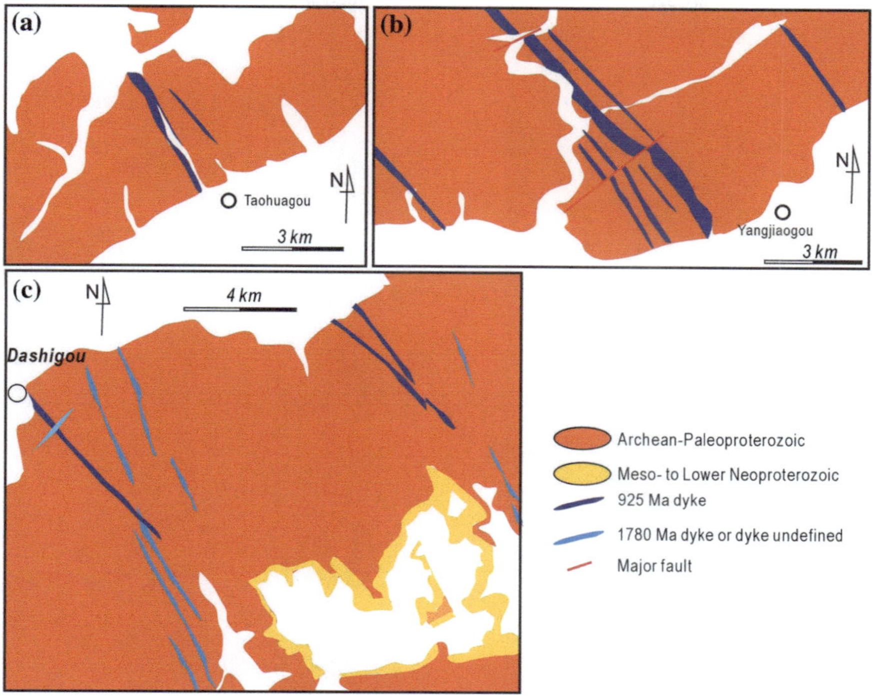

Fig. 11 Representative ~925 Ma dykes in Liangcheng (Taohuagou, **a**), Huai'an (Yangjiaogou, **b**), and Yingxian (Dashigou, **c**). After Peng et al. (2011a). Locations of these maps are shown in Fig. 3

epidote, chlorite, Na-rich plagioclase, and hornblende. They have been slightly deformed. The metamorphism and deformation could have occurred at ca. 400 Ma, resulting from an orogenic process affecting the cratonic margin (Peng et al. 2011b).

Peng et al. (2011a) suggest that these sills are cogenetic with the Dashigou dykes. And as these ~925 Ma dykes are rare in other continents, their global match could potentially suggest a paleogeographic affinity. Based on the geometry reconstruction (Fig. 10) and geological comparison, it is proposed that North China may be paleogeographically neighboring the Congo–São Francisco craton prior to Neoproterozoic (Fig. 6).

1.4.3 The 810 Ma Qianlishan Dyke Swarm

There are tens of ~810 Ma mafic dykes in the Helan and Qianli Mts. (Fig. 3). They are generally several meters wide and within a couple of kilometers long and composed of diabase rocks. They locally cut Mesoproterozoic sediments and unconformably overlain by the Carboniferous strata. More than half of them are

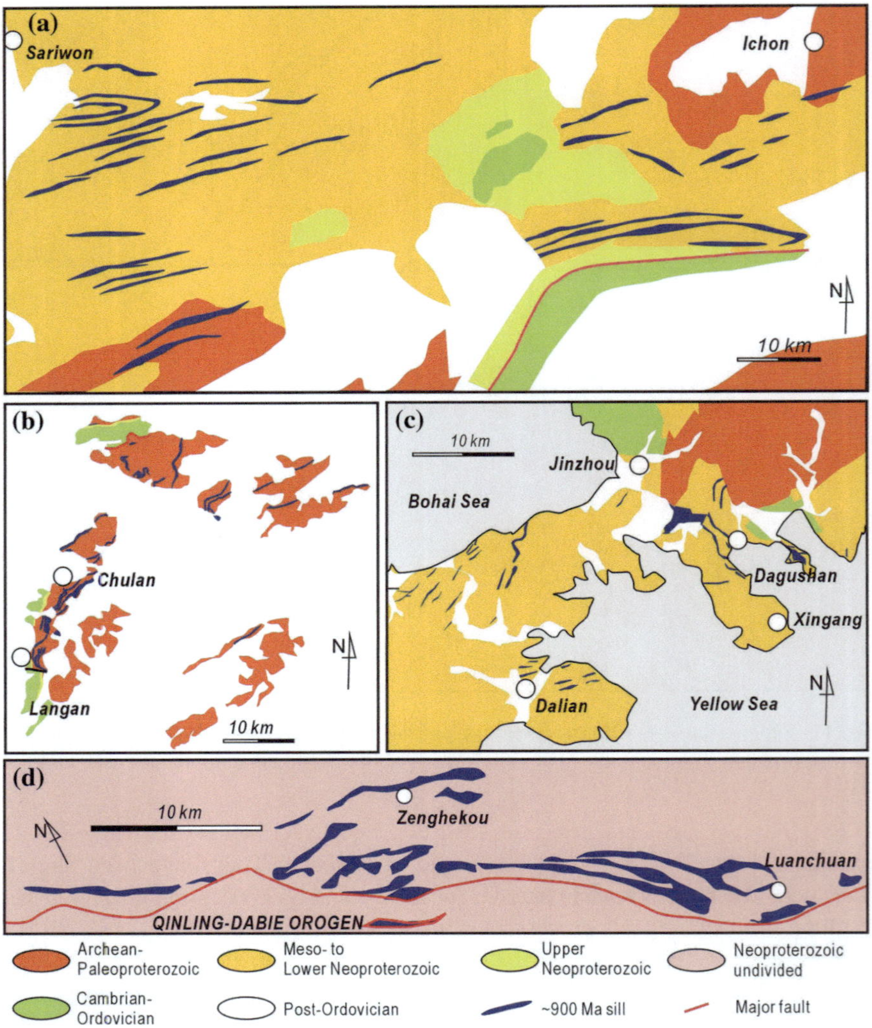

Fig. 12 The representative ~900 Ma sill swarms in the Pyongnam basin (Korea, **a**), the Xu-Huai basin (**b**, Chulan), the Lv-Da basin (**c**, Dalian), and the Luanchuan basin (**d**). After Peng et al. (2011a). Locations of these maps are shown in Fig. 3

NW–trended and the rest is NE-trended. One NE-trended dyke gives an 813 ± 7 Ma baddeleyite $^{207}Pb/^{206}Pb$ age. Peng et al. (2010) reported minor ~810 Ma rhyolitic lavas from the Zhaertai Group in the western Alashan (Alxa) block. Ultramafic intrusions from Jinchuan also have a similar age (Li et al. 2004).

2 Strata and Rift Systems

Figure 13 shows the distribution of the main Late Paleoproterozoic to Neoproterozoic strata in different rift systems. Here in this section, the description of the strata comes from the provincial geology volumes, including Hebei (1991), Inner Mongolia (Neimenggu) (1991), Henan (1989), Shandong (1991), and Liaoning (1989).

The Late Paleoproterozoic to Neoproterozoic rift systems of North China could be summarized as four: the Xiong'er, Yan-Liao, Bayan Obo, and Xu-Huai rift systems (Fig. 13; Peng 2015). Among them, the Xiong'er and Yan-Liao rifts may have evolved from the Late Paleoproterozoic to the Mesoproterozoic, whereas the Xu-Huai rift may be active in the Early Neoproterozoic. The evolution of the Bayan Obo rift system is controversial.

2.1 Strata of the Xiong'er Rift System

The Xiong'er rift system distributes along the southern parts of the eastern North China Craton (Fig. 13). It has three branches with one extended inside the central and two along the southern margin of the craton. It contains the Xiong'er Group, Guandaokou Group, Ruyang (and Luoyu) Group(s), Luanchuan Group, and the Sinian strata of the Huanglianduo, Dongjia, and Luoquan Fms. (Fig. 14). The Xiong'er rift initiated at ~1800 Ma, followed by volcanism mainly at ~1780 Ma

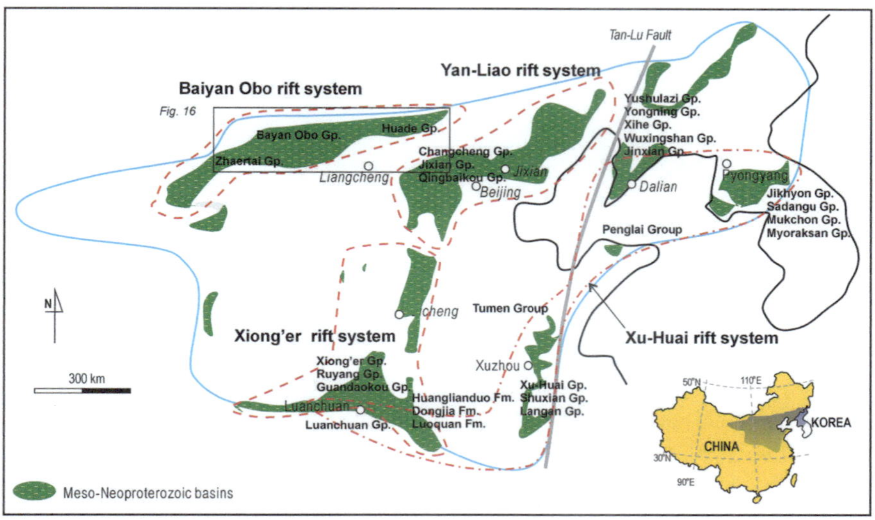

Fig. 13 Distribution of the Late Paleoproterozoic to the Neoproterozoic strata and rift systems in North China. *Gp.* Group; *Fm.* Formation

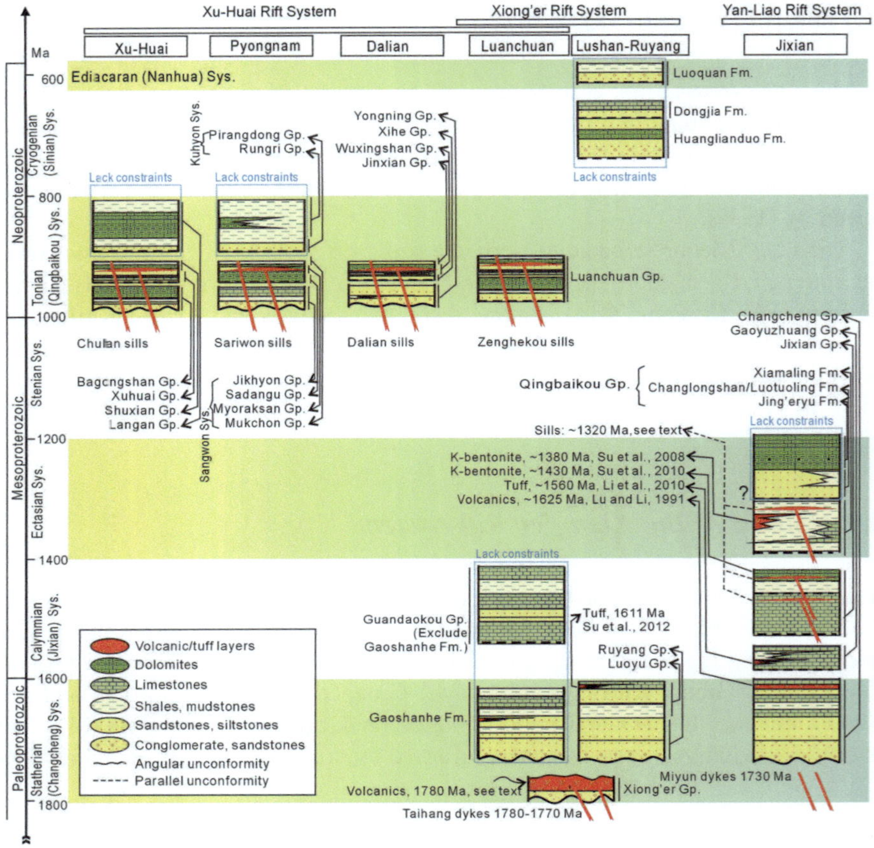

Fig. 14 Representative strata in the Xu-Huai, Pyongnam, Dalian, Luanchuan, Lushan–Ruyang, and Jixian basins of the Xu-Huai, Xiong'er, and Yan-Liao rift systems. After Peng (2015)

(the Xiong'er Group volcanic province) and minor younger events at ~1600 Ma (Su et al. 2012) and ~1400 Ma (Zhao et al. 2009a). The strata are dominantly 1780–1600 Ma clastic sediments (the Ruyang Group and the Luoyu Group) and younger carbonate (the Guandaokou Group) (Fig. 14).

2.1.1 The Xiong'er Group

The Xiong'er Group is distributed in the southern part of the eastern North China Craton (Fig. 13), and it can be subdivided into 4 formations, named Dagushi, Xushan, Jidanping, and Majiahe from bottom to top (Fig. 15a). The Dagushi Fm. is dominated by clastic sediments, but the other three formations are dominated by volcanic rocks (predominantly andesite; Fig. 15b, c). It has been well-constrained that the Dagushi Fm. was deposited after 1800 Ma, the volcanism in the other three

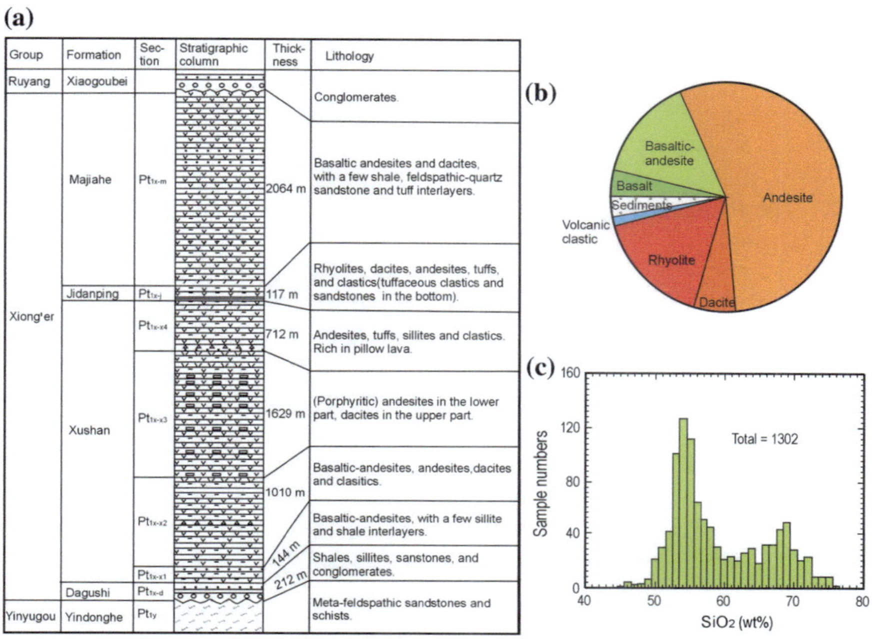

Fig. 15 Strata of the Xiong'er Group (**a**) and the compositions of volcanic rocks (**b**, **c**). After Zhao et al. (2002)

formations is ~1780 Ma (Zhao et al. 2004; He et al. 2009; Cui et al. 2010). It has also been revealed that the uppermost formation of the Majiahe Fm. was intruded by ~1780 Ma dioritic stocks, and thus, the Xiong'er Group volcanism is short-lived (Cui et al. 2010).

The lower part of the Dagushi Fm. comprises of yellowish, yellow-greenish, and amaranthine pebbly feldspathic quartz sandstone; while the upper part is dominated by amaranthine sandstone and shale. The sandstone in this formation is poorly sorted with largely varied composition and thickness. The Dagushi Fm. is 40–289 m thick, and it deposited in a relatively dry and unstable environment from a provenance which was not far away. The Xushan Fm. is composed of andesite, pyroxene andesite, and andesitic basalt, with a few rhyolite and volcaniclastic rock. The volcanic rocks are characterized by plagioclase megacryst and pyroxene. In the lowermost part, there are a few thin feldspathic quartz sandstone and amaranthine shale layers. It is conformably overlain on the Dagushi Fm. or unconformably sitting on the Archean gneisses. The thickness is 2400–3000 m. The Jidanping Fm. is a series of acidic volcanic rocks, including rhyolite, dacite, quartz porphyry, and minor volcaniclastic rocks. Pillow lavas can be found in some places. And there is also lentoid marble. It is conformably lying on the Xushan Fm. and has a thickness varied from 100 to ~1000 m. The Majiahe Fm. is predominantly andesite, basaltic andesite and pyroxene andesite with minor volcaniclastic rock, sandstone, shale, and limestone layers. It has a thickness of about 2000 m, up to 3910 m in the Xiong'er Mts.

2.1.2 The Ruyang (Ruyang + Luoyu) Group and the Guandaokou Group

The Ruyang Group is limited in the Xiong'er rift (Fig. 13), and it is composed by the Bingmagou (Xiaogoubei), Yunmengshan, Baicaoping, Beidajian, Cuizhuang, Sanjiaotang, and Luoyukou Fms. Sometimes, the Cuizhuang, Sanjiaotang, and Luoyukou Fms. are specially referred as the Luoyu Group. Basically, the Ruyang Group is a slightly metamorphosed series of carbonates and clastic sediments. It is unconformably overlying on the Xiong'er Group or locally on the Archean gneisses; and it is unconformably under the Luoquan and Xinji Fms. The total thickness of this group is about 13,070 m. Among them, the Yunmengshan, Baicaoping, and Beidajian Fms. contain microfossil plants; and the upper part of the Beidajian Fm. has stromatolites.

The Bingmagou Fm. (also called as the Xiaogoubei Fm. in Songshan area) comprises conglomerate, pebbly sandstone, sandstone, sandy shale, and glauconite- and iron-concretion-bearing sandstone. It has a thickness of 880 m. The Yunmengshan Fm. is composed by quartz sandstone and shale, with some iron-ore-bearing conglomerate in the bottom. It is ~276 m thick. The Baicaoping Fm. is mainly composed by red sandy shale with thin layers of quartz sandstone and locally conglomerate and dolomite. It is ~200 m thick. The Beidajian Fm. is dominated by quartz sandstone, feldspathic quartz sandstone and glauconite-bearing quartz sandstone with dolomite on the top. There are also iron-ore layers on the upper part. The thickness is ~198 m. The Cuizhuang Fm. is predominantly shale with minor quartz sandstone. There are oolitic hematite- and siderite-bearing layers in the bottom and middle parts. It is ~213 m in thickness. The Sanjiaotang Fm. is a series of quartz sandstone, with some glauconite-bearing layers in the top part. It has a thickness of 231 m. The Luoyukou Fm. comprises ~146-m-thick carbonates with minor clastic sediment interlayers.

The Guandaokou Group is distributed in the southern parts of the Xiong'er rift (Fig. 13) and is divided into the Gaoshanhe Fm., the Longjiayuan Fm., the Xunjiansi Fm., the Duguan Fm., and the Fengjiawan Fm. It is dominated by a series of clastic rocks and carbonate, rich in microfossil plants and stromatolites. The thickness is varied from 1950 to 5440 m. It is unconformably overlain on the Xiong'er Group.

The ages of the Ruyang and Guandaokou Groups are not well-constrained (Fig. 13), though they were thought to be comparable with the Changcheng Group or the Jixian Group. Recently, Su et al. (2012) discovered a ~1611 Ma tuff layer on the top of the Ruyang (Luoyu) Group, which indicates that it should be the Late Paleoproterozoic in age.

2.2 *Strata of the Yan-Liao Rift System*

The Yan-Liao rift system distributes in the northern parts of the eastern North China Craton (Fig. 13). It contains the Changcheng Group, Jixian Group, and Qingbaikou Group (Fig. 16), which were thought to compose the stratotype section of the Mesoproterozoic to the Early Neoproterozoic in China (Chen et al. 1980). However, recent work suggests that parts of the Qingbaikou Group are older than 1300 Ma (Li et al. 2009; Zhang et al. 2009), and thus, whether there is Neoproterozoic strata in the Yan-Liao rift or not is questionable. In addition, as they

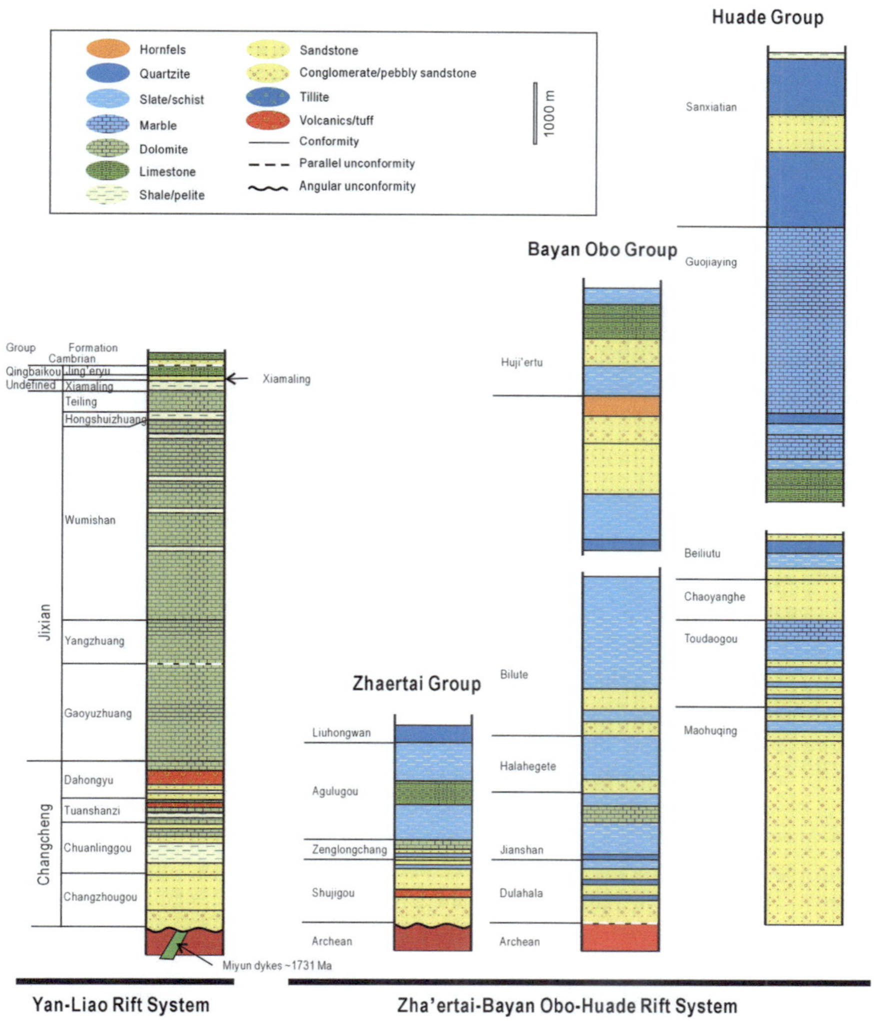

Fig. 16 Strata in the Yan-Liao rift system and the Bayan Obo rift system

were thought to be the cover of the craton (Chen et al. 1980), the lower time limit of the Mesoproterozoic is represented by the initiation of the Changcheng Group at 1800–1700 Ma. However, we think that the cratonization of the eastern North China Craton was completed at ~2500 Ma and the Changcheng–Jixian–Qingbaikou sediments were intra-continental rift deposits rather than covers (Peng 2015).

The Yan-Liao rift initiated at ~1730 Ma (Peng et al. 2012) and contains magmatism and volcanism mainly at 1730–1680, ~1620 and ~1320 Ma (Rämö et al. 1995; Yang et al. 2005; Zhang et al. 2007, 2009, 2012a; Li et al. 2009; Zhao et al. 2009b; Jiang et al. 2011; Wang et al. 2013). It is a long-lived rift system (Zhai et al. 2015).

2.2.1 The Changcheng Group

The Changcheng Group is the initial deposition of the Yan-Liao rift system, and it comprises the Changzhougou, Chuanlinggou, Tuanshanzi, and Dahongyu Fms. from bottom to top (Figs. 14 and 16). The Changzhougou Fm. is unconformably sitting on the crystallized basement and is composed by conglomerate, pebbly sandstone, siltstone, and shales, with locally argillaceous dolomite and stromatolites. The Chuanlinggou Fm. is dominated by black shales or sandy shales, with minor siltstone and dolomite. In the west of the Yan-Liao rift (Xuanhua–Chicheng area), there are so-called Xuanlong-type iron deposits in the sandstone. The thickness of the Chuanlinggou Fm. varies from 10 to 1000 m, and in the middle–upper parts, there are some microfossil plants, e.g., Margominuscula and Dictyosphaera. The Tuanshanzi Fm. is dominated by thick dolomite with minor sandy shale, sandy dolomite, and argillaceous dolomite. The thickness is about ten to several hundred meters. The Dahongyu Fm. is composed by sandstone and feldspathic sandstone with minor potassium-rich shales in the middle–lower parts, and dolomite in the upper part. There are also potassium-rich volcanic layers and volcaniclastic layers in this formation. Meng et al. (2011) suggest that the Changcheng Group was deposited in rift basins that experienced strong subsidence in association with volcanism, and the unconformity above the Changcheng Group may indicate a breakup event.

The deposition age of the Changcheng Group has long been debated, though the age of volcanic layers in the Dahongyu Fm. is well-constrained to be at ~1625 Ma (Lu et al. 2008; Gao et al. 2008a). Wang et al. (1995) revealed that the maximum deposition age of the Changcheng Group is about 1700 Ma based on the Ar–Ar age dating on basement rocks. Wan et al. (2003) suggested that the maximum deposition age is ~1800 Ma based on the ages of detrital zircon. Li et al. (1995a, b) and Gao et al. (2009)'s ages from volcanic rock or diabase dyke suggest that the initial deposition age of the Changcheng Group should be older than 1683–1638 Ma.

Recently, Li et al. (2011, 2013)'s work from a granitic vein, which was seemly cut the Changcheng Group, gave an age of 1673 ± 10 Ma. If this relationship is verified, it suggests that the Changcheng Group is younger than 1673 Ma.

Nevertheless, Peng et al. (2012) have dated a ~40-m-wide ~1730 Ma dyke that has covered by basal conglomerate of the Changcheng Group, which constrains the initiation age to be younger than ~1730 Ma.

2.2.2 The Jixian Group

The Jixian Group is dominated by carbonate, and it comprises five formations, the Gaoyuzhuang, Yangzhuang, Wumishan, Hongshuizhuang, and Tieling from bottom to top (Figs. 14 and 16).

The Gaoyuzhuang Fm. is composed by thick dolomite and chert-bearing argillaceous dolomite. There are plenty of stromatolites in the lower part. The thickness of this formation is up to 2000 m. The Yangzhuang Fm. is dominated by amaranth- and white-colored dolomites with locally sandstone or conglomerate. It is conformably or deceptive conformably contacted with the Gaoyuzhuang Fm. and has a thickness of several ten meters to 1000 m. The Wumishan Fm. is also dominated by dolomite but generally gray in color. The Hongshuizhuang Fm. is composed by black to green illite-bearing shales, with minor thin dolomite in the lower part and thick sandstone in the upper part. It has several ten meters to more than 100 m. The Tieling Fm. contains mainly dolomite and stromatolites rich limestone and dolomite, with some shale. The thickness is 200–300 m.

It has long been thought that the Jixian Group deposited between 1600–1000 Ma; however, recent discovery of ~1350 Ma sills from the Wumishan Fm. reveals that this group ended earlier (Zhang et al. 2009). In addition, Li et al. (2010) got 1559–1560 Ma ages from tuff layers in the Gaoyuzhuang Fm.; Su et al. (2010) obtained ~1437 Ma age from potassium bentonite in the Tieling Fm. These works suggest the deposition age of the Jixian Group to be older than ~1400 Ma.

2.2.3 The Qingbaikou Group

The Qingbaikou Group comprises three formations, i.e., the Xiamaling, Changlongshan (or Luotuoling), and Jing'eryu Fms. (Figs. 14 and 16). The Xiamaling Fm. is dominated by multicolor sandy shale, with minor siltstone and marlstone. There is a parallel unconformity to micro-angle unconformity between the Xiamaling Fm. and the Tieling Fm. The thickness is about one hundred to several hundred meters. The Changlongshan Fm. is made of pebbly feldspathic sandstone, quartz sandstone, glauconite-bearing sandstone, and shale. It is parallel unconformably contact with the Xiamaling Fm. and is locally overriding the Wumishan Fm. The thickness is about 100 m. The Jing'eryu Fm. (also called as the Luotuoling Fm.) contains dolomitic limestone, with some glauconite-bearing feldspathic sandstone or fine conglomerate. It is about several 10 m to >100 m thick.

It has long been suggested that the Qingbaikou Group belongs to the Early Neoproterozoic in age (e.g., Chen et al. 1980). However, two suites of K-bentonite

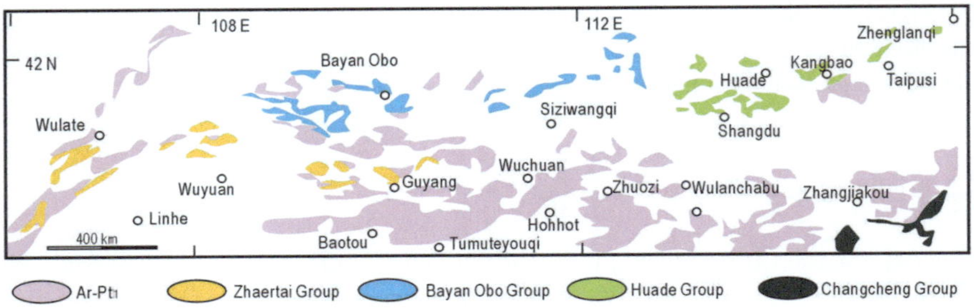

Fig. 17 Distribution of the Zhaertai–Bayan Obo rift strata

beds were discovered in the Tieling and Xiamaling Formations, and they yielded U–Pb zircon SHRIMP ages of ~1440 Ma (Su et al. 2010) and 1380–1366 Ma (Gao et al. 2007, 2008a, b; Su et al. 2008, 2010), respectively. These tuff layers probably originated from the volcanism events around the Yan-Liao rift. Li et al. (2009) and Zhang et al. (2009) discovered some ~1320 Ma sills from the Xiamaling Fm. Thus, the Xiamaling Fm. could be actually 1400–1200 Ma in age; however, the age of the Jing'eryu and Changlongshan Fms. is not well-constrained (Figs. 14 and 16).

2.3 *Strata of the Bayan Obo Rift System*

The Bayan Obo rift distributes in the central Inner Mongolia (Neimenggu) province (Fig. 13). It contains the Zhaertai, Bayan Obo, and Huade Groups; however, there are no direct contacts between the three groups, or with the Changcheng Group in the Yan-Liao rift (Fig. 17). It is in debating whether the four groups have similar ages. The Bayan Obo rift probably has magmatic events at ~1300–1400 Ma (Fan et al. 2006; Le Bas et al. 2007; Yang et al. 2011; Zhang et al. 2012a, b) and ~810 Ma (Peng et al. 2010; Peng 2015).

2.3.1 The Zhaertai Group

The Zhaertai Group is unconformably sitting on the Archean basement and is composed by the Shujigou, Zenglongchang, Agulugou, and Liuhongwan Fms. (Fig. 16). The Shujigou Fm. comprises meta-conglomerate, meta-feldspathic quartz sandstone and quartzite; the Zenglongchang Fm. is composed of dolomitic slate, stromatolite-bearing crystalline limestone and dolomite; the Agulugou Fm. is dominated by carbonatic slate; and the Liuhongwan Fm. is prominently quartzite. There are minor alkaline basaltic volcanic layers in the top of the Shujigou Fm. (Wang et al. 1992).

Li et al. (2007a) dated a mafic volcanic layer in the Shujigou Fm., and it gives a ~1743 Ma age. In the western parts of the Bayan Obo rift in Langshan area, Peng et al. (2010) have obtained 817–805 Ma acidic lavas, which may indicate some Neoproterozoic strata in this area.

2.3.2 The Bayan Obo Group

The Bayan Obo Group distributes east to the Zhaertai Group (Fig. 17) and is dominated by clastic sediments and clay, and it can be divided into six formations (Fig. 16). The bottom formation is the Dulahala Fm., which comprises quartzite, quartz granule conglomerate, and pebbly feldspathic quartz sandstone. The Jianshan, Halahuogete, and Bilute Fms. compose mainly of pelite and turbidite. The Baiyinbaolage and Hujiertu Fms. comprise clastic sediment and carbonate, with minor volcanic rocks. The Bayan Obo Group is the host of the world-class REE-Nb-Fe-deposit (mainly within the Dulahala–Bilute Fms.: Fan et al. 2006; Le Bas et al. 2007; Yang et al. 2011).

Zhong et al. (2015)'s detrital zircon ages constrained a Late Paleoproterozoic to Mesoproterozoic age for the Bayan Obo Group. The basalt from the lower parts of the Bayan Obo Group gave U–Pb zircon age of ~1730 Ma (Lu et al. 2002), whereas the carbonatite from the Dulahala Fm. gave zircon U–Pb ages of ~1300–1400 Ma (Fan et al. 2006; Yang et al. 2011).

2.3.3 The Huade Group

The Huade Group distributes east to the Bayan Obo Group (Fig. 17) and is characterized by a series of slightly metamorphosed sandstone, graywacke, feldspathic sandstone, pelite, phyllite, schist, slate, marble and limestone, up to ~10,000 m in thickness. In Shangdu area, the lower parts of the Huade Group are comparable with the 6 formations of the Bayan Obo Group. In addition, there is another formation on the top, named the Ayadeng Fm., which is dominated by crystalline limestone with minor sandy slate. The Huade Group in Zhangbei–Kangbao areas is divided into another 6 formations, named the Maohuqing, Toudaogou, Chaoyanghe, and Beiliutu Fms. in the Zhangbei area, and the Gejiaying and Sanxiatian Fms. in Kangbao area. There is no contact between the Huade Group in the two areas. The Maohuqing Fm. is dominated by metamorphosed thick pebbly feldspathic sandstone with minor quartz sandstone, carbon slate, and siltstone. The Toudaogou Fm. composes of metamorphosed pebbly quartz sandstone and quartz sandstone in the lower part, siltstone and carbon slate in the middle part, and carbonate in the upper part. The Chaoyanghe Fm. is dominated by quartz schist and two-mica schist with carbon phyllite and garnet slate. The Beiliutu Fm. is composed by medium- to fine-grained quartz sandstone in the lower part and andalusite-bearing quartzite, meta-feldspathic quartzite and sericite phyllite. The Gejiaying Fm. is a series of calc-silicate rock, metamorphosed carbonate, pelite, and clastic sediments. The Sanxiatian Fm. is a series of metamorphosed clastic rocks, including meta-quartz sandstone, meta-feldspathic quartz sandstone, and two-mica schist.

It has long been thought that the Huade Group was deposited in the Mesoproterozoic. Zhang et al. (2012a) reported ~1331–1313 Ma granites which have intruded the Toudaogou–Beiliutu Fms. Hu et al. (2009)'s detrital zircon ages constrained the maximum deposit age of the Huade Group to be 1800–1600 Ma. Chen (1993), and Tan and Shi (2000) have claimed Cambrian fossils in some strata. On the contrary, Zheng et al. (2004) and Li et al. (2005) proposed Paleoproterozoic deposit ages based also on the detrital zircon ages.

2.4 *Strata of the Xu-Huai Rift System*

The Xu-Huai rift system lies on the eastern parts of North China and partly in North Korea. It contains strata of the Xu-Huai, Langan, Tumen, Yongning, Xihe, Wuxingshan, Jinxian, Jikhyon, Sadangu, Mukchon, Myoraksan, and Penglai Groups (Fig. 14). The Xu-Huai rift system mainly evolved in the Early Neoproterozoic, characterized by the development of the 925–900 Ma sills (Peng et al. 2011a, b).

2.4.1 The Xuhuai, Shuxian, and Langan Groups

The strata in the Xuzhou, Huaibei, and adjacent areas are divided into three groups—the Xuhuai, Shuxian, and Langan Groups (Figs. 13 and 14). The Xuhuai Group has 9 formations, i.e., the Lanling, Xinxing, Jushan, Jiayuan, Zhaoxu, Niyuan, Jiudingshan, Zhangqu, and Weiji Fms. from bottom to top. The first three formations are made of mainly conglomerate, quartz sandstone, siltstone, and shales, whereas the rest six formations are made of sandy marlstone, limestone, and dolomite. Locally, the bottom formations (the Lanling and Xinxing Fms.) of the Xuhuai Group are referred as the Bagongshan Group and are subdivided into the Xiaodian, Wushan (Lanling), and Liulaobei (Xinxing) Fms. On the other hand, the Xuhuai Group is included into the Shuxian Group in some other regional maps. The Shuxian Group (the Shijia and Wangshan Fms.) and Langan Group (the Jinshanzhai and Gouhou Fms.) comprise shale, quartz sandstone, siltstone, marlstone, limestone, and dolomite. The Xuhuai Group is unconformably sitting on the Archean basement; and the three groups are unconformably covered by the Cambrian strata of the area.

These strata are thought to be the Neoproterozoic in age (Cao 2000; Xue et al. 2001); and some thought the three groups could have overlapped deposition ages (Qiao et al. 1996; Zhang, 2001). Recently, Wang et al. (2012) reported ~900 Ma sills from the Xuhuai Group, which supports an Early Neoproterozoic age for this group. However, the sedimentation age of the Shuxian and Langan Groups is not well-constrained.

2.4.2 The Tumen Group and the Penglai Group

The Tumen Group is mainly distributed in west Shandong province (Fig. 13), and it can be subdivided into five formations, i.e., the Heishanguan, Erqingshan, Tongjiazhuang, Fulaishan, and Shiwangzhuang Fms. The Tumen Group is dominated by sandstone, shale, and limestone, and it is unconformably sitting on the Archean gneisses and is parallel unconformably covered by the Cambrian strata (the Changqing Group). The Heishanguan Fm. is composed of glauconite-bearing quartz sandstone and fuchsia shale. The Erqingshan Fm. is parallel unconformably sitting on the Heishanguan Fm. and is made of glauconite-bearing quartz sandstone, thin limestone, calcic shale, and marlstone. The Tongjiazhuang Fm. is parallel unconformably sitting on the Erqingshan Fm. and is dominated by dark shale, with some quartz sandstone, algal limestone, and marlstone. In some localities, the Tongjiazhuang Fm. is sitting on the Archean basement. The Fulaishan Fm. is dominated by siltstone with minor shale and sandy marlstone, and it is conformably sitting on the Tongjiazhuang Fm. The Shiwangzhuang Fm. is composed of sandy limestone, limestone, and dolomite, with minor shale. It is conformably sitting on the Fulaishan Fm. In the provincial geological report (Shandong 1991), the Tumen Group is thought to be the Neoproterozoic in age. There are some ~910 Ma Rb–Sr isochron ages from limestone (Zhou and Hu 1998). Hu et al. (2012) suggest that the maximum deposit age is 1200–1000 Ma based on the ages from detrital zircons.

The Penglai Group is distributed in Qixia–Penglai area in eastern Shandong province. It is a series of low-grade metamorphosed sediments, including phyllite, slate, quartzite, crystalline limestone, and marble. The deposition can be divided into 4 formations—the Baoshankou, Fuzikuang, Nanzhuang, and Xiangkuang Fms. The Baoshankou Fm. is unconformably sitting on the Paleoproterozoic Fenzishan Group and is composed by phyllite, slate, marble, and phyllitic slate. The Fuzikuang Fm. is composed by quartzite, silicic slate, and phyllite. The Nanzhuang Fm. is composed by slate, phyllitic slate, marble, calcic slate, and marlstone. The Xiangkuang Fm. is composed by marlstone with minor limestone. Zhou et al. (2008) suggest that the maximum deposit age to be 2000–1800 Ma, whereas Li et al. (2007b) suggest that it is younger than 1200 Ma. It also needs to be mentioned that Zhou et al. (2008) think the provenance to be from the North China Craton, whereas Li et al. (2007b) favor a provenance from the South China craton.

2.4.3 The Yushulazi, Yongning, Xihe, Wuxingshan, and Jinxian Groups

The Yushulazi, Yongning, Xihe, Wuxingshan, and Jinxian Groups are mainly distributed in Liaodong Peninsula (Fig. 13). Specifically, the first four groups are mainly distributed in Fuzhou area, whereas the Jinxian Group is mainly outcropped in Dalian area. The Yushulazi Group is parallel unconformably sitting on the Paleoproterozoic Liaohe Group and is composed mainly by quartzite with phyllite, sericite quartz schist, and meta-sandstone. The Yongning Group/Formation is parallel unconformably sitting on the Yushulazi Group and is composed by feldspathic quartz sandstone and feldspathic sandstone, with minor conglomerate and pebbly sandstone. The Xihe Group is parallel unconformably sitting on the Yongning Group and comprises the Diaoyutai, Nanfen, and Qiaotou Fms. The Diaoyutai Fm. is composed by fine quartz sandstone, pebbly quartz sandstone, siltstone, and shale. The Nanfen Fm. is dominated by dark sandy shale with minor siltstone in the lower part, crystalline limestone, sandy micrite with minor calcic shale, sandy shale, and calcic sandy shale with minor quartz sandstone in the upper part. The Qiaotou Fm. is composed by quartz sandstone, feldspathic quartz sandstone, with minor glauconite-bearing quartz sandstone and sandy shale.

The Wuxingshan Group comprises the Changlingzi, Nanguanling, and Ganjinzi Fms. from bottom to top. The Changlingzi Fm. is dominated by shale, sandy shale, sandstone, glauconite-bearing quartz sandstone, siltstone, and micrite. The Nanguanling Fm. is dominated by clastic limestone with stromatolite limestone. The Ganjinzi Fm. is composed by stromatolite dolomite and clastic dolomite. The Jinxian Group includes the Yingchengzi, Shisanlitai, Majiatun, Cuijiatun, Xingmincun, and Dalinzi Fms. The Yingchengzi Fm. is composed by micrite and stromatolite limestone. The Shisanlitai Fm. is composed by micrite, oomicrite, stromatolite limestone, and yellow-greenish to purple shales. The Majiatun Fm. is composed of micrite, stromatolite limestone, and calcic shale. The Cuijiatun Fm. is dominated by sandy shale, siltstone, and stromatolite limestone. The Xingmincun Fm. is dominated by siltstone, sandy shale, quartz sandstone, micrite, and calcic shale. The Dalinzi Fm. is composed by black shale, quartz sandstone, micrite, and dolomite, with minor sandy shale, iron–manganese-bearing mudstone, and limestone.

Liaoning (1989) suggests that the Yushulazi Group belongs to the Paleoproterozoic, the Yongning Group and the Diaoyutai and Nanfen Fms. of the Xihe Group belong to the Early Neoproterozoic, and the Qiaotou Fm. of the Xihe Group, the Wuxingshan Group and the Jinxian Group belong to the Late Neoproterozoic. However, some recent works reveal detrital zircon ages of $\sim$1100 Ma from the Yushulazi and Xihe Group (Luo et al. 2006); thus, they could belong to the Early Neoproterozoic (Fig. 14).

2.4.4 The Luanchuan Group and the Huanglianduo–Dongjia–Luoquan Formations

The Luanchuan Group is distributed in the southern edge of the eastern North China Craton (Fig. 13), and it comprises the Baishugou, Sanchuan, Nannihu, and Meiyaogou Fms. It is unconformably overlain on the Guandaokou Group, but unconformably covered by the Sinian or the Cambrian strata (Fig. 14). The thickness is 1700–3100 m. The Baishugou Fm. is composed by carbon sericite phyllite, sericite quartz schist, and feldspathic quartzite, with carbon schist in the lower part, thick fine-grained K-feldspar-bearing quartzite in the middle part, and black slaty carbon phyllite, thin carbon sericite quartzite and carbon marble in the upper part. The Sanchuan Fm. contains pebbly quartz sandstone, sandstone, siltstone, marble, and sericite schist. It is 320–452 m thick. The Nannihu Fm. is 309–509 m thick and is made of fine-grained quartzite, two-mica schist, carbon sericite schist, and calcic two-mica schist, with minor quartzite and marble. The Meiyaogou Fm. is 855–1154 m thick and comprises meta-siltstone, mica schist, marble, stromatolite- or bone coal-bearing marble, quartzite, and magnetite mica schist.

The Huanglianduo, Dongjia and Luoquan Fms. distribute east to the Luanchuan Group (Fig. 13). The Huanglianduo Fm. composes of conglomerate, pebbly sandstone, sandstone, silicic banded dolomite (contain stromatolites), and banded chert. It is about 252–449 m thick. The Dongjia Fm. composes conglomerate, pebbly sandstone, feldspathic quartz sandstone, quartz sandstone, siltstone, and shales, with minor glauconite-bearing sandstone, dolomite, and micrite. This formation varies from 28 to 353 m. The Luoquan Fm. composes of mudstone, dolomitic tillite (locally changed to dolomite), sandy tillite, shale, slate, and siltstone, with phosphate nodule. It is about 100 m thick. These are thought to be the Late Neoproterozoic strata but with little constraints (Henan 1989).

2.4.5 The Jikhyon, Sadangu, Mukchon, and Myoraksan Groups

In the Pyongnam basin in North Korea, there is a series of 8000–1000-m-thick greenschist facies sediments. The North Korea geologists divided it into two systems, the Sangwon System and the Kuhyon System (Paek et al. 1993). The Sangwon System is composed by the Jikhyon, Sadangu, Mukchon, and Myoraksan Groups, while the Kuhyon System composes of the Pirangdong and Rungri Groups. The Jikhyon Group is subdivided into the Jangbong, Obongri, Jangsusan, and Ansimryong Fms. and is made of mainly conglomerate, quartz sandstone, schist, and phyllite. The Sadangu Group is subdivided into the Unjoksan, Tokjaesan, and Chongsokturi Fms. and is dominated by stromatolite limestone and dolomite. The Mukchon Group is subdivided into the Solhwasan, Okhyonri, and Mukchon Fms. and is dominated by quartz sandstone, phyllite, and micrite. The Myoraksan Group is composed of limestone, dolomite, and sandy phyllite. The Kuhyon System is unconformably sitting on the Sangwon System. The Pirangdong Group of the Kuhyon System is composed by conglomerate, schist, dolomite, pebbly limestone, and phyllite, whereas the Rungri Group of this system is dominated by pebbly phyllite, phyllite, and minor siltstone.

Paek et al. (1993) have suggested that the Jikhyon and Sadangu Groups are Mesoproterozoic in age, whereas the Mukchon and Myoraksan Groups should belong to Neoproterozoic strata. Hu et al. (2012) reported detrital zircon ages which indicate that the Sangwon System could be initiated at the Late Mesoproterozoic to the Early Neoproterozoic. Peng et al. (2011b) have obtained ~900 Ma sills, which have been metamorphosed at ~400 Ma in the Kuhyon System and have suggested the Sangwon and Kuhyon Systems to be the Early Neoproterozoic strata (Fig. 14).

3 Summary: A Prolonged Stepwise Rifting Lasting for Ten Billion Years

Zhai et al. (2015) suggest that the Late Paleoproterozoic to Neoproterozoic sedimentation and rifting in North China are featured as the Earth's 'middle age' by a long-term continuous stable platform or para-platform tectonic setting without any record of the Grenville orogenic event. And it is possible that the Earth's 'middle age' represents a particular tectonic evolution period, during which the Earth had a stable lithosphere with underlying hot mantle that resulted in multistage magmatism and rifting from the Late Paleoproterozoic to the Neoproterozoic.

Although the Proterozoic rifting event in North China is prolonged, it is stepwise as the magmatic events and the strata can be divided into different stages.

Stage 1 (1800–1730 Ma, Early Statherian/Changcheng System): It is represented by the Taihang dykes (1780–1770 Ma), the Miyun dykes ($\sim$1730 Ma), and the strata of the Xiong'er Group with volcanism at $\sim$1780 Ma in the Xiong'er rift system. It is dominated by volcanism with a little clastic sedimentation (conglomerate, sandstone, and shale). It is likely in an intra-continental environment.

Stage 2 (1730/1710–1600 Ma, Late Statherian/Changcheng System): It is represented by the Damiao–Shachang anorthosite–rapakivi granite–dyke complexes (1710–1680 Ma), the Laiwu dykes ($\sim$1680 Ma), the Taishan dykes ($\sim$1620 Ma), and the strata of the Changcheng Group in the Yan-Liao rift, the Ruyang (Ruyang + Luoyu) Group in the Xiong'er rift, probably the Zhaertai Group, lower parts of the Bayan Obo Group, and the Huade Group in the Bayan Obo rift, with the $\sim$1620 Ma Dahongyu volcanic rocks. The strata are dominated by clastic rocks (conglomerate, sandstone, and shale) with a little limestone. There were a few volcanic rocks evolved in the end of this period.

Stage 3 (1600–1400 Ma, the Calymmian/Jixian System): It is represented by the strata of the Jixian Group in the Yan-Liao rift, the Guandaokou Group in the Xiong'er rift, and probably upper parts of the Bayan Obo Group and the Huade Group, with little volcanism and magmatism. The sediments are dominated by the dolomites with a little clastic sedimentation.

Stage 4 (1400–1200 Ma, the Ectasian System): It is represented by the $\sim$1320 Ma dykes/sills/granites, the $\sim$1230 Ma Licheng dyke swarm, and the strata of the Qingbaikou Group in the Yan-Liao rift. The sedimentation is dominated by clastic rocks (sandstone, shale, and siltstone) with a little carbonates (marlstone and limestone).

Stage 5 (1200–1000 Ma, the Stenian System): It is not certain yet if there were any magmatism and deposition during this period. It is probably a period of 'quiet' time with little geological record as from the data available.

Stage 6 (1000–800 Ma, the Tonian System): It is represented by the Dashigou dyke swarm ($\sim$925 Ma), the Sariwon (Dalian–Chulan–Zenghekou) sills, the Qianlishan dyke swarm ($\sim$810 Ma), and the Bagongshan/Xuhuai/Shuxian, Langan, Jikhyon,

Sadangu, Mukchon, Yongning, Xihe, Wuxingshan, Jinxian, and Luanchuan Groups in the Xu-Huai rift. The sedimentation is dominated by clastic rocks and carbonates. Stage 7 (800–541 Ma, the Cryogenian–Ediacaran/Nanhua–Sinian System): It is uncertain yet if there were any igneous events or sedimentation during this period. Some formations, e.g., the Huanglianduo, Dongjia, and Luoquan Fms. in Henan province thought to be deposited during this time period.

The rifting center was in the Xiong'er rift (1780–1730 Ma), and it then moved to the Yan-Liao rift, which was prolonged and lasted till ~1200 Ma. Then in the Early Neoproterozoic, the rifting was mainly in the Xu-Huai rift (1000–800 Ma). The evolution of the Bayan Obo rift system is likely similar with the Yan-Liao rift, but with uncertainty. This means that there was a stepwise and prolonged rifting lasting 1000 Ma in North China, with the rifting centers shifted from the south (the Xiong'er rift, 1780–1730 Ma) to the north (the Yan-Liao rift, 1730–1200 Ma) and to the southeast (the Xu-Huai rift, 1000–800 Ma).

Acknowledgements This review has incorporate publications from many experts who have worked in this area. Drs. Mingguo Zhai, Jinghui Guo, Taiping Zhao, Yusheng Wan, Guochun Zhao, Shuwen Liu, Shuanhong Zhang, Hongrui Fan, Guiting Hou, Minli Cui, Linzhi Gao, H.C. Halls, Yanhong He, Baofu Han, Bo Hu, Neng Jiang, T. Kusky, Huaikun Li, Xianhua Li, Qiuli Li, Linzhi Gao, Huichu Wang, Xianghui Li, Songnian Lu, Qingren Meng, Runmin Peng, Xiaolei Wang, Yuejun Wang, Jinhui Yang, Kuifeng Yang, Yan Zhong, and Jianbo Zhou are specially thanked. Our own research is currently supported by the 973 (2012CB416601) and NFSC (41210003 and 41322018) projects.

References

Cao RJ (2000) Discussion on some problems in the Mesoproterozoic and Neoproterozoic stratigraphical study in China. J Stratigr 24:1–7 (in Chinese)

Chen CY (1993) The ages of the Bayan Obo Gorup, Zhaertai Gorup and Huade Group. Reg Geol China 1:59–67 (in Chinese)

Chen J, Zhang H, Zhu S, Zhao Z, Wang Z (1980) Research on the Sinian System. Sinian System of China. Tianjing Science and Technology Publication, Tianjin, pp 1–114

Cui ML, Zhang BL, Peng P, Zhang LC, Shen XL, Guo ZH, Huang XF (2010) Zircon/baddeleyite U-Pb dating for the Paleoproterozoic intermediate-acid intrusive rocks in Xiaoshan Mountains, west of Henan province and their constraints on the age of the Xiong'er volcanic province. Acta Petrologica Sinica 26:1541–1549 (in Chinese)

Fan HR, Hu FF, Chen FK, Yang KF, Wang KY (2006) Intrusive age of No. 1 carbonatite dyke from Bayan Obo REE-Nb-Fe deposit, Inner Mongolia: with answers to comment of Dr. Le Bas. Acta Petrologica Sinica 22:519–520 (in Chinese)

Gao L, Zhang C, Yin C, Shi X, Wang Z, Liu Y, Liu P, Tang F, Song B (2008a) SHRIMP zircon ages: basis for refining the chronostratigraphic classification of the Meso-and Neoproterozoic strata in North China Old Land. Acta Geoscientica Sinica 29:366–376 (in Chinese)

Gao LZ, Zhang CH, Liu PJ, Ding XZ, Wang ZQ, Zhang YJ (2009) Recoginiation of Meso- and Neoproterozoic stratigraphic framework in North and South China. Acta Geoscientica Sinica 30:433–446 (in Chinese)

Gao LZ, Zhang CH, Shi XY, Song B, Wang ZQ, Liu YM (2008b) Mesoproterozoic age for Xiamaling Formation in North China Plate indicated by zircon SHRIMP dating. Chin Sci Bull 53:2665–2671

Gao LZ, Zhang CH, Shi XY, Zhou HR, Wang ZQ (2007) Zircon SHRIMP U-Pb dating of the tuff bed in the Xiamaling Formation of the Qingbaikouan System in North China. Geol Bull China 26:249–255 (in Chinese)

Guo JH, Sun M, Chen FK, Zhai MG (2005) Sm-Nd and SHRIMP U-Pb zircon geochronology of high-pressure granulites in the Sanggan area, North China Craton: timing of Paleoproterozoic continental collision. J Asian Earth Sci 24:629–642

Guo J, Peng P, Chen Y, Jiao S, Windley BF (2012) UHT sapphirine granulite metamorphism at 1.93–1.92 Ga caused by gabbronorite intrusions: Implications for tectonic evolution of the northern margin of the North China Craton. Precambr Res 222–223:124–142

Halls HC, Li J-H, Davis D, Hou G-T, Zhang B-X, Qian X-L (2000) A precisely dated Proterozoic paleomagnetic pole form the North China Craton, and its relevance to paleocontinental construction. Geophys J Int 143:185–203

Han BF, Zhang L, Wang YM, Song B (2007) Enriched mantle source for Paleoproterozoic high Mg and low Ti-P mafic dykes in central part of the North China craton: constraints from zircon Hf isotopic compositions. Acta Petrologica Sinica 23:277–284 (in Chinese)

He YH, Zhao GC, Sun M, Xia XP (2009) SHRIMP and LA-ICP-MS zircon geochronology of the Xiong'er volcanic rocks: implications for the Paleo-Mesoproterozoic evolution of the southern margin of the North China craton. Precambr Res 168:213–222

Hebei (Bureau of Geology and Mineral Resources of Hebei) (1991) Geology of Hebei. Geology Publishing House, Beijing, pp 1–741 (in Chinese)

Henan (Bureau of Geology and Mineral Resources of Henan) (1989) Geology of Shandong. Geology Publishing House, Beijing, pp 68–124 (in Chinese)

Hou G-T, Li J-H, Qian X-L (2001b) Geochemical characteristics and tectonic setting of Mesoproterozoic dyke swarms in northern Shanxi. Acta Petrologica Sinica 17:352–357 (in Chinese)

Hou G-T, Li J-H, Qian X-L, Zhang B-X, Halls HC (2001a) The paleomagnetism and geological significance of Mesoproterozoic dyke swarms in the central North China craton. Sci China Ser D Earth Sci 44:185–192

Hou GT, Santosh M, Qian XL, Lister GS, Li JH (2008a) Tectonic constraints on 1.3–1.2 Ga final breakup of Columbia supercontinent from a giant radiating dyke swarm. Gondwana Res 14:561–566

Hou GT, Santosh M, Qian XL, Lister GS, Li JH (2008b) Configuration of the late Paleoproterozoic supercontinent Columbia: Insights from radiating mafic dyke swarms. Gondwana Res 14:395–409

Hou GT, Wang CC, Li JH, Qian XL (2006) Late Paleoproterozoic extension and a paleo-stress field reconstruction of the North China Craton. Tectonophysics 422:89–98

Hu B, Zhai MG, Guo JH, Peng P, Liu F, Liu S (2009) LA-ICP-MS U-Pb geochronology of detrital zircons from the Huade Group in the northern margin of the North China Craton and its tectonic significance. Acata Petrologica Sinica 25:193–211

Hu B, Zhai M, Li T, Li Z, Peng P, Guo J, Kusky T (2012) Mesoproterozoic magmatic events in the eastern North China Craton and their tectonic implications: geochronological evidence from detrital zircons in the Shandong Peninsula and North Korea. Gondwana Res 22(3–4):828–842

Jiang N, Guo JH, Zhai MG (2011) Nature and origin of the Wenquan granite: implications for the provenance of Proterozoic A-type granites in the North China craton. J Asian Earth Sci 42:76–82

Kusky TM, Li JH (2003) Paleoproterozoic tectonic evolution of the North China Craton. J Asian Earth Sci 22:383–397

Kusky TM, Li JH, Santosh M (2007) The Paleoproterozoic North Hebei Orogen: North China Craton's collisional suture with the Columbia supercontinent. Gondwana Res 12:4–28

Le Bas MJ, Yang XM, Taylor RN, Spiro B, Milton JA, Zhang PS (2007) New evidence from a calcite-dolomite carbonatite dyke for the magmatic origin of the massive Bayan Obo ore-bearing dolomite marble, Inner Mongolia, China. Mineral Petrol 91:223–248

Li CD, Zheng JM, Zhang YL, Zhang K, Hua YQ (2005) Redefinition of the Huade Group and its tectonic significance. Geol China 32:354–362

Li HK, Li HM, Lu SN (1995a) Grain zircon U-Pb age for volcanic rocks from Tuanshanzi Formation of Changcheng System and their geological implication. Geochimica 24:43–48 (in Chinese)

Li HK, Li HM, Lu SN (1995b) Grain zircon U-Pb ages for volcanic rocks from Tuanshanzi Formation of Changcheng System and their geological implications. Geochemica 24:43–48 (in Chinese)

Li HK, Lu SN, Li HM, Sun LX, Xiang ZQ, Geng JZ, Zhou HY (2009) Zircon and baddeleyite U-Pb precision dating of basic rock sills intruding Xiamaling Formation, North China. Geol Bull China 28:1396–1404 (in Chinese)

Li HK, Lu SN, Su WB, Xiang ZQ, Zhou HY, Zhang YQ (2013) Recent advances in the study of the Mesoproterozoic geochronology in the North China Craton. J Asian Earth Sci 72:216–227

Li HK, Su WB, Zhou HY, Geng JZ, Xiang ZQ, Cui YR, Liu WC, Lu SN (2011) The base age of the Changchengian System at the northern North China Craton should be younger than 1670 Ma: constraints from zircon U-Pb LA-MC-ICP-MS dating of a granite-porphyry dike in Miyun County, Beijing. Earth Sci Front 18:108–120 (in Chinese)

Li HK, Zhu SX, Xiang ZQ, Su WB, Lu SN, Zhou HY, Geng JZ, Li S, Yang FJ (2010) Zircon U-Pb dating on tuff bed from Gaoyuzhuang Formation in Yanqing, Beijing: further constraints on the new subdivision of the Mesoproterozoic stratigraphy in the northern North China Craton. Acta Petrologica Sinica 26:2131–2140 (in Chinese)

Li QL, Chen FK, Guo JH, Li XH, Yang YH, Siebel W (2007a) Zircon ages and Nd–Hf isotopic composition of the Zhaertai Group (Inner Mongolia): evidence for early Proterozoic evolution of the northern North China Craton. J Asian Earth Sci 30:573–590

Li XH, Su L, Song B, Liu DY (2004) SHRIMP U-Pb zircon age of the Jinchuan ultramafic intrusion and its geological significance. China Sci Bull 49:420–422

Li XH, Chen FK, Guo JH, Li QL, Xie LW, Siebel W (2007b) South China provenance of the lower-grade Penglai Group north of the Sulu UHP orogenic belt, eastern China: evidence from detrital zircon ages and Nd-Hf isotopic composition. Geochem J 41:29–45

Li Y, Peng P, Wang X, Wang H (2015) Nature of 1800-1600 Ma mafic dyke swarms in the North China Craton: Implications for the rejuvenation of the sub-continental lithospheric mantle. Precambr Res 257:114–123

Liaoning (Bureau of Geology and Mineral Resources of Liaoning) (1989) Geology of Liaoning. Geology Publishing House, Beijing, pp 1–393 (in Chinese)

Liu SW, Pan YM, Xie QL, Zhang J, Li QG, Yang B (2005) Geochemistry of the Paleoproterozoic Nanying granitic gneisses in the Fuping Complex: implications for the tectonic evolution of the Central Zone, North China Craton. J Asian Earth Sci 24:643–658

Lu SN, Yang CL, Li HK, Li HM (2002) A Group of Rifting Events in the Terminal Paleoproterozoic in the North China Craton. Gondwana Res 5:123–131

Lu SN, Zhao GC, Wang HC, Hao GJ (2008) Precambrian metamorphic basement and sedimentary cover of the North China craton: a review. Precambr Res 160:77–93

Luo Y, Sun M, Zhao GC (2006) LA-ICP-MS U-Pb Zircon Geochronology of the Yushulazi Group in the Eastern Block, North China Craton. Int Geol Rev 48:828–840

Meng QR, Wei HH, Qu YQ, Ma SX (2011) Stratigraphic and sedimentary records of the rift to drift evolution of the northern North China craton at the Paleo- to Mesoproterozoic transition. Gondwana Res 20:205–218

Neimenggu (Bureau of Geology and Mineral Resources of Neimenggu/Inner Mongolia) (1991) Geology of Neimenggu. Geology Publishing House, Beijing, pp 1–725 (in Chinese)

Paek RJ, Kan HG, Jon GP, Kim YM, Kim YH (eds) (1993) Geology of Korea. Foreign languages books publishing house, Pyongyang, p 619

Peng P, Zhai MG, Zhang HF, Guo JH (2005) Geochronological constraints on the Paleoproterozoic evolution of the North China Craton: SHRIMP zircon ages of different types of mafic dikes. Int Geol Rev 47:492–508

Peng P, Zhai MG, Ernst R, Guo JH, Liu F, Hu B (2008) A 1.78 Ga large igneous province in the North China craton: the Xiong'er Volcanic Province and the North China dyke swarm. Lithos 101:260–280

Peng P (2010) Reconstruction and interpretation of giant mafic dyke swarms: a case study of 1.78 Ga magmatism in the North China craton. In: Kusky TM, Zhai MG, Xiao WJ (eds) The evolving continents: understanding processes of continental growth. Geological Society, London, Special Publications, vol 338, pp 163–178

Peng P, Bleeker W, Ernst RE, Söderlund U, McNicoll V (2011a) U-Pb baddeleyite ages, distribution and geochemistry of 925 Ma mafic dykes and 900 Ma sills in the North China craton: evidence for a Neoproterozoic mantle plume. Lithos 127:210–221

Peng P, Zhai M, Li Q, Wu F, Hou Q, Li Z, Li T, Zhang Y (2011b) Neoproterozoic (~900 Ma) Sariwon sills in North Korea: geochronology, geochemistry and implications for the evolution of the south-eastern margin of the North China Craton. Gondwana Res 20:243–254

Peng P, Liu F, Zhai M, Guo J (2012) Age of the Miyun dyke swarm: constraints on the maximum depositional age of the Changcheng System. Chin Sci Bull 57:105–110

Peng P, Wang X, Windley BF, Guo J, Zhai M, Li Y (2014) Spatial distribution of ~1950–1800 Ma metamorphic events in the North China Craton: implications for tectonic subdivision of the craton. Lithos 202–203:250–266

Peng P (2015) Precambrian mafic dyke swarms in the North China Craton and their geological implications. Sci China: Earth Sci 58:649–675

Peng RM, Zhai YS, Wang JP, Chen XF, Liu Q, Lv JY, Shi YX, Wang G, Li SW, Wang LG, Ma YT, Zhang P (2010) Discovery of Neoproterozoic acid volcanic rock in the western section of Langshan, Inner Mongolia, and its geological significance. Chin Sci Bull 55:2611–2620 (in Chinese)

Peng TP, Wilde SA, Fan WM, Peng BX, Mao YS (2013) Mesoproterozoic high Fe–Ti mafic magmatism in western Shandong, North China Craton: petrogenesis and implications for the final breakup of the Columbia supercontinent. Precambr Res 235:190–207

Qiao XF, Ji Q, Gao LZ, Zhang YX, Gao ZJ (1996) Stratigraphic correlation of the Sinian in the eastern part of the northern China plate. Reg Geol China 2:135–142 (in Chiense)

Qian X-L, Chen Y-P (1987) Late Precambrian mafic dyke swarms of the North China craton. In: Halls HC, Fahrig WF (eds) Mafic Dykes Swarms. Geology Association of Canada Special Paper 34, pp 385–391

Rämö OT, Haapala I, Vaasjoki M, Yu JH, Fu HQ (1995) 1700 *Ma* Shachang complex, northeast China: Proterozoic rapakivi granite not associated with Paleoproterozoic orogenic crust. Geology 23:815–818

Shandong (Bureau of Geology and Mineral Resources of Shandong) (1991) Geology of Shandong. Geology Publishing House, Beijing, pp 1–595 (in Chinese)

Su W, Li H, Xu L, Jia S, Geng J, Zhou H, Wang Z, Pu H (2012) Luoyu and Ruyang Group at the south margin of the North China Craton (NCC) should belong in the Mesoproterozoic Changchengian System: Direct constraints from the LA-MC-ICPMS U-Pb age of the tuffite in the Luoyukou Formation, Ruzhou, Henan, China. Geol Survey Res 35:96–108

Su WB, Zhang SH, Huff WD, Li HK, Ettensohn FR, Chen XY, Yang HM, Han YG, Song B, Santosh M (2008) SHRIMP U-Pb ages of K-bentonite beds in the Xiamaling Formation: implications for revised subdivision of the Meso- to Neoproterozoic history of the North China Craton. Gondwana Res 14:543–553

Su WB, Li HK, Warren DH, Ettensohn FR, Zhang SH, Zhou HY, Wan YS (2010) Zircon SHRIMP U-Pb ages of tuff in the Tieling Formation and their geological significance. Chin Sci Bull 55:3312–3323

Tan LK, Shi TZ (2000) Discovery and significance of samll shelly fossils in the Bayan Obo Group in Shangdu, Inner Mongolia. Geol Rev 46:573–583 (in Chinese)

Wan YS, Zhang QD, Song TR (2003) SHRIMP detrial zircon ages of the Changzhougou Fm. Of the Changcheng System in Shisanling, Beijing: implications for the provenance and maximum deposition age of the cover of the North China Craton. Chin Sci Bull 48:1970–1975 (in Chinese)

Wang J, Li SQ, Wang BL, Li JJ (1992) The Langshan-Bayan Obo rift system. Peking University Publishing House, Beijing, pp 1–132 (in Chinese)
Wang QH, Yang DB, Xu WL (2012) Neoproterozoic basic magmatism in the southeast margin of North China Craton: Evidence from whole-rock geochemistry, U-Pb and Hf isotopic study of zircons from diabase swarms in the Xuzhou-Huaibei area of China. Sci China: Earth Sci 55:1416–1479
Wang SS, Sang HQ, Qiu J, Chen ME, Li MR (1995) The metamorphic age of pre-Changcheng System in Beijing-Tianjin area and a discussion about the lower limit age of Changcheng System. Scientia Geologica Sinica 30:348–354 (in Chinese)
Wang W, Liu S, Bai X, Li Q, Yang P, Zhao Y, Zhang S, Guo R (2013) Geochemistry and zircon U-Pb-Hf isotopes of the late Paleoproterozoic Jianping diorite-monzonite-syenite suite of the North China Craton: implications for petrogenesis and geodynamic setting. Lithos 162:175–194
Wang XL, Jiang SY, Dai BZ, Griffin WL, Dai MN, Yang YH (2011) Age, geochemistry and tectonic setting of the Neoproterozoic (ca. 830 Ma) gabbros on the southern margin of the North China Craton. Precambr Res 190:35–47
Wang YJ, Fan WM, Zhang YH, Guo F, Zhang HF, Peng TP (2004) Geochemical, 40Ar/39Ar geochronological and Sr–Nd isotopic constraints on the origin of Paleoproterozoic mafic dikes from the southern Taihang Mountains and implications for the ca. 1800 Ma event of the North China Craton. Precambr Res 135:55–77
Wang YJ, Zhao GC, Cawood PA, Fan WM, Peng TP, Sun LH (2008) Geochemistry of Paleoproterozoic ($\sim$1770 Ma) mafic dikes from the Trans-North China Orogen and tectonic implications. J Asian Earth Sci 33:61–77
Xue YS, Cao RJ, Tang TF, Yin LM, Yu CL, Yang JD (2001) The Sinian stratigraphic sequence of the Yangtze region and correlation to the late Precambrian strata of North China. J Stratigr 25:207–216
Yang JH, Wu FY, Liu XM (2005) Zircon U-Pb ages and Hf isotopes and their geological significance of the Miyun rapakivi granites from Beijing, China. Acta Petrologica Sinica 21:1633–1644 (in Chinese)
Yang KF, Fan HR, Santosh M, Hu FF, Wang KY (2011) Mesoproterozoic carbonatitic magmatism in the Bayan Obo deposit, Inner Mongolia, North China: constraints for the mechanism of super accumulation of rare earth elements. Ore Geol Rev 40:122–131
Zhai MG, Guo JH, Li JH, Li YG, Yan YH, Zhang WH (1996) Retrograded eclogites in the Archaean North China craton and their geological implication. Chin Sci Bull 41:315–326
Zhai MG, Peng P (2007) Paleoproterozoic events in the North China Craton. Acta Petrologica Sinica 23:2665–2682 (in Chinese)
Zhai MG, Hu B, Zhao T-P, Meng Q-R (2015) Late Paleoproterozoic-Neoproterozoic multi-rifting events in the North China Craton and their geological significance: a review. Tectonophysics. doi:10.1016/j.tecto.2015.01.019
Zhang PF (2001) Division and correlation of late Precambrian strata in southern Liaoning, northern Jiangsu and Anhui and eastern Shandong. Geol Resour 10:11–17 (in Chinese)
Zhang SH, Liu SW, Zhao Y, Yang JH, Song B, Liu XM (2007) The 1.75–1.68 Ga anorthosite-mangerite-alkali granitoid-rapakivi granite suite from the northern North China Craton: magmatism related to a Paleoproterozoic orogen. Precambr Res 155:287–312
Zhang SH, Zhao Y, Yang ZY, He ZF, Wu H (2009) The 1.35 Ga diabase sills from the northern North China craton: implications for breakup of the Columbia (Nuna) supercontinent. Earth Planet Sci Lett 288:588–600
Zhang SH, Zhao Y, Santosh M (2012a) Mid-Mesoproterozoic bimodal magmatic rocks in the northern North China Craton: implications for magmatism related to breakup of the Columbia supercontinent. Precambr Res 222–223:339–367
Zhang SH, Li ZX, Evans DAD, Wu HC, Li HY, Dong J (2012b) Pre-Rodinia supercontinent Nuna shaping up: a global synthesis with new paleomagnetic results from North China. Earth Planet Sci Lett 353–354:145–155

Zhao GC, Wilde SA, Cawood PA, Sun M (2001) Archean blocks and their boundaries in the North China Craton: lithological, geochemical, structural and P-T path constraints and tectonic evolution. Precambr Res 107:45–73

Zhao GC, He YH, Sun M (2009a) Xiong'er volcanic belt in the North China Craton: Implications for the outward accretion of the Paleo-Mesoproterozoic Columbia Supercontinent. Gondwana Res 16:170–181

Zhao GC (2014) Precambrian evolution of the North China Craton. Elsevier, Oxford, p 194

Zhao TP, Zhou MF, Zhai MG, Xia B (2002) Paleoproterozoic rift–related volcanism of the Xiong'er group, North China Craton: implications for the breakup of Columbia. Int Geol Rev 44:336–351

Zhao TP, Zhai MG, Xia B, Li HM, Zhang YX, Wang YS (2004) Zircon U-Pb SHRIMP dating for the volcanic rocks of the Xiong'er Group: Constraints on the initial formation age of the cover of the North China craton. Chin Sci Bull 49:2495–2502

Zhao TP, Chen W, Zhou MF (2009b) Geochemical and Nd–Hf isotopic constraints on the origin of the ~1.74 Ga Damiao anorthosite complex, North China Craton. Lithos 113:673–690

Zheng JM, Liu YS, Chen YF, Gao X (2004) Zircon U-Pb ages of Paleopterozoic granite and the age of the Huade Gorup in Kangbao area, north Hebei province. Geol Survey Res 27:14–17 (in Chinese)

Zhong Y, Zhai M, Peng P, Santosh M, Ma X (2015) Detrital zircon U-Pb dating and whole-rock geochemistry from the clastic rocks in the northern marginal basin of the North China Craton: Constraints on depositional age and provenance of the Bayan Obo Group. Precambr Res 258:133–145

Zhou JB, Hu K (1998) Tectonic activities of the Yihe-Shuhe Fault during the Jinning movement period. Seismol Geol 20:208–212 (in Chinese)

Zhou JB, Wilde SA, Zhao GC, Zheng CQ, Jin W, Zhang XZ, Cheng H (2008) SHRIMP U-Pb zircon dating of the Neoproterozoic Penglai Group and Archean gneisses from the Jiaobei Terrane, North China, and their tectonic implications. Precambr Res 160:323–340

Part III
The South China Craton

Early Precambrian Geological Signatures in South China Craton

Yuan-Sheng Geng

Abstract The South China Craton is formed through coalescence of the Yangtze and Cathaysian blocks in the Neoproterozoic and with 3 tectonic units included, the Yangtze and Cathaysia blocks and the intervening Jiangnan orogenic belt. As relatively young in formation time, the Yangtze craton is only sporadically distributed with Early Precambrian terranes. Some occurrences of Archean outcrop in northern Yangtze craton, such as the localities at Kongling, Yichang city, Huji, Zhongxiang city, of Hubei Province, and Yudongzi locality, Mian-Lue area, of Shaanxi Province. Among them, the Kongling terrane is well exposed, including the Meso–Neoarchean metamorphic supracrustal rocks of khondalite feature, the TTG gneiss series of 3.3, 2.9 and 2.7–2.6 Ga, and the metamorphism grade reached upper amphibolite to granulite facies. The Paleoproterozoic terranes are the Tangdan Group at southwestern margin, Houhe Complex at the northern margin of the Yangtze craton, the Badu Group, and the S-, I-granites at the northern margin of the Cathaysian block. Though the Archean and Paleoproterozoic rocks are sporadically exposed in South China Craton, the ubiquitous presence of the older detrital and inherited zircons suggests the once widespread occurrence of the Early Precambrian terranes on the craton. The zircon Hf-isotope compositions demonstrate three major magmatic events which were manifested substantially by the recycling of the crustal materials, accompanied by minor juvenile crustal accretion.

Keywords Archean · Paleoproterozoic · South China Craton · Yangtze block · Cathaysian block · Jiangnan (orogenic) belt · Dongchonghe complex

Y.-S. Geng (✉)
Institute of Geology, Chinese Academy of Geology Sciences, Beijing 100037, China
e-mail: ys-geng@cags.ac.cn

M. Zhai (ed.), *Precambrian Geology of China*,
Springer Geology, DOI 10.1007/978-3-662-47885-1_5

1 Introduction (Distribution of Precambrian Basement in South China)

South China Craton (SCC), located in the southeast part of China mainland, is formed through coalescence of the Yangtze and Cathaysian blocks in the Neoproterozoic and with 3 tectonic units included, the Yangtze and Cathaysia blocks and the intervening Jiangnan orogenic belt. The SCC is separated from the North China Craton by the central orogenic belt to the north and from the Tibetan Plateau by the Sanjiang Orogen to the west, and was developed from three Precambrian basements (or paleoplates) of the Yangtze basement, the Cathaysian basement, and the Indochina basement (Cheng 1994).

The SCC was developed on the Neoproterozoic-Caledonian orogeny, and therefore (Early) Neoproterozoic geological bodies constitute its major basement, and Early Precambrian rocks are rarely documented. Therefore, the SCC is

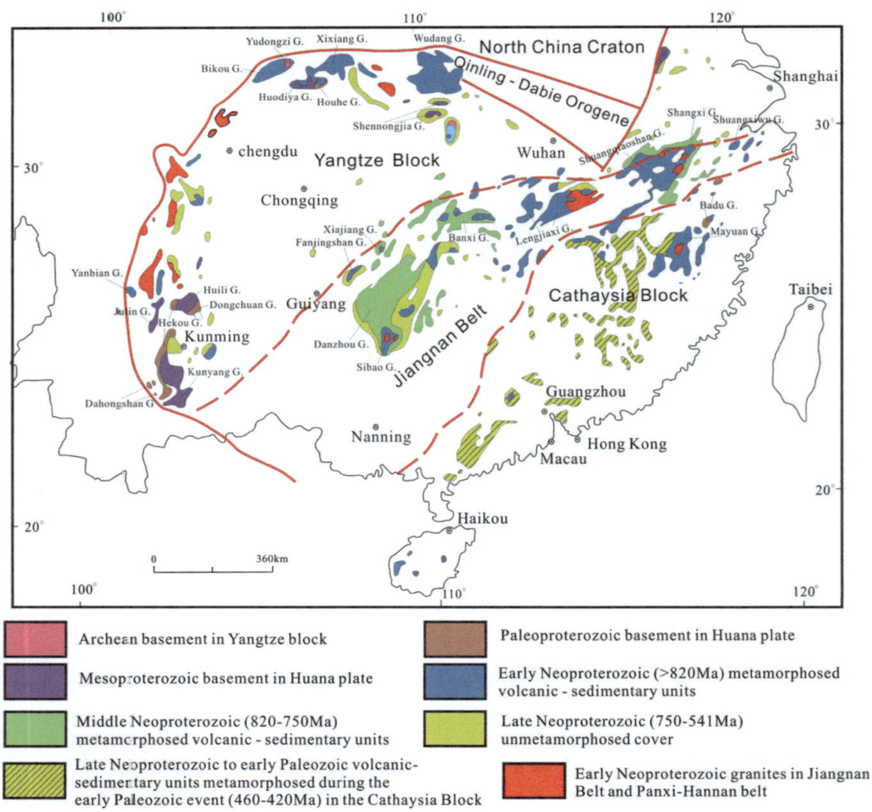

Fig. 1 Geological map showing the distribution of Precambrian geological bodies in the South China. (Revised after Zhao and Cawood 2012) Abbreviations: G = Group

apparently distinct from the North China Craton (NCC) because the NCC basement is dominantly made of Early Precambrian geological bodies. Paleoproterozoic geological bodies only were identified from the Huangling area in Hubei Province, the southwestern Zhejiang Province, and the northern Fujian Province (Fig. 1). In the Dongchuan area of the northeastern Yunnan Province, the Tangdan Group was also possibly formed during the Paleoproterozoic era. Moreover, the Kangding Group, located at the southwestern margin of the Yangtze Block, was previously considered to be formed during Archean to Paleoproterozoic (Li et al. 1988; Feng 1990; BGMRSC 1991; Cheng 1994). However, recent SHRIMP zircon U–Pb dating revealed that the Kangding Group is dominantly of Neoproterozoic magmatic complexes, with minor Meso- to Neoproterozoic metamorphosed strata (Zhou 2002a, b, 2006a; Geng et al. 2007a, 2008).

In this contribution, the Precambrian geological features of the Yangtze Block, the Jiangnan belt, and the Cathaysian Block are described in order. In convenience of work correlations in China, an age of 1.8 Ga is invoked as the lower limited age of Mesoproterozoic era.

2 Archean Geological Signatures of the SCC

Rare Archean terranes are exposed in South China, and minor outcrops are present in the northern Huangling, Huji town in Zhongxiang County, Hubei Province and Yudongzi locality in Lueyang, Shaanxi Province. Up to now, no substantial Archean materials have been observed in Jiangnan belt and Cathaysian block, but some detrital and inherited information of Archean can be disserved in the clastics and magmatic rocks in these Yangtze block, Jiangnan belt, and Cathaysian block.

2.1 Archean Geological Signatures of the Yangtze Block

In northern Yangtze block, some high-grade metamorphic terranes outcrop at the Zigui–Xingshan area in the Huangling region, Yichang city, Hubei Province. The rocks are mainly present at the core of the Huangling Dome or Anticline, and originally called as Kongling schist (Lee and Chao 1924). Geological survey has demonstrated that the Huangling dome is bounded at the northern margin by the Late Proterozoic–Paleozoic sequence, the core of medium- to high-grade metamorphics and granitic gneisses (Fig. 2). With the total area of some 300 km^2, the dome is actually an anticline along the NS strike. Further research has shown that the metamorphics of the southern dome are different from that of the northern part of the dome, and the former dominated by the granitoids of Neoproterozoic, with minor metamorphic sequence of Kongling Group, and the latter mainly consists of Archean TTG gneisses and supracrustal rocks, which were called as Dongchonghe complex (Li and Nie 1987; Shen et al. 1996) or Kongling complex (Zhang et al. 2006a). The complex

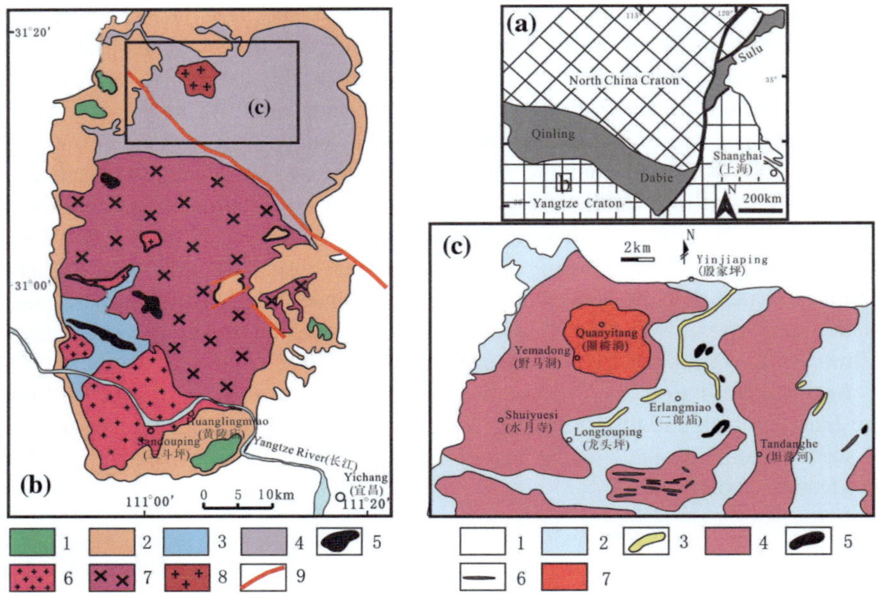

Fig. 2 Tectonic setting of the Kongling area (**a**), geological map of the Kongling anticline and Dongchonghe complex (**b**, **c**) (Modified from Qiu et al. (2000) and Zhang et al. (2006b)), **b** *1* Cambrian, *2* Neoproterozoic Sinian, *3* Meso-Neoproterozoic Kongling Group, *4* Archean Dongchonghe Complex, *5* Mafic-ultramafic complexes, *6* Neoproterozoic potassic granites, *7* Neoproterozoic Huangling granites, *8* Paleoproterozoic Quanyitang granites, *9* major faults; **c** *1* Neoproterozoic and later covers, *2* Archean metamorphosed supracrustal rocks, *3* Marble layers, *4* Archean TTG gneisses, *5* Mafic-ultramafic rocks, *6* mafic dykes, *7* Paleoproterzoic Quanyitang granites

has been intruded by the Paleoproterozoic Quanyitang potassic granite and Neoproterozoic Huangling granite (Fig. 2b, c).

The Dongchonghe complex is distributed at the Quchangsai, Yemadong, Dongchonghe, and Yanluoping area of Yichang, Hubei Province, in northern dome, and some places near Maoya in northern dome. The complex is composed of both the TTG and supracrustal rocks (Fig. 2c). The TTG is the dominant and includes the diorite, tonalite, trondhjemite, and granitic gneisses. The supracrustal rocks are graphite-bearing sillimanite-garnet biotite–plagioclase gneiss, (garnet) biotite-bio-plagioclase gneiss, migmatitic-bio-plagioclase gneiss, and minor bands of amphibolite, quartzite, graphite schist, biotite schist, siliceous rock, and marble. Some index minerals like andalusite, sillimanite, staurolite, and garnet can be found in the schists and gneisses, showing the feature of khondalite (Lu et al. 1996). Some diabase dykes intruded in the TTG and supracrustal rocks. The Dongchonghe complex has reached amphibolite facies in metamorphism, with the peak condition of 700 °C, 0.4 GPa, and then isothermal decompression process of dockwise *P-T* path (Lu et al. 1996). The complex has been intensively migmatized and undergone multiple phases of deformation.

The Dongconghe complex mainly consists of the Mesoarchean TTG gneisses and Neoproterozoic granitic (granodioritic) rocks. The Mesoarchean rocks are basically situated at the northern dome and the Neoproterozoic granite (granodiorite) at the eastern dome. Three stages of magmatism in the Archean can be discerned in the complex, the early phase of ~3.4–3.2 Ga, with a few exposures of TTG gneisses, such as the gray gneiss at Maoya (Wang et al. 2001), trondhjemitic gneiss at Yemadong (Zhang et al. 2006a), and granitic gneisses (Chen et al. 2013). The ~2.9 Ga phase of magmatism is the dominant part of the Dongchonghe complex and is widely distributed, including the trondhjemitic, tonalitic-dioritic, and migmatitc gneisses (Li and Nie 1987; Zheng et al. 1991; Ma et al. 1997; Ling et al. 1998; Qiu et al. 2000; Gao et al. 2001; Wang et al. 2001; Zhang et al. 2006a; Chen et al. 2013). The Neoarchean phase (~2.7–2.6 Ga) of magmatism is manifested in the eastern part of the northern dome, forming the magmatic complex dominated by the granitic and granodioritic schists (Chen et al. 2013). The magmatic rocks of the 3 phases are the skeleton of the Dongchonghe complex, but the setting and features of the magmatism are poorly understood and need further research. The metasediments of the Dongchonghe complex have shown plenty of 3.2–2.8 Ga detrital zircon records (Gao et al. 2001; Zhang et al. 2006b). One sample (KH21) has given the 2.75 Ga age of a metamorphic zircon, constraining the lowest deposition age of the complex (Gao et al. 2001). Recent study shows that the oldest rocks in the Kongling terrain were formed at ~3.4 Ga and metamorphosed after a few tens of million years (Guo et al. 2014).

All the geochronological data above suggest that there are three stages of magmatism in Archean in the Kongling region: the first 3.3–3.2 Ga, the second 3.0–2.9 Ga, and the third ~ 2.6 Ga. However, the 2.75 Ga of metamorphic zircon can give preliminary constraint on the deposition age older than 2.8 Ga and metamorphism reactivation at c. 2.75 Ga.

With the area of some 8 km^2, the Yangpo Group is located some 200 km NE of the Huangling dome and is present as a narrow NNW band of metamorphic sequence at Huji town, Zhongxiang city, Hubei Province (Fig. 3). The group is mainly composed of various schists, leptynites, meta-quartzite, meta-quartz-graywacke, intercalated with minor amphibolite, and intruded by beads of granites. The quartz schist produces many groups of zircon ages, the oldest one 3057 ± 4 Ma ($n = 4$, MSWD = 0.78), suggesting the presence of the Mesoarchean continental materials in the source. The youngest group is 2801 ± 24 Ma ($n = 33$, MSWD = 2.7) and can be used to limit the maximum deposition age (Wang et al. 2013a). However, the granite intruding the Yangpo Group is measured to have the zircon U–Pb age of 2655 ± 9 and 2652 ± 21 Ma (Wang et al. 2013a, 2013b). Thus, the Yangpo Group is deduced to have formed between ~2.8 and 2.65 Ga.

A suite of metamorphic strata sequence was exposed along the Gelaoling–Yudongzi area of the Lueyang County, Shaanxi Province (Fig. 4), and was traditionally named as the Bikou Group. Subsequently, the generalized Bikou Group was separated into three lithological units, from bottom to top in turn, of the Late Archean Yudongzi Group, the Mesoproterozoic Bikou Group (sensu stricto), and the upper Sinian to lower Cambrian strata units, respectively (Qin et al. 1990, 1992).

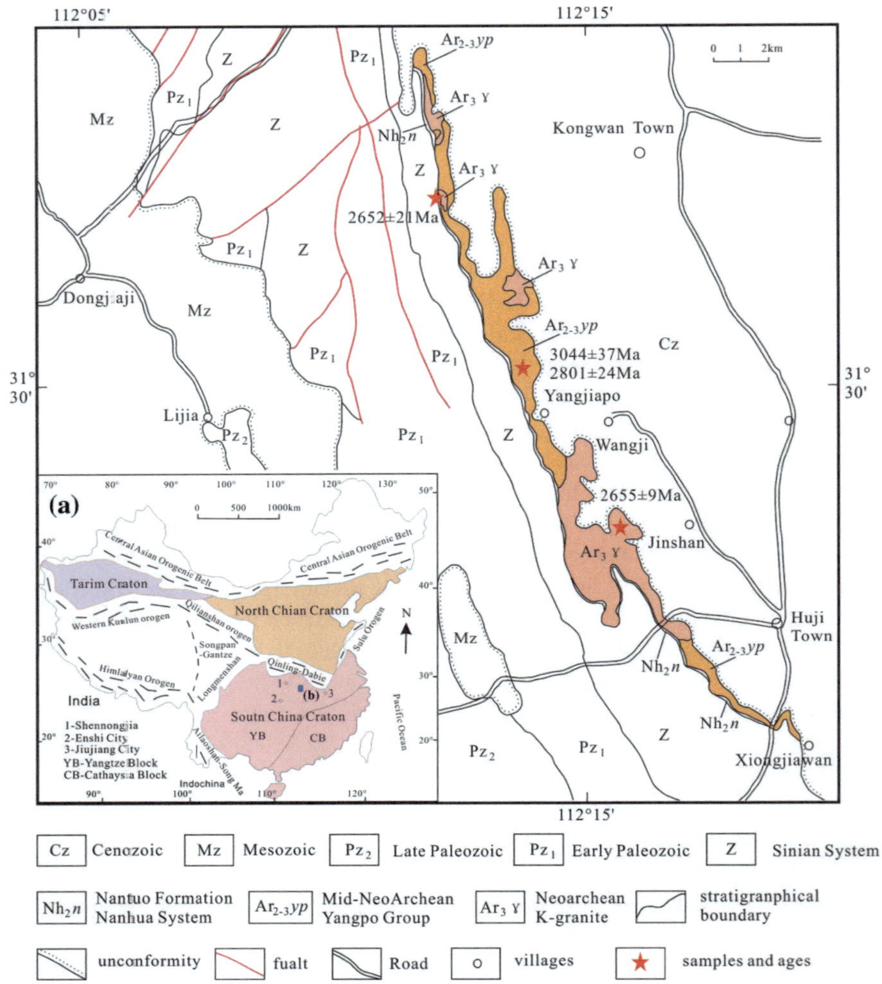

Fig. 3 (**a**) Schematic tectonic map of China showing the major Precambrian blocks connected by Phanerozoic fold belts (Zhao and Cawood 2012). **b** Geological map of the Huji region of Zhongxiang City. (after Wang et al. 2013a)

Although the Yudongzi Group is located in the southern Qinling Orogenic belt (Fig. 4a), the similar Nd isotopic features of its metamorphic rocks to those of the Archean rocks in the Kongling area imply that the Yudongzi Group should be attributed to the basement debris of the Yangtze Block (Zhang et al. 2001, 2002). Recent 1:50000 geological survey suggests that the original Late Archean Yudongzi Group can be further divided into the Yudongzi Group and the Archean TTG gneisses (Fig. 4b) (Huangniping gray gneisses and Longwanggou leuco-gneisses).

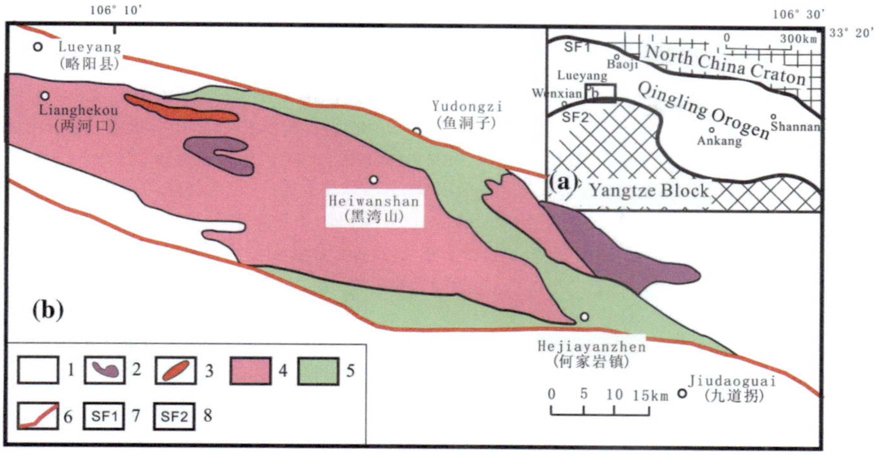

Fig. 4 Sketch geological map of the Yudongzi Complex in the Lueyang County, Shaanxi Province. *1* Neoproterozoic and later covers, *2* Altered mafic-ultramafic rocks, *3* Biotite monzogranites, *4* Archean TTG gneisses, *5* Archean Yudongzi Group, *6* Major faults, *7* Shangdan suture zone, *8* Mianlue suture zone

The newly defined Yudognzi Group is located in the northern and southern flanks of the TTG gneisses, and some occur as enclaves within the TTG gneisses, and they together constitute a granite–greenstone terrane (Wang et al. 1998). Narrow sense Yudongzi Group is mainly distributed in the Heishangou–Majiagou and Shuilinshu areas in the east of the Lueyang County, with a lithological assemblage of banded magnetite quartzites, amphibolites, felsic leptites, leptynites, magnetite actinolite schists, chlorite epidote actinolite schists, albite chlorite schists, and chlorite sericite schists. Moreover, metamorphosed volcano-sedimentary-type iron deposits and ductile shear-zone-type gold deposits locally occur within the Yudongzi Group (Wang et al. 1998). Conventional zircon U–Pb dating for a migmatic amphibolite yields an upper intercept age of 2657 ± 9 Ma, which was interpreted as a metamorphic age (Qin et al. 1992). In the Majiagou iron deposit field to the north of the Yudongzi–Gelaoling terrane, thirteen amphibolites and gneisses from the Yudongzi Group yield a comparable whole-rock Sm–Nd isochron age of 2688 ± 100 Ma (Zhang et al. 2002). A pinky fine-grained granite intruded into the Yudongzi Group gives an upper intercept age of 2693 ± 9 Ma, obtained by conventional zircon dating methods (Zhang et al. 2002). Recently, precise LA–ICPMS zircon U–Pb dating of a mylonitized fine-grained biotite granite and a strongly foliated biotite granite yields two ages of 2661 ± 17 Ma and 2703 ± 26 Ma, respectively (Zhang et al. 2010a, b). Albeit the lack of reliable ages for the Yudongzi Group, the accompanied deformed granites yield Neoarchean ages, implying that rocks of the Yudongzi Group should be formed in Neoarchean, and earlier than the deformed granites.

2.2 Age Information from the South China Craton

Except for the minor Archean geological bodied exposed in the Kongling area, Huji area of Hubei Province and the Yudongzi area of Shaanxi Province, no other reliable Neoarchean rocks, have been documented from the South China Craton. However, detrital zircons from Mesoproterozoic to Late Paleozoic (even modern rivers) sediments and zircon xenocrysts from some Mesozoic to Cenozoic magmatic rocks contain amounts of Archean age information, which are listed in Table 1. From these in situ zircon dating results, the following three features are observed. Firstly, Archean detrital zircons were identified not only in the Neoproterozoic Liantuo Formation near the Kongling Complex and in the Bikou Group around the Yudongzi Group, but also from the Yunnan Province, Western Sichuan, the Jiangnan belt and the Cathaysian Block, which are far away from the exposure regions of Archean rocks or even without exposed Archean rocks. Therefore, Archean terranes may have been widely distributed in the South China Craton. Secondly, host rocks with the Archean age information show a large age range, suggesting that certain amounts of Archean rocks outcropped in the source region of both Meso- to Neoproterozoic and recent sediments. Moreover, large amounts of Archean zircons captured by the Mesozoic to Cenozoic magmatic rocks reveal the existence of Archean basement in the lower or middle continental crust. Thirdly, hidden Archean basement may be existed widely in some areas, as indicated by ~3856 Ma zircon xenocrysts from the Meso- to Cenozoic volcanic rocks in the southeastern Guangxi Province (Zheng et al. 2008, 2011), ~3802 Ma detrital zircon grains from the Neoproterozoic Liantuo Formation in Hubei Province (Zhang et al. 2006b), ~3778, 3732 and 3775 Ma detrital zircons from Proterozoic sedimentary rocks in Dongchuang area of Yunnan Province (Zhu et al. 2011a, b; Zhao et al. 2010; Li et al. 2013), ~3817 and ~3959 Ma detrital zircons from the Ordovician sandstones in Jiangxi Province (Yao et al. 2011), and the ~3755 Ma inherited zircons from Tanxi gneisses in the Nanxiong area of Guangdong Province (Yu et al. 2007). These ancient zircons provide important clues to trace the early formation and evolution history of the South China Craton. Until now, 2.5 Ga rocks have not been identified from the South China Craton, and the large amounts of 2.5 Ga detrital or inherited zircons in the sedimentary or magmatic rocks led to the conclusion that intense ~2.5 Ga tectonothermal events may also occur in the South China Craton, similar to the North China Craton.

Although minor Archean geological bodies exposed in the South China Craton, the numerous Archean detrital zircons in sedimentary rocks indicate that large Archean terranes may have served as a major source region. Meanwhile, the large amounts of Archean inherited zircons imply the possible presence of Archean basement during Meso- to Cenozoic in the lower continental crust of the South China Craton. It is noteworthy that the South China Craton experienced crustal evolution in the early earth history and was significantly modified during the terminal Archean, as suggested by the 3.7 and 2.5 Ga ancient zircons.

Table 1 Archean age information of the South China Craton

Geological unit	Location	Lithology or zircon types	Analytical method	Age/Ma	Citation
Formation ages of Archean rocks					
Yangtze Block	Kongling, Yichang of Hubei	Granitic gneiss	SIMS concordia age (6)	3437 ± 12	Guo et al. 2014
	Kongling, Yichang	Granitic gneiss	SIMS concordia age (2)	3426 ± 69	Guo et al. 2014
	Kongling, Yichang	Granitic gneiss	LA-ICPMS concordia age (7)	3329 ± 42	Chen et al. 2013
	Kongling, Yichang	Biotite–plagioclase gneiss	LA-ICPMS concordia age (24)	3218 ± 13	Jiao et al. (2009)
	Kongling, Yichang	Trondhjemitic gneiss	SHRIMP concordia age (13)	2947 ± 75	Qiu et al. (2000)
	Kongling, Yichang	Trondhjemitic gneiss	SHRIMP concordia age (6)	2903 ± 10	Qiu et al. (2000)
	Kongling, Yichang	Migmatite	SHRIMP upper intercept age	2916 ± 31	Zhang et al. (2006a)
	Kongling, Yichang	Migmatite	LA-ICPMS upper intercept age	2936 ± 28	Zhang et al. (2006a)
	Kongling, Yichang	Migmatite	LA-ICPMS upper intercept age	2930 ± 44	Zhang et al. (2006a)
	Kongling, Yichang	Migmatite	LA-ICPMS upper intercept age	2947 ± 28	Zhang et al. (2006a)
	Kongling, Yichang	Migmatite	LA-ICPMS upper intercept age	2947 ± 28	Zhang et al. (2006a)
	Kongling, Yichang	Trondhjemitic gneiss	LA-ICPMS concordia age (9)	2909 ± 30	Chen et al. 2013
	Kongling, Yichang	Trondhjemitic gneiss	LA-ICPMS concordia age (16)	2937 ± 16	Chen et al. 2013
	Kongling, Yichang	Trondhjemitic gneiss	LA-ICPMS concordia age (20)	2907 ± 15	Chen et al. 2013
	Kongling, Yichang	Granitic gneiss	LA-ICPMS concordia age (9)	2691 ± 32	Chen et al. 2013
	Kongling, Yichang	Granitic gneiss	LA-ICPMS concordia age (15)	2707 ± 24	Chen et al. 2013

(continued)

Table 1 (continued)

Geological unit	Location	Lithology or zircon types	Analytical method	Age/Ma	Citation
	Kongling, Yichang	A-type granitic gneiss	LA-ICPMS concordia age (9)	2645 ± 15	Chen et al. 2013
	Kongling, Yichang	A-type granitic gneiss	LA-ICPMS concordia age (15)	2622 ± 14	Chen et al. 2013
	Kongling, Yichang	A-type granitic gneiss	LA-ICPMS concordia age (14)	2640 ± 18	Chen et al. 2013
	Kongling, Yichang	A-type granitic gneiss	LA-ICPMS concordia age (17)	2671 ± 17	Chen et al. 2013
	Huji, Zhongxiang	K-granitic gneiss	LA-ICPMS concordia age (15)	2652 ± 21	Wang et al. 2013a
	Huji, Zhongxiang	K-granitic gneiss	SHRIMP concordia age (12)	2655 ± 9	Wang et al. 2013b
	Lueyang, Shaanxi	Granite of Yudongzi Complex	LA-ICPMS upper intercept age	2703 ± 26	Zhang et al. (2010)
	Lueyang, Shaanxi	Granite of Yudongzi Complex	LA-ICPMS upper intercept age	2661 ± 17	Zhang et al. (2010)
Archean age information from detrital zircons					
Yangtze Block	Kongling, Yichang	Inherited zircons from trondhjemitic gneiss	SHRIMP concordia spot age	3051 ± 12	Qiu et al. (2000)
	Kongling, Yichang	Inherited zircons from trondhjemitic gneiss	SHRIMP concordia spot age	2738 ± 18, 2727 ± 8	Qiu et al. (2000)
	Kongling, Yichang	Meta-pelite	SHRIMP single-grain 7/6 age	3275 ± 11, 3213 ± 16, 3169 ± 6, 3133 ± 14	Qiu et al. (2000)
	Kongling, Yichang	Meta-pelite	SHRIMP concordia spot age	3234 ± 6, 2949 ± 4	Qiu et al. (2000)
	Kongling, Yichang	Meta-pelite	SHRIMP upper intercept age	2974 ± 49	Qiu et al. (2000)
	Kongling, Yichang	Inherited zircons from migmatite	LA-ICPMS concordia age (2)	3182 ± 175	Zhang et al. (2006a)
	Kongling, Yichang	Inherited zircons from migmatite	LA-ICPMS concordia age (2)	3242 ± 40	Zhang et al. (2006a)
	Kongling, Yichang	Inherited zircons from migmatite	LA-ICPMS concordia age (5)	3123 ± 36	Zhang et al. (2006a)
	Kongling, Yichang	Biotite–plagioclase gneiss	LA-ICPMS concordia age (24)	3218 ± 13	Jiao et al. (2009)

(continued)

Table 1 (Continued)

Geological unit	Location	Lithology or zircon types	Analytical method	Age/Ma	Citation
	Kongling, Yichang	Metamorphic zircons from biotite–plagioclase gneiss	LA-ICPMS concordia age (5)	2732 ± 16	Jiao et al. (2009)
	Huji, Zhongxiang	Detritao zircons from Quartze schist, Yangpo Group	LA-ICPMS concordia age (4) and Concordia age (33)	3057 ± 41 and 2801 ± 24	Wang et al. (2013a, b)
	Liantuo, Yichang	Red sandstone, Liantuo Formation	LA-ICPMS 7/6 age	3508 ± 20, 3369 ± 21, 3321 ± 26, 3319 ± 18, 3267 ± 21, 3235 ± 17	Liu et al. (2006)
	Gaojiayan, Changyang	Tillite, Nantuo Formation	LA-ICPMS 7/6 age	3502 ± 16, 3437 ± 15, 3086 ± 18	Liu et al. (2006)
	Liantuo, Yichang	Sandstone, Liantuo Formation	SHRIMP concordia spot age	3802 ± 8, 3445 ± 10	Zhang et al. (2006b)
	Liantuo, Yichang	Sandstone, Liantuo Formation	SHRIMP concordia age (5)	2942 ± 42	Zhang et al. (2006b)
	Liantuo, Yichang	Sandstone, Liantuo Formation	SHRIMP concordia age (3)	3306 ± 15	Zhang et al. (2006b)
	Liantuo, Yichang	Sandstone, Liantuo Formation	SHRIMP concordia age (10)	2951 ± 18	Zhang et al. (2006b)
	Yanbian, Sichuan	Sandstone, Neoproterozoic Zhagu Formation	LA-ICPMS 7/6 age	2649 ± 22, 2517 ± 24, 2943 ± 23	Sun et al. (2008)
	Yanbian, Sichuan	Arkose, Neoproterozoic Xiaoping Formation	LA-ICPMS 7/6 age	2669 ± 20, 2594 ± 20, 2680 ± 22, 2728 ± 24	Sun et al. (2008)
	Hekou, Sichuan	Quartzite, Mesoproterozoic Hekou Group	LA-ICPMS concordia spot age	2668 ± 8	Greentree et al. (2006)
	Hekou, Sichuan	Quartzite, Mesoproterozoic Hekou Group	SHRIMP concordia spot age	2798 ± 5, 3051 ± 9	Greentree et al. (2006)
	Dongchuan, Yunnan	Shale, Yinmin Formation of Mesoproterozoic Dongchuan Group	SHRIMP concordia age (12)	2736 ± 5	Greentree et al. (2006)
	Dongchuan, Yunnan	Shale, Mesoproterozoic Meidang Formation	SHRIMP concordia age (2)	3575 ± 9	Greentree et al. (2006)
	Dongchuan, Yunnan	Shale, Mesoproterozoic Meidang Formation	SHRIMP concordia spot age	3364 ± 8, 3034 ± 8	Greentree et al. (2006)

(continued)

Table 1 (Continued)

Geological unit	Location	Lithology or zircon types	Analytical method	Age/Ma	Citation
	Kunyang, Yunnan	Fine-grained sandstone, Mesoproterozoic Laowushan Formation	SHRIMP concordia age (2)	2707 ± 4	Greentree et al. (2006)
	Xide, Sichuan	Quatz schist, Mesoproterozoic Dengxiangying Group	SHRIMP 7/6 age	3308 ± 8, 2563 ± 41, 2551 ± 11	Geng et al. (2008)
	Dongchuan, Yunnan	Conglomerate, Proterozoic Wangchang Formation	LA-ICPMS 7/6 age	3778 ± 14	Zhu et al. (2011a)
	Dongchuan, Yunnan	Conglomerate, Proterozoic Wangchang Formation	LA-ICPMS concordia age (8)	2855 ± 14	Zhu et al. (2011a)
	Dongchuan, Yunnan	Phyllite, Yinmin Formation of Mesoproterozoic Dongchuan Group	LA-ICPMS 7/6 age	3755 ± 12, 3405 ± 13, 3409 ± 14	Li et al. (2013)
	Kunyang, Yunnan	Sandstone, Heishantou Formation of Mesoproterozoic Kunyang Group	LA-ICPMS 7/6 age	2690 ± 27, 2446 ± 29	Sun et al. (2009)
	Huili, Sichuan	Sandstone, Mesoproterozoic Tongan Formation	LA-ICPMS 7/6 age	2521 ± 26, 2488 ± 27, 2485 ± 27	Sun et al. (2009)
	Mianxian, Shaanxi	Turbidite, Proterozoic Bikou Group	LA-ICPMS 7/6 age	2529 ± 29, 2492 ± 26, 2592 ± 32	Sun et al. (2009)
	Kunyang, Yunnan	Siltstone Mesoproterozoic Heishantou Formation	LA-ICPMS 7/6 age	2526 ± 37	Wang et al. (2012)
	Kunyang, Yunnan	Sandstone, Mesoproterozoic Yinmin Formation	LA-ICPMS 7/6 age	2734 ± 23, 2855 ± 21	Wang et al. (2012)
	Kunyang, Yunnan	Sandstone, Mesoproterozoic Yinmin Formation	LA-ICPMS upper intercept age	2766	Wang et al. (2012)
	Kunyang, Yunnan	Sandstone, Mesoproterozoic Yinmin Formation	LA-ICPMS 7/6 age	2783 ± 23, 2888 ± 23	Wang et al. (2012)
	Kunyang, Yunnan	Sandstone, Mesoproterozoic Yinmin Formation	LA-ICPMS upper intercept age	2800	Wang et al. (2012)
	Kunyang, Yunnan	Sandstone, Mesoproterozoic Yinmin Formation	LA-ICPMS 7/6 age	2642 ± 24, 3030 ± 26, 2943 ± 24, 2830 ± 26	Wang et al. (2012)

(continued)

Table 1 (Continued)

Geological unit	Location	Lithology or zircon types	Analytical method	Age/Ma	Citation
	Kunyang, Yunnan	Sandstone, Mesoproterozoic Yinmin Formation	LA-ICPMS upper intercept age	2800	Wang et al. (2012)
	Kunyang, Yunnan	Sandstone, Neoproterozoic Chengjiang Formation	LA-ICPMS 7/6 age	2849 ± 29, 2702 ± 37, 2988 ± 27	Wang et al. (2012)
	Dongchuan, Yunnan	Sandstone, Mesoproterozoic Yinmin Formation	LA-ICPMS upper intercept age (6)	2849 ± 21	Zhao et al. (2010)
	Dongchuan, Yunnan	Sandstone, Mesoproterozoic Yinmin Formation	LA-ICPMS 7/6 age	2596 ± 33, 2574 ± 23	Zhao et al. (2010)
	Dongchuan, Yunnan	Breccia, Mesoproterozoic Yinmin Formation	LA-ICPMS 7/6 age	3732 ± 26, 3511 ± 26	Zhao et al. (2010)
Jiangnan Belt	Sibao, Guangxi	Sandstone, Neoproterzoic Sibao Group	LA-ICPMS concordia spot age	2547 ± 8, 2536 ± 9, 2631 ± 13, 2528 ± 8, 2508 ± 9, 2513 ± 13, 2512 ± 12, 2883 ± 40, 2624 ± 10, 2512 ± 10, 2509 ± 11	Wang et al. (2007)
	Sibao, Guangxi	Siltstone, Neoproterzoic Sibao Group	LA-ICPMS concordia spot age	2521 ± 7	Wang et al. (2007)
	Sibao, Guangxi	Quartz schist, Neoproterozoic Sibao Group	LA-ICPMS concordia spot age	2707 ± 29, 2710 ± 24, 2622 ± 7	Wang et al. (2007)
	Wenjiashi, Hunan	Sandy pelite, Neoproterozoic Lengjiaxi Group	LA-ICPMS concordia spot age	2543 ± 7, 2520 ± 7	Wang et al. (2007)
	Tianli, Jiangxi	Quartz mica schist, Mesoproterozoic Tianli Formation	SHRIMP 7/6 age	3505 ± 3	Li et al. (2007)
	Tianli, Jiangxi	Quartz mica schist, Mesoproterozoic Tianli Formation	SHRIMP peak age	3300	Li et al. (2007)
	Tianli, Jiangxi	Quartz mica schist, Mesoproterozoic Tianli Formation	SHRIMP peak age	2800	Li et al. (2007)
	Tianli, Jiangxi	Quartz mica schist, Mesoproterozoic Tianli Formation	SHRIMP peak age	2750	Li et al. (2007)
	Wuyuan, Jiangxi	Sandy slate, Neoproterozoic Xikou Group	LA-ICPMS concordia spot age	2535 ± 35, 2545 ± 38, 2530 ± 34	Zhang et al. (2010)

(continued)

Table 1 (continued)

Geological unit	Location	Lithology or zircon types	Analytical method	Age/Ma	Citation
Cathaysian Block	Taoxi, Fujian	Proterozoic pelitic granulite	LA-ICPMS upper intercept age	2523 ± 26	Yu et al. (2005)
	Longchuan, Guangdong	Late Neoproterozoic paragneiss	LA-ICPMS 7/6 age	3012 ± 15, 2915 ± 116	Yu et al. (2006)
	Longchuan, Guangdong	Late Neoproterozoic paragneiss	LA-ICPMS upper intercept age	2577 ± 48	Yu et al. (2006)
	Tanxi, northern Guangdong	Neoproterozoic paragneiss	LA-ICPMS 7/6 age	3755 ± 15	Yu et al. (2007)
	Tanxi, northern Guangdong	Neoproterozoic paragneiss	LA-ICPMS upper intercept age	3284 ± 86	Yu et al. (2007)
	Tanxi, northern Guangdong	Neoproterozoic paragneiss	LA-ICPMS 7/6 age	2650 ± 16	Yu et al. (2007)
	Tanxi, northern Guangdong	Neoproterozoic paragneiss	LA-ICPMS upper intercept age	2518 ± 38	Yu et al. (2007)
	Zengcheng, Guangdong	Neoproterozoic migmatite	LA-ICPMS 7/6 age	3315 ± 9, 3294 ± 8, 2838 ± 8	Yu et al. (2008)
	Zengcheng, Guangdong	Neoproterozoic migmatite	LA-ICPMS upper intercept age	2517 ± 30	Yu et al. (2008)
	Chongyi, Jiangxi	Ordovician feldspathic quartz sandstone	LA-ICPMS concordia spot age	3817 ± 17, 3959 ± 21, 3232 ± 13, 3257 ± 20, 3327 ± 21, 3379 ± 11, 3353 ± 13	Yao et al. (2011)
	Ji'an, Jiangxi	Permian medium-grained sandstone	Cameca SIMS 7/6 age	3278	Li et al. (2012a)
	Ji'an, Jiangxi	Permian medium-grained sandstone	Cameca SIMS peak age	2520	Li et al. (2012a)
	Leiyang, Hunan	Permian fine-grained sandstone	Cameca SIMS 7/6 age	2962	Li et al. (2012a)
	Leiyang, Hunan	Permian fine-grained sandstone	Cameca SIMS peak age	2500	Li et al. (2012a)
	Yongding, Fujian	Permian medium-grained sandstone	Cameca SIMS 7/6 age	2865	Li et al. (2012a)
	Yongding, Fujian	Permian medium-grained sandstone	Cameca SIMS peak age	2510	Li et al. (2012a)
	Jiangle, Fujian	Permian coarse-grained sandstone	Cameca SIMS 7/6 age	3248	Li et al. (2012a)
	Jiangle, Fujian	Permian coarse-grained sandstone	Cameca SIMS peak age	2540	Li et al. (2012a)
	Zhoutan, Jiangxi	Garnet mica schist, Zhoutan Group	LA-ICPMS concordia spot age	2501 ± 38	Li et al. (2012a)
	Lanhe river, Guangdong	Mesoproterozoic Lanhe gneiss	LA-ICPMS 7/6 age	2607 ± 29, 2517 ± 9, 2691 ± 9, 2659 ± 9	Xu et al. (2005)

(continued)

Table 1 (continued)

Geological unit	Location	Lithology or zircon types	Analytical method	Age/Ma	Citation
	Lanhe river, Guangdong	Devonian sedimentary rock	LA-ICPMS 7/6 age	2669 ± 9	Xu et al. (2005)
	Shunchang, Fujian	Quartz mica schist, Proterozoic Wanquan Group	LA-ICPMS 7/6 age	2686 ± 16, 2653 ± 11, 2642 ± 12, 2562 ± 14	Xu et al. (2010)
	Jian'ou Fujian	Biotite gneiss, Proterozoic Mamianshan Group	LA-ICPMS 7/6 age	2595 ± 14, 2591 ± 14, 2532 ± 12	Xu et al. (2010)
	Jian'ou Fujian	Fine-grained gneiss, Proterozoic Mayuan Group	LA-ICPMS 7/6 age	2802 ± 12	Xu et al. (2010)
	Qujiang River, Fujian	Modern river sediments	LA-ICPMS 7/6 age	3054 ± 23	Xu et al. (2007)
	Beijiang river, Guangdong	Modern river sediments	LA-ICPMS 7/6 age	2696 ± 7, 2625 ± 7, 2730 ± 8, 2556 ± 30, 2566 ± 27, 3550 ± 15, 2752 ± 16	Xu et al. (2007)
Archean age information from zircon xenocrysts or inherited zircons from magmatic rocks					
Yangtze Block	Huili, Sichuan	Triassic patassic felspar syenite	SHRIMP 7/6 age	2818 ± 14, 2692 ± 12	Liu et al. (2004)
	Ningxiang, Hunan	Xenocrysts from lamprophyre	LA-ICPMS concordia spot age	2980 ± 7, 2835 ± 10, 2751 ± 8, 2740 ± 9, 2525 ± 7	Zheng et al. (2006)
	Ma'anshan, Anhui	Xenocrysts from Cretaceous trachyandesite	SHRIMP 7/6 age	3098 ± 1, 2592 ± 10	Zhang et al. (2003)
	Tongling, Anhui	Mesozoic quartz diorite	SHRIMP 7/6 age	2670, 2598	Wang et al. (2004)
	Kunming, Yunnan	Cambrian bentonite	SHRIMP 7/6 age	2955 ± 24, 2914 ± 6	Compston et al. (1997)
	Yiyang, Hunan	Proterozoic basaltic komatiite	SHRIMP 7/6 age	2636 ± 12, 3122 ± 47, 2925 ± 45, 2640 ± 13	Shen et al. (2005)
Jiangnan Belt	Pingnan, Guangxi	Xenocryst from Cenozoic olivine basalts	LA-ICPMS concordia spot age	3178 ± 20	Zheng et al. (2011)
	Pingnan, Guangxi	Xenocrysts from Cenozoic olivine basalts	LA-ICPMS 7/6 age	3856 ± 18, 3313 ± 20	Zheng et al. (2011)
	Pingnan, Guangxi	Xenocrysts from Cenozoic olivine basalts	LA-ICPMS concordia spot age	2879 ± 20, 2716 ± 19	Zheng et al. (2011)
	Pingnan, Guangxi	Xenocrysts from Cenozoic olivine basalts	LA-ICPMS concordia age (20)	2716 ± 19	Zheng et al. (2011)

(continued)

Table 1 (continued)

Geological unit	Location	Lithology or zircon types	Analytical method	Age/Ma	Citation
	Pingnan, Guangxi	Xenocrysts from Cenozoic olivine basalts	LA-ICPMS concordia spot age	2500 ± 50, 2890 ± 1[illegible] 2798 ± 17	Zheng et al. (2011)
	Pingnan, Guangxi	Xenocrysts from Cenozoic olivine basalts	LA-ICPMS concordia age (4)	2530 ± 30	Zheng et al. (2011)
	Pingle, Guangxi	Xenocryst from Jurassic minette	LA-ICPMS concordia spot age	2525 ± 21	Zheng et al. (2011)
	Zhenyuan, Guizhou	Xenocrysts from lamprophyre	LA-ICPMS concordia spot age	2676 ± 9, 2632 ± 1[illegible]	Zheng et al. (2006)
Cathaysian Block	Jianning, Fujian	Inherited zircons from Proterozoic amphibolite of Tianjingping Formation	SHRIMP 7/6 age	2770 ± 27, 2818 ± [illegible] 2696 ± 41	[illegible] et al. (1998)
	Guzhai, Guangdong	Inherited zircon from Caledonian granodiorite	LA-ICPMS 7/6 age	3102 ± 20	Ding et al. (2005)
	Guzhai, Guangdong	Inherited zircon from Caledonian granodiorite	LA-ICPMS upper intercept age	2708 ± 100	Ding et al. (2005)

Note Concordia age—weighted average age of a set of data along the concordia curve
Digits in ()—the number of data for weighted mean age calculation
Concordia spot age—a single concordia age on the concordia curve from one analysis
7/6 age—$^{207}Pb/^{206}Pb$ age of one analysis with certain Pb loss

3 Paleoproterozoic Era of South China Craton

In South China, Paleoproterozoic geologic body is distributed sporadically in the southwestern Zhejiang and northern Fujian. In addition, Paleoproterozoic geologic bodies in Yangtze Block also spread scatteredly in Kongling of Hubei, Dahongshan of Yunnan, and Dongchuan of Yunnan. Mayuan Group in Fujian, Xingzi Group in Jiangxi, Ailaoshan Group in Yunnan, Susong Group in Anhui, and Kangding complex were previously considered to be Paleoproterozoic (Cheng et al. 1994; Jin et al. 1996; Li et al. 1988). However, recently geochronologic data suggest these groups might be formed in Meso-Neoproterozoic (Wan et al. 2007; Dong et al. 2010; Gao et al. 2012a; Jiang et al. 2003; Geng et al. 2008).

3.1 Paleoproterozoic Era of the Yangtze Block

Houhe complex, as a metamorphic rock series in the northwestern Yangtze Block, consists of Houhe Group and Bajiaoshu Gneiss. Houhe Group is mainly made up of banded garnet-bearing biotite-plagioclase gneiss, two-mica plagioclase gneiss, amphiolite, and few diopside-bearing amphiolite and marble and is metamorphosed in amphibolite facies. Bajiaoshu Gneiss consists of biotite-plagioclase gneiss, amphibole plagioclase gneiss, and a few gray gneisses which are made up from granitic gneiss. All of these rock series construct the Houhe complex, as the crystalline basement of this area (He et al. 1997), which is uncomfortably covered by low-grade metamorphic Huodiya Group (Ling et al. 2003). Using LA–ICPMS zircon U–Pb method, investigators have obtained a $^{207}Pb/^{206}Pb$ weighted average age of 2081 ± 9 Ma from the gray gneiss in Houhe complex, which represented the emplacement age of the parent rock of the gray gneiss (Wu et al. 2012). These gray gneisses are characteristic of high Fe (Fe_2O_3 = 2.86–6.69 %), high Al (Al_2O_3 = 16.01–18.88 %), high Y (12.9–32.7 ppm), high Yb (0.95–2.25 ppm), Low Sr (149–390 ppm), Cr (9.07–45.1 ppm), and Ni (4.97–21.3 ppm), similar to the arc-related calc-alkaline granite, suggesting these rocks might formed in the arc-subduction environment (Wu et al. 2012).

Although there is no Paleoproterozoic metamorphic strata outcrop in Kongling area, Hubei Province, Paleoproterozoic magmatic rocks have been founded, including the 1.85 Ga Quanyitang granites that intruded into Archean gneisses (Xiong et al. 2009; Peng et al. 2012) and the 1.85 Ga mafic dikes that emplaced into Archean gneisses and metamorphic strata sequence (Peng et al. 2009). The Quanyitang granites show geochemistry characteristics of A2-type granites (Xiong et al. 2009; Peng et al. 2009, 2012). 1851 ± 18 Ma (zircon LA–ICPMS U–Pb) Rapakivi granites have been found in Huashanguan area of Zhongxiang, Hubei Province, in south to the Kongling area (Zhang et al. 2011). The A-type granites, Rapakivi granites, and mafic dikes in the northern margin of the Yangtze Block show similar forming ages of ~1.85 Ga, suggesting at least partial South China Craton had stabilized into the rigid block before 1.85 Ga because of the existence of the ~1.85 Ga A-type granites, Rapakivi granites, and mafic dikes that formed in rigid block under extensional environment (Whalen et al. 1987; Eby 1992; Bonin 2007; Halls et al. 2000; Halls and Zhang 2003; Hou et al. 2006; Peng et al. 2009, 2012). However, whether the whole South China Block had already finished the cratonization before 1.85 Ga similar to the North China Craton is still a question and needs more lines of evidence.

Paleoproterozoic geological bodies might also exist in Dongchuan area, Yunnan Province, the southwest part of Yangtze Block. The metamorphic strata sequence in Dongchuan area is usually attributed to Mesoproterozoic Kunyang Group; however, these strata sequence is made up of Tangdan Group (including Shaihaigou Formation, Wangchang Formation, Caiyuanwan Formation, and Pingdingshan Formation) in the lower part and Dongchuan Group (including Yinmin Formation, Luoxue Formation, Heishan Formation, and Qinglongshan Formation) in the upper part, and the former are uncomfortably overlain by the latter (Yin et al. 2011a, b). Zircon grains of the volcanic tuff from Yinmin Formation of Dongchuan Group yield a U–Pb age of 1740 ± 15 Ma, which represents the forming age of the tuff (Zhao et al. 2010). Thus, the Tangdan Group that lies below the Dongchuan Group must be older than 1740 Ma. Two groups of zircon ages were obtained from the welded tuff in the Shaihaigou Formation of Tangdan Group. One group of zircons shows Pb loss, yielding an upper intercept age of 2742 ± 48 Ma. The other group contains 12 analyses plotted close to concordia curve and yielded $^{207}Pb/^{206}Pb$ weighted mean age of 2285 ± 12 Ma. The former was interpreted as inherited zircon age, and the later suggested as a forming age of the tuff (Zhu et al. 2011b). The zircon grains from the tuff in Wangchang Formation gave SHRIMP U–Pb age of 2299 ± 14 Ma (Zhou et al. 2012). Additionally, two ages of 2855 ± 14 Ma and 1838 ± 10 Ma were obtained from detrital zircon of the metamorphic sedimentary rocks (Zhu et al. 2011a). In spite of some paradoxes in the chronological data, the Tangdan Group may be deduced as Paleoproterozoic due to its tuff ages of 2285–2299 Ma and the constraints of uncomfortably overlying Dongchuan Group.

3.2 *Paleoproterozoic Era of the Cathaysian Block*

The Proterozoic metamorphic geological bodies in Southwestern Zhejiang can be divided into Chencai Group in Zhuji, Badu Group, and Longquan Group in Suichang-Longquan (Fig. 5). The believable Paleoproterozoic geological bodies in

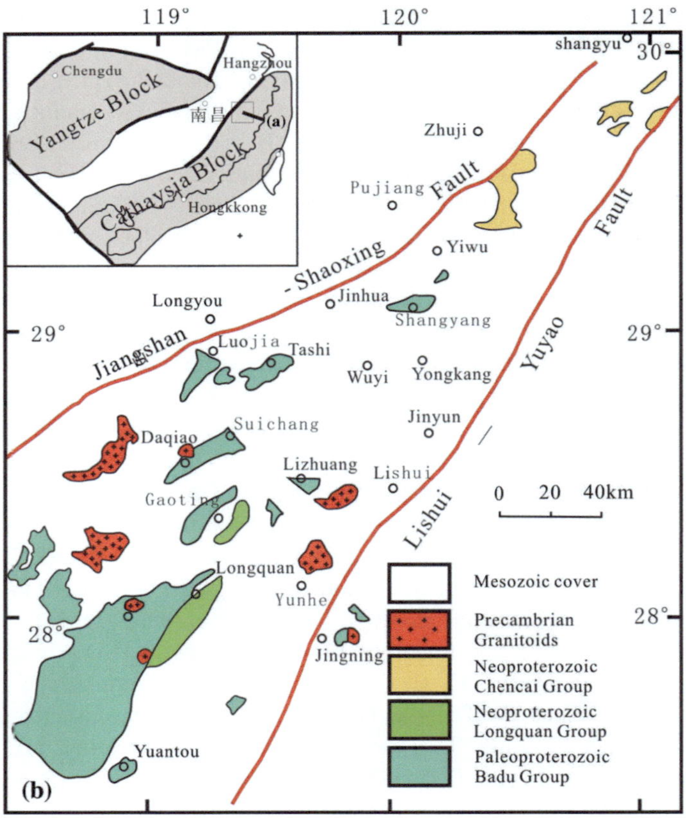

Fig. 5 Geological map of the South China Block (**a**) and the southwestern Zhejiang Province (**b**)

this area mainly comprise medium-grade metamorphic Badu Group and the granitoid plutons that emplaced into the Badu Group (Yu et al. 2009, 2010, 2012; Ding et al. 2005; Kong et al. 1995; Gan et al. 1995; Li et al. 1996; Li et al. 1998). However, Chencai Group and Longquan Group were controversially considered to be Paleoproterozoic (Hu et al. 1991, 1992) or Meso-Neoproterozoic (Kong et al. 1995; Jin et al. 1997). A zircon SHRIMP U–Pb dating of the Metamorphic gabbro from Chencai Group yielded an emplacement age of 1781 ± 21 Ma (Li et al. 2009), but this is the only sample so it is hard to confirm their distribution range. Moreover, new zircon SHRIMP and LA–ICPMS dating indicated that the major lithological assemblages of Chencai Group and Longquan Group belong to Neoproterozoic, and these results will be discussed later.

Badu Group was divided into Tangyuan Formation, Qiantou Formation, Zhangyan Formation, Siyuan Formation, and Dayanshan Formation from the bottom to top. Tangyuan Formation consists of meta-mafic rocks and leptynites, mainly comprising the lithological assemblage of amphibolite, amphibole anorthosite, sahlite-bearing amphibole anorthosite, garnet-bearing biotite–plagioclase leptynite, etc., and their protoliths are mafic igneous rocks and volcanic graywacke. Qiantou Formation is mainly composed of biotite–plagioclase leptynite, partial amphibole plagioclase leptynite, pyroxene-bearing amphibolites, etc., and their protoliths are graywackes and intermediate-acid volcanic rocks. The major lithological assemblage of Zhangyan Group is biotite quartz schist, biotite schist, and biotite leptynite, undergoing strong migmatization, and their protoliths are argillaceous-half argillaceous clastic rocks. Siyuan Formation contains biotite leptynite, arkose quartzite, tint leptite, biotite quartz schist, etc., generally undergoing migmatization, and their protoliths are sandstones and argillaceous-half argillaceous clastic rocks. Dayanshan Formation consists chiefly of biotite schist, biotite quartz schist, locally changing into biotite–plagioclase gneiss, and their protoliths are mainly terrigenous clay, with less granitoid rocks that were derived from migmatization (Jin et al. 1997; Zhao 2012).

The metamorphic rock series in Badu Group generally contain garnet, sillimanite, biotite, K-feldspar, and crystalline graphite. These rocks were metamorphosed under granulite facies condition (Zhao 2012), showing complicatedly superimposed fold deformation, and developing abundant migmatization genetic granitoid rocks.

Previous investigators reported some whole-rock Sm–Nd method and Sm–Nd internal isochron ages of 1735 ± 55 to 2199 ± 95 Ma from the metamorphic rocks of the Badu Group (Wang et al. 1992; BGMRZJ 1989; Li et al. 1996). Recently, Zhao et al. (2014a) obtained a metamorphic age of 1869 ± 19 Ma (zircon Th/U = 0.009–0.093) and a forming age of 1923 ± 8 Ma from the garnet-bearing felsic gneiss in Badu Group. Two groups of detrital zircon ages, 2451 ± 63 Ma and 2002 ± 29 Ma, were obtained from sillimanite amphibole plagioclase gneiss. Garnet amphibole plagioclase gneiss, garnet two-pyroxene granulite, and marble revealed their metamorphic ages of 1884 ± 18 to 1852 ± 12 Ma. Yu et al. (2012) obtained metamorphic ages of 1887 ± 26 and 1885 ± 9 Ma from the migmatitic biotite gneiss and quartz-enriched gneiss revealed, respectively, and the two dated samples have abundant Archean detrital zircons and suggest the Badu Group had deposited since

2.5 Ga. However, in these samples, there were also some ~2.1 Ga detrital zircons that plotted on or close to the concordia, indicating the Badu Group actually started deposition after 2.2 Ga. Integrated all the information above, forming time of the Badu Group may be determined from 2.2 to 1.9 Ga, following the granulite facies-amphibolite facies metamorphism in 1880–1850 Ma.

Paleoproterozoic granitoids intruded Badu Group are extensive in southwest Zhejiang Province and always exhibit gneissic structure with the same structural orientation as the Badu Group. Most of these granitoids were regarded as S-type granites, but some of them might be A-type granites (Hu et al. 1991; Yu et al. 2009; Liu et al. 2009; Xia et al. 2012; Zhao et al. 2014a; Liu et al. 2014). The representative granitoids include the Xiaji, Danzhu Quankeng and Huaqiao plutons in Longquan County, Lizhuang (Jinluohou) and Jingju plutons in Songyang County, Tianhou (Dazhe) pluton in Suichang County, and Chimushan (Sanzhishu) and Wengkeng plutons in Jingning County (Fig. 6). On the basis of previous geochronological data through conventional multi-grain or single-grain zircon U–Pb (TIMS), Rb–Sr and Sm–Nd methods, these granitoids were thought to have a wide age span of 2080–1755 Ma (Hu et al. 1991, 1992; Wang et al. 1992; Gan et al. 1993, 1995). Recent precise zircon U–Pb ages (SHRIMP and LA–ICP–MS) have revealed the most of the granitoids formed on 1832 and 1888 Ma, and the two samples of granite yield the age results of the 1912 and 1925 Ma (Fig. 6).

Except for the Badu Group and the concomitant granites in the southwestern Zhejiang, Paleoproterozoic geological bodies also outcrop sporadically in other regions of the Cathaysian Block. For example, Mayuan Group in Wuyishan area, Fujian. A SHRIMP zircon U–Pb age of 1766 ± 19 Ma was obtained from an amphibolite sample of Tianjingping Formation in Mayuan Group (Li et al. 1998). Similarly, another SHRIMP U–Pb age of 1790 ± 19 Ma from a gneiss sample of Tianjingping Formation was documented by Wan et al. (2007). The gneissic granites that intruded into the Mayuan Group show their forming ages of 1851–1857 Ma (Li et al. 2011). Although some Neoproterozoic or even younger strata were involved locally (Wan et al. 2007), major geological bodies may be determined as Paleoproterozoic, but their distribution and scale are still unclear and need to be further investigated.

In the southwestern Cathaysian Block, a suit of metamorphic series was reported in western Guangdong–southeastern Guangxi and is divided into Paleoproterozoic Tiantangshan Group, Meso-Neoproterozoic Yunkai Group, and Neoproterozoic plutons (Qin et al. 2006). Of which the Tiantangshan Group is mainly composed of garnet sillimanite feldspar biotite (quartz) schists, garnet sillimanite biotite gneisses, biotite leptynites, pyroxene (amphibole) plagioclase leptynites, feldspar quartzites, garnet pyroxenites, and diopsidites, etc. These rocks generally underwent amphibolite facies metamorphism (partial granulite facies), with a certain degree of migmatization. Zircon SHRIMP U–Pb dating reveals that the garnet pyroxenite emplaced at 1817 ± 36 Ma (Qin et al. 2006), indicating the Tiantangshan Group is Paleoproterozoic.

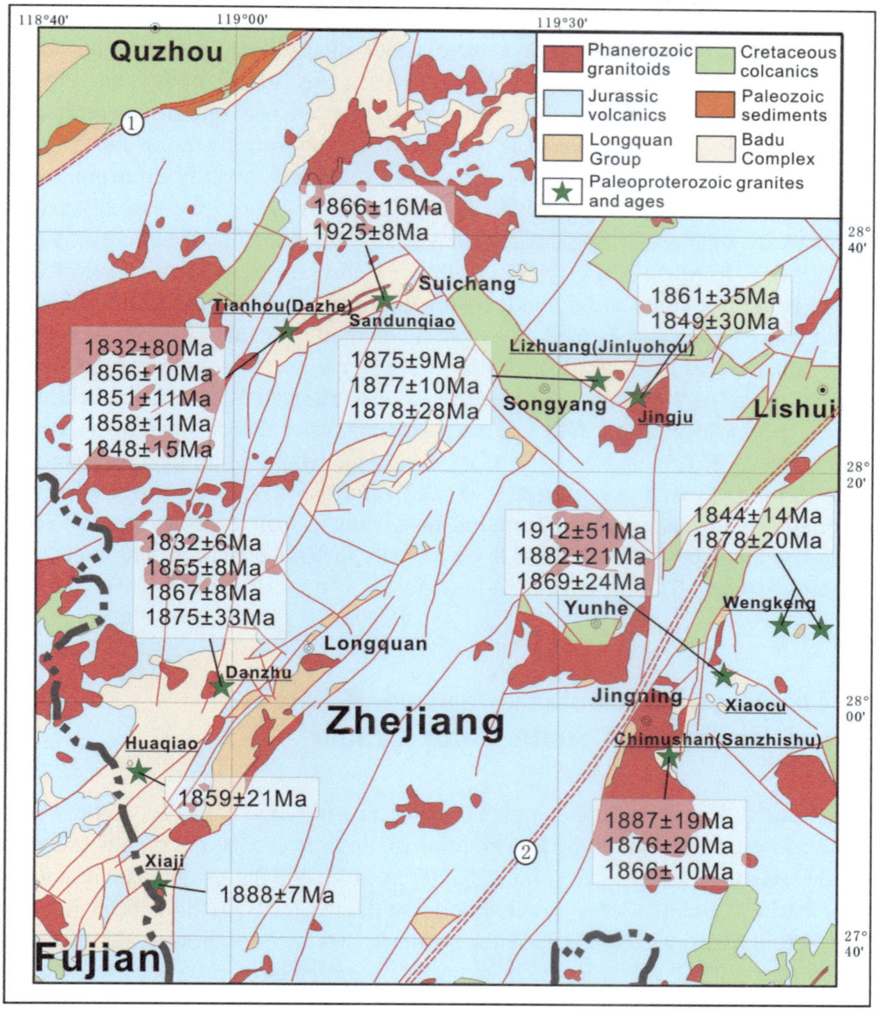

Fig. 6 Geological map of southwestern Zhejiang Province (after Liu et al. 2014). Zircon SHRIMP and LA–ICP–MS U–Pb ages are from Li and Li (2007), Wang et al. (2008), Yu et al. (2009), Liu et al. (2009), Xia et al. (2012), Zhao et al. (2014b), Liu et al. (2014). ① Jiangshan-Shaoxing Fault, ②-Zhenghe-Dabu Fault

3.3 *Information of Paleoproterozoic Ages in the South China Craton*

Except for these identified Paleoproterozoic geological bodies, detrital zircons from post-Mesoproterozoic sedimentary rocks and inherited zircons from magmatic rocks contain a large number of Paleoproterozoic age information (Li et al. 2012b;

Yao et al. 2011; Zheng et al. 2008, 2011; Yu et al. 2007), revealing Paleoproterozoic geological bodies have wider distribution in the South China Craton. In Mesozoic–Cenozoic, Paleoproterozoic geological bodies still spread widely in the deep crust. In the histogram of detrital and inherited zircon ages (Fig. 7), two Paleoproterozoic peaks can be obviously recognized in the Yangtze Block (2.3 and 1.85 Ga), suggesting the two events had strongly influence to the Yangtze Block. In the Jiangnan Belt, the age peaks of 1.8–1.9 Ga and 2.0–2.1 Ga are relatively small, revealing this region was also influenced by the two events, but was different from the Yangtze Block. In the Cathaysian Block, the peak of 1.8–1.9 Ga is very obvious and a relatively smaller peak of 2.3 Ga, indicating that the Cathaysian Block was reformed by a tectonic thermal event at $\sim$2.3 Ga. According to the statistics by Li et al. (2012b), the western Yangtze Block appears relatively obvious peaks of 1.85 and 2.32 Ga; the eastern Yangtze Block shows relatively obvious peaks of 2.0 and 2.48 Ga; and the Cathaysian Block displays relatively obvious peaks at 1.85 and 2.48 Ga. Considering the whole South China Craton, the three tectonic thermal events (at 2.48, 2.3, and 1.8–1.9 Ga, respectively) had important significance to the formation and evolution of the basement of South China Craton, especially the 1.8–1.9 Ga tectonic thermal events play a crucial role to the formation of the South China basement.

4 The Major Precambrian Geological Events and Evolution of South China Craton

The exposed Precambrian geological bodies are infrequent in the South China Craton. In order to discuss the Precambrian geological events and the evolution of South China Craton, we have collected 3297 in situ U–Pb dating analyses of detrital and inherited zircons. Some low harmonious degree analyses have been rejected (<80 % for zircons >1.0 Ga; <90 % for zircons <1.0 Ga) from these analyses. Using the remaining 2711 analyses, the histograms are produced in order to discuss some important Precambrian events. Combined with some exposed geological bodies, we discussed the Precambrian geological events and the evolution of South China Craton as below.

As shown in Fig. 7, South China Craton, whether the Yangtze Block, the Jiangnan Belt, or the Cathaysian Block, has some detrital and inherited zircon grains showing >3.5 Ga ages, indicating the existence of some ancient geological bodies with older than 3.5 Ga in South China Craton. These zircons are characterized by oscillation zonings and high Th/U ratios, suggesting they came from intermediate to acid magmatic rocks (Pidgen 1992; Hanchar and Miller 1993; Hoskin and Schaltegger 2003; Wu and Zheng 2004). Therefore, the Geological events before 3.5 Ga dominated intermediate to acid volcanic magmatic events. The $\varepsilon_{Hf}(t)$ values of the old zircon grains are around 0 instead of getting close to the depleted mantle evolution curve (Fig. 8), suggesting these zircons came from the

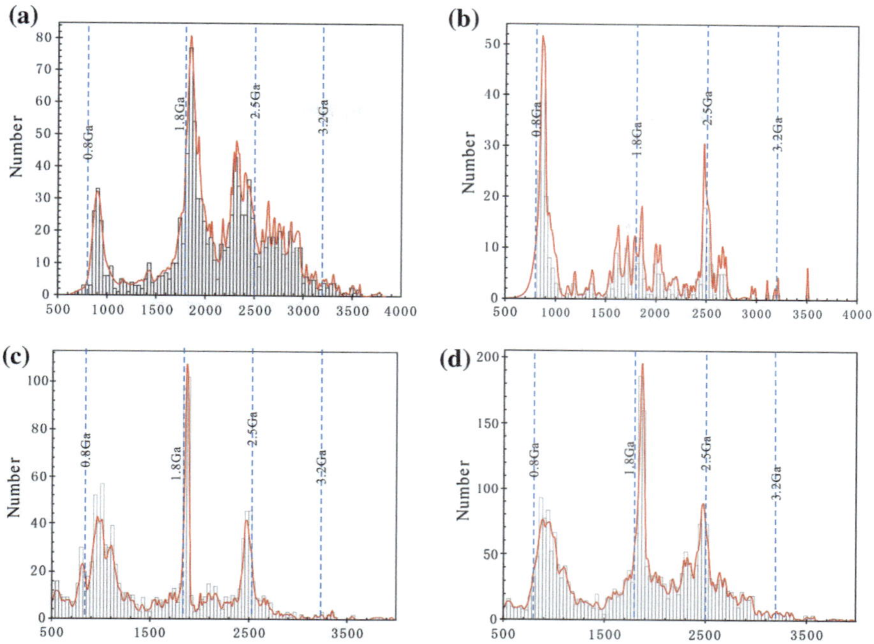

Fig. 7 Histogram of detrital zircon ages in the South China Block (**a**-Yangtze Block, **b**-Jiangnan Belt, **c**-Cathaysian Block, **d**-the whole South China). These data come from Zhao et al. (2010), Greetree et al. (2006), Greetree and Li (2008), Zhu et al. (2011a), Geng et al. (2008), Sun et al. (2008, 2009), Wang et al. (2012), Zhang et al. (2006b), Zhang et al. (2010), Liu et al. (2006), Gao et al. (2001), Wang et al. (2007), Xu et al. (2005, 2007), Yu et al. (2007, 2008, 2009), Zheng et al. (2006, 2011), Liu et al. (2009), Li et al. (2007), Yao et al. (2011), Li et al. (2012a), Xu et al. (2010), Zhao (2012), and unpublished data of the author

magma undergoing crustal contamination, suggesting South China Craton have thicker crust during >3.5 Ga. The zircon grains from the 2.9–3.3 Ga old metamorphic migmatites display Hf-isotopic model ages from 3.6–3.4 Ga, up to 4.0 Ga in the Kongling area (Gao et al. 2001; Zhang et al. 2006b; Jiao et al. 2009; Zhang and Zheng 2013). The crustal residual ages of zircon Hf isotope also shows the existence of the reconstructive 4.0 Ga crustal materials (Li et al. 2012b). These data reveal that South China Craton indeed preserves Paleoarchean, even Hadean, crustal materials.

Present older rocks with 3.2–3.3 Ga U–Pb ages only have been found in Kongling area (Jiao et al. 2009; Gao et al. 2011). Integrated with the geological facts, the 2.9 Ga TTG gneisses intruded into the Dongchonghe supracrustal rock series, and the Dongchonghe supracrustal rocks were metamorphosed at ~2.75 Ga; therefore, the Dongchonghe supracrust rocks are most likely to be formed during 3.3–2.9 Ga. This supracrust rock series dominate migmatic biotite–plagioclase gneisses, intercalated Al-enriched gneisses, garnet sillimanite quartzites, calcium and magnesium silicate rocks and marbles, and all of these rocks have graphite

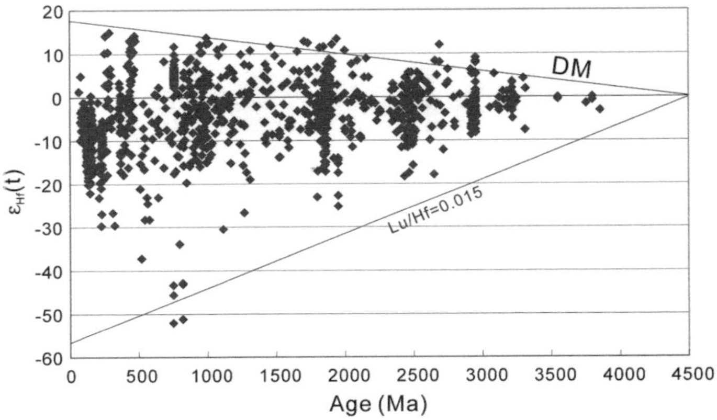

Fig. 8 εHf(t)-Age diagram of detrital zircons from South China. Data resource: Peng et al. (2009), Zhang et al. (2006a, 2006b, 2006c), Wang et al. (2011), Xu et al. (2005, 2007), Yu et al. (2007, 2008, 2009), Zheng et al. (2011), Liu et al. (2009), Li et al. (2012a), Yao et al. (2011), Xiang et al. (2008), Zhao et al. (2008)

mineral, indicating that this association has the features of Khondalite series, which is a suit of organic terrigenous sedimentary formation under the depression-rift zones located in neritic region of ancient continental margin (Jiang 1986; Lu et al. 1996). These features suggest that the basement of the South China Craton had already larger scale, where can provide stable material sources for the deposition of Khondalites. From the age histogram, we can see only a very small peak around 3.2 Ga, suggesting the crust at that time was seldom preserved because of the strong denudation before Mesoproterozoic.

During 2.9–3.0 Ga, large-scale TTG magmatism occurred in Hubei Kongling area (Qiu et al. 2000; Zhang et al. 2006a). In the age histogram, detrital zircons from the Yangtze block and the Jiangnan Belt have the signatures of this event, while zircons from the Cathaysian Block do not have (Fig. 7), suggesting this event affected mainly the Yangtze block and the Jiangnan Belt. The zircon $\varepsilon_{Hf}(t)$ values of 2.9 Ga TTG gneisses in Kongling area range from −8.7 to −0.1, with an average valve of −4.03, weighted average values (T_{DM1}) of 3.35 Ga, and weighted T_{DM2} average values of 3.46 Ga (Zhang et al. 2006a), suggesting these TTG gneisses are the productions of the partial melting of ancient crust, suggesting the crust-mantle differentiation ages were Paleoarchean or Hadean. Detrital zircons formed at 2.9 Ga show $\varepsilon_{Hf}(t)$ from −10 to +10. Some spots plot around the depleted mantle evolutionary line, while most analyses are actually far from the line (Fig. 8), suggesting that 2.9 Ga magmatic event was mainly produced by of the partial melting of the pre-existent crust, with little involvement of the mantle materials.

An important magmatic tectonothermal event occurred around 2.7–2.6 Ga, mainly displaying the emplacement of magmatic rocks in the northern margin of the Yangtze Block (Zhang et al. 2010a, b; Wang et al. 2013a, b; Chen et al. 2013), and the magmatic zircon rims display 2.7 Ga metamorphic growth rims from the 3.2 Ga biotite–plagioclase gneisses in Kongling area, Hubei. A smaller age peak of 2.7–2.6 Ga is shown in the zircon age histogram (Fig. 7) from the Yangtze Block, the Cathaysian Block, and the Jiangnan Belt (see Table 1), suggesting that this tectonothermal event has affected a certain range of the South China Craton.

No 2.5 Ga geological body has been found in South China Craton, but detrital and inherited zircons reveal that this event affect extensively on the Yangtze Block, the Cathaysian Block, and the Jiangnan Belt. The xenolith from the Cenozoic lamprophyre contains ~2.5 Ga inherited zircons (Zheng et al. 2006, 2011). The fluviatile sands in Beijiang also discovered ~2.5 Ga zircons (Xu et al. 2007), suggesting preserving 2.5 Ga geological bodies in the Cenezoic (even present) deep crust. Based on the analyses of the detrital zircons from Badu Group metamorphic sedimentary rocks in the Cathaysian Block, Yu et al. (2012) suggested that Badu Group might deposit in a sedimentary basin that was connected with arc around 2.5 Ga. Based on zircon Hf isotopes, they suggest that the 2.5 Ga magmatic event led a mass of the juvenile crust growth and the reworking of the ancient crust (Yu et al. 2012). In Figs. 6, 7, and 8, these zircons show $\varepsilon_{Hf}(t)$ from −10 to +8, inferring that the partial melting of pre-existent crust was the main part of this event, coupled with the involvement of mantle material. However, the scale and evolution process of this event still need to be more investigated.

The 1.8–1.9 Ga geological events were recorded in the Yangtze Block and the Cathaysian Block. In Kongling area, the Yangtze Block, only records of the Late Paleoproterozoic breakup event are preserved; however, a full orogenic process was retained in southwestern Zhejiang region of the Cathaysian Block. Some 1.85 Ga geological bodies, for example, simultaneously Quanyitang A-type granite, Huashanguan rapakivi granite, and mafic dikes in the northern margin of the Yangtze Block, suggest a Paleoproterozoic extensional environment (Whalen et al. 1987; Eby 1992; Bonin 2007; Halls et al. 2000; Halls and Zhang 2003; Hou et al. 2006; Peng et al. 2009, 2012). There are no believable >1.85 Ga geological bodies in the northern margin of the Yangtze Block to be reported, so we can only discuss from the related zircon dating information. Qiu et al (2000) obtained SHRIMP ages of 1.99 and 1.93 Ga from TTG gneisses and metamorphic pelites, respectively. Zhang et al. (2006a, b) documented metamorphic ages of 1.94–1.98 Ga from metamorphic pelites and amphibolites and metamorphic ages of 1.98–2.01 Ga from migmatites. Based on the summarization of the Paleoproterozoic magmatic and metamorphic events, Zheng and Zhang (2007) provided that the Paleoproterozoic continental substances are widely distributed in different regions of the Yangtze Block and the Cathaysian Block. In addition, Hf-isotopic data of the Paleoproterozoic zircons suggest that the reworking of the ancient crust is the main Paleoproterozoic event of South China Craton. Combining all of these data, we can be inferred that the northern margin of the Yangtze Block occurred widely regional metamorphism during 2.0–1.9 Ga in the northern margin of the Yangtze Block, and the magmatism under extensional environment occurred around 1.85 Ga.

In the southwestern Zhejiang region of the Cathaysian Block, there are Badu Group and many Late Paleoproterozoic granitoid rock outcrops (Gan et al. 1995), for example, 1866 Ma Sanzhishu granite (Liu et al. 2009),1832–1867 Ma Danzhu granite (Li and Li 2007; Liu et al. 2009; Yu et al. 2009), 1856 Ma Tianhou granodiorite (Yu JH et al., 2009), 1875 Ma Lizhuang biotite granite (Yu et al. 2009), 1887 Ma Xiaji two-mica monzonitic granite (Yu et al. 2009), which were attributed to S-type granites under the collisional orogenic environment (Hu et al. 1991).The current researches show that earlier (1875–1887 Ma) Xiaji two-mica monzonitic granites and the Lizhuang biotite granites show high SiO_2, K_2O, and Rb, low Sr, REE and mafic compositions, high A/CNK value (1.09–1.40), and high Rb/Sr ratios, which are similar to those of S-type granites in geochemistry (Yu et al. 2009). However, the later (1832–1867 Ma) Danzhu, Houtian, and Sanzhishu plutons have relatively low-SiO_2 content, metaluminous to peraluminous (A/CNK = 0.80–1.07) features, high Ga/Al and FeO/(FeO + MgO) ratios, and these magmatic rocks have formation temperature of 885–920 °C, belonging to the high-temperature A2-type granite (Liu et al. 2009; Yu et al. 2009). S-type granites are usually thought to be mainly formed in the syn-collision tectonic environment (Pearce et al. 1984; Harris et al. 1986), while the A-type granites form in anorogenic environment or the extensional environment after orogenic period (Anderson and Thomas 1985; Whalen et al. 1987; Eby 1992; Bonin 2007), so the granites in this area suggest the evolution from the syn-collision S-type granites to the high-temperature A-type granites in the extensional environment after orogenic period. Moreover, Badu Group Al-rich gneiss peak metamorphic temperature and pressure are 800–850 °C, 0.60–0.70 GPa, respectively, and the retrograde metamorphic peak temperature and confining pressure are 560–590 °C, 0.25–0.33 GPa, respectively; Badu Group garnet two-pyroxene granulite shows its peak metamorphic temperature and pressure of 850–900 °C, 0.92–1.10 GPa, and the retrograde metamorphic peak temperature and confining pressure of 650–700 °C, 0.58–0.65 GPa, reflecting a clockwise P–T path (Zhao and Zhou 2012). Nearly, isothermal decompression clockwise PT path is usually associated with rapid uplift after the collision (Bohlen and Mezger 1989; Harley 1989). The peak metamorphism occurred during 1858–1884 Ma in this region (Zhao 2012; Yu et al. 2009), being consistent with the formation age of S-type granites. Both the transition from S-type granites to A-type granites, and metamorphism from the >1.1 GPa granulites to the clockwise P–T path reflect a more complete collision orogenic process. It can be said that the 1.9–1.8 Ga event was a collision orogenic event. Due to limited field outcrops, it is still difficult to determine the scale and direction of the collision orogenic event.

References

Anderson JL, Thomas WM (1985) Proterozoic anorogenic two-mica granites: silver Plume and St Vrain batholiths of Colorado. Geology 13:177–180

BGMRSC (Bureau of Geology and Mineral Resources Sichuan Province) (1991) Regional Geology of Sichuan Province. Geological Publishing House, Beijing, p 730 (in Chinese with English abstract)

BGMRZJ (Bureau of Geology Mineral Resources of Zhejing Province) (1989) Regional geology of Zhejiang Province. Geological Publishing House, Beijing, p 688 (in Chinese with English summary)

Bohlen SR, Mezger K (1989) Origin of granulite terranes and the formation of the lowermost continental crust. Science 244:326–329

Bonin B (2007) A-type granites and related rocks: evolution of a concept, problems and prospects. Lithos 97:1–29

Chen K, Gao S, Wu Y-B, Guo J-L, Hu Z-C, Liu Y-S, Zong K-Q, Liang Z-W, Geng X-L (2013) 2.6–2.7 Ga crustal growth in Yangtze craton, South China. Precambr Res 224:472–490

Cheng Y-Q (1994) An introduction to regional geology of China. Geological Publishing House, Beijing, p 517 (in Chinese)

Compston W, Williams IS, Kirschvink JL (1997) Zircon U–Pb ages for the early Cambrian time-scale. J Geol Soc London 149:171–184

Ding X, Zhou X-M, Sun T (2005) The Episodic Growth of the Continental Crustal Basement in South China: Single Zircon LA-ICP-MS U-Pb Dating of Guzhai Granodiorite in Guangdong. Geol Rev 51(4):382–392 (in Chinese with English abstract)

Dong S-W, Xue H-M, Xiang X-K, Ma L-C (2010) The discovery of Neoproterozoic pillow lava in spilite-ceratophyre of Lushan area, northern Jiangxi Province, and its geological significance. Geol China 37(4):1021–1033 (in Chinese with English abstract)

Eby GN (1992) Chemical subdivision of the A-type granitoids: petrogenetic and tectonic implications. Geology 20:641–644

Feng B-Z (1990) The Presinian Geology and Metallogeny of the Kangdian Area. Geological Publishing House, Beijing, p 202 (in Chinese with English abstract)

Gan X-C, Li H-M, Sun D-Z, Zhuang J-M (1993) Geochronological study on the Precambrian metamorphic basement in northern Fujian. Geol Fujian 12(1):17–32 (in Chinese with English abstract)

Gan X-C, Li H-M, Sun D-Z, Jin W-S, Zhao F-Q (1995) A geochronological study on early Proterozoic granitic rocks, southwestern Zhejiang. Acta Petrologica et Mineralogica 14(1):1–8 (in Chinese with English abstract)

Gao L-Z, Huang Z-Z, Ding X-Z, Liu Y-X, Zhang C-H, Wang Z-Q, Pang J-F, Han K-Y (2012) The Geochronological Relationship between the Shaojiwa Formation and the Xingzi complex group in Northwestern Jiangxi and the Constraints on Zircon SHRIMP U–Pb Age. Acta Geoscientica Sinica 33(3):295–304 (in Chinese with English abstract)

Gao S, Qiu Y-M, Ling W-L, McNaughton NJ, Groves DI (2001) Single zircon U–Pb dating of the Kongling high-grademetamorphic terrain: evidence for >3.2 Ga old continental crust in the Yangtze craton. Sci China Ser D 44:326–335

Gao S, Yang J, Zhou L, Li M, Hu Z-C, Guo J-L, Yuan H-L, Gong H-J, Xiao G-Q, Wei J-Q (2011) Age and growth of the Archean Kongling terrain, South China, with emphasis on 3.3 Ga granitoid gneisses. Am J Sci 311:153–182

Geng Y-S, Yang C-H, Wang X-S, Ren L-D, Du L-L, Zhou X-W (2007) Age of crustalline basement in western margin of Yangtze terrane. Geol J China Univ 13(3):429–441 (in Chinese with English abstract)

Geng Y-S, Yang C-H, Wang X-S, Du L-L, Ren L-D, Zhou X-W (2008) Metamorphic basement evolution in Western Margin of Yangtze Block. Geological Publishing House, Beijing, p215 (in Chinese)

Greentree MR, Li Z-X, Li X-H, Wu H-C (2006) Late Mesoproterozoic to earliest Neoproterozoic basin record of the Sibao orogenesis in western South China and relationship to the assembly of Rodinia. Precambr Res 151:79–100

Greentree MR, Li Z-X (2008) The oldest known rocks in south western China: SHRIMP U–Pb magmatic crystallization age and detrital provenance analysis of the Paleoproterozoic Dahongshan Group. J Asian Earth Sci 33:290–302

Guo J-L, Gao S, Wu Y-B, Li M, Chen K, Hu Z-C, Ling Z-W, Liu Y-S, Zhou L, Zong K-Q, Zhang W, Chen H-H (2014) 3.45 Ga granitic gneisses from the Yangtze Craton, South China: implications for early Archean crustal growth. Precambr Res 242:82–95

Halls HC, Li J-H, Davis D, Hou G-T, Qian X-L (2000) A precisely dated Proterozoic palaeomagnetic pole from the North China Craton, and its relevance to palaeocontinental reconstruction. Geophys J Int 143:185–203

Halls HC, Zhang B (2003) Crustal uplift in the southern Superior Province, Canada, revealed by paleomagnetism. Tectonophysics 362(4):123–136

Hanchar JM, Miller CF (1993) Zircon zonation patterns as revealed by cathodoluminescence and backscattered electron images: implications for interpretation of complex crustal histories. Chenmical Geol 110:1–13

Harley SL (1989) The origin of granulite: a metamorphic perspective. Geol Mag 126:215–247

Harris NBW, Pearce JA, Tindle AG (1986) Geochemical characteristics of collision zone magmatism. In: Coward MP, Reis AC (eds) Collision Tectonics. Special Publications Geological Society, London, vol 19, pp 67–81

He Z-W, Liu Y-C, Wei X-G, Xiao Y-P, Ma R-Z, Wu D-C (1997) Isotopic geochronology of basement metamorphic rocks series in the Micangshan area along the northern margin of Yangtze Craton, China. J Miner Petrol 17(Suppl):83–87 (in Chinese with English abstract)

Hoskin PWO, Schaltegger U (2003) The composition of zircon and igneous and metamorphic petrogenesis. Rev Miner Geochem 53:27–55

Hou G-T, Liu Y-L, Li J-H (2006) Evidence form 1.8 Ga extension of the Eastern Block of the North China Craton from SHRIMP U–Pb dating of mafic dyke swarms in Shandong Province. J Asian Earth Sci 27:392–401

Hu X-J, Xu J-K, Tong C-X, Chen C (1991) The Precambrian geology of southwestern Zhejiang Province. Geological Publishing House, Beijing p 278 (in Chinese with English abstract)

Hu X-J, Xu J-K, Chen C-H, Li C-Z, Li H-M (1992) Single-grain zircon U–Pb ages for the early Proterozoic granite and pegmatite from southwestern Zhejiang. Chin Sci Bull 37:1016–1018 (in Chinese)

Jiang J-S (1986) Isotope geological age and crustal evolution in Huangling metamorphic area. J Changchun Geol Coll 3:1–11 (in Chinese with English abstract)

Jiang L-L, Wu W-P, Liu Y-C, Li H-M (2003) U–Pb zircon and Ar–Ar hornblende ages of the Susong complex of the southern Dabie orogen and their geological implication. Acta Petrologica Sinica 19(3):497–505 (in Chinese with English abstract)

Jiao W-F, Wu Y-B, Peng M, Wang J, Yang S-H (2009) The oldest basement rock in the Yangtze Craton revealed by zircon U–Pb age and Hf isotope composition. Science China: Earth Sciences, doi: 10.1007/s11430-009-0135-7

Jin W-S, Wang R-Z, Sun D-Z, Wu C-H (1996) The Paleoproterozoic Erathem. In: Stratigraphical Lexicon of China. Geological Publishing House, Beijing, pp 85–175 (in Chinese)

Jin W-S, Sun D-Z, Zhao F-Q (1997) Deep crustal structure and its evolution of South China. Geological Publishing House, Beijing, p 175 (in Chinese)

Kong X-S, Li Z-D, Feng B, Guo M, Ma J (1995) The Precambrian Geology of Chencai Region in Zhejiang Province, Precambrian Geology. Geological Publishing House, Beijing, p 136 (in Chinese with English summary)

Lee J-S, Chao Y-T (1924) Geology of the Gorges district of the Yangtze from Ichang to Tzekuei with special reference to development of the Gorges. Bulle Geol Soc China 3:351–391

Li F-X, Nie X-W (1987) The geological age and stratigraphic division of Kongling group in Norther part of Huangling faulted upwarpin. Hubei Geol 1(1):28–41 (in Chinese with English abstract)

Li F-H, Tan J-M, Shen Y-L, Wang F-X, Zhou G-F, Pan X-N, Li X-Z (1988) The Kangdian region Pre-Sinian. Chongqing Publishing House, Chongqing, p 396 (in Chinese)

Li H-K, Zhang C-L, Yao C-Y, Xiang Z-Q (2013) U–Pb zircon age and Hf isotope compositions of Mesoproterozoic sedimentary strata on the western margin of the Yangtze massif. Sci China Earth Sci 56:628–639. doi:10.1007/s11430-013-4590-9

Li L-M, Sun M, Wang Y, Xing G-F, Zhao G-C, Lin S, Xia X, Chan L, Zhang F, Wong J (2011) U–Pb and Hf isotopic study of zircons from migmatised amphibolites in the Cathaysia Block: implications for the early Paleozoic peak tectonothermal event in Southeastern China. Gondwana Res 19(1):191–201

Li S-G, Chen Y-Z, Ge N-J, Hu X-J, Liu D-L (1996) Isotopic ages of metavolcanic rocks and metacrust mylonite in the Badu group in southwestern Zhejiang Province and their implication for tectonics. Acta Petrol Sinica 12(1):79–87 (in Chinese with English abstract)

Li X-H, Wang Y-X, Zhao Z-H, Chen D-F, Zhang H (1998) SHRIMP U–Pb zircon geochronology for amphibolete from the Precambrian basement in SW Zhejiang and NW Fujian provinces. Geochemica 27(4):327–334 (in Chinese with English abstract)

Li X-H, Li W-X, He B (2012a) Bulding of the South China Block and its relevance to assembly and breakup of Rodinia supercontinent: observations, interpretations and tests. Bull Mineral Petrol Geochem 31(6):543–559 (in Chinese with English abstract)

Li X-H, Li Z-X, He B, Li W-X, Li Q-L, Gao Y, Wang X-C (2012b) The Early Permian activecontinental margin and crustal growth of the Cathaysia Block: In situ U–Pb, Lu-Hf and O isotope analyses of detrital zircons. Chem Geol 328:195–207

Li Z-X, Wartho JA, Occhipinti S, Zhang C-L, Li X-H, Wang J, Bao C-M (2007) Early history of the eastern Sibao Orogen (South China) during the assembly of Rodinia: New mica 40Ar/39Ar dating and SHRIMP U–Pb detrital zircon provenance constraints. Precambr Res 159:79–94

Li Z-X, Li X-H (2007) Formation of the 1300 km wide intracontinental orogen and postorogenic magmatic province in Mesozoic South China: a flat-slab subduction model. Geology 35 (2):179–182. doi:10.1130/G23193A.1

Li Z-X, Li X-H, Wartho JA, Clark C, Li W-X, Zhang C-L, Bao C (2009) Magmatic and metamorphic events during the early Paleozoic Wuyi-Yunkai orogeny, southeastern South China: new age constraints and pressure-temperature conditions. Geol Soc Am Bull 122(5–6):772–793

Ling W-L, Gao S, Zheng H-F, Zhou L, Zhao Z-B (1998) An Sm–Nd isotopic dating study of the Archean Kongling complex in the Huangling area of the Yangtze Craton. Chin Sci Bull 43 (14):1187–1191

Ling W-L, Gao S, Zhang B-R, Li H-M, Liu Y, Cheng J-P (2003) Neoproterozoic tectonic evolution of the northwestern Yangtze craton, South China: implications for amalgamation and break-up of the Rodinia Supercontinent. Precambr Res 122:111–140

Liu H-Y, Xia B, Zhang Y-Q (2004) Zircon SHRIMP dating of sodium alkaline rocks from Maomaogou area of Huili County in Panxi, SW China and its geological implications. Chin Sci Bull 49(16):1750–1757

Liu Q, Yu J-H, O'Reilly SY, Zhou M-F, Griffin WL, Wang L-J, Cui X (2014) Origin and geological significance of Paleoproterozoic granites in thenortheastern Cathaysia Block, South China. Precambr Res 248:72–95

Liu R, Zhou H-W, Zhang L, Zhong Z-Q, Zeng W, Xiang H, Jin S, Lu X-Q, Li C-Z (2009) Paleoproterozoic reworking of ancient crust in the Cathaysia Block, South China: evidence from zirconc trace elements, U–Pb and Lu–Hf isotopes. Chin Sci Bull 54(9):1543–1554. doi:10.1007/s11434-009-0096-4

Liu X-M, Gao S, Ling W-L, Yuan H-L, Hu Z-C (2006) Identification of 3.5 Ga detrital zircons from Yangtze craton in south China and the implication for Archean crust evolution. Prog Nat Sci 16(6):663–666

Lu L-Z, Xu X-C, Liu F-L (1996) Early Precambrian Khondalite Series in North China. Changchun Publishing House, Changchun, p 210 (in Chinese)

Ma D-Q, Li Z-C, Xiao Z-F (1997) The constitute, geochronology and geologic evolution of the Kongling complex, western Hubei. Acta Geoscientia Sinica 18(3):233–241 (in Chinese with English abstracta)

Pearce JA, Harris NBW, Tindle AG (1984) Trace element discrimination diagrams for the tectonic interpretation of granitic rocks. J Petrol 25:956–983

Peng M, Wu Y-B, Wang J, Liu X-C, Jiao W-F, Yang, S-H (2009) Paleoproterozoic mafic dyke from Kongling terrain in the Yangtze Craton and its implication. Chin Sci Bull 54:1098–1104

Peng M, Wu Y-B, Gao S, Zhang H-F, Wang J, Liu X-C, Gong H-J, Zhou L, Hu Z-C, Liu Y-S, Yuan H-L (2012) Geochemistry, zircon U–Pb age and Hf isotope compositions of Paleoproterozoic aluminous A-type granites from the Kongling terrain, Yangtze Block: constraints on petrogenesis and geologic implications. Gondwana Res 22:140–151

Pidgen RT (1992) Rerystallisation of oscillatory-zoned zircon: some geochronological and petrological implications. Contrib Miner Petrol 110:463–472
Qin K-L, Shao X-H, He S-P (1990) The sequence and age division of Bikou Group in Motianling region bounded by Shaanxi, Gansu and Sichuan, China. Bulletin Xi'an Institute of Geology and Mineral Resources, CAGS 30:1–60 (in Chinese with English abstract)
Qin K-L, He S-P, Song S-G (1992) Isotopic geochronology and its significance in Bikou terrain, China. Northwest Geosci 13(2):97–109 (in Chinese with English abstract)
Qin X-F, Pan Y-M, Li J, Li R-S, Zhou F-S, Hu G-A, Zhong F-Y (2006) Zircon SHRIMP U-Pb geochronology of the Yunkai metamorphic complex is shoutheastern Guangxi, China. Geol Bull China 25(5):553–559 (in Chinese with English abstract)
Qiu Y-M, Gao S, McNaughton NJ, Groves DI, Ling W-L (2000) First evidence of > 3.2 Ga continental crust in the Yangtze craton of south China and its implications for Archean crustal evolution and Phanerozoic tectonics. Geology 28(1):11–14
Shen Q-H, Geng Y-S, Liu G-H, Gao J-F (1996) The Paleoproterozoic Erathem. In: Stratigraphical Lexicon of China. Geological Publishing House, Beijing, pp 85–175 (in Chinese)
Shen Q-H, Geng Y-S, Song B, Wan Y-S (2005) New information from the surface outcrops and deeo crust of Archean rocks of the Noeth China and Yangtze Blocks, and Qinling-Dabie orogenic belt. Acta Geol Sinica 79(5):616–627 (in Chinese with English abstract)
Sun W-H, Zhou M-F, Yan D-P, Li J-W, Ma Y-X (2008) Provenance and tectonic setting of the Neoproterozoic Yanbian Group, western Yangtze Block (SW China). Precambr Res 167:213–236
Sun W-H, Zhou M-F, Gao J-F, Yang Y-H, Zhao X-F, Zhao J-H (2009) Detrital zircon U–Pb geochronological and Lu–Hf isotopic constraints on the Precambrian magmatic and crustal evolution of the western Yangtze Block, SW China. Precambr Res 172:99–126
Wan Y-S, Liu D-Y, Xu M-H, Zhuang J-M, Song B, Shi Y-R, Du L-L (2007) SHRIMP U–Pb zircon geochronology and geochemistry of metavolcanic and metasedimentary rocks in Northwestern Fujian, Cathaysia block, China: Tectonic implication and the need to redefine lithostratigraphic units. Gondwana Res 12:166–183
Wang L-J, Yu J-H, Griffin WL, O'Reilly SY (2012) Early crustal evolution in the western Yangtze Block: Evidence from U–Pb and Lu–Hf isotopes on detrital zircons from dedimentary rocks. Precambr Res. doi:10.1016/j.precamres.2011.08.001
Wang M, Dai C-G, Wang X-H, Chen J-S, Ma H-Z (2011) In-situ zircon geochronology and Hf isotope of muscovite-bearing leucogranites from Fanjingshan, Guizhou Province, and constraints on continental growth of the southern China block. Earth Sci Front 18(5):213–223 (in Chinese with English abstract)
Wang X, Chen J, Luo D (2008) Study on petrogenesis of zircons from the Danzhugranodiorite and its geological implication. Geol Rev 54(3):387–398 (in Chinese with English abstract)
Wang X-F, Ma D-Q, Chen X-H, Wang C-S (2001) Precambrian evolution of the Kuangling Arch, China, and its relation to the Rodinia breakup and assembly. Gondwana Res 4(4):816–817
Wang X-L, Chang H, Cui J-G (1998) The feature of the crystal basement in Mian-Lue-Ning triangular area during late Archean era. Geol Shaanxi 16(1):58–66 (in Chinese with English abstract)
Wang X-L, Zhou J-C, Griffin WL, Wang R-C, Qiu J-S, O'Reilly SY, Xu X-S, Liu X-M, Zhang G-L (2007) Detrital zircon geochronology of Precambrian basement sequences in the Jiangnan orogen: dating the assembly of the Yangtze and Cathaysia Blocks. Precambr Res 159:117–131
Wang Y-B, Liu D-Y, Zeng P-S, Yang Z-S, Meng Y-F, Tian S-H (2004) SHRIMP U–Pb geochronology of Xiaotongguanshan quartz dioritic intrusions in Tongling district and its petrogenetic implications. Acta Petrologica et Mineralogica 23(4):298–312 (in Chinese with English abstract)
Wang Y-X, Yang J-D, Guo L-Z, Shi Y-S, Hu X-J, Wang X (1992) The discovery of the lower Proterozoic granite in Longquan, Zhejiang Province, and the age of the basement. Geol Rev 38 (6):525–531 (in Chinese with English abstract)

Wang Z-J, Wang J, Du Q-D, Deng Q, Yang F (2013a) The evolution of the Central Yangtze Block during early Neoarchean time: evidence from geochronology and geochemistry. J Asian Earth Sci 77:31–44

Wang Z-J, Wang J, Du Q-D, Deng Q, Yang F (2013b) Mature Archean continental crust in the Yangtze craton: evidence from petrology, geochronology and geochemistry. Chinese Science Bulletin 58. http://dx.doi.org/10.1007/s11434-013-5668-7

Whalen JB, Currie KL, Chappell BW (1987) A-type granites: geochemical characteristics, discrimination and petrogenesis. Contrib Miner Petrol 95:407–419

Wu Y-B, Zheng Y-F (2004) Genesis of zircon and its constraints on interpretation of U–Pb age. Chin Sci Bull 49(15):1554–1569

Wu Y-B, Gao S, Zhang H-F, Zheng J-P, Liu X-C, Wang H, Gong H-J, Zhou L, Yuan H-L (2012) Geochemistry and zircon U–Pb geochronology of Paleoproterozoic arc related granitoid in the Northwestern Yangtze Block and its geological implications. Precambr Res 200–203:26–37

Xia Y, Xu X-S, Zhu K-Y (2012) Paleoproterozoic S-and A-type granites in southwestern Zhejiang: Magmatism, metamorphism and implications for the crustal evolution of the Cathaysia basement. Precambr Res 216–219:177–207

Xiang H, Zhang L, Zhou H-W, Zhong Z-Q, Zeng W, Liu R, Jin S (2008) U–Pb zircon geochronology and Hf isotope study of metamorphosed basic-ultrabasic rocks from metamorphic basement in southwestern Zhejiang: the response of the Cathaysia Block to Indosinian orogenic event. Sci China Ser D(Earth Sciences), 51(6):788–800

Xiong Q, Zheng J-P, Yu C-M, Su Y-P, Tang H-Y, Zhang Z-H (2009) Zircon U–Pb age and Hf isotope of Quanyishang A-type granite in Yichang: Signification for the Yangtze continental cratonization in Paleoproterozoic. Chin Sci Bull 54(3):436–446

Xu X-B, Zhang Y-Q, Shu L-S, Jia D, Wang R-R, Xu H-Z (2010) Precambrian Geochronology and stratigraphy in the Wuyishan area, south China. J Stratigr 34(3):254–267 (in Chinese with English abstract)

Xu X-S, O'Reilly SY, Griffin WL, Deng P, Pearson NJ (2005) Relict Proterozoic basement in the Nanling Mountains (SE China) and its tectonothermal overprinting. Tectonics 24:1–16

Xu X-S, O'Reilly SY, Griffin WL, Wang X-L, Pearson NJ, He Z-Y (2007) The crust of Cathaysia: age, assembly and reworking of two terranes. Precambr Res 158:51–78

Yao J-L, Shu LS, Santosh M (2011) Detrital zircon U–Pb geochronology, Hf-isotopes and geochemistry: New clues for the Precambrian crustal evolution of Cathaysia Block, South China. Gondwana Res 20:553–567

Yin F-G, Sun Z-M, Bai J-K (2011a) Stratigraphic framework of the Mesoproterozoic in Dongchuan and central Yunnan region. J Stratigr 35(1):49–54 (in Chinese with English abstract)

Yin F-G, Sun Z-M, Zhang Z (2011b) Mesoproterozoic stratigraphic – structure framework in Huili – Dongchuan area. Geol Rev 57(6):770–778 (in Chinese with English abstract)

Yu J-H, Zhou X-M, O'Reilly YS, Zhao L, Griffin WL, Wang R-C, Wang L-J, Chen X-M (2005) Formation history and protolith characteristics of granulite facies metamorphic rock in Central Cathaysia. Chin Sci Bull 50(18):2080–2089

Yu J-H, Wang L-J, Zhou X-M, Jiang S-Y, Wang R-C, Xu X-S, Qiu J-S (2006) Compositions and formation history of the basement metamorphic rocks in northeastern Guangdong Province. Earth Sci 31(1):38–48 (in Chinese with English abstract)

Yu J-H, O'Reilly YS, Wang L-J, Griffin WL, Jiang S-Y, Wang R-C, Xu X-S (2007) Finding of ancient materials in Cathaysia and implication for the formation of Precambrian crust. Chin Sci Bull 52(1):13–22

Yu J-H, O'Reilly SY, Wang L-J, Griffin WL, Zhang M, Wang R-C, Jiang S-Y, Shu L-S (2008) Where was South China in the Rodinia supercontinent? Evidence from U–Pb geochronology and Hf isotopes of detrital zircons. Precambr Res 164:1–15

Yu J-H, Wang L-J, O'Reilly SY, Griffin WL, Zhang M, Li C, Shu L-S (2009) A Paleoproterozoic orogeny recorded in a long-lived cratonic remnant (Wuyishan terrane), eastern Cathaysia Block, China. Precambr Res 174(3–4):347–363

Yu J-H, O'Reilly SY, Wang L-J, Griffin WL, Zhou M-F, Zhang M, Shu L-S (2010) Components and episodic growth of Precambrian crust in the Cathaysia Block, South China: Evidence from U–Pb ages and Hf isotopes of zircons in Neoproterozoic sediments. Precambr Res 181:97–114

Yu J-H, O'Reilly SY, Zhou M-F, Griffin WL, Wang L (2012) U–Pb geochronology and Hf–Nd isotopic geochemistry of the Badu Complex, Southeastern China: implications for the Precambrian crustal evolution and paleogeography of the Cathaysia Block. Precambr Res 222–223:424–449

Zhang L-J, Ma C-Q, Wang L-X, Yu Z-B, Wang S-M (2011) Discovery of Paleoproterozoic rapakivi granite on the northern margin of the Yangtze block and its geological significance. Chin Sci Bull 56:306–318

Zhang Q, Jian P, Liu D-Y, Wang Y-L, Qian Q, Wang Y, Xue H-M (2003) SHRIMP dating of volcanic rocks from Ningwu area and its geological implications. Sci China (Ser D) 46:830–837

Zhang S-B, Zheng Y-F, Wu Y-B, Zhao Z-F, Gao S, Wu F-Y (2006a) Zircon isotope evidence for ≥3.5 Ga continental crust in the Yangtze Craton of China. Precambr Res 146:16–34

Zhang S-B, Zheng Y-F, Wu Y-B, Zhao Z-F, Gao S, Wu F-Y (2006b) Zircon U–Pb age and Hf isotope evidence for 3.8 Ga crustal remnant and episodic reworking of Archean crust in South China. Earth and Plane Sci Lett 252:56–71

Zhang S-B, Zheng Y-F, Wu Y-B, Zhao Z-F, Gao S, Wu F-Y (2006c) Zircon U–Pb age and Hf–O isotope evidence for Paleoproterozoic metamorphic event in South China. Precambr Res 151:265–288

Zhang S-B, Zheng Y-F (2013) Formation and evolution of Precambrian continental lithosphere in South China. Gondwana Res 23:1241–1260

Zhang X, Xu X-Y, Song G-S, Wang H-L, Chen J-L, Li T (2010a) Zircon LA-ICP-MS U–Pb dating and significance of Yudongzi Group deformation granite from Lueyang area, western Qinling, China. Geol Bull China 29(4):510–517 (in Chinese with English abstract)

Zhang Y-J, Zhou X-H, Liao S-B, Zhang X-D, Wu B, Wang C-Z, Yu M-G (2010b) Neoproterozoic crustal composition and orogenic process of the Zhanggongshan area, Anhui–Jiangxi. Acta Geol Sinica 84(10):1401–1427 (in Chinese with English abstract)

Zhang Z-Q, Zhang G-W, Tang S-H, Wang J-H (2001) On the age of metamorphic rocks of the Yudongzi Group and the Archean crystalline basement of the Qinling Orogen. Acta Geol Sinica 75:198–204 (in Chinese with English abstract)

Zhang Z-Q, Zhang G-W, Tang S-H (2002) Isotopic Geochronology of Metamorphic Strata in the South Qinling. Geological Publishing House, Beijing, p 256 (in Chinese)

Zhao X-F, Zhou M-F, Li J-W, Sun M, Gao J-F, Sun W-H, Yang J-H (2010) Late Paleoproterozoic to early Mesoproterozoic Dongchuan Group in Yunnan, SW China: implications for tectonic evolution of the Yangtze Block. Precambr Res 182:57–69

Zhao G-C, Cawood PA (2012) Precambrian geology of China. Precambr Res 222–223:13–54

Zhao J-H, Zhou M-F, Yan D-P, Yang Y-H, Sun M (2008) Zircon Lu–Hf isotopic constraints on Neoproterozoic subduction-related crustal growth along the western margin of the Yangtze Block, South China. Precambr Res 163:189–209

Zhao L (2012) Study on the early precambrain metamorphic evolution and geochronology of the Badu group in southwestern Zhejiang. Dissertation submitted to Chinese Academy of Geological Science for master's degree. pp 1–111(in Chinese with English abstract)

Zhao L, Zhou X-W (2012) The metamorphic evolution and *PT* path of politic granulite from Badu group in southwestern Zhejiang Province. Acta Petrologica et Mineralogica 31(1):61–72 (in Chinese with English abstract)

Zhao L, Zhou X-W, Zhai M-G, Santosh M, Geng Y-S (2014a) Zircon U–Th–Pb–Hf isotopes of the basement rocks in northeastern Cathaysia block, South China: Implications for Phanerozoic multiple metamorphic reworking of a Paleoproterozoic terrane. Gondwana Research. http://dx.doi.org/10.1016/j.gr.2014.03.019

Zhao L, Zhou X-W, Zhai M-G, Santosh M, Ma X-D, Shan H-Q, Cui X-H (2014b) Paleoproterozoic tectonic transition from collision to extension in the eastern Cathaysia

Block, South China: evidence from geochemistry, zircon U–Pb geochronology and Nd–Hf isotopes of a granite- charnockite suite in southwestern Zhejiang. Lithos 184–187:259–280

Zheng Y-F, Zhang S-B (2007) Formation and evolution of Precambrian continental crust in South China. Chin Sci Bull 52:1–12

Zheng J-P, Griffin WL, O'Reilly SY, Zhang M, Pearson N, Pan Y-M (2006) Widespread Archean basement beneath the Yangtze craton. Geology 34:417–420

Zheng J-P, Griffin WL, Tang H-Y, Zhang Z-H, Su Y-P, Liu G-L (2008) Archean basemen similar to the North China and Yangtze continents may be existed benesth the western Cathaysia. Geol J China Univ 14(4):549–557 (in Chinese with English abstract)

Zheng J-P, Griffin WL, Li L-S, O'Reilly SY, Pearson NJ, Tang H-Y, Liu G-L, Zhao J-H, Yu C-M, Su Y-P (2011) Highly evolved Archean basement beneath the western Cathaysia Block, South China. Geochimica et Cosmochimica Acta 75(1):242–255

Zheng W-Z, Liu G-L, Wang X-W (1991) Archean information from the Kongling Complex, northern Huangling anticline. Yichang Inst Geol Min Res Bull 16:97–105 (in Chinese with English abstract)

Zhou B-G, Wang S-W, Sun X-M, Liao Z-W, Guo Y, Jiang X-F, Zhu H-P, Luo M-J, Liu C-Z, Ma D, Shen Z-W, Zhang H (2012) SHRIMP U–Pb age and its significance of zircons in welded tuff of Wangchang formation in Dongchuan area, Yunnan Province, SW China. Geol Rev 58(2):359–368 (in Chinese with English abstract)

Zhou M-F, Yan D-P, Kennedy AK, Li Y-Q, Ding J (2002a) SHRIMP U–Pb zircon geochronological and geochemical evidence for Neoproterozoic arc-magmatism along the western margin of the Yangtze Block, South China. Earth Planet Sci Lett 196:51–67

Zhou M-F, Kennedy AK, Sun M, Malpas J, Lesher CM (2002b) Neoproterozoic arc-related mafic intrusions along the Northern Margin of South China: implications for the Accretion of Rodinia. J Geololy 110:611–618

Zhou M-F, Ma Y-X, Yan D-P, Xia X-P, Zhao J-H, Sun M (2006) The Yanbian terrane (Southern Sichuan province, SW China): a Neoproterozoic arc assemblage in the western margin of the Yangtze block. Precambr Res 144:19–38

Zhu H-P, Zhou B-G, Wang S-W, Luo M-J, Liao Z-W, Guo Y (2011a) Detrital zircon U–Pb dating by LA-ICP-MS And its geological significance in western margin of Yangtze terrane. J Miner Petrol 31(1):70–74 (in Chinese with English abstract)

Zhu H-P, Fan W-Y, Zhou B-G, Wang S-W, Luo M-J, Liao Z-W, Guo Y (2011b) Assessing Precambrian stratigraphic sequence of Dongchuan area: evidence from zircon SHRIMP and LA-ICP-MS dating. Geol J China Univ 17(3):452–461 (in Chinese with English abstract)

Mesoproterozoic Era of South China Craton

Yuan-Sheng Geng

Abstract The Mesoproterozoic terranes are mainly distributed at the northern and western margin of the Yangtze Block and can be discerned into 2 major stages, early Pt_2 (1.8–1.5 Ga) and late Pt_2 (1.5–1.0 Ga), according to the formation time. The sequence of early Pt_2 stage is present as the Dahongshan, Dongchuan, and Hekou Groups at the southwestern margin of the Yangtze Block, consisting of sandstone, siltstone, slate, phyllite, dolomitic marble, and marble, with volcanic intercalation in which usually produce the Fe–Cu deposits. In addition, some magmatic rocks of early Pt_2 stage occur as stocks and dykes intruding the early Pt_2 sequence in the area. The magmas can be subdivided into two events: the early ca 1.7 Ga dominated by dolerites and the late ca 1.5 Ga gabbro stocks and dykes. The sequence of late Pt_2 stage crop out as the Kunyang, Huili Groups at the south-western margin and the Shennongjia Group at the northern margin of the Yangtze Block. Magmatic events are obviously manifested in the period (1.1–1.0 Ga) in the block, but magmatic rocks (including the eruptives and intrusives) in different areas vary in formation setting according to their association and geochemical features. While in the Cathaysian Block and Jiangnan orogenic belt, the Mesoproterozoic terranes are only rarely exposed, such as the 1.5–1.4 Ga Baoban Group (including part of the Shilu Group) in Hainan Island of the Cathaysian Block, and the Tianli schist and Tieshajie Formation of Late Mesoproterozoic in the eastern segment of the Jiangnan belt.

Keywords Mesoproterozoic · Dahongshan Group · Hekou Group · Dongchuan Group · Kunyang Group · Huili Group · Shennongjia Group · 1.7 and 1.5 Ga magmatic event · 1.0 Ga magmatic event

---

Y.-S. Geng (✉)
Institute of Geology, Chinese Academy of Geology Sciences,
100037 Beijing, China
e-mail: ys-geng@cags.ac.cn

M. Zhai (ed.), *Precambrian Geology of China*,
Springer Geology, DOI 10.1007/978-3-662-47885-1_6

1 Mesoproterozoic Era of Yangtze Block

The Mesoproterozoic era on international stratigraphic chart (IUGS 2008) is the formation strata systems between 1600 and 1000 Ma, including Calymmian (1600–1400 Ma), Ectasian (1400–1200 Ma), and Stenian (1200–1000 Ma). However, Chinese regional chrono-stratigraphic scale (China National Commission of Stratigraphy, CNCS, 2001) defined the period of Mesoproterozoic era as 1800s to 1000 Ma, comprising of Statherian (1800–1600 Ma), Calymmian, Ectasian, and Stenian. Previous Chinese stratigraphic chart divided the Mesoproterozoic era into the Changcheng System (1.8–1.4 Ga) and the Jixianian System (1.4–1.0 Ga). Nevertheless, recent researches of isotope chronology (Gao et al. 2007, 2008a; Su et al. 2008; Li et al. 2009ab, 2011; He et al. 2011) indicated that previous Changcheng System and Jixianian System in the standard Mesoproterozoic Changcheng–Jixianian Section were limited to 1.68–1.40 Ga, lacking a lot of upper strata sequence. Therefore, the CNCS proposed that Chinese Mesoproterozoic era can be divided into Changcheng System (1800–1600 Ma), Jixianian System (1600–1400 Ma), and Unnamed System (1400–1000 Ma, which maybe can be divided into two systems furthermore). This paper adopts the scheme proposed by the CNCS. Mesoproterozoic era distributed widely in the South China Block, but mainly in the Yangtze Block, with minor in the Cathaysian Block and Jiangnan belt.

1.1 The Early Mesoproterozoic Era of Yangtze Block

Mesoproterozoic era of Yangtze Block distributed mainly in the southwestern and northern parts of the Block. The southwestern part is comprised of the Early Mesoproterozoic Dahongshan Group, Hekou Group, and Dongchuan Group; and the Mid–Late Mesoproterozoic Kunyang Group and Huili Group. The northern part is mainly composed of Shennongjia Group.

Dahongshan Group is exposed chiefly in the Dahongshan area at the both sides of the Lüzhijiang River, Xinping County of Yunnan Province, and partially in the Yuanjiang County and Xiaoliukou and Dongchuan counties. From the bottom to the top, they can be divided into Laochanghe Formation, Manganghe Formation, Hongshan Formation, Feiweihe Formation, and Potou Formation (Qian and Shen 1990). Many ages were obtained from the Manganghe Formation. In the early 1990s, a formation age of 1665 + 14/−11 Ma was documented from the meta-sodic lava by TIMS zircon dating method, and an isochron age of 1657 ± 82 Ma (Hu et al. 1991) was reported by the whole-rock Sm–Nd isotopic dating. Greentree and Li (2008) got the SHRIMP U–Pb isotopic age of 1675 ± 8 Ma from the meta-tuff (Greentree and Li 2008), and Zhao and Zhou (2011) got the LA-ICPMS zircon U–Pb isotopic age of 1681 ± 13 Ma from the imandrite and suggested that the dolerites intruded into the Manganghe Formation at 1659 ± 16 Ma. Yang et al. (2012) reported the ages of 1711 ± 4 Ma and 1688 ± 16 Ma from meta-acid volcanics and meta-basic volcanics, respectively. These data indicate that the Manganghe Formation was formed mainly between 1711 and 1675 Ma. Inasmuch as the Manganghe Formation is the lower part of the Dahongshan Group, it was deduced that the Dahongshan Group was formed between 1750 and 1600 Ma, namely the Early Mesoproterozoic.

The thickness of the Laochanghe Formation is about 1284 m, locally copper mineralization, which can be subdivided into five segments from the bottom to the top: The first segment is composed of quartzites interbedded with biotite–quartz schists, all of them yield cross-bedding; the second segment comprises chiefly garnet–quartz schists, muscovite–quartz schists, and muscovite schists interbedded with minor hornblende–biotite schists; the third segment consists of marbles interbedded with argillaceous siltstones and carbon mudstones; the fourth segment is composed mainly of garnet–quartz–muscovite schists and bambo-leaf marbles interbedded with carbon mudstones; and the fifth segment is composed of garnet–muscovite schists and mica schists. Manganghe Formation is composed mainly of hornblende–albite schists, albite–hornblende schists, magnetite–albite–hornblende schists, meta-volcanic breccias, and chlorite schists interbedded with dolomite marbles in the middle and the upper segments, and the protoliths is chiefly composed of volcanics and volcanic tuff, having unconformable contact with the underlying Laochanghe Formation. Hongshan Formation has the thickness of greater than 319 m and shows unconformable contact with the underlying Manganghe Formation. The lower segment of Hongshan Formation comprises amygdaloidal and globular albite leptites, and magnetite–albite leptites interbedded with magnetite deposit, and the upper part of the lower segment is the essential Dahongshan-type copper and iron deposit ore-bearing horizon, which is composed of silica–magnetites interbedding with garnet chlorite hornblende schists. The upper segment of Hongshan Formation is composed of garnet–hornblende–chlorite schists and hornblende–albite leptites. The thickness of the Feiweihe Formation is about 259 m, which can be divided into three segments from the bottom to the top. The first segment has a petrological assemblage of banded dolomite marbles, with minor scapolites, quartzite, and carbonatite. The second segment is chiefly composed of dolomite marbles in the lower part, and carbonaceous slates, marbles, and pyrites in the upper part. The third segment comprises dolomite marbles interbedded with meta-gabbro–diabases, with copper-bearing carbonaceous slates in the lower part which has the thickness of about 3 m. Potou Formation has the thickness of about 577 m: the lower segment is composed of banded carbon-bearing garnet two mica schists, the middle segment comprises banded dolomite marbles and carbonaceous slates and interbedded with albite–quartzites, and the upper segment is composed of carbon-bearing biotite–quartz schists and carbonaceous sericite schists interbedded with marbles and carbonaceous slates.

Numerous Mesoproterozoic strata distributed in the southwest part of the Yangtze Block, including Hekou Group, Tangdan Group, Dongchuan Group, Huili Group and Tong'an Formation, and Neoproterozoic Yanbian Group (Fig. 1). However, some researchers proposed that the Tangdan Group was formed in Paleoproterozoic. The Hekou Group in the Lalachang, Huili County, and

Fig. 1 A simplified geological map of the southwest margin of Yangtze Block *1* Neoproterozoic Yanbian Group, *2* Mesoproterozoic Huili Group, *3* Dongchuan Group, *4* Tong'an Formation, *5* Hekou Formation, *6* Tangdan Group, *7* Amphogneiss, *8* Dioritic gneiss, *9* Granite, *10* Granodiorite, *11* Ultramafic rock, *12* Major fault

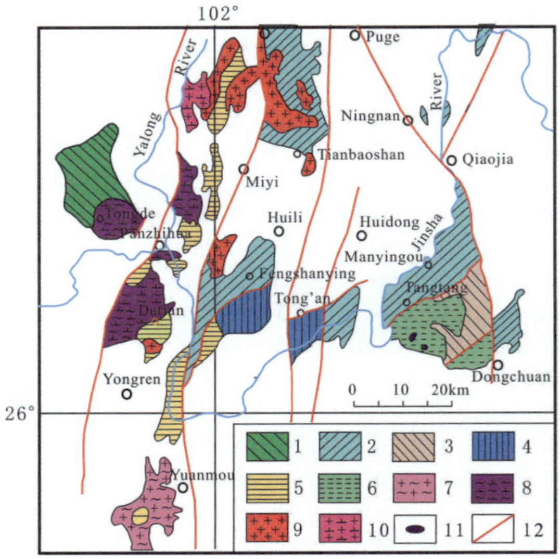

southwestern Sichuan Province has the same age with the Dahongshan Group and Dongchuan Group in Yunnan Province.

The Hekou Group can be subdivided into three formations from bottom to the top, comprising Dayingshan Formation, Luodang Formation, and Changchong Formation (Li et al. 1988; BGMRSC 1991). Previous researchers obtained some ages from 1987 to 858 Ma using the whole rock Rb–Sr isochron dating, conventional zircon U–Pb isotopic dating, molybdenite Re–Os isotopic dating, the whole Pb–Pb dating (Li et al. 1988, 2003a, b), and the various ages cannot clearly confirm the formation age of the Hekou Group. Greentree et al. (2006) got the oldest age of 3051 Ma from the detrital zircons of the metamorphic quartzites from the Hekou Group, indicating that Archean materials were involved in the sediments of the Hekou Group, and a group of younger ages of 1699 to 1759 Ma, implying the sedimentary time is younger than 1699 Ma. Geng et al. (2008) got the ages of 1785–2932 Ma from the detrital zircons of metamorphic feldspathic quartz sandstones from the Hekou Group using SHRIMP U–Pb isotopic dating, and most ages distributed between 1800 and 2400 Ma, and the youngest $^{207}Pb/^{206}Pb$ weighted average age is 1817 ± 10 Ma from six zircons, so he proposed that the Hekou Group was formed in Mesoproterozoic. Recently, the ages of 1680 ± 13 Ma (Zhou et al. 2011), 1722 ± 25 Ma (Wang et al. 2012a, b) from the soda-volcanic rocks, and 1703 ± 5, 1708 ± 7, 1679 ± 13 Ma from tuff (Chen et al. 2013) of the Hekou Group have been obtained by zircon LA-ICP-MS or SHRIMP U–Pb methods. The data suggest that Hekou Group was mainly formed in Early Mesoproterozoic, which is comparable with the Dahongshan Group and Dongchuan Group in Yunnan Province.

The Dayingshan Formation is composed mainly of light-colored metamorphic sandstones with minor muscovite–quartz schists in the lower part, and imandrites interbedded with muscovite-quartz schists in the middle and upper parts. Luodang Formation can be subdivided into three segments from bottom to the top. The lower part is composed mainly of muscovite–quartz schists interbedded with dolomites and imandrites. The middle part comprises the imandrites interbedded with two mica schists, metamorphic quartz sandstones, and marble lenticles, which is the most important copper-bearing position in Lixi. The upper part is composed of albite–quartz schists interbedded with imandrites and mica schists. The mica–quartz schists in this formation contain garnets ordinarily. Changchong Formation consists of imandrites, muscovite–albite schists, garnet–mica schists, and garnet–hornblende schists interbedded with carbonaceous slates and copper-bearing marble lenticles. Changchong Formation suffered greenschist facies to low-amphibolite-facies metamorphism.

Protoliths of imandrites and some albite–quartz schists in the Hekou Group were keratophyres and quartz–keratophyres, and they have lower SiO_2 contents (<60 %) and K_2O contents of (0.09–0.18 %), but higher contents of CaO (8.15–9.43 %) and Na_2O (6.22–6.95 %), but lower and higher Na_2O/K_2O ratios. The contents of rare earth elements vary from right oblique to left oblique. The high contents of Hf and Zr and lower contents of La, Ce, and Nd show the similarities with the typical spilitic keratophyres. The geochemical characteristics reveal that the Hekou Group formed in a continental rifting setting, similar to the contemporary Dahongshan and Dongchuan Groups in the Yunnan Province (Chen et al. 2013).

The Dongchuan Group is located in the northeast part of Yunnan Province and formed during Early to Middle Mesoproterozoic. Previous studies attributed the state of the group into the Kunyang Group in the central part of Yunnan Province, but its affinity is still ambiguous in the stratigraphic division (Li et al. 1984, 1988; Wu et al. 1990; Dai 1997; Lü and Dai 2001). Recent studies suggested Central Yunnan and Dongchuan areas are different blocks before Neoproterozoic (Li and Powell 1996; Li 1999; Zhou et al. 2002a; Li et al. 2002; Wu 2006; Zhang et al. 2007), and the Yinmin Formation, Luoxue Formation, Heishan Formation, and Qinglongshan Formation in Dongchuan area belonged to Dongchuan Group, and Huangcaoling Formation, Heishantou Formation, Fuliangpeng Formation, Dalongkou Formation, and Meidang Formation belonged to Kunyang Group (Lu et al. 2010; Yin et al. 2011a, b).

Dongchuan Group can be divided into four formations from the bottom to top, comprising Yinmin Formation, Luoxue Formation, Heishan Formation, and Qinglongshan Formation. The Yinmin Formation is composed of purple slates, dolarenites, sandstones, breccias, and intermediate–basic volcanics (mainly of tuff with minor lavas). The Luoxue Formation consists mainly of siliceous-band-bearing siliceous dolomites. The Heishan Formation is a suit of black slates interbedded with tufaceous volcanic systems. Qinglongshan Formation is a suite of carbonatites. Zircons of tuff from Yinmin Formation yield the age of 1740 ± 15 Ma, representing the formation age of the tuff. Sandstones, breccias, and trachybasalt dykes yield the youngest zircon isotopic ages of 1842 ± 18, 1838 ± 17, and 1690 ± 32 Ma, respectively, indicating that the dykes intruded into Luoxue Formation at 1690 ± 32 Ma (Zhao et al. 2010). These data imply that the Yinmin Formation and Luoxue Formation were formed during 1838–1690 Ma, showing the resemble time to Dahongshan Group and Hekou Group. Zircons of tuff from Heishan Formation yield the SHRIMP U–Pb isotopic age of 1503 ± 17 Ma (Sun et al. 2009), and Lu et al. (2010) got the LA-ICPMS U–Pb isotopic ages of 1500 ± 4 and 1499 ± 15 Ma (Li et al. 2013a); therefore, Heishan Formation and Qinglongshan Formation were formed during Middle Mesoproterozoic Jixianian Period.

Tong'an Formation in Huili County, Sichuan Province, has the same formation age with Dongchuan Group and can be divided into five segments. The lower part of the first segment is composed of sandy slates, the middle part is composed of dolomite, and the upper part is chiefly composed of shales. The lower part of the second segment consists chiefly of argillaceous dolomites, dolomites, and dolomitic limestones, and the middle part comprises shales and slates interbedded with marbles. The lower part of the third segment consists chiefly of siltites and sandstones sediments with minor shales, and the middle part comprises mainly of slates interbedded with tuff. The lower part of the fourth segment is a suite of limestones and mainly composed of medium-thick-bedded limestones interbedded with minor argillaceous dolomites and argillaceous limestones, with minor calcareous shale locally. The lower part of the fifth segment consists chiefly of sandstones and siltites, the middle part comprises chiefly slates and shales, and the upper part is composed of slates. The author got the LA-ICPMS U–Pb isotopic ages from the edge and the middle part of the doleritic-gabbro veins which intruded into the third part of Tong'an Formation of 1513 ± 13 and 1531 ± 18 Ma, respectively (Geng et al. 2012), according to the formation age of 1503 ± 17 Ma of tuff from Heishan Group in Dongchuan. However, Yin et al. (2011b) got the formation age of the fifth segment of 1082 ± 13 Ma and suggested a fault contact relationship between the fourth and the fifth segment, perhaps the fifth segment is not belong to the Tong'an Formation.

The Dahongshan, Dongchuan, and Hekou Groups and Tong'an Formation were intruded by a large number of mafic intrusions (diabase, dolerite, and gabbro) and a little of acidic intrusions (granite–porphyry) in Dahongshan, Dongchuan, and Wuding areas of Yunnan Province and Hekou and Tong'an areas of Sichuan Province. The geochronological and geochemical data indicated that these intrusions emplaced at ~1.7 and ~1.5 Ga and formed in continental rifting setting (Zhao et al. 2010; Zhao and Zhou. 2011; Guan et al. 2011; Zhu et al. 2011b; Chen et al. 2013; Hou et al. 2013; Wang et al. 2013a, b, c, d; Ye et al. 2013; Geng et al. 2012; Fan et al. 2013).

The Early Mesoproterozoic (1.75–1.50 Ga) Dahongshan, Dongchuan, and Hekou Groups and related mafic intrusions in the southwestern margin of Yangtze Block formed in a rift-related tectonic setting were associated with the breakup of the supercontinent Columbia (Wang and Zhou 2014; Wang et al. 2013a, b, c, d, 2014; Fan et al. 2013; Chen et al. 2013; Zhou et al. 2014)

1.2 The Later Mesoproterozoic Era of Yangtze Block

Middle–Late Mesoproterozoic strata of Yangtze Block are represented by Kunyang Group in Central Yunnan Province, Huili Group in the south of Sichuan Province and Shennongjia Group in the north margin of Yangtze Block.

Standard section of Kunyang Group is located in the Yimen County, Yunnan Province, which is subdivided into Huangcaoling Formation, Heishantou Formation, Fuliangpeng Formation, Dalongkou Formation, and Meidang Formation from bottom to top. Huangcaoling Formation is composed mainly of banded slates interbedded with sandstones, with minor hematites in the lower part. Heishantou Formation comprises thick-bedded rhythmic quartz sandstones interbedded with sandstones and slates. Fuliangpeng Formation is made up of volcanic rocks. Dalongkou Formation consists of medium-thick-bedded limestones, with thin-bedded limestones, marlstones, slates, and siderites in the lower and upper parts. Meidang Formation comprises celadon banded slates interbedded with siltites, sandstones, limestones, and cavity-bearing slates. The SHRIMP zircon U–Pb isotopic age from the tuff of Fuliangpeng Formation is 1032 ± 9 Ma (Zhang et al. 2007). Zircons of tuff from Heishantou Formation (which is named as Laowushan Formation by Greentree) yield 1142 ± 16 Ma age (SHRIMP U–Pb, Greentree and Li 2008) and 1043 ± 7 Ma age (LA-ICP-MS U–Pb, Li et al. 2013a, b), Late Mesoproterozoic.

Huili Group occurs mainly in the Huili County and Huidong County of Sichuan Province, which is subdivided into Limahe Formation, Fengshanying Formation, and Tianbaoshan Formation from bottom to top. Limahe Formation is subdivided into two segments from bottom to top: the first segment is composed mainly of slates, phyllites, and shales in the lower part, and metamorphic quartz sandstones and quartzites interbedded with thin-bedded volcanic tuff in the upper part; and the second segment comprises quartz sandstones and quartzites interbedded with thick-bedded porphyres and phyllites, and the protoliths are sedimentary clastic rocks with the granularity turns coarser from bottom to top. Fengshanying Formation is made up of gray thin- and medium-bedded banded limestones, argillaceous–sandy dolomites, with few calcareous clasolites in the bottom, and siderite bed occurs locally in the medium to lower part, and the protoliths are carbonatites. Tianbaoshan Formation is composed mainly of slates and phyllites interbedded with volcanic tuff and carbon slates in the lower part; the middle part is made up of metamorphic siltstones, sandstones, and phyllites; and the upper part consists of slates interbedded with acid volcanics, and the SHRIMP zircon U–Pb isotopic ages from the upper part are 1028 ± 13 and 1036 ± 12 Ma, respectively, representing the formation age of the acid volcanics (Geng et al. 2007; Yin et al. 2011b). The author recently got the LA-ICPMS U–Pb isotopic ages of 1021 ± 7 and 1034 ± 9 Ma, respectively, for tuff from the lower part of Tianbaoshan Formation near Kongmingzhai and Hongchuanqiao (Fig. 2). These data indicate that Huili Group was formed before 1.0 Ga, in Late Mesoproterozoic.

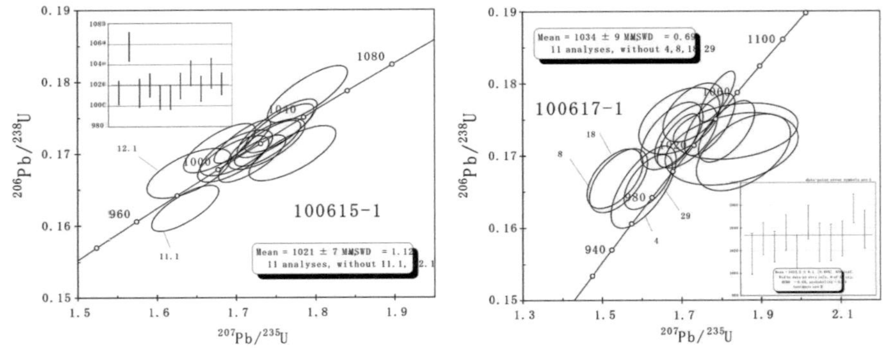

Fig. 2 U–Pb concordia diagrams of zircons from the tuffs of the Tianbaoshan Formation of the Huili Group

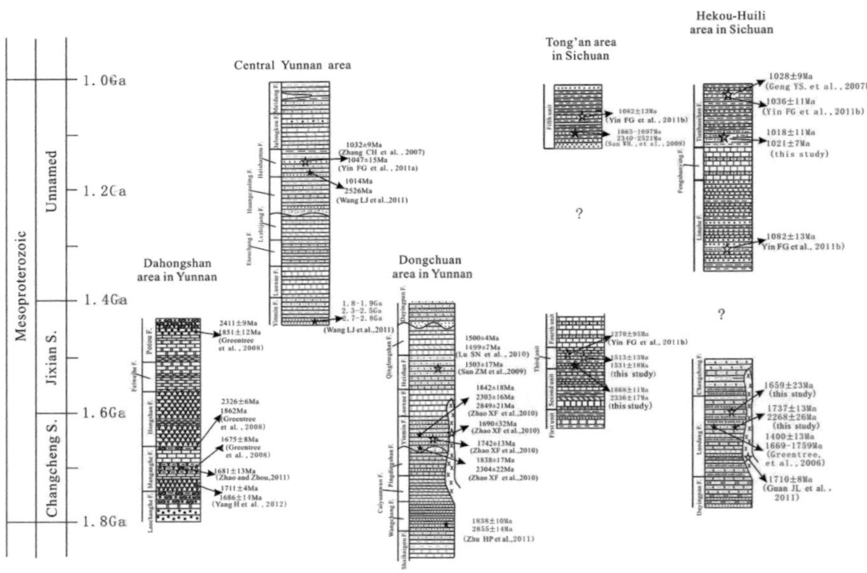

Fig. 3 Correlation diagram of columnar section of Mesoproterozoic strata in the southwestern margin of the Yangtze Block

Stratigraphic sequences, petrographic compositions, and zircon in situ dating results of Mesoproterozoic strata which occur in the southwest margin of Yangtze Block are shown in Fig. 3. Dahongshan Group is comparable with Hekou Group and the lower part of Dongchuan Group. The upper part of Dongchuan Group is comparable with the medium–lower part of Tong'an Formation. Kunyang Group in Central Yunnan is comparable with Huili Group in western Sichuan and the fifth segment of Tong'an Formation generally.

Shennongjia Group is located in the north margin of Yangtze Block, exposing mainly in Shennongjia, northwest of Hubei Province, and Dahongshan area, Central Hubei Province, was subdivided into eleven lithological segments, and mainly consists of magnesian carbonates, clasolites, and volcanics, with stromatolites and microfossil plants in the carbonatites (Fig. 3). Shennongjia Group is overlain unconformably by Macaoyuan Group (Li and Leng 1987; Chen et al. 2000). A few ages were got for the bad exposed condition of Shennongjia Group. The zircon U–Pb isotopic age from the tholeiitic andesites is 1103 ± 8 Ma (Qiu et al. 2011), and the zircon from the tuff of the Yemahe Formation in the upper part of the Shennongjia Group yields an age of 1224 ± 87 Ma. The baddeleyite and zircon of U–Pb ages of the mafic dike intruding the Shicaohe Formation are 1115 ± 89 and 1083 ± 46 Ma, respectively (Li et al. 2013b). So, the age span of the Shennongjia Group can be precisely put in 1.4–1.1 Ga. An age of 1157 ± 819 Ma has been obtained from the tuff of Macaoyuan Group (Wang et al. 2013a, b, c, d). Therefore, the controversy on the mutual relations between Shennongjia Group and Macaoyuan Group remained.

2 Mesoproterozoic Era of Cathaysian Block

2.1 Mesoproterozoic Era in Hainan Island

Longquan Group, southwestern Zhejiang Province; Mayuan Group, northwestern Fujian Province; Mamianshan Group, northern Fujian Province; and Yunkai Group, Guangdong Province, were attributed into Mesoproterozoic in previous investigations (Cheng 1994). However, high-precision zircon in situ dating revealed that they were formed in Neoproterozoic (Wan et al. 2007, 2010; Shu et al. 2011; Yao et al. 2011). Precambrian basement in Hainan Island consists of Baoban Group and Shilu Group (Zhang et al. 1997). The Baoban Group shows obvious dual structure, with the lower Gezhencun Formation comprising migmatized plagiogneisses, which have undergone upper amphibolite-facies metamorphism with granulite-facies metamorphism locally (Xu et al. 2007). The upper Ewenling Formation of Baoban Group is composed of schists interbedded with quartzites and leptites, which have undergone low-amphibolite-facies metamorphism. Calc-alkali granites intruded into Baoban Group (Long et al. 2005). Shilu Group is made up of phyllites, mica schists, diopside–tremolite schists, dolomites, quartzites, tuffs, and iron formation, which suffered greenschist-facies metamorphism. Previous investigators reported 1824–1364 Ma ages of Baoban Group of using conventional zircon isotopic dating and granule zircon isotopic dating (Yu et al. 1992; Ma et al. 1997; Zhang et al. 1999), 975 and 841 Ma ages of Shilu Group using whole rock Sm–Nd method (Tan et al. 1991; Zhang et al. 1992), and discovered the macroscopic alga fossils in Shilu Group (Zhang et al., 1989). These chronological data and alga fossils reveal that the Baoban Group formed in the Early–Middle Mesoproterozoic and the Shilu Group formed in the Neoproterozoic. Li et al. (2008a) got the SHRIMP U–Pb isotopic weighted mean $^{207}Pb/^{206}Pb$ ages of 1433 ± 6 Ma for meta-volcanics. Li et al. (2002) and Xu et al. (2006) obtained the formation ages of deformational granites, 1436 ± 7, 1431 ± 5 and 1455 ± 12 Ma. These data imply that the Baoban Group in Hainan Island was formed in Mesoproterozoic. Li et al. (2008a) got the SHRIMP zircon U–Pb isotopic age of 1439 ± 9 Ma from meta-tuff from Shilu Group and suggested that Baoban Group and Shilu Group were contemporary. However, the former suffered high-grade metamorphism, but the latter was metamorphosed under the low-grade conditions. On account of only one reliable chronological datum, the Shilu Group needed to be further constrained for its exact formation time.

2.2 *Mesoproterozoic Era in Yunkai Area*

Yunkai Group occurs in Yunkai mountains range crossing Guangdong and Guangxi provinces where are at the southwest margin of Cathaysian Block and was overlain unconformably on the Paleoproterozoic Tiantangshan Group. It consists of phyllitic mica (quartz) schists, iron-bearing mica feldspar quartzites, marbles with minor calc-siliceous rocks, and metamorphic basic volcanics, with the protolith being a suite of flysh-like and flysh formation (Qin et al. 2006, 2007), which suffered low-grade metamorphism, such as greenschist facies, with amphibolite facies locally. The SHRIMP zircon U–Pb isotopic dating revealed a weighted $^{207}Pb/^{206}Pb$ mean age of 1462 ± 28 Ma at the cores having clear oscillatory zoning, and another age of 455 ± 10 Ma at the rims (Qin et al. 2006), indicating Yunkai Group was formed in Mesoproterozoic and was metamorphosed in Early Paleozoic.

3 Mesoproterozoic Era of Jiangnan Orogenic Belt

Traditionally, numerous strata in Jiangnan belt such as Sibao Group in Guangxi Province, Fanjingshan Group in Guizhou Province, Lengjiaxi Group in Hunan Province, Shuangqiaoshan Group and Xingzi Group in Jiangxi Province, and Shangxi Group and Shuangxiwu Group in the border land of Zhejiang and Anhui and Jiangxi provinces were regarded as metamorphosed Mesoproterozoic strata (Shui et al. 1988; Cheng. 1994). However, recent data of zircon in situ dating indicated that most of these strata were formed in Neoproterozoic (Gao et al. 2008b, 2009, 2010a, b, c, 2011a, b, 2012a, b, c; Wang et al. 2004, 2006, 2008; Ding et al.2008; Dong et al. 2010). Perhaps, Tianli schists and Tieshajie Formation in the eastern segment of the Jiangnan belt were the only Mesoproterozoic sequence in Jiangnan orogenic belt.

Tianli schists are distributed chiefly at Tianli of Guangfeng County, Jiangxi Province, where is located in the Jiangshan–Shaoxing fault zone, and the fault zone of northeastern stretching has outcrop area of Tianli schists about 6 × 2.5 km^2. The

Tianli schists thrusted onto the Jurassic system in northern part and were overlain high angle unconformably by Neoproterozoic Wengjialing Formation in the southern part (Fan et al. 1997). Tianli schists are composed of mica–quartz schists, calc-muscovite quartz schists, quartz marbles, quartzites, and meta-sandstones, all of them suffered greenschist-facies metamorphism and strong deformation. Based on Tianli schists were overlain high angle unconformably by Neoproterozoic Wengjialing Formation and the muscovite $^{40}Ar/^{39}Ar$ plateau ages of 1108–1019 Ma, Tianli schists were usually regarded to be formed in Mesoproterozoic (Fan et al. 1997). Li et al. (2007) got the SHRIMP zircon U–Pb isotopic ages from two schists of Tianli, excepting one grain with the metamorphic age of 1029 Ma, other 82 grains yielded ages older than 1500 Ma, and they also got two groups of muscovite $^{40}Ar/^{39}Ar$ peak ages of 1015 ± 4 and 968 ± 4 Ma, indicating that Tianli schists were formed in Mesoproterozoic between 1040 and 1530 Ma. The Tieshajie Formation in the segment of the Jiangnan belt consists of schist, slate, and rhyolite, and metamorphosed up to high green schist facies. The ages of 1132 ± 8, 1140 ± 7, 1143 ± 9, and 1172 ± 10 Ma have been obtained from the rhyolites in the formation by zircon SHRIMP U–Pb method (Gao et al. 2013), indicating the Tieshajie Formation was formed in Late Mesoproterozoic (Fig. 4).

4 Discussions with Respect to Geological Events Between Late Mesoproterozoic and Early Neoproterozoic, South China Plate

Grenville orogenic event is the significant continental collision event to form Rodinia supercontinent, which occurred between 1300 and 900 Ma (Hoffman 1991; Clarke et al. 1995; Boger et al. 2001; Jacobs et al. 2003). Li et al. (2002, 2003a, b) proposed that the South China Plate was located between the Laurentia and Australian–Eastern Pacific Ocean continent, and Sibao orogenic event through which the Yangtze Block amalgamated with the Cathaysian Block together was comparable with the Grenville orogenic event (Li et al. 2007). However, Zhou et al. (2002b) suggested that South China Block was not a part of Rodinia supercontinent from Late Mesoproterozoic to Early Neoproterozoic, but the periphery of Rodinia supercontinent. Obviously, geological events during Late Mesoproterozoic to Early Neoproterozoic in South China Block play an essential role in understanding the properties of South China Block and the relationships between South China Block and Rodinia supercontinent. Geological signatures during Late Mesoproterozoic to Early Neoproterozoic are listed in Table 1, and besides, detrital zircons also recorded the geological events around 1.0 Ga (see Fig. 7 in Chap. 5). Late Mesoproterozoic to Early Neoproterozoic geologic bodies are mainly distributed in four areas (Table 1) of (1) granites and volcanoclastic rocks in southwestern Sichuan Province, Central Yunnan, southwestern Yangtze Block; (2) amphibolites and quartzites interbedded with serpentinization–dunites, harzburgites, gabbros and

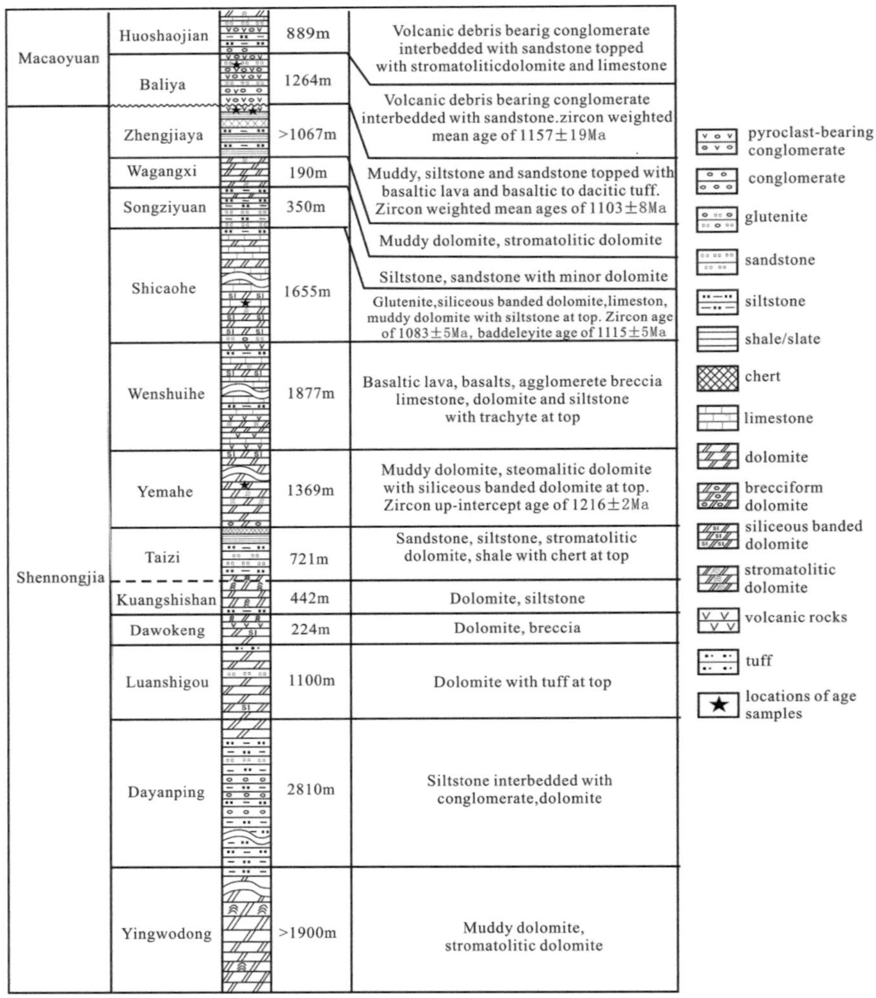

Fig. 4 Stratigraphic column for the Late Mesoproterozoic to the Early Neoproterozoic sedimentary successions in the Shennongjia region. Modified after Qiu et al. 2011, zircon ages from Qiu et al. (2011), Li et al. (2013b), Wang et al. (2013a)

diabases in Miaowan, Kongling, north margin of Yangtze Block; (3) meta-basalts and rhyolites in western Yunkai–Wuyi, southwestern Cathaysian Block, (4) volcanics, volcanoclastic rocks interbedded with felsic tuff, tuff sandstones, and siltites in Shuangxiwu, northeastern Jiangnan (orogenic) belt. The different rock associations imply various tectonic environments. Granites and acid volcanics in southwestern Yangtze Block display features of calc-alkali rocks from crust remelting and were formed in an island-arc collision setting (Geng et al. 2008; Yang et al. 2009); however, some granites (e.g., Tangtang granites) record the characteristics of

Table 1 Geochronology of Late Mesoproterozoic to Early Neoproterozoic geological events of South China Plate

Location	Recodes of geological events	Isotopic age (Ma)	Data source
Central Yunnan (Yangtze Block)	Volcanic tuff (Laowushan Formation)	1142 ± 16	Greentree et al. (2006))
Central Yunnan (Yangtze Block)	Volcanic tuff (Heishantou Formation)	1043 ± 7	Li et al. (2013a)
Central Yunnan (Yangtze Block)	Volcanic tuff (Fuliangpeng Formation)	1032 ± 15	Zhang et al. (2007)
Central Yunnan (Yangtze Block)	Volcanic tuff (Fuliangpeng Formation)	1047 ± 15	Yin et al. (2011a)
Central Yunnan (Yangtze Block)	Meta-basalt (Pudeng Formation)	1043 ± 19	Chen et al. (2014)
Central Yunnan (Yangtze Block)	Meta-basalt (Pudeng Formation)	1050 ± 14	Chen et al. (2014)
Southwestern Sichuan (Yangtze Block)	Volcanic tuff (Tong'an Formation)	1082 ± 13	Yin et al. (2011b)
Southwestern Sichuan (Yangtze Block)	Dacitic porphyrite (Tianbaoshan Formation)	1036 ± 11	Yin et al. (2011b)
Southwestern Sichuan (Yangtze Block)	Volcanic tuff (Tianbaoshan Formation)	1019 ± 10	Li et al. (2013a)
Southwestern Sichuan (Yangtze Block)	Meta-dacite (Dengxiangying Group)	1030 ± 19	Geng et al. (2008)
Western Sichuan (Yangtze Block)	Granitic gneisses of Huijinggou	1007 ± 14	Li et al. (2002)
Western Sichuan (Yangtze Block)	Acid volcanic rock (Tianbaoshan Formation)	1028 ± 9	Geng et al. (2007)
Western Sichuan (Yangtze Block)	Monzonitic granite of Huoshangou	1014 ± 8	Yang et al. (2009)
Western Sichuan (Yangtze Block)	Tangtang potassic granite	1063 ± 7	Wang et al. (2012a, b)
Western Sichuan (Yangtze Block)	Caiyuanzi potassic granite	1040 ± 6	Wang et al. (2013a, b, c, d)
Kongling (Yangtze Block)	Foliated gabbro	1118 ± 24	Peng et al. (2012)
Kongling (Yangtze Block)	Meta-gabbro	1001 ± 16	Peng et al. (2012)
Kongling (Yangtze Block)	Tabular gabbro	974 ± 11	Peng et al. (2012)
Kongling (Yangtze Block)	Diabase	978 ± 12	Peng et al. (2012)
Yunkai (Cathaysian Block)	Amphibolite	997 ± 21	Zhang et al. (2012)
Yunkai (Cathaysian Block)	Meta-basalt	978 ± 19	Zhang et al. (2012)

(continued)

Table 1 (continued)

Location	Recodes of geological events	Isotopic age (Ma)	Data source
Xingning (Cathaysian Block)	Meta-rhyolite	972 ± 8	Shu et al. (2008)
Shuangxiwu (Jiangnan (orogenic) belt)	Rhyolite	926 ± 15	Li et al. (2009a, b)
Shuangxiwu (Jiangnan (orogenic) belt)	Rhyolite	891 ± 12	Li et al. (2009a, b)
Shuangxiwu (Jiangnan (orogenic) belt)	Tonalitic intrusion intruding Shuangxiwu Group	913 ± 15	Ye et al. (2007)
Shuangxiwu (Jiangnan (orogenic) belt)	Granodioritic intrusion intruding Shuangxiwu Group	905 ± 14	Ye et al. (2007)
Xiwan (Jiangnan (orogenic) belt)	Jadeite-kyanited anorthosites	968 ± 23	Li et al. (1994)
Tianli schists (Jiangnan (orogenic) belt)	Metamorphic $^{40}Ar/^{39}Ar$ age of micas	1029 ± 12	Li et al. (2007)
Tieshajie (Jiangnan belt)	Rhyolite (Tieshajie Formation)	1132 ± 8	Gao et al. (2013)
Tieshajie (Jiangnan belt)	Rhyolite (Tieshajie Formation)	1140 ± 7	Gao et al. (2013)
Tieshajie (Jiangnan belt)	Rhyolite (Tieshajie Formation)	1143 ± 9	Gao et al. (2013)
Tieshajie (Jiangnan belt)	Rhyolite (Tieshajie Formation)	1172 ± 10	Gao et al. (2013)

post-collision extension environment (Wang et al. 2012a, b). Amphibolites, serpentinization–dunites, harzburgites, and gabbros in Kongling area of Hubei Province constructed an ophiolite suite, namely Miaowan ophiolite (Peng et al. 2012). Petrological and geochemical features indicate that they were suprasubduction ophiolites in the fore-arc subduction zone of Shennongjia (Peng et al. 2012). The ~1.0 Ga Yunkai geological bodies are composed of meta-basalts, and their geochemical features imply that they were formed in an oceanic island-arc setting (Zhang et al. 2012). The volcanics, volcanoclastic rocks, and Jadeite-kyanited anorthosites in eastern Jiangnan belt show features of typical calc-alkali volcanics at active continental margin (Li et al. 2009a, b). Therefore, there are ~1.0 Ga geological bodies in Yangtze Block, Cathaysian Block, and Jiangnan (orogenic) belt, and they have different petrological and geochemical features and therefore were formed in different tectonic environments. Peng et al. (2012) regarded the ophiolites in northern Yangtze Block as the record of the Yangtze Block pieced with Australian Block. However, Song et al. (2012) regarded them as a part of Grenville orogenic belt in Yangtze–Chaidamu–Qilian–Talimu belt, which formed the South-West China United Continent. The ~1.0 Ga (or younger than 1.0 Ga) tectonic events in eastern Jiangnan (orogenic) belt were regarded as the significant component of the Sibao orogenic belt between Yangtze Block and Cathaysian Block, and they amalgamated together to form the Yangtze Block (Li et al. 2002, 2008b, 2009a, b). Whether the ~1.0 Ga geological bodies in southwestern Yangtze Block and Yunkai range of Cathaysian Block are related to

Grenville orogenic event or not, what they played roles in Grenville orogenic event, and their exact locations in Rodinia supercontinent still need to be solved. It is hard to say that the uniform South China Block because ~1.0 Ga geological events in different locations display various tectonic settings.

References

BGMRSC (Bureau of Geology and Mineral Resources Sichuan Province) (1991) Regional Geology of Sichuan Province. Geological Publishing House, Beijing, p 730 (in Chinese with English abstract)

Boger SD, Carson CJ, Wilson CJL, Fanning CM (2001) Neoproterozoic deformation in the Radok Lake region of the northern Prince Charles Mountains, East Antarctica: evidence for a single protracted orogenic event. Precambr Res 104:1–24

Chen J-B, Zhang P-Y, Gao Z-J, Sun S-F (2000) The Mesoproterozoic Erathem. Stratigraphical Lexicon of China. Geological Publishing House, Beijing, pp 177–294

Chen WT, Zhou M-F, Zhao X-F (2013) Late Paleoproterozoic sedimentary and mafic rocks in the Hekou area, SW China: Implication for the reconstruction of the Yangtze Block in Columbia. Precambr Res 231:61–77

Chen WT, Sun W-H, Wang W, Zhao J-H, Zhou M-F (2014) Grenvillian intra-plate mafic magmatism in the southwestern Yangtze Block, SW China. Precambr Res 242:138–153

Cheng Y-Q (1994) An introduction to regional geology of China. Geological Publishing House, Beijing, p 517 (in Chinese)

China National Committee of Stratigraphy (2001) Guide to stratigraphy in China and Explanation of the guide (revised edition). Geological Publishing House, Beijing, pp 1–59 (in Chinese)

Clarke GL, Sun SS, White RW (1995) Grenville age belts and adjacent older terranes in Australia and Antarctica. J Aust Geol Geophys 16:25–39

Dai H-G (1997) On the strata, structure and prospecting target area of Kunyang Group and Huili Group in Kang-Dian region. Yunnan Geol 16(1):1–39 (in Chinese)

Ding B-H, Shi R-D, Zhi X-C, Zheng L, Chen L (2008) Neoproterozoic (~850 Ma) subduction in the Jiangnan orogen: evidence from the SHRIMP U-Pb dating of the SSZ-type ophiolite in southern Anhui Province. Acta Petrol Mineral 27(5):375–388 (in Chinese with English abstract)

Dong S-W, Xue H-M, Xiang X-K, Ma L-C (2010) The discovery of Neoproterozoic pillow lava in spilite-ceratophyre of Lushan area, northern Jiangxi Province, and its geological significance. Geol China 37(4):1021–1033 (in Chinese with English abstract)

Fan G-M, Xue C-S, Yang X-S (1997) Characteristics and significance of ductile zones in Tianli, Jiangxi Province and Youxi, Zhejiang Province. J East China Geol Inst 20(4):301–308 (in Chinese with English abstract)

Fan H-P, Zhu W-G, Li Z-X, Zhong H, Bai Z-J, He D-F, Chen C-J, Cao C-Y (2013) Ca. 1.5 Ga mafic magmatism in South China during the break-up of the supercontinent Nuna/Columbia: the Zhuqing Fe-Ti-V oxide ore-bearing mafic intrusions in western Yangtze Block. Lithos 168–169:85–98

Gao L-Z, Zhang C-H, Shi X-Y, Zhou H-R, Wang Z-Q, Song B (2007) A new SHRIMP age of the Xiamaling Formation in the North China Plate and its geological significance. Acta Geol Sinica (English edition) 81(6):1103–1109

Gao L-Z, Zhang C-H, Shi X-Y, Song B, Wang Z-Q, Liu Y-M (2008a) Meso–proterozoic age for Xiamaling Formation in North China Plate indicated by zircon SHRIMP dating. Chin Sci Bull 53(17):2665–2671

Gao L-Z, Yang M-G, Ding X-Z, Liu Y-X, Liu X, Ling H-L, Zhang C-H (2008b) SHRIMP U-Pb zircon dating of tuff in the Shuangqiaoshan and Heshangzhen groups in South China: constraints on the evolution of the Jiangnan Neoproterozoic orogenic belt. Geol Bull China 27 (10):1744–1751 (in Chinese with English abstract)

Gao L-Z, Zhang C-H, Liu P-J, Ding X-Z, Wang Z-Q, Zhang Y-J (2009) Recognition of Meso-and Neoproterozoic stratigraphic framework in North and South China. Acta Geosci Sin 30(4): 433–446 (in Chinese with English abstract)

Gao L-Z, Dai C-G, Liu Y-X, Wang M, Wang X-H, Chen J-S, Ding X-Z (2010a) Zircon SHRIMP U-Pb dating of the tuffaceous bed of Xiajiang Group in Guizhou Province and its stratigraphic implication. Geol China 37(4):1071–1080 (in Chinese with English abstract)

Gao L-Z, Dai C-G, Liu Y-X, Wang M, Wang X-H, Chen J-S, Ding X-Z, Zhang C-H, Cao Q, Liu J-H (2010b) Zircon SHRIMP U-Pb dating of tuff bed of the Sibao Group in southeastern Guizhou-northern Guangxi area, China and its stratigraphic implication. Geol Bull China 29(9):1259–1267 (in Chinese with English abstract)

Gao L-Z, Ding X-Z, Cao Q, Zhang C-H (2010c) New geological time scale of Late Precambrian in China and geochronology. Geol China 37(4):1014–1020 (in Chinese with English abstract)

Gao L-Z, Ding X-Z, Pang W-H, Liu Y-X, Lu S-N, Liu Y-R, Chen J, Zhang Y-H (2011a) SHRIMP zircon U-Pb dating of metamorphic tuff from the Precambrian Cangxi Complex-Group in northeastern Hunan. Geol Bull China 30(10):1479–1484 (in Chinese with English abstract)

Gao L-Z, Chen J, Ding X-Z, Liu Y-R, Zhang C-H, Zhang H, Liu Y-X, Pang W-H, Zhang Y-H (2011b) Zircon SHRIMP U-Pb dating of the tuff bed of Lengjiaxi and Banxi groups, northeastern Hunan: constraints on the Wuling Movement. Geol Bull China 30(7):1001–1008 (in Chinese with English abstract)

Gao L-Z, Liu Y-X, Ding X-Z, Zhang C-H, Wang Z-Q, Chen J, Liu Y-R (2012a) SHRIMP dating of Cangshuipu Group in the middle part of the Jiangnan Orogen and its implications for tectonic evolutions. Geol China 39(1):12–20 (in Chinese with English abstract)

Gao L-Z, Huang Z-Z, Ding X-Z, Liu Y-X, Zhang C-H, Wang Z-Q, Pang J-F, Han K-Y (2012b) The geochronological relationship between the Shaojiwa Formation and the Xingzi complex group in northwestern Jiangxi and the constraints on zircon SHRIMP U-Pb Age. Acta Geosci Sin 33(3):295–304 (in Chinese with English abstract)

Gao L-Z, Ding X-Z, Zhang C-H, Lu S-N, Liu Y-X, Pang W-H (2012c) A revised chrono-stratigraphic dating of metamorphosed basement strata of Jiangnan old land and its implication for Wuling tectonic movement. Resour Surv Environ 33(2):71–76 (in Chinese with English abstract)

Gao L-Z, Liu Y-X, Ding X-Z, Song Z-R, Huang Z-Z, Zhang C-H, Zhang H, Shi Z-G (2013) Geochronographic dating of the Tieshajie Formation in the Jiang-Shao fault zone and its implicatios. Geol Bull China 32(7):30–39 (in Chinese with English abstract)

Geng Y-S, Yang C-H, Du L-L, Wang X-S, Ren L-D, Zhou X-W (2007) Chronology and tectonic environment of the Tianbaoshan Formation: new evidence from zircon SHRIMP U-Pb age and geochemistry. Geol Rev 53:556–563 (in Chinese with English abstract)

Geng Y-S, Yang C-H, Wang X-S, Du L-L, Ren L-D, Zhou X-W (2008) Metamorphic basement evolution in western margin of yangtze block. Geological Publishing House, Beijing, pp 215 (in Chinese)

Geng Y-S, Liu Y-Q, Gao L-Z, Peng N, Jiang X-J (2012) Geochronology of the Mesoproterozoic Tong'an formation in southwestern margin of Yangtze Craton: new evidence from zircon LA-ICPMS U-Pb ages. Acta Geol Sinica 86(9):1479–1490 (in Chinese with English abstract)

Greentree MR, Li Z-X, Li X-H, Wu H-C (2006) Late Mesoproterozoic to earliest Neoproterozoic basin record of the Sibao orogenesis in western South China and relationship to the assembly of Rodinia. Precambr Res 151:79–100

Greentree MR, Li Z-X (2008) The oldest known rocks in south western China: SHRIMP U-Pb magmatic crystallization age and detrital provenance analysis of the Paleoproterozoic Dahongshan Group. J Asian Earth Sci 33:290–302

Guan J-L, Zheng L-L, Liu J-H, Sun Z-M, Cheng W-H (2011) Zircons SHRIMP U-Pb dating of diabase from Hekou, Sichuan Province, China and Its geological significance. Acta Geol Sinica 85(4):482–490 (in Chinese with English abstract)

He Z-J, Niu B-G, Zhang X-Y, Zhao L, Liu R-Y (2011) Discovery of the Paleo-weathered mantle of the rapakivi granite covered by the Proterozoic Changzhougou formation in the Miyun area, Beijing and their detrital zircon dating. Geol Bull China 30(5):798–802 (in Chinese with English abstract)

Hoffman PF (1991) Did the breakout of Laurentia turn Gondwanaland inside-out? Science 252:1409–1412

Hou L, Ding J, Deng J, Liao Z-W, Peng H-J (2013) Zircon LA-ICP-MS dating of the Magmatic breccia from the Yinachang iron-copper deposit in Wuding County of Yunnan Province and its geological significance. Geol Bull China 32(4):580–588 (in Chinese with English abstract)

Hu A-Q, Zhou B-Q, Mao C-X, Zhu N-J, Huang R-S (1991) Geochronology of the Dahongshan Group. Chin J Geochem 10(3):195–203

IUGS, International Commission Stratigraphy (2008) International Stratigraphic Chart, 1

Jacobs J, Fanning CM, Bauer W (2003) Timing of Grenville-age vs. Pan-African medium- to high grade metamorphism in western Dronning Maud Land (East Antarctica) and significance for correlations in Rodinia and Gondwana. Precambr Res 125:1–20

Li X-H (1999) U-Pb zircon ages of granites from the southern margin of the Yangtze Block: timing of Neoproterozoic Jinning Orogeny in SE China and implications for Rodinia Assembly. Precambr Res 97:43–57

Li Q, Leng J (1987) The upper Precambrian in the Shennongjia region. Tianjin: Science and Technology Publishing House, pp 565 (in Chinese)

Li Z-X, Powell CM (1996) Break of Rodinia and Gondwanaland andassembly of Asian. Aust Earth Sci 43(4):591–592

Li Z-X, Li X-H (2007) Formation of the 1300-km wide intracontinental orogen and postorogenic magmatic province in Mesozoic South China: a flat-slab subduction model. Geology 35(2): 179–182. doi:10.1130/G23193A.1

Li X-J, Wu M-D, Duan J-S (1984) The stratigraphic sequence of the Kunyang Group and its top and bottom boundaries. Geol Rev 30(5):399–408 (in Chinese with English abstract)

Li F-H, Tan J-M, Shen Y-L, Wang F-X, Zhou G-F, Pan X-N, Li X-Z (1988) The Kangdian Region Pre-Sinian. Chongqing Publishing House, Chongqing, pp 396 (in Chinese)

Li X-H, Zhou G-Q, Zhao J-X, Fanning CM, Compston W (1994) SHRIMP Ion microprobe zircon U-Pb age of the NE Jiangxi ophiolite and its tectonic implications. Geochemica 23(2):125–131 (in Chinese with English abstract)

Li Z-X, Li X-H, Zhou H-W, Kinny PD (2002) Grenvillian Continental collisionin south China: new SHRIMP U- Pb zircon results and implications for the configuration of Rodinia. Geology 30(2):163–166

Li Z-Q, Wang J-Z, Liu J-J, Li C-Y, Du A-D, Liu Y-P, Ye L (2003a) Re-Os dating of molybdenite from Lala Fe-Oxide-Cu-Au-Mo-REE deposit, southwest China: implications for ore genesis. Contrib Geol Miner Resour Res 18(1):39–42 (in Chinese with English abstract)

Li Z-X, Li X-H, Kinny PD, Wang J, Zhang S, Zhou H-W (2003b) Geochronology of Neoproterozoic syn-rift magmatism in the Yangtze Craton, South China and correlations with other continents: evidence for a mantle superplume that broke up Rodinia. Precambr Res 122:85–109

Li Z-X, Wartho JA, Occhipinti S, Zhang C-L, Li X-H, Wang J, Bao C-M (2007) Early history of the eastern Sibao Orogen (South China) during the assembly of Rodinia: New mica $^{40}Ar/^{39}Ar$ dating and SHRIMP U-Pb detrital zircon provenance constraints. Precambr Res 159:79–94

Li Z-X, Li X-H, Li W-X, Ding S-J (2008a) Was Cathaysia part of Proterozoic Laurentia?—new data from Hainan Island, south China. Terra Nova 20:154–164

Li Z-X, Bogdanova SV, Collins AS, Davidson A, De Waele B, Ernst RE, Fitzsimons ICW, Fuck RA, Gladkochub DP, Jacobs J, Karlstrom KE, Lu S, Natapov LM, Pease V, Pisarevsky SA, Thrane K, Vernikovsky V (2008b) Assembly, configuration, and break-up history of Rodinia: a synthesis. Precambrian Res 160:179–210

Li H-K, Lu S-N, Li H-M, Sun L-X, Xiang Z-Q, Geng J-Z, Zhou H-Y (2009a) Zircon and beddeleyite U-Pb precision dating of basic rock sills intruding Xiamaling Formation North China. Geol Bull China 28(10):1396–1404 (in Chinese with English abstract)

Li X-H, Li W-X, Li Z-X, Lo C-H, Wang J, Ye M-F, Yang Y-H (2009b) Amalgamation between the Yangtze and Cathaysia Blocks in South China: constraints from SHRIMP U-Pb zircon ages, geochemistry and Nd-Hf isotopes of the Shuangxiwu volcanic rocks. Precambr Res 174:117–128

Li H-K, Su W-B, Zhou H-Y, Geng J-Z, Xiang Z-Q, Cui Y-R, Liu W-C, Lu S-N (2011) The base age of the Changchengian System at the northern North China Craton should be younger than 1670 Ma: constraints from zircon U-Pb LA-MC-ICPMS dating of a granite-porphyry dike in Miyun County Beijing. Earth Sci Front 18(3):108–120 (in Chinese with English abstract)

Li H-K, Zhang C-L, Yao C-Y, Xiang Z-Q (2013a) U-Pb zircon age and Hf isotope compositions of Mesoproterozoic sedimentary strata on the western margin of the Yangtze massif. Sci China: Earth Sci 56:628–639. doi:10.1007/s11430-013-4590-9

Li H-K, Zhang C-L, Xiang Z-Q, Lu S-N, Zhang J, Geng J-Z, Qu L-S, Wang Z-X (2013b) Zircon and baddeleyite U-Pb geochronologe of the Shennongjia Group in the Yangtze Craton and its tectonic significance. Acta Petrol Sinica 29(2):673–697 (in Chinese with English abstract)

Long W-G, Ding S-J, Ma D-Q, Lin Y-H, Mo W-M, Zhou J-B (2005) Formation and Evolution of the Precambrian Basement in Hainan Island. Earth Sci 30(4):421–429 (in Chinese with English abstract)

Lu S-N, Li H-K, Xiang Z-Q (2010) Advances in the study of Mesoproterozoic geochronology in China: a review. Geology in China 37(4): 1002–1013 (in Chinese with English abstract)

Lü S-K, Dai H-G (2001) A review of the set-up of Kunyang Group's sequence and the discovery of important ore-bearing horizons in Kang-Dian area. Yunnan Geol 20(1):1–24 (in Chinese with English abstract)

Ma D-Q, Huang X-D, Chen Z-P, Xiao Z-F, Zhang W-C, Zhong S-Z (1997) New advance in the study of the Baoban Group in Hainan Province. Reg Geol China 16(2):130–136 (in Chinese with English abstract)

Peng S-B, Kusky TM, Jiang X-F, Wang L, Wang J-P, Deng H (2012) Geology, geochemistry, and geochronology of the Miaowan ophiolite, Yangtze craton: implications for South China's amalgamation history with the Rodinian supercontinent. Gondwana Res 21:577–594

Qian J-H, Shen Y-R (1990) The Dahongshan volcanogenic Fe-Cu deposit in Yunnan Province. Geological Publishing House, Beijing, pp 1–183 (in Chinese)

Qin X-F, Pan Y-M, Li J, Li R-S, Zhou F-S, Hu G-A, Zhong F-Y (2006) Zircon SHRIMP U-Pb geochronology of the Yunkai metamorphic complex is shoutheastern Guangxi China. Geol Bull China 25(5):553–559 (in Chinese with English abstract)

Qin X-F, Xia B, Zhou F-S, Hu G-A, Li G-N, Xie L-F, Zhou K-H, Pan Y-W (2007) Confirm of Proterozoic unconformity event in the Yunkai area of Guangxi and its tectonic significance. Geoscience 21(1):22–30 (in Chinese with English abstract)

Qiu X-F, Ling W-L, Liu X-M, Kusky TM, Berkana W, Zhang Y-H, Gao Y-J, Lu S-S, Kuang H, Liu C-X (2011) Recognition of Grenvillian volcanic suite in the Shennongjia region and its tectonic significance for the South China Craton. Precambr Res 191:101–119

Shu L-S, Faure M, Yu J-H, Jahn B-M (2011) Geochronological and geochemical features of the Cathaysia block (South China): new evidence for the Neoproterozoic breakup of Rodinia. Precambr Res 187:263–276

Shui T (1988) Tectonic framework of the continental basement of southeast China. Scientia Sinica (series B) 31(1):885–896

Song S-G, Su L, Li X-H, Niu Y-L, Zhang L-F (2012) Grenville-age orogenesis in the Qaidam-Qilian block: the link between South China and Tarim. Precambr Res 220–221:9–22

Su W-B, Zhang S-H, Huff WD, Li H-K, Ettensohn FR, Chen X-Y, Yang H-M, Han Y-G, Song B, Santosh M (2008) SHRIMP U-Pb ages of K-bentonite beds in the Xiamaling Formation: implications for revised subdivision of the Meso- to Neoproterozoic history of the North China Craton. Gondwana Res 14:543–553

Shu L-S, Deng P, Yu J-H, Wang Y-B, Jiang S-Y (2008) The age and tectonic environment of the rhyolitic rocks on the western side of Wuyi Mountain South China. Sci China (series D) 51(8):1053–1063

Sun Z-M, Yin F-G, Guan J-L, Liu J-H, Li J-M, Geng Q-R, Wang L-Q (2009) SHRIMP U-Pb dating and its stratigraphic significance of tuff zircon from Heishan formation of Kunyang group, Dongchuan area, Yunnan Province China. Geol Bull China 28(7):896–900 (in Chinese with English abstract)

Tan Z-F, Jiang D-H, Ma G-G, Chen S-M, Chen P, Huang Z-X (1991) The isotopic ages of the Baoban Group in Hainan and its geological significance. Gegional Geol China 3:241–245 (in Chinese with English abstract)

Wan Y-S, Liu D-Y, Xu M-H, Zhuang J-M, Song B, Shi Y-R, Du L-L (2007) SHRIMP U-Pb zircon geochronology and geochemistry of metavolcanic and metasedimentary rocks in Northwestern Fujian, Cathaysia block, China: tectonic implication and the need to redefine lithostratigraphic units. Gondwana Res 12:166–183

Wan Y-S, Liu D-Y, Wilde SA, Cao J, Chen B, Dong C-Y, Song B, Du L-L (2010) Evolution of the Yunkai Terrane, South China: evidence from SHRIMP zircon U-Pb dating, geochemistry and Nd isotope. J Asian Earth Sci 37:140–153

Wang D-B, Sun Z-M, Yin F-G, Wang L-Q, Wang B-D, Zhang W-P (2012a) Geochronology of the Hekou Group on the western margin of the Yangtze Block: evidence from zircon LA-ICP-MS U-Pb dating of volcanic rocks. J Stratigr 36(3):630–635 (in Chinese with English abstract)

Wang Z-Z, Zhou B-G, Guo Y, Yang B, Liao Z-W, Wang S-W (2012b) Geochemistry and zircon U-Pb dating of Tangtang granite in the western margin of the Yangtze Platform. Acta Petrol Mineral 31(5):652–662 (in Chinese with English abstract)

Wang D-B, Yin F-G, Sun Z-M, Wang L-Q, Wang B-D, Liao S-Y, Tang Y, Ren G-M (2013a) Zircon U-Pb age and Hf isotope of Paleoproterozoic mafic intrusion on the western margin of the Yangtze Block and their implications. Geol Bull Chian 32(4):617–630 (in Chinese with English abstract)

Wang J, Deng Q, Wang Z-J, Qiu Y-S, Duan T-Z, Jiang X-S, Yang Q-X (2013b) New evidences for sedimentary attributes and timing of the Macaoyuan conglomerates on the northern margin of the Yangtze block in southern China. Precambr Res 235:58–70

Wang S-W, Liao Z-W, Sun X-M, Zhou B-G, Guo Y, Jiang X-F, Zhu H-P, Sun Z-M, Luo M-J, Ma D, Shen Z-W, Zhang H (2013c) Age and Geochemistry of the Caiyuanzi granite in Sichuan, SW China: Mwchanism of the Grenvillian Orogenic Movement in the western margin of Yangtze Block. Acta Geol Sinica 87(1):55–70 (in Chinese with English abstract)

Wang Z-Z, Guo Y, Yang B, Wang S-W, Sun X-M, Hou L, Zhou B-G, Liao Z-W (2013d) Discovery of the 1.73 Ga Haizi anorogenic type granite in the western margin of Yangtze Craton, and its geological significance. Acta Geol Sinica 87(7):931–942 (in Chinese with English abstract)

Wang W, Zhou M-F (2014) Provenance and tectonic setting of the Paleo- to Mesoproterozoic Dongchuan Group in the southwestern Yangtze Block, South China: implication for the breakup of the supercontinent Columbia. Tectonophysics 610:110–127

Wang W, Zhou M-F, Zhao X-F, Chen WT, Yan D-P (2014) Late Paleoproterozoic to Mesoproterozoic rift successions in SW China: Implication for the Yangtze Block-North Australia-Northwest Laurentia connection in the Columbia supercontinent. Sed Geol 309:33–47

Wang X-L, Zhou J-C, Qiu J-S, Gao J-F (2004) Geochemistry of the Meso- to Neoproterozoic basic-acid rocks from Hunan Province, South China: implications for the evolution of the western Jiangnan orogen. Precambr Res 135:79–103

Wang X-L, Zhou J-C, Qiu J-S, Zhang W-L, Liu X-M, Zhang G-L (2006) LA-ICP-MS U-Pb zircon geochronology of the Neoproterozoic igneous rocks from Northern Guangxi, South China: implications for tectonic evolution. Precambr Res, 145(1–2):111–130

Wang X-L, Zhao G, Zhou J-C, Liu Y, Hu J (2008) Geochronology and Hf isotopes of zircon from volcanic rocks of the Shuangqiaoshan Group, South China: implications for the Neoproterozoic tectonic evolution of the eastern Jiangnan orogen. Gondwana Res 14:355–367

Wu M-D, Duan J-S, Song X-L (1990) Geology of Kunyang Group in Yunnan. Kunming: Yunnan Science and Technology Publishing House, pp 1–265 (in Chinese)
Wu G-Y (2006) Division of the Precambrian in south China in the light of key geological events. J Stratigr 30(3):271–286 (in Chinese with English abstract)
Xu D-M, Ma D-Q, Zhang Y-M, Xie C-F (2007) Origin and geological significance of granulites from Qiongzhong complex. Acta Petrol Mineral 26(1):35–41 (in Chinese with English abstract)
Xu D-R, Xia B, Li P-C, Zhang Y-Q, Chen G-H, Ma C (2006) SHRIMP U-Pb Dating of zircon from the Precambrian granitoids in northwestern Hainan Island and its geological implications. Geotectonica et Metallogenia 30(4):510–518 (in Chinese with English abstract)
Yang C-H, Geng Y-S, Du L-L, Ren L-D, Wang X-S, Zhou X-W, Yang Z-S (2009) The identification of the Grenvillian granite on the western margin of the Yangtze Block and its geological implications. Geol China 36(3):647–657 (in Chinese with English abstract)
Yang H, Liu F-L, Du L-L, Liu P-H, Wang F (2012) Zircon U-Pb dating for metavolcanites in the Laochanghe Formation of the Dahongshan Group in southwestern Yangtze block, and its geological significance. Acta Petrol Sinica 28(9):2994–3014 (in Chinese with English abstract)
Yao J-L, Shu LS, Santosh M (2011) Detrital zircon U-Pb geochronology, Hf-isotopes and geochemistry: new clues for the Precambrian crustal evolution of Cathaysia Block, South China. Gondwana Res 20:553–567
Ye M-F, Li X-H, Li W-X, Liu Y, Li Z-X (2007) SHRIMP zircon U-Pb geochronological and whole-rock geochemical evidence for an early Neoproterozoic Sibaoan magmatic arc along the southeastern margin of the Yangtze Block. Gondwana Res 12:144–156
Ye X-T, Zhu W-G, Zhong H, He D-F, Ren T, Bai Z-J, Fan H-P, Hu W-J (2013) Zircon U-Pb and chalcopyrite Re-Os geochronology, REE geochemistry of the Yinachang Fe-Cu-REE deposit in Yunnan Province and its geological significance. Acta Petrol Sinica 29(4):1167–1186 (in Chinese with English abstract)
Yin F-G, Sun Z-M, Bai J-K (2011a) Stratigraphic framework of the Mesoproterozoic in Dongchuan and central Yunnan region. J Stratigr 35(1):49–54 (in Chinese with English abstract)
Yin F-G, Sun Z-M, Zhang Z (2011b) Mesoproterozoic stratigraphic—structure framework in Huili—Dongchuan area. Geol Rev 57(6):770–778 (in Chinese with English abstract)
Yu S-J, Xia P, Deng T-Y, Li Q, Chen Y-W (1992) Characteristics of the accessory mineral zircon in Mesoproterozoic granitoids and its measurements of U-Pb isotopic in Baoban area, Hainan Province. Geochemica 3:213–219 (in Chinese with English abstract)
Zhang A-M, Wang Y-J, Fan W-M, Zhang Y-Z, Yang J (2012) Earliest Neoproterozoic (ca. 1.0 Ga) arc–back-arc basin nature along the northern Yunkai Domain of the Cathaysia Block: geochronological and geochemical evidence from the metabasite. Precambr Res 220–221:217–233
Zhang C-H, Gao L-Z, Wu Z-J, Shi X-Y, Yan Q-R, Li D-J (2007) SHRIMP U-Pb zircon age of tuff from the Kunyang Group in central Yunnan: evidence for Grenvillian orogeny in South China. Chin Sci Bull 52(11):1517–1525
Zhang R-J, Feng S-N, Xu G-H, Yang D-L, Yan D-P, Li Z-H, Jiang D-H, Wu W (1989) Discovery of Chuarian-tawuia assemblage in Shilu Group, Hainan Island and its significance. Sci China (Series B) 33(2):211–220
Zhang R-J, Ma G-G, Feng S-N, Yan D-P (1992) Sm -Nd age of the Shilu iron ore deposirs on Hainan Island and its significance. Sci Geol Sinica 1:38–43 (in Chinese with English abstract)
Zhang Y-M, Zhang R-J, Yao H-Z, Ma G-G (1997) Proterozoic tectonic evolution of Hainan island. Earth Sci J China Univ Geosci 22(4):395–400 (in Chinese with English abstract)
Zhang Y-M, Zhang R-J, Hu N, Zhang S-H, Duan Q-F (1999) High grade metamorphic complexes in Middle Hainan Island: age of the Pb-Pb single zircond and their geological significance. Acta Geosci Sinica 20(3):284–288 (in Chinese with English abstract)
Zhao X-F, Zhou M-F, Li J-W, Sun M, Gao J-F, Sun W-H, Yang J-H (2010) Late Paleoproterozoic to early Mesoproterozoic Dongchuan Group in Yunnan, SW China: Implications for tectonic evolution of the Yangtze Block. Precambr Res 182:57–69

Zhao X-F, Zhou M-F (2011) Fe-Cu deposit in the Kangdian region, SW China: a Proterozoic IOCG (iron oxide copper gold) metallogenic province. Miner Deposita 46(7): 731–747

Zhou J-Y, Mao J-W, Liu F-Y, Tan H-Q, Shen B, Zhu Z-M, Chen J-B, Luo L-P, Zhou X, Wang Y (2011) SHRIMP U-Pb zircon chronology and geochemistry of albitite from the Hekou Group in the western Yangtze Block. J Miner Petrol 31(3):66–73 (in Chinese with English abstract)

Zhou M-F, Yan D-P, Kennedy AK, Li Y-Q, Ding J (2002a) SHRIMP U-Pb zircon geochronological and geochemical evidence for Neoproterozoic arc-magmatism along the western margin of the Yangtze Block, South China. Earth Planet Sci Lett 196:51–67

Zhou M-F, Kennedy AK, Sun M, Malpas J, Lesher CM (2002b) Neoproterozoic arc-related mafic intrusions along the Northern Margin of South China: implications for the Accretion of Rodinia. J Geololy 110:611–618

Zhou M-F, Zhao X-F, Chen W-T, Li X-C, Wang W, Yan D-P, Qiu H-N (2014) Proterozoic Fe–Cu metallogeny and supercontinental cycles of the southwestern Yangtze Block, southern China and northern Vietnam. Earth Sci Rev 139:59–82

Zhu H-P, Fan W-Y, Zhou B-G, Wang S-W, Luo M-J, Liao Z-W, Guo Y (2011b) Assessing Precambrian stratigraphic sequence of Dongchuan area: Evidence from zircon SHRIMP and LA-ICP-MS dating. Geol J China Univ 17(3):452–461 (in Chinese with English abstract)

Neoproterozoic Era of South China Craton

Yuan-Sheng Geng

Abstract The Neoproterozoic of the Yangtze block may be subdivided into early Pt_3 (1000–850 Ma), middle Pt_3 (850–780 Ma), and late Pt_3 (780–541 Ma). The Neoproterozoic sequence is most developed in the Jiangnan orogenic belt. The early Pt_3 strata, or the lower structure sequence of Neoproterozoic, include the Sibao (Guangxi Province), Fanjingshan (Guizhou Province), Lengjiaxi (Hunan Province), Xiko (Anhui Province), Shuangxiwu (Zhejiang Province), and Pingshui (Zhejiang Province) Groups. The middle Pt_3 strata, or the upper structure sequence of Neoproterozoic, include the Danzhou (Guangxi Province), Xiajiang (Guizhou Province), Banxi (Hunan Province), Likou (Anhui Province), and Heshangzhen (Zhejiang Province) Groups. The upper sequence unconformably overlies the lower sequence. The tectonic movement between the two sequences was responsible for the amalgamation of the Yangtze and Cathaysian blocks, thus formed the South China craton. But it is still controversy on the time and mechanism of the movement. The early and middle Pt_3 strata are represented by the Yanbian Group, the Suxiong, and Kaijianqiao Formations at the western Yangtze block. The Huodiya and Xixiang Groups occur at the northern margin. Besides the Pt_3 strata, voluminous magmatic rocks of the period are present from Panxi (Panzhihua and Xichang regions in western Sichuan Province) to Hannan region in southern Shaanxi Province, including the basic gabbros to intermediate-acid granites during 850–780 Ma ago, so forming the Panxi-Hannan magmatic belt. In the Cathaysian block, the early and middle Pt_3 sequences are characterized by the Chencai and Longquan Groups, while the late Pt_3 strata in the South China Craton are manifested by the Nanhua System (corresponding to the Cryogenian System) and the Sinian System (corresponding to the Ediacaran System), which are widely distributed as the cover sequence in southern China after the consolidation of the South China Craton.

Keywords South China Craton · Neoproterozoic · Jiangnan (orogenic) belt · Nanhua system · Sinian system

Y.-S. Geng (✉)
Institute of Geology, Chinese Academy of Geology Sciences,
100037 Beijing, China
e-mail: ys-geng@cags.ac.cn

M. Zhai (ed.), *Precambrian Geology of China*,
Springer Geology, DOI 10.1007/978-3-662-47885-1_7

1 Introduction

Conventionally, the Neoproterozoic strata sequences of the South China Block are categorized into the Qingbaikou System (1000–800 Ma), Nanhua System (800–680 Ma), and Sinian System (680–543 Ma), with geochronological boundaries of 800 and 680 Ma (National Commission of Stratigraphy, China, 2001; NCS). However, recent geological investigations published a lots of high-quality in situ zircon U-Pb isotopic dating data that revealed obvious angular-unconformity boundary within the Qingbaikou System of the South China Craton, indicating that the lower boundary of Nanhua System was not so old as the previous classification, and then promoted intensively debates on lower boundary of the Nanhua Systems (Zhang et al. 2003; Wang 2005; Wang et al. 2008), leading some researchers to suggest building a new system between the Qingbaikou and Nanhua systems (Wang 2008). Consequently, the NCS (National Commission of Stratigraphy, China) organized some related specialists to systematic investigation for the boundary, and then suggested that the boundary between Qingbaikou and Nanhua Systems need to be revised although a new system can be not built, therefore, redivided the Neoproterozoic strata of the South China Craton into the Qingbaikou System (1000–780 Ma), Nanhua System (780–635 Ma), and Sinian System (635–541 Ma). In the South China Craton, the Qingbaikou System consists mainly of strongly deformed, greenschist phase metamorphic rocks in the lower structural layer (e.g., Sibao Group in Guangxi, Fanjingshan Group in Guizhou, Lengjiaxi Group in Hunan, Shuangqiaoshan Group in Jiangxi, Xikou Group in southern Anhui, and Shuangxiwu Group in western Zhejiang), overlain by the weakly deformed and low-greenschist phase metamorphic rocks of the upper structural layer with unconformable contacts (Danzhou Group in Guangxi, Xiajiang Group in Guizhou, Banxi Group in Hunan, Xiushui Group in Jiangxi and Likou, and Heshangzhen Groups straddling border of Anhui, Zhejiang, and Jiangxi). Therefore, in this contribution, we categorize the Neoproterozoic strata systematics of the South China Craton into the Early Neoproterozoic (1000–820 Ma), Middle Neoproterozoic (820–780 Ma), and Late Neoproterozoic (780–541 Ma), and the Late Neoproterozoic contains the Nanhua System (780–635 Ma) and Sinian System (635–541 Ma). The Nanhua System is coincident with the international Cryogenian with a set of aqueoglacial deposits, and the Sinian System is synchronous with the Ediacaran with a set of carbonates and abundant fossil records during marine transgression.

The interior of the Yangtze Block is composed mainly of terranes of Late Neoproterozoic Nanhua System and Sinian System, and the western margin of the Yangtze Block is dominated by the Early Neoproterozoic volcanic sedimentary sequences and deformed granites, and the Late Neoproterozoic Sinian System strata. The Neoproterozoic strata of the Jiangnan (orogenic) belt consists of the Early Neoproterozoic greenschist phase metamorphosed volcanic sedimentary rocks and S-type granites, the Middle Neoproterozoic low-greenschist phase metamorphic sedimentary sequences and synchronous undeformed granites, and the

Late Neoproterozoic aqueoglacial deposits of Nanhua System and carbonates of Sinian System during widely marine transgression. The Cathaysian Block is composed of the Paleo–to Mesoproterozoic amphibolite phase metamorphic volcanic and sedimentary rocks, overlain by Sinian covers. Stratigraphic division and correlation of the Neoproterozoic strata for blocks of the South China Craton are listed in Table 1, and we can figure out that the unconformity between the lower and upper structural layers at ca. 820 Ma and the boundary between the metamorphic basement and Nanhua System at ca. 780 Ma.

2 Early to Middle Neoproterozoic of South China Craton

The Early Neoproterozoic geological bodies in the South China Craton have been constrained well at 860–820 Ma, and the 1000–860 Ma old geological records exposed limitedly in the Shuangxiwu area—straddling the Anhui, Zhejiang, and Jiangxi Province—and the Shennongjia area—northwestern Hubei Province. The Neoproterozoic tectonic environment of the South China Craton, especially the Yangtze Block and the Jiangnan (orogenic) belt, remains still controversial. Some workers considered that the Yangtze Block and Cathaysian Block amalgamated before 850 Ma, and the later mantle plume upwelling results in the formation of the NE–SW striking Huanan rift, the south–north striking Kangdian rift, and east–west striking Bikou-Hannan rift (Li et al. 1999, 2002a, 2003, 2008a, b; 2002a, 2008, 2009; Wang and Li 2003; Zhu et al. 2008a, b; Wang et al. 2006, 2011). Other researchers, on the basis of the Early Neoproterozoic (mostly 850–820 Ma) volcanic rocks with island-arc-type or back-arc-type geochemical features, proposed existence of Jiangnan and Panxi-Hannan island arc belts in the margin of the Yangtze Block (Zhou et al. 2002a, b, 2006a, b; Zhao and Zhou 2008; Zhao et al. 2011; Liu et al. 2013; Zhang et al. 2013; Yao et al. 2014). Zhao and Cawood (2012) provided systematic assessments on the previous models for the Jiangnan belt and proposed that each model has its own lines of geological evidence but the unexplainable problems still existed. Consequently, Zhao and Cawood (2012) put forward the bifurcation two-way subduction model and renamed the Jiangnan belt as the Jiangnan folding belt and the Neoproterozoic tectonic belt in the northern margin of the Yangtze Block as the Panxi-Hannan folding belt. Before the final determination of the tectonic environment (rift or orogenic belt?), we consider that the nomenclature of "folding belt" is much more suitable for the tectonic belts above. Therefore, in this contribution, the tectonic belts will be nominated as Jiangnan folding belt (abbreviated for the Jiangnan belt) and Panxi-Hannan folding belt (abbreviated for the Panxi-Hannan belt).

Table 1 Stratigraphic division and correlation of Neoproterozoic strata in the South China Plate (Revised after Gao et al. 2010a, b)

		桂北 North Guangxi	黔东北 Northeast Guizhou	湖南地区 Hunan	赣西北 Northwest Jiangxi	皖南 South Anhui	赣东北 Northeast Jiangxi	浙西南 Southwest Zhejiang	
Paleozoic		Cambrian	Cambrian	Cambrian	Cambrian	Cambrian	Cambrian	Cambrian	541Ma
Neoproterozoic E.	Sinian P.	Laopu F.	Liuchapo F.	Liuchapo F.	Dengying F.	Piyuancun F.	Xifengsi F.	dengying F.	
	Sinian P.	Doushantuo F.	Doushantuo F.	Jinjiadong F.	Doushantuo F.	Lantian F.		Doushantuo F.	635Ma
	Nanhua P.	Nantuo F. Fulu F.	Nantuo F. Datangpo F. Tiesiao F.	Nantuo F. Datangpo F. Dongshanfeng F	Nantuo F.	Leigongwu F.	Nantuo F.	Leigongwu F. Yang'an F. Xiayabu F.	
	Nanhua P.	Chang'an F.	Liangjiehe F.	Xieshuihe F.	Dongmen F.	Xiuning F.	Xiuning F.	Zhitang F.	780Ma (760Ma)
	Qingbaikou P.	Danzhou Group: 765±14[6]	Xiajiang Group: 780±9[17], 782±8[16], 785±8[15], 814±6[14]	Banxi Group: 780±28[26], 803±8[25], 809±8[24], 814±12[23]	Xiushui Group	Likou Group: 777±9[45], 783±8[44], 773±7[43], 779±7[42], 820±16[41]	Heshangzhen G.: 767±5	Likou G.	820Ma
	Qingbaikou P.	Sibao Group.: 823±4[5], 824±13[4], 827±6[3], 836±3[2], 842±6[1]	Fanjingshan G.: 821±4[13], 823±2[12], 838±2[11], 814±6[10], 827±24[9], 831±6[8], 840±10[7]	Lengjiaxi Group: 806±9[22], 816±5[21], 822±10[20], 828±10[19], 855±5[18]	Shuangqiaoshan G.: 801±4[33], 818±10[32], 819±9[31], 829±5[30], 831±5[29], 878±4[28], 879±5[27]	Xikou Group: 823±8[40], 825±7[39], 826±6[38], 827±7[37], 837±14[36], 827±9[35], 848±12[34]	Shuangxiwu G.: 818±6[56], 844±3[55], 902±5[54], 905±14[53], 916±6[51], 913±5[52], 932±7[50], 891±21[48], 926±15[47], 968±23[46]	Shangxi G.: 866±9[49]	1000Ma

Note the isotopic ages above are listed with unit of Ma, data from (1): tuff from Sibao Group, Gao et al. (2010b); (2): Zhaigun granite, Wang et al. (2006); (3): Motianling granite, Gao et al. (2010b); (4): Dongma granite, Wang et al. (2006); (5): Bendong granite, Wang et al. (2006); (6): tuff from Sanmenjie Formation of Danzhou Group, Zhou et al. (2007); (7): tuff from Fanjingshan Group, Gao et al. (2010b); (8): basalt from the Huixiangping Formation of Fanjingshan Group, Zhou et al. (2009); (9): basalt from the Huixiangping Formation of Fanjingshan Group, Zhou et al. (2009); (10): basalt from the Xiaojiahe Formation of Fanjingshan Group, Zhou et al. (2009); (11): muscovite granite intruded into Fanjingshan Group and overlain by Xiajiang Group, Wang et al. (2011); (12): Gangbian granite unconformity overlain by Xiajiang Group, Chen et al. (2007); (13): gabbro intruded into Fanjingshan Group, Xue et al. (2012); (14): tuff from the Hongzixi Formation of Xiajiang Group, Gao et al. (2010a); (15)–(17): tuff from the Ejiaao Formation of Xiajiang Group, Wang et al. (2010); (18): tuff from Cangxi Group, northeastern Hunan Province, Gao et al. (2011b); (19): volcanic rock from the Yunjunli of Lengjiaxi Group; Bai et al. (2010); (20): bentonite from Lengjiaxi Group, and Linxiang, Hunan Provinces, Gao et al. (2011a); (21): Zhangbangyuan granite in northeastern Hunan Province, Ma et al. (2009); (22): Yexijiang granodiorite in Bucheng country of Hunan Province, Bai et al. (2010); (23): dacitic agglomerate from the Cangshuipu Formation of Banxi Group, Wang et al. (2003); (24): tuff from the Wuqiangxi Formation of Banxi Group, Zhang et al. (2008b); (25): bentonite from the Zhangjiawan Formation of Banxi Group, Gao et al. (2011a); (26): trachybasalt intruded into Banxi Group and overlain by Nanhua System, Zhou et al. (2007); (27): rhyolitic tuff from the Hengyong Formation of Shuangqiaoshan Group, Wang et al. (2008); (28): quartz-keratophyre from the Hengyong Formation of Shuangqiaoshan Group, Wang et al. (2008); (29): fuff from the Hengyong Formation of Shuangqiaoshan Group, Gao et al. (2008a); (30): tuff from the Anlelin Formation of Shuangqiaoshan Group, Gao et al. (2008a); (31): Jiuling cordierite-bearing granodiorite intruded into Shuangqiaoshan Group, Li et al. (2003a); (32): cordierite-bearing granodiorite,Li et al. (2001); (33): gabbro intruded into Shuangqiaoshan Group,Wang et al. (2008); (34): gabbro from ophiolites in southern Anhui Province, Ding et al. (2008); (35): wehrlite from ophiolites in southern Anhui Province, Ding et al. (2008); (36): Shexian granite, Xue et al. (2010); (37): Xucun granite, Wu et al. (2006); (38): Xiuning granite, Xue et al. (2010); (39): Xiuning granite, Wu et al. (2006); (40): Xucun granite, Li et al. (2003a); (41): tuff from the Jingtan Formation, Wu RX et al. (2007); (42)–(43): dacites from the Jingtan Formation, Wu et al. (2007);(44)–45: Shiershan granite, Xue et al.

(2010); Wu et al. (2005a); (46): Jadeite-kyanited anorthosites in Xiwan area, Li et al. (1994b); (47): rhyolite from the Beiwu Formation of Shuangxiwu Group, Li et al. (2009); (48): rhyolite from the Zhangcun Formation of Shuangxiwu Group, Li et al. (2009); (49): dacite from the Niuwu Formation of Shangxi Group, Gao et al. (2009); (50): high-Mg diorite intruded into Shuangxiwu Group, Chen et al. (2009); (51): basaltic porphyrite intrude into Shuangxiwu Group, Chen et al. (2009); (52): Taohong granite intruded into Shuangxiwu Group, Ye et al. (2007); (53): Xiqiu granite intruded into Shuangxiwu Group, Ye et al. (2007); (54): plagioclase granite, Chen et al. (2009); (55): pyroxenolite,Wangs et al. (2012; (56): quartz diorite, Wang et al. (2012); (57): tuff from the Heshangzhen Group, Gao et al. (2008a)

2.1 The Early to Middle Neoproterozoic Jiangnan Belt

Abundant Early to Middle Neoproterozoic geological bodies exposed in Jiangnan belt, which are composed of extremely thick volcanic sedimentary sequences, except for some granitic intrusions. This volcanic sedimentary suite shows bilayer structure: the lower structural layer comprises Sibao Group exposed in Guangxi, Fanjingshan Group in Guizhou, Lengjiaxi Group in Hunan, Xikou Group in Anhui, Shuangxiwu and Pingshui groups in Zhengjia, etc.; the upper structural layer includes Danzhou Group exposed in Guangxi, Xiajiang Group in Guizhou, Banxi Group in Hunan, Likou Group in Anhui, and Heshangzhen Group in Zhejiang. The lower structural layer was considered to be the Mesoproterozoic strata sequence in the published papers (Cheng 1994; BGMRGX 1985; BGMRGZ 1987; BGMRHN 1988; BGMRJX 1984;

BGMRZJ 1989; Xu et al. 2008). However, recent SHRIMP and LA-ICP-MS zircon U-Pb dating revealed these complexes formed at the Early Neoproterozoic. The upper structural layer was overlain on the lower structural layers with unconformable contacts (Fig. 1b), and distinct orientations of them and boundaries with uneven surfaces could be observed locally (Fig. 1a–c). The tectonic event of Wuling Orogeny was considered to result in the unconformity between two structural layers. In Lucheng, Linxiang City of the Hunan Province, it can be recognized that the Banxi Group of the upper structural layer was overlain on the Lengjiaxi Group of the lower structural layer with unconformable contact. The SHRIMP zircon U-Pb dating revealed that the tuffs in the Lengjiaxi Group gave an age of 822 ± 10 Ma and tuffs in the overlain Banxi Group yielded an age of 803 ± 8 Ma (Gao et al. 2011a), which constrained the Wuling Orogeny during 820–803 Ma. The lower structural layers are characterized by the developments of high-angle closed upright folds and chevron folds (Fig. 1d), and the upper structural layers are characterized by developments of relax folds (Fig. 1c). Conglomerates usually develop at the bottom of the upper structural layers (Danzhou, Xiajiang, and Banxi Groups) (Fig. 1f), and these conglomerates contain mainly gravels of quartzite (quartz vein), phyllite, slate, and granite. The stratigraphic division and correlation of the Early to Middle Neoproterozoic strata for diverse areas of the Jiangnan belt and respective isotopic ages are listed in Table 1.

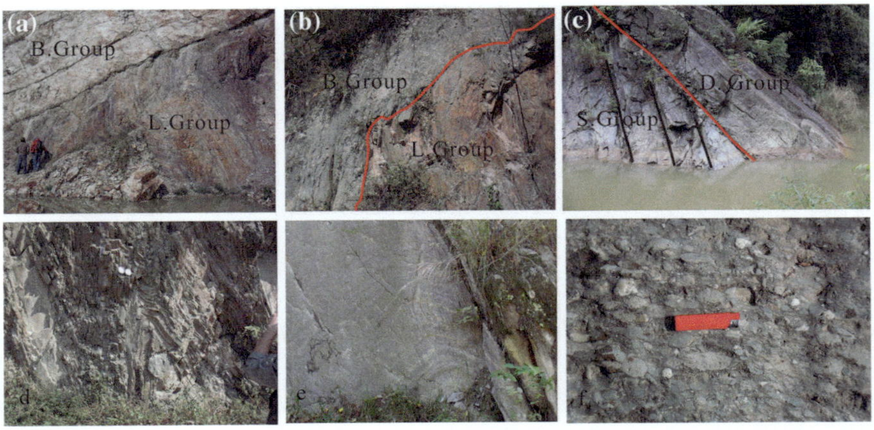

Fig. 1 Unconformities and tectonic features between the lower and upper structural layers **a** Hunan Linxiang-Lucheng Banxi group (*B.Group*) unconformably blanketed on the Lengjiaxi group (*L.Group*); **b** Hunan Yuanling-Madiyi Banxi group (*B.Group*) unconformably blanketed on the Lengjiaxi group (*L.Group*) with uneven surface; **c** Guangxi Luocheng-Huangjin Danzhou group (*D.Group*) unconformably blanketed on the Sibao group (*S.Group*); **d** Hunan Yuanling-Madiyi Lengjiaxi group, tight homoclinal folding; e-Guizhou Taihe Xiajiang group of upper structural layer with rolling fold with axial plane cleavages; **f** Hunan Yuanling-Madiyi Banxi group of the upper structural layer with conglomerates at the bottom

In the Jiangnan belt, terranes of the lower structural layers, e.g., Sibao, Fanjingshan, Lengjiaxi, Xikou, and Shuangxiwu Groups, consist mainly of a set of low-grade metamorphic volcanic sedimentary rocks, and these rocks show distinct lithological associations and timings in diverse areas.

In the border land of Anhui–Zhejiang–Jiangxi Provinces, the Shuangxiwu Group of the Jiangnan belt consists of a set of basic–intermediate and intermediate–acidic lavas, and pyroclastic rocks interbedded by the sandy, siliceous, carbonaceous shale and limestone lens. Some researchers suggested that the combination of these rocks was similar to ophiolitic melange (Zhou et al. 1989, 1990; Zhou and Zhao 1991; Zhou 1997; Shen et al. 1992; Li et al. 1994a; Zhao et al. 1995), and later blue schists were discovered in this complex (Zhou et al., 1989; Gao 2001). The high-pressure metamorphic rocks consist mainly of aragonite–jadeite blue schists (Zhou 1989), jadeite-bearing aegirine albite hornblende schists, jadeite-bearing aegirine quartz albite rocks, glaucophane quartz albite schists, and torendrikite quartz schists (Gao 2001), and the pressure of the peak metamorphism has over 12 kbar (Zhou et al. 1989, Zhou 1997; Gao 2001). The SHRIMP zircon U-Pb dating revealed the rhyolites of the Shuangxiwu with ages of 891–926 Ma, which was interpreted to represent their crystallization ages (Li et al. 2009). The K-Ar isotopic dating revealed the glaucophane of these blue schists with ages at 866 ± 14 Ma (Shu et al. 1994) and the $^{40}Ar/^{39}Ar$ isotopic dating constrained the crossites with ages of 799 Ma (Hu et al. 1992). These metamorphic ages have provided well constraints on the timing of the high-pressure metamorphism (Shu et al. 1994; Hu et al. 1992;

Gao 2001; Shen and Geng 2012). The zircons from gabbro in Fuchua ophiolite belt yielded two groups of age of 891 ± 13 and 824 ± 3 Ma, the former were interpreted as inherited zircon age, and the latter were interpreted as crystallization age of the gabbro (Zhang et al. 2012). Based on the dating and geochemical feature of gabbro, authors consider that the Fuchuan ophiolite was formed in the back-arc basin at about 825 Ma due to partial melting of the enriched mantle wedge (Zhang et al. 2012).

The Shuangqiaoshan Group (Wannian Group) in the north of Jiangnan belt, northern Jiangxi Province, consists dominantly of the extreme thick muddy, sandy sedimentary rocks, interbedded by minor volcanic rocks. This volcanic suite is composed mainly of splites and quartz keratophyres (Wang et al. 2008; Zhao and Cawood 2012) The Shuangqiaoshan Group, from bottom to top, was categorized into the Hengyong Formation, Jilin Formation, Anlelin Formation, and Xiushui Formation within successions (Gao et al. 2008; Wang et al. 2008) These rocks experienced widely greenschist phase metamorphism and strong deformations (dominantly folding deformations with slaty cleavages but few synchronous faults developed) (Xue et al. 2010). The sedimentary environment for the Shuangqiaoshan Group remains disputed recent years. On the basis of the sedimentary formation, Huang et al. (2003) proposed the gravity flow deposit system of the littoral and bathyal environments for the Shuangqiaoshan Group; however, Chen et al. (2002) suggested turbidity current phase of bathyal and abyssal environments for these sedimentary rocks. The LA-ICPMS zircon U-Pb dating revealed the quartz keratophyres and the rhyolitic tuffs of the Hengyong Formation of lower Shuangqiaoshan Group, respectively, at 878 ± 4 and 879 ± 5 Ma (Wang et al. 2008). The SHRIMP an LA-ICP-MS zircon U-Pb dating has constrained the Shuangqiaoshan Group (Wannian Group) formed during 880–830 Ma (Gao et al. 2008a; Wang et al. 2008; Liu et al. 2013; Yao et al. 2014).

In the southwest of Jiangnan belt, the lower structural layers consist of mainly Sibao Group in Guangxi, Fanjingshan Group in Guizhou, and Lengjiaxi Group in Hunan. These groups are composed dominantly of low-grade, mostly greenschist facies, metamorphic sandstones, siltstones, and slates with interbedded mafic-ultramafic and basaltic volcanic rocks with strong deformations (Fig. 1d). The volcanic sequences, in Sibao area of Guangxi, Fanjingshan area of Guizhou, and Yiyang area of Hunan, are relatively thick and composed mainly of basalts with developments of the pillow structure (Fig. 2), indicating an underwater eruption environment. The ultramafic rocks in Sibao area of Guangxi were previously considered to be komatiites (Yang 1990; Tang et al. 1992). However, lack of spinifex structure and low contents of MgO (<30 wt%) suggested that these ultramafic rocks should belong to the komatiitic basalts and high-Mg basalts with cumulative characteristics (Ge et al. 2001). The mafic-ultramafic volcanic rocks were usually accompanied by the intrusions of mafic gabbros. The published geochronological data have constrained the Sibao Group during 842–830 Ma (Gao et al. 2010b; Wang et al. 2012), the Fanjingshan Group during 840–815 Ma (Gao et al. 2010b; Zhou et al. 2009), and the Lengjiaxi Group during 855–822 Ma (Gao et al. 2011a, b; Bai et al. 2010). The forming environment for

Fig. 2 The pillow lava in Guizhou Fanjingshan group (**a**); Guangxi Sibao group (**b**); and in Hunan Yiyang area (**c**)

these mafic-ultramafic volcanic rocks remains disputed. Some workers considered similar incompatible element distribution characteristics and εNd(t) values between the ultramafic rocks in Sibao area of Guangxi and the dike swarms at Gairdner of Australia, and then suggested these volcanic rocks correlated with the mantle plume resulting in the break off of the Rodinia supercontinent during Neoproterozoic (Ge et al. 2001). Others researchers, on the basis of the enrichments of strong incompatible elements (e.g., LREEs, Rb, Ba, Th, and U), the strong depletion of HFSEs (e.g., Nb and Ta), and the low εNd(t) values, proposed back-arc limited oceanic basin during subduction for these volcanic rocks (Xue et al. 2012; Yao et al. 2014).

In general, the Early to Middle Neoproterozoic strata of the Jiangnan belt show distinct lithologic associations in the diverse area. The eastern Shuangxiwu Group is composed of mainly volcanic sequence with the lithologic combination similar to the ophiolitic melanges. The Shuangqiaoshan Group contains minor volcanic rocks in the north-center of Jiangxi, but more basaltic volcanic rocks with pillow structure in the southwest of Guizhou and Guangxi Province. The distinction in the lithologic associations of the Groups, in diverse areas, demonstrates their possible differences in forming environment. In addition, the isotopic dating constrained the Shuangxiwu Group in the east at 891–926 Ma, the Shuangqiaoshan Group in the north-center at 880–830 Ma, and the Lengjiaxi Group, the Fanjingshan Group, and the Sibao Group in the southwest, respectively, at 855–822, 840–815, and 842 ± 6 Ma, showing younger isotopic ages from the east to the southwest.

Except for the Early to Middle Neoproterozoic strata mentioned above, the Jiangnan belt also contains mounts of granites, which intrude into the lower structural layers and are overlain by the upper structural layers with unconformable contacts, indicating their synchronous emplacements with the lower structural layers. Some granites intruded into the lower structural layer and are exposed without any covers of the upper structural layers. As mentioned above, the Wuling Orogeny (movement), which resulted in the unconformity between the lower and upper structural layers, has been suggested to happen at ~820 Ma, and then we consider the granites with ages older than 820 Ma synchronous with the lower structural layers. In the border of Anhui, Zhejiang, and Jiangxi Province, geochronological data have constrained the high-Mg diorite intruding into the Shuangxiwu Group at 932 Ma (Chen et al. 2009), the Taohong granite at 913 Ma and the Xiqiu granite at 905 Ma (Ye et al. 2007), the plagiogranite at 902 Ma (Chen et al. 2009), and the amphibole pyroxenite at 844 Ma (Wang et al. 2012). A lot of granites exposed along the Qimen-Shexian-Sanyangkeng arc-shaped zone in the southern Anhui Province, which were categorized into two Groups, the S-type and A-type granite. The S-type granites have been considered to be part of the lower structural layer, and the A-type granites, intruding into the Shangshu Formation in the Heshangzhen Group or the Jingtan Formation in the Likou Group of Late Neoproterozoic (Xue et al. 2010), are obviously attributed to the upper structural layer. In this fracture zone, a mass of granites was recognized in the lower structural layer, including the Xucun pluton, Shexian pluton, and Xiuning pluton. The Xucun pluton is composed of gneissic granodiorites, and the SHRIMP zircon U-Pb dating revealed their crystallization age at 823 ± 12 Ma (Li et al. 2002b) and LA-ICPMS zircon U-Pb dating obtained isotopic ages ranging from 852 ± 6 to 820 ± 10 Ma (Xue et al. 2010; Wu et al. 2005b; Li et al. 2003a). The Shexian pluton is composed of granodiorites with weak deformations, and the LA-ICPMS zircon U-Pb isotopic dating obtained their crystallization ages of 823 ± 10 Ma (Wu et al. 2005b) and 837 ± 14 Ma (Xue et al. 2010). The Xiuning pluton consists mainly of the undeformed granodiorites, and the LA-ICPMS zircon U-Pb isotopic dating obtained their crystallization ages of 826 ± 6 and 824 ± 6 Ma for these granodiorites (Xue et al. 2010; Wu et al. 2005b). The Early Neoproterozoic Jiuling pluton, as the largest Neoproterozoic pluton in Jiangnan belt, exposed in the northern Jiangxi with outcrop of ca. 2500 km^2, intruding into the Shuangqiaoshan Group and covered by the Middle to Late Neoproterozoic Dongmen Formation (as section of the Banxi Group in Hunan) with unconformable contacts (Li et al. 2003a). This batholith is composed mainly of cordierite-bearing muscovite diorites. Many synchronous plutonic records were also reported by the previous researchers, such as the muscovite granite at 823 ± 2 Ma, intruding into the Fanjingshan Group and covered by Xiajiang Group with unconformable contacts (Wang et al. 2011); the gabbros at 821 ± 4 Ma (Xue et al. 2012) in Guizhou Province; and the Xiutang pluton and Gangbian pluton, respectively, at 836 ± 5 Ma (Fan et al. 2010) and 823 ± 2 Ma (Chen et al. 2007), intruding into the Sibao Group and covered by the Jialu Formation of the Xiajiang Group. In Guangxi Province, amount of plutons, such as the Gunzhai pluton (836 ± 3 Ma), the Bendong pluton (823 ± 4 Ma), the Dongma pluton (824 ± 13 Ma) (Wang et al. 2006), the Motianling pluton (827 ± 6 Ma, Gao et al. 2010b), and Yanbaoshan pluton (822 ± 5 Ma ~ 833 ± 6 Ma, Yao et al. 2014), intruded into the Sibao Group and were overlain by the Danzhou Group of the upper structural layer with unconformable contacts. These granites show mostly alkali-rich peraluminous to strongly peraluminous features similar to S-type granite (Li et al. 2002b; Wang et al. 2006, 2011; Xue et al. 2010; Fan et al. 2010). These granites show the zircon $\varepsilon_{Hf}(t)$ ranging from −10 to +10 (see Fig. 8 of chapter "Late Archean—Mesoproterozoic geology of the Tarim Craton"), indicating their derivations from partial melting of the crustal rocks. The tectonic environment for these granites remains disputation in recent years. Some researchers suggested syn-collisional emplacements (Xue et al. 2010; Bai et al. 2010; Yao et al. 2014); however, others proposed post-collisional extensional emplacements for these granites (Fan et al. 2010), and some workers considered these granites derived from partial melting of crustal rocks caused by the upwelling mantle plume (Li et al. 2002b; Wang et al. 2006, 2011).

In Jiangnan belt, the upper structural layer of the Early to Middle Neoproterozoic are composed mainly of Danzhou Group in Guangxi, Xiajiang Group in Guizhou, Banxi Group in Hunan, Xiushui Group in northern Jiangxi, Likou Group in southern Anhui, and Heshangzhen Group in northern Zhejiang. These groups consist of low-grade metamorphic sandstone, siltstone, slate, and phyllite, interbedded by volcanic rocks. The lithologic associations are different in diverse area, and various stratigraphic divisions have been proposed, respectively. The Heshanzhen Group, in the northeast of the Jiangnan belt, was divided into the Luojiamen Formation, Hongchicun Formation and Shangshu Formation from the bottom to top, which covered on the intermediate-acidic volcanic rocks of the pre-Sinian Shuangxiwu Group and was overlain by the Zhitang Formation of the Sinian with unconformable contacts. The Luojiamen Formation consists of a set of pyroclastic rocks with dominantly greywackes and sand slates, and the conglomerates show glaciations-related characteristics. The Hongchicun Formation is composed of purple thick sandstones, and the Shangshu Formation consists of mainly terrestrial basic and acidic volcanic associations, which contain basalts in the lower segment and the rhyolites in the upper segment, showing bimodal volcanic features. The Banxi Group in the center of Jiangnan belt is divided into the Cangshuipu Formation, Madiyi Formation, and Wuqiangxi Formation from the bottom to top. The Cangshuipu Formation is composed of meta-andesite, dacitic agglomerate, andesitic–dacitic conglomerate, and dusty tuff within limited outcrop (Gao et al. 2012). The Madiyi Formation consists dominantly of conglomerate, sandy conglomerate, siltstone, muddy slate, calcic slate with interbedded limestone, carbonaceous slate, and sandy slate. The Wuqiangxi Formation contains mainly pebbly sandstone interbedded by sandy conglomerate, sandstone, and siltstone with muddy slate bedding in the lower segment; the silty slate, rhyolitic sandstone, and slate interbedded by crystal pyroclast-bearing sedimentary tuff in the middle segment; and the pebbly sandstone, feldspathic quartz sandstone, silty slate, and sedimentary tuff in the upper segment. The Madiyi Formation and Wuqiang Formation is characterized by the heather intraclastic Formation with purple-gray fuchsia and gray-sage green interbeddings and named as “Hongbanxi,” which stands for red slaty feature in Chinese. The Danzhou Group, in the southwestern Jiangnan belt, is divided into Baizhu Formation, Hetong Formation, Sanmenjie Formation, and Gongdong Formation from the bottom to top. The Baizhu Formation consists of mainly pebbly sandstone, siltstone, sericite schist, and calcic schist, and the Hetong Formation is composed of sericite schist and calcic schist. The Sanmenjie Formation contains dominantly basic volcanic rocks, and the Gongdong Formation consists of mainly sericite quartz schist and slate. The Danzhou Group, Xiajiang Group, and Gaojian Group in the southwestern margin of Jiangnan belt show mostly sage green and rubricans features, and then are named as “Heibanxi,” which stands for black slaty characteristics in Chinese.

The LA-ICPMS zircon U-Pb dating revealed the dacites at 820 ± 16 and 773 ± 7 Ma, and the volcanic tuff at 779 ± 7 Ma in the Jingtan Formation of Likou Group at the northeastern segment of the Jiangnan belt (Wu et al. 2007). SHRIMP zircon U-Pb dating revealed the tuff in Heshangzhen Group at 767 ± 5 Ma (Gao et al. 2008a). In the center of Jiangnan belt, SHRIMP zircon U-Pb dating revealed tuffs in Wuqiangxi Formation of Banxi Group at 809 ± 12 Ma (Zhang et al. 2008b), the dacitic agglomerate of the Cangshuipu Formation at 814 ± 12 Ma (Wang et al. 2003), and bentonites of Zhangjiawan Formation at 803 ± 8 Ma (Gao et al. 2011b). In the southwestern Jiangnan belt, the SHRIMP zircon U-Pb dating revealed the tuff in the Hongzixi Formation of Xiajiang Group at 814 ± 6 Ma (Gao et al. 2010a); three samples of pyroclastic crystal bearing tuffs in the Ejiaao Formation of Banxi Group in nearby Tongren, respectively, at 780 ± 93, 782 ± 8, and 785 ± 8 Ma (Wang et al. 2010); and the volcanic rocks in Sanmenjie Formation of Danzhou Group in northern Guangxi at 765 ± 14 Ma (Zhou et al. 2007). Recently, Gao et al. (2013a) obtained SHRIMP zircon U-Pb ages of 801 ± 3 and 787 ± 6 Ma for the tuffs in the middle segment of the Hetong Formation and upper segment of the Gongdong Formation of Danzhou Group, respectively. But detrital zircon dating demonstrates that the Danzhou Group formed between ~770 and 730 Ma. Geochronological data listed above are mostly older than 780 Ma, and then we consider forming of the Early to Middle Neoproterozoic upper structural layer during 820–780 Ma, suggesting the bottom of the Nanhua System at 780 Ma. However, there are still some geochronological data younger than 780 Ma, such as the 765 Ma old volcanic rocks in Sanmenjie that obviously younger than the bottom age of Nanhua System, and some detrital zircon ages younger than 780 Ma. In summary, the bottom of the Nanhua System remains still debated and deserves considerably studies.

2.2 *Early to Middle Neoproterozoic Strata in Panxi-Hannan Belt*

Early to middle Neoproterozoic strata in Panxi-Hannan belt are composed of Yanbian Group, Suxiong Formation, and Kaijianqiao Formation in the western Sichuan, Bikou Group in the northwestern of Yangtze Block, and Hannan Complex in the northern Yangtze Block.

Yanbian Group is mainly distributed in the southwestern Yangtze Block and subdivided into Huangtian Formation, Yumen Formation, Xiaoping Formation, and Zhagu Formation from bottom to top (Fig. 3). Huangtian Formation consists of basalts, breccia-bearing volcanic rocks, and breccia lavas, with thin siliceous rocks as intercalated beds. In addition, basalts occurred with pillow structure. The lower segment of Yumen Formation is composed of light-gray to dark-gray carbonaceous slates, sericite slates, and siliceous slates, interbedded with crystalline limestones as lenses, metamorphosed maristones, and sandy limestones. Light-gray sandstones

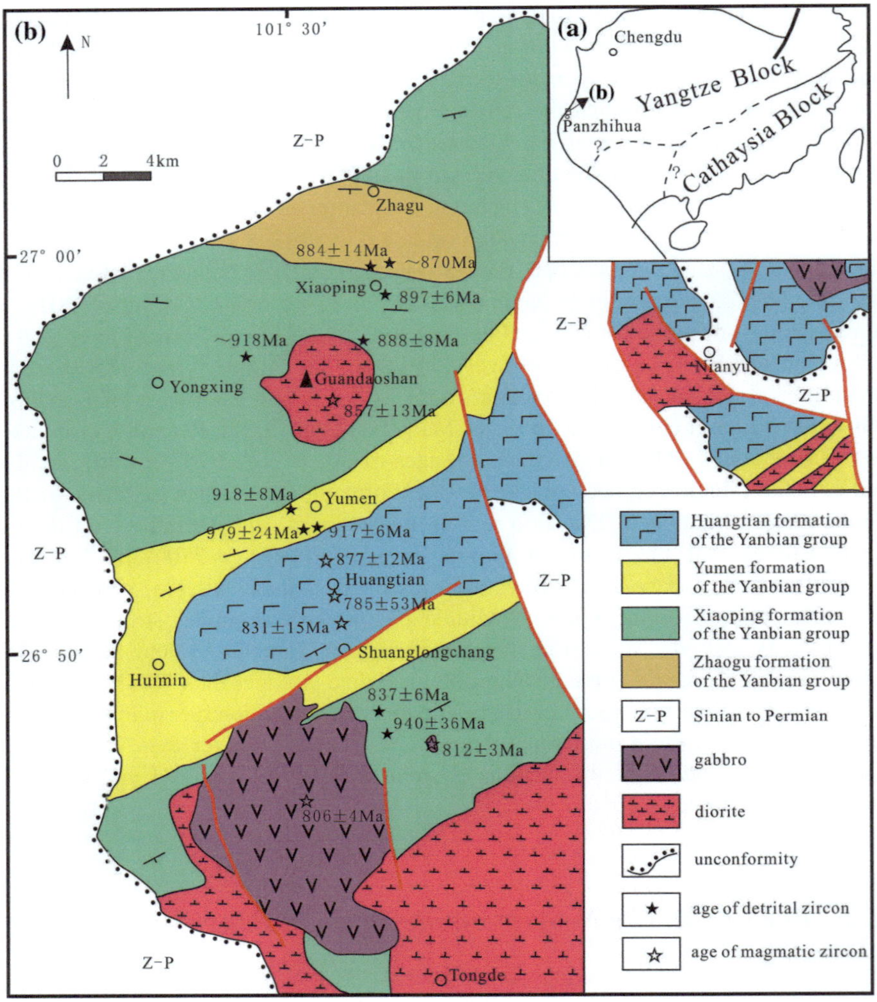

Fig. 3 Geological map of Yanbian area in Sichuan Province *Note* Geochronological data from Du et al. (2005, 2013), Zhou et al. (2006a), Li et al. (2003b), Sun and Zhou (2008), Sun et al. (2009), Du (2010)

occur as parallel bedding and graded bedding structure. The middle-top segment comprises a rhythmite which consists of metamorphosed tuffaceous slates, sandy slates, and slates with developed folds. Yumen Formation with thickness of about 1,700 m displays a conformable contact with underlying Huangtian Formation. Xiaoping Formation contains light-gray, dark-gray sericite slates, sandy slates, and carbonaceous slates, which are interbedded with metamorphosed sandstones and carbonaceous slates. Thick-bedded metamorphosed tuffaceous fine conglomerate and sandstones constitute the lower segment, and carbonaceous slates constitute the upper segment with increasing thickness, and certain sedimentary tectonics such as convolute bedding and wave erosion surface. Locally, the whole Bouma sequence occurs (Sun et al. 2008) and granitic veins invade. Xiaoping Formation, 2, 260 m thick, occurred in conformable contact with underlying Yumen Formation (BGMRSC, 1991). Zhagu Formation is primarily composed of sericite slates, siltstone, and slate, and its bottom consists of metamorphosed tuffaceous conglomerate or sandy conglomerate as lenses. The gavels are mainly volcanic lavas. The lower segment of Zhagu Formation is composed of carbonaceous slates, metamorphosed fine sandstones, and siltstones, and the upper segment is interbedded with dolomite limestones and slates. Brecciaous dolomite limestones were locally exposed. Zhagu Formation shows parallel uncomfortable contact with underlying Xiaoping Formation. Some researchers (Li et al. 1983; Li 1984; Sun et al. 1994) suggested that volcanic rocks in Huangtian Formation and south Gaojiacun mafic intrusion might be ophiolite suite in last 1980s; however, later investigations negated the "ophiolite", so it still is a hot-debated issue on the tectonic setting of Yanbian Group. Some researchers argued that both Yanbian Group and contemporary magmatic rocks formed in an extensional environment caused by mantle plume (Li et al. 2003b, 2006). Some researchers believed they formed in arc environment (Zhou et al. 2002a, 2006a, b). Other proposed that Yanbian Group was a back-arc basin which was related to the subduction induced by western oceanic crust of the Yangtze Block (Du et al. 2005; Sun et al. 2007). And some argue the lower segments of Yanbian Group formed in back-arc basin-related setting while the upper Zhagu Formation in foreland basin environment (Jiang et al. 2005).

Yanbian Group was used to consider as the Mesoproterozoic fold basement of Yangtze Block (Li et al. 1983, 1988; BGMRSC, 1991; Cheng 1994). Figure 3 exhibits isotopic age data measured by in situ zircon dating and reveals that the peak ages of detrital zircons range from 979 to 837 Ma, the age of Lengshuiqing pluton which intruded into Yanbian Group is 812 ± 3 Ma, and the age of Gaojiacun pluton which intruded Xiaoping Formation and Yumen Formation is 806 ± 4 Ma (Zhou et al. 2006a). The ages of Guandaoshan pluton which intruded Xiaoping Formation in the north are 857 ± 13, 857 ± 7, 856 ± 6, and 856 ± 8 Ma (Li et al. 2003b; Du et al. 2014). Recent study reveals that the formation ages of basalts from Yanbian Group vary from 877 to 831 Ma (Du 2010; Du et al. 2013). According to various ages mentioned above, the ages of Yanbian Group is likely to be limited between 880 and 830 Ma. In terms of the lithologies and formation ages, Yanbian Group resembles the lower structural layers in Jiangnan belt such as Sibao Group and Fanjing Group.

Except Yanbian Group, early to middle Neoproterozoic low-degree metamorphic strata in the west of Sichuan also consist of Suxiong Formation and Kaijianqiao Formation. Suxiong Formation is primarily composed of extraordinarily thick acid volcanic rocks (mainly rhyolites and dacites, thickness varies from hundreds meters to ten thousands meters), which is interbedded with minor mafic volcanic rocks and pyroclastic rocks. In addition, the proportion of mafic rocks to acid rocks is about 1:9 (Li et al. 2002a). Kaijianqiao Formation mainly comprises purple or gray-green acid pyroclastic rocks and tuffaceous sandstone. Geochemically, the acid volcanic rocks from Suxiong Formation resemble A_2-type granite forming in extensional tectonic setting. SHRIMP zircon U-Pb isotopic dating revealed that an age of acid volcanic rocks from Suxiong Formation is 803 ± 12 Ma (Li et al. 2001b, 2002a) and age of tuff from Kaijianqiao Formation is 801 ± 7 Ma (from Jiang et al. 2012), which are similar to early to middle Neoproterozoic upper structural layer in Jiangnan belt such as Danzhou Group, Xiajiang Group, and Banxi Group.

Various degrees of metamorphic rocks were exposed within a north-south trend belt, which is 700 km long and several ten thousands meters wide and distributed in western Sichuan Province to Yuanmou of Yunnan Province. Researchers used to name them Kangding Group or Kangding Complex, which constitutes an Archean to Paleoproterozoic crystalline basement (BGMRSC 1991; Cheng 1994). For a decade years, however, amounts of studies revealed that abundant deformed rocks occurred within Kangding Complex, such as granitic gneisses with ages of 797 ± 10, 795 ± 11, and 796 ± 13 Ma in Kangding area (SHIRMP U-Pb dating method, Zhou et al. 2002a); biotite trondhjemites in Gezhong area with age of 864 ± 11 Ma (SHIRMP U-Pb dating method, Zhou et al. 2002a); diorites and granodiorites in Guzan area with ages of 768 ± 7 and 755 ± 6 Ma (SHIRMP U-Pb dating method, Li et al. 2003); mozogranite ing Kangding area with an age of 767 ± 24 Ma (SHIRMP U-Pb dating method, Liu et al. 2009); diorites in Tianwan area with age of 823 ± 12 Ma and granites 876 ± 40 Ma in Pianlugang area within Kangding-Luding area (SHIRMP U-Pb dating method, Guo et al. 1998); granites in Huangcaoshan with age of 786 ± 36 Ma and Xiasuozi granites with an age of 805 ± 15 Ma (TIMS zircon U-Pb dating method, Shen et al. 2000); gabbros yielding ages of 752 ± 11 and 752 ± 12 Ma (Li et al. 2003); and granitic gneiss 772 ± 15 Ma (Chen et al. 2005) in Shaba area. We also recognized plenty of early to middle Neoproterozoic deformed magmatic rocks (Geng et al. 2007, 2008; Fig. 4). Figure 4 illustrates Neoproterozoic complex from the western Yangtze Block formed from 746 to 864 Ma, especially focusing during 840–780 Ma. Furthermore, the lithologies of complex vary from granitic, intermediate mafic to ultramafic rocks, and mafic dykes (Zhu WG et al. 2008). It is still not clear that weather these distinguished rocks formed by same geological event.

Neoproterozoic Bikou Group is located in Shaanxi, Gansu, Sichuan Provinces' junction, northwestern Yangtze, from Mianxian in Shaanxi to Pingwu in Sichuan and many places between them like Kangxian, Bikou in Gansu, with an exposed area of about 10,000 km^2. There have been controversies on the constituent and formation age of Bikou Group. One clastic sedimentary suit from above Bikou Group was named Hengdan Group (Zhang et al. 1993; Yan et al. 2004a). Now most researchers proposed that Bikou Group consists of only low-degree metamorphosed volcanic suit (Pei 1989; Xia et al. 1989; Xu et al. 2002; Yan et al. 2003). Bikou Group primarily comprises mafic to intermediate volcanic rocks and pyroclastic rocks. Volcanic rocks are composed of spilites, basalts, andesites, keratophyres, and minor rhyolite. Pyroclastic rocks consist of breccia lava, tuff lava, volcanic breccia, and tuff. The pyroclastic sedimentary rock suit comprises sedimentary volcanic

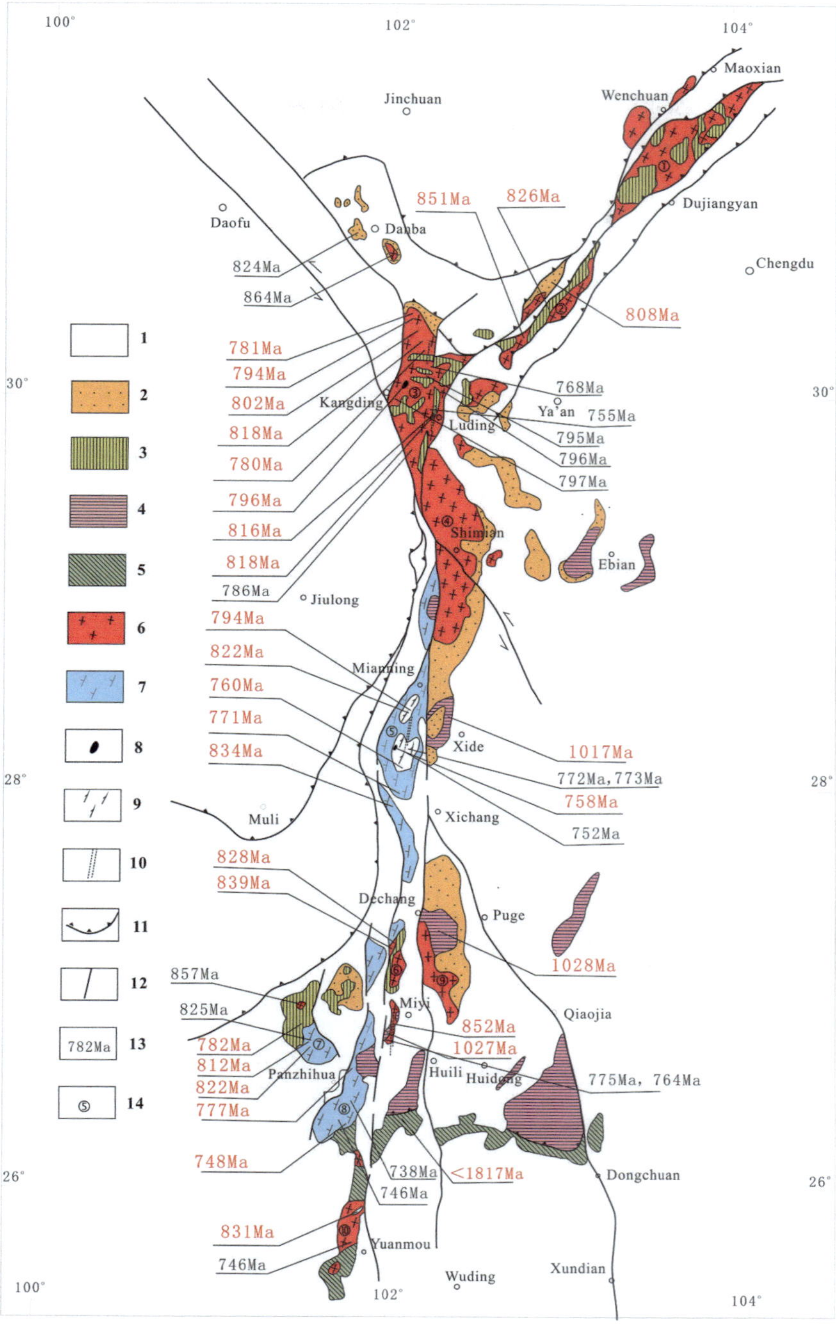
100°
102°
104°
30°
28°
26°
Maoxian
Wenchuan
Jinchuan
Dujiangyan
Chengdu
Daofu
Danba
851Ma
826Ma
824Ma
864Ma
808Ma
1
2
3
4
5
6
7
8
9
10
11
12
13
14
782Ma
781Ma
794Ma
802Ma
818Ma
780Ma
796Ma
816Ma
818Ma
786Ma
Kangding
Luding
Ya'an
768Ma
755Ma
795Ma
796Ma
797Ma
Shimian
Ebian
Jiulong
794Ma
822Ma
760Ma
771Ma
834Ma
Mianning
Xide
1017Ma
772Ma, 773Ma
758Ma
752Ma
Muli
Xichang
828Ma
839Ma
Dechang
Puge
1028Ma
857Ma
825Ma
782Ma
812Ma
822Ma
777Ma
Panzhihua
Miyi
852Ma
1027Ma
Qiaojia
Huili
Huidong
775Ma, 764Ma
748Ma
738Ma
<1817Ma
746Ma
Dongchuan
831Ma
746Ma
Yuanmou
Wuding
Xundian

◀ **Fig. 4** The distribution of lithologies in the western basement of Yangtze terrane and obtained zircon U-Pb age by SHRIMP dating method 1-Geological body younger than Sinian system, 2-Sinian system, 3-Neoproterozoic stratum, 4-Late Mesoproterozoic stratum, 5-Early Mesoproterozoic rocks, 6-Neoproterozoic granites, 7-Neoproterozoic intermediate -mafic intrusions, 8-super mafic rocks, 9-two-pyroxene gneiss, 10-ductile shear zone, 11-thrust fault, 12-fault, 13-SHRIMP zircon U-Pb age, 14-the numbers of magmatic complexes:① Pengguan Complex;② Baoxing Complex;③ Xiasuozi- Kangding Complex; ④ Shimian Complex; ⑤Mianning Complex; ⑥ Mopanshan-Miyi Complex; ⑦ Tongde Complex; ⑧ Datian Complex; ⑨ Moshaying Complex; ⑩ Longchuanjiang Complex. The red age in the figure is from Geng et al. 2008 and the black one from Zhou et al. 2002a, Li et al. 2002a, 2003, Chen et al. 2005

breccia, sedimentary tuff, tuffaceous sandstone, tuffaceous siltstone, and phyllite (Yan et al. 2004a). Moreover, Bikou Group could be subdivided into three volcanic eruption cycles: the lower segment of the first cycle is mainly composed of spilites, the upper primarily quartz keratophyric tuffs which interbedded with thin lavas. Spilites, spilitic porphyrites, and spilitic tuffs constituted the second cycle, upper of which is composed of quartz keratophyric tuffs, with thin lavas as intercalated beds. The lower segment of the third cycle consists of spilites with pillow structure, and the upper part is mainly composed of quartz keratophyric tuffs and interbedded with metamorphosed sedimentary tuffs or siltites as intercalated beds (Xu et al. 2002). The first cycle is characterized by alkaline volcanic rocks, and the second and third cycle by tholeiitic basaltic rocks. Bikou Formation widely underwent low-greenschist face (Wei 1993). The SHRIMP zircon U-Pb dating ages of three mafic volcanic rocks reveal at 840 ± 10, 846 ± 19, and 876 ± 17 Ma, separately (Yan et al. 2003, 2004b), and the ages of 790 ± 15 and 776 ± 13 Ma for two acid volcanic rocks, separately (Yan et al. 2003). Diorite intruded Bikou Group in Guankouya yielded an age of 884 ± 14 Ma and diorite in Pingtoushan of 884 ± 6 Ma, gabbro with an age of 877 ± 13 Ma by LA-ICPMS U-Pb dating method (Xiao et al. 2007), and granodiorite gave an age of 791 ± 13 Ma and diabase of 689 ± 24 Ma by SHRIMP U-Pb dating method (Yan et al. 2004b) in Liujiaping. Monzonitic granites intruded into Daomuliang Group (equal to Bikou Group) gave ages of 793 ± 11 and 792 ± 11 Ma (Pei et al. 2009). All data are accessed to reveal that Bikou Group and related magmatic rocks mainly formed in the early to middle Neoproterozoic era, though these data are to some extent in conflict, for example, the age of diorite which invaded Bikou Group is older than it of Bikou Group. If turbidites from Hengdan Group (Druschke et al. 2006), which are contemporaneous relationship with Bikou Group, and Bikou volcanic rocks are considered as a whole volcanic sedimentary rock suit, then their constituent resembles Yanbian Group (Sun et al. 2008).

Early to middle Neoproterozoic strata in the north of Panxi-Hannan belt comprise Huodiya Group and Xixiang Group. Huodiya Group is an unconformity relationship with the underlying Houhe Group, which is covered by Sinian System in an unconformity relationship. Houdiya Group was subdivided into three Formations from bottom to top, Shangliang Formation, Mawozi Formation, and Tiechuanshan Formation. Shangliang Formation and Mawozi Formation were metamorphic sedimentary rock suit, which is primarily composed of metamorphosed conglomerates, quartzites, carbonatites, siliceous rocks, and slates. Tiechuanshan Formation is volcanic rock suit and mainly composed of red alkaline rhyolites, dacite–rhyolites, tholeiitic basalts, ignimbrite, and pyroclastic rocks, which is characterized by bimodal volcanic rocks geochemically (Ling et al. 1996a). On the basis of the whole-rock Sm–Nd isotopic ages, Huodiya Group was considered as Mesoproterozoic rock series (Ling et al. 1996a). The rhyolite with an age of 817 ± 5 Ma in Tiechuanshan Formation was recognized later by TIMS zircon U-Pb dating method (Ling et al. 2003). According to the comparison with Xixiang Group, Huodiya Group is identified as early to middle Neoproterozoic strata sequence. Xixiang Group is a volcanic sedimentary rock suit and traditionally subdivided into six Formations from bottom to top, Baimianxia Formation, Sanwan Formation, Sanhuashi Formation, Sunjiahe Formation, Dashigou Formation, and Sanlangpu Formation. Some researchers also divide Xixiang Group into two suits: the lower suit is composed of low-K basalts erupting undersea and basaltic andesites, with metamorphosed sedimentary rocks as intercalated beds, while the upper suit consists of calc-alkaline to alkaline basaltic andesites, dacites, rhyolites which erupted under water, above of which is molasse formation (Ling et al. 2003). The ages during 950–895 Ma were obtained from volcanic rocks of Xixiang Group by single-zircon TIMS U-Pb dating method. However, ages new acquired by LA-ICPMS and SHRIMP zircon U-Pb dating method are much younger recent years. For example, basalts and dacites in Sunjiahe Formation yield ages of 845 ± 17 and 833 ± 5 Ma separately by LA-ICPMS U-Pb dating method (Xia et al. 2009; Xu et al. 2010), and dacites 815 ± 5 Ma by SHRIMP U-Pb dating method (Cui et al. 2010). Rhyolites and porphyrites in Dashigou Formation give ages of 803 ± 5 and 776 ± 6 Ma separately, and basalts of 730 ± 13 Ma by LA-ICPMS U-Pb dating method (Xia et al. 2009). All those ages reveal that Xixiang Group mainly formed between 845 and 770 Ma. Except Neoproterozoic Huodiya Group and Xixiang Group, there are large abundant early to middle Neoproterozoic magmatic complexes in Hannan area, which includes Beiba mafic intrusive complex, Tianpinghe granodiorite, Yangjiahe granodiorite, Maoerzhai moyite, Tiechuanshan aegirine granite, Wangjiangshan mafic complex, and Liunan complex (Ling et al. 1996b). According to tremendous study on isotopic ages recent years, all magmatic complex mainly formed between 820 and 700 Ma, except Liushudian gabbro with age of 898 ± 10 Ma (Zhou et al. 2002a, b; Zhao et al. 2010; Dong et al. 2011, 2012; Xu et al. 2010, 2011). The geochemistry of volcanic rocks in Xixiang Group and Hannan complex revealed that they both formed in continental arc tectonic setting (Xu et al. 2010). On the basis of regional analyses, Dong et al. (2012) proposed that the Neoproterozoic magmatic evolution migrated from south to north in the north Yangtze Block. Arc-related magmatic activity in Michangshan range of the south occurred between 870 and 820 Ma, and in Huijiaba of the middle Yangtze Block occurred from 840 to 820 Ma, and in Hannan area of the north Yangtze Block from 825 to 706 Ma. Such migration from south to north supports a tectonic model of accretionary orogenesis around continental marginal arc in the north Yangtze Block.

2.3 Early to Middle Neoproterozoic in the Cathaysian Block

The Neoproterozoic geologic bodies in the Cathaysian Block chiefly exposed in the southwestern Zhejiang Province and Wuyi mountain range in Fujian Province, while there is also sporadic exposures in the area of Nanling and Yunkai areas.

The early to middle Neoproterozoic strata comprise Chencai Group and Longquan Group in the southwestern Zhejiang Province and Mayuan Group, Mamianshan Group, Wanquan Group in Wuyi mountain range, Louziba Group, Jiaoxi Formation in the western Wuyi mountains, and Taoxi Formation at the Wuping area in the southwestern Fujian Province. According to the conventional lithostratigraphic unit, previous investigators analyzed and sorted for the regional strata sequences, and divided these strata as different groups and formations (Gong and Lin 1987; Hu et al. 1991; Gan et al. 1993; Zhuang et al. 2000; Zhang et al. 2005b; Xu et al. 2010; Fu et al. 2010). However, recent researches show that these lithological assemblages underwent a complicated structural deformation, with complicated contact relationships among them, especially showing generally structural contact between Formation and Formation instead of the conventional stratigraphic units (Wan et al. 2007; Li et al. 2010). From these strata, Paleoproterozoic age information was obtained (Li et al. 1998; Wan et al. 2007; Li et al. 2009), while vast Neoproterozoic age information also was documented (Li et al. 2005; Shu 2006; Shu et al. 2006, 2011; Wan et al. 2007; Li et al. 2008a; Li et al. 2009; Xu et al. 2010; Yu et al. 2010). Meanwhile, some Paleozoic age information was discovered from the migmatitic gneisses in these strata (Wan et al. 2007; Zeng et al. 2008). The new obtained data one side indicate a complexity of these lithological assemblages and suggested that these strata were mainly formed in the Neoproterozoic era. These rocks may be roughly divided into two structural layers. The lower structural layer is represented by Chencai Group and Mayuan Group, while the upper structural layer is represented by Longquan Group, Mamianshan Group, and Wanquan Group (Hu et al. 1991; Jin et al. 1997; Jin et al. 2008).

In the traditional ideas, high-grade metamorphic Chencai Group in the southwestern Zhejiang Province and Mayuan Group in the northern Fujian Province may be basically contrasted, of which the Chencai Group in the southwestern Zhejiang Province is mainly composed of aluminum-enriched gneisses, leptites, amphibolites, garnet-mica schists, calc-silicate rocks, and marbles. The Chencai Group was usually divided into four rock assemblages (Kong et al. 1994) and was suggested as khondalite series (Lan et al. 1995), which underwent high amphibolite facies metamorphism (Zhao and Sun 1994; Kong et al. 1994; Lan et al. 1995). Mayuan Group in the northern Fujian Province is mainly composed of sillimanite-bearing mica quartzite schists, garnet biotite plagioclase gneisses, amphibolites, and marbles. Some researches consider that it has undergone a high amphibolite to granulite facies metamorphism (Mei et al. 1993; Zhao and Cawood 1999). Chencai Group and Mayuan Group were traditionally regarded as the metamorphosed basement of the Cathaysian Block, forming in the Paleoproterozoic or Mesoproterozoic (Shui

1988; Shui et al. 1988; Zhao 1999; Kong et al. 1994; Fu et al. 2010). However, recent SHRIMP zircon U-Pb chronological data indicate that these two groups of rocks were mainly formed in the Neoproterozoic. For example, A SHRIMP zircon U-Pb age of 857 ± 7 Ma was obtained for the basalts of Chencai Group (Shu et al. 2011) and 838 ± 5 Ma for a metamorphic rhyolite of this group (Li et al. 2009). Similarly, a SHRIMP zircon U-Pb age of 807 ± 5 Ma was obtained for the volcanic rock in upper Mayuan Group (Wan et al. 2007), and the youngest detrital zircon U-Pb age of 879 Ma was obtained by LA-ICPMS dating method for the paragneiss in lower Mayuan Group (Xu et al. 2010), which indicates that the formative period of Mayuan Group was formed after 879 Ma. Furthermore, the SHRIMP zircon U-Pb dating got 858 ± 11 and 836 ± 7 Ma ages for the metamorphic gabbros in Chencai area(Shu 2006; Shu et al. 2011), and similarly, 841 ± 6 Ma for the gabbro-diorites (Li et al. 2009). All of these data indicate that Chencai Group and Mayuan Group were mainly formed in the early to middle Neoproterozoic era.

Longquan Group, Mamianshan Group, and Wanquan Group in the southwestern Zhejiang Province and the Wuyi mountain range are attributed into the upper structure layer. The Longquan Group is divided into three formations, namely Nannong Formation, Qingkeng Formation, and Wanshan Formation (Xu et al. 2010). The Longquan Group is composed mainly of fine-grained garnet-bearing biotite gneisses, mica-quartz schists, amphibolites, epidote amphibolites and marbles, and their protolith of which are mainly volcanic formation and clastic rock formation. The volcanic rocks of Wanshan Formation exhibit the characteristics of the bimodal volcanic rocks (Xu et al. 2010). Mamianshan Group was divided into three formations, namely Longbeixi Formation, Dongyan Formation, and Daling Formation (Xu et al. 2010). The Mamianshan Group is mainly composed of fine-grained gneisses, mica-quartz schists, actinolite schists, amphibolites, and a bit of marbles. The volcanic rocks of Dongyan Formation also display the characteristic of the bimodal volcanic rocks (Zhang et al. 2005a). Wanquan Group is divided into three formations, which are Dutan Formation, Huangtan Formation, and Xiafeng Formation (Xu et al. 2010), and is composed mainly of fine-grained biotite gneisses, biotite–albite gneisses, and mica-quartz schists, with protolith being intermediate-acid volcanic rocks and pelitic-arenaceous sedimentary rocks. Lithological assemblages in the three groups may be contrasted, and all of these three groups underwent high greenschist to amphibolite facies metamorphism (Mei et al. 1993; Jin et al. 1997; Zhao and Cawood 1999). Except for the above three groups, Louziba Group, Dikou Formation, and Jiaoxi Formation in Fujian Province also were attributed to the upper structure layer of the early to middle Neoproterozoic. Based on whole-rock Sm–Nd isochron ages, these rock groups and formations were considered to be attributed to the Mesoproterozoic to Neoproterozoic in previous investigations (Shui 1988; Gan et al. 1993; Jin et al. 1997; Zhou 1997; Fu et al. 2010). However, recent zircon U-Pb chronological data using SHRIMP and LA-ICPMS dating methods indicate that these rock groups and formations were mainly formed in the Neoproterozoic. For example, SHRIMP zircon U-Pb age of 818 ± 9 Ma was obtained for the acid volcanic rock in Mamianshan Group (Li et al. 2005), while the LA-ICPMS zircon U-Pb age is

818 ± 14 Ma (Xu et al. 2010). The SHRIMP zircon U-Pb age of the fine-grained biotite gneiss in this group is 751 ± 7 Ma (Wan et al. 2007) and the 853 ± 4 and 797 ± 7 Ma for basalts (Shu 2006). LA-ICPMS zircon U-Pb ages are 825 ± 18 and 746 ± 6 Ma for metamorphic volcanic rocks in Wanquan Group (Xu et al. 2010), while the SHRIMP zircon U-Pb ages are 728 ± 8 Ma (Wan et al. 2007), 800 ± 14, and 788 ± 27 Ma (Li et al. 2009). SHRIMP zircon U-Pb ages are 841 ± 12 Ma and 837 ± 8 Ma for the gabbros in Zhenghe County of Fujian Province (Shu et al. 2011). All of these data indicate that the Longquan Group, Mamianshan Group, and Wanquan Group were mainly formed in the early to middle Neoproterozoic and were metamorphosed in Caledonian (Zeng et al. 2008; Li et al. 2010; Hu et al. 2011; Yao et al. 2012).

There exists different understanding about the formation environment of geologic bodies in early to middle Neoproterozoic in the Cathaysian Block. According to their characteristic of bimodal volcanic rocks, some researches considered that the eruptions of volcanic rocks in Longquan Group and Mamianshan Group were simultaneous with the Neoproterozoic magmatism which is very widespread in the Yangtze Block and are related to the rifting caused by the mantle plume concomitant with the breakup of Rodinia supercontinent (Li et al. 2005, 2008a, 2010; Shu et al. 2011). Some researchers suggested that the mafic-ultramafic rocks in Longquan Group and Mamianshan Group along Shangyu-Zhenghe Fault Belt might be the components of ophiolitic melange (Nie and Wang 1992; Ren et al. 1997; Wang et al. 1988; Wang and Shu 2007). These mafic-ultramafic rocks were formed in an island arc environment, which reflect a history from arc–arc collision to continental–continental collision (Wang and Shu 2007). Through the study of their geochemical characteristics, some researchers suggested that the metamorphic sedimentary rocks in Longquan Group, Mamianshan Group, and Wanquan Group were formed in an island arc-active continental margin environment (Jin et al.2008).

2.4 Discussions About Tectonic Environment of Jiangnan Belt

The many bifurcations are preserved with respect to cognitions for early to middle Neoproterozoic tectonic environment in South China Plate, and some tectonic questions have been hotly debated, for example, for long time, and whether a controversial tectonic environment of continental accretion, back-arc basin or rift in the northwestern margin of Yangtze Block (Zhou et al. 2002a, 2006a, b; Du et al. 2005; Sun et al. 2007; Xu et al. 2010; Dong et al. 2011, 2012; Li et al. 2003b, 2006; Wang and Li 2003). Moreover, the tectonic rifting environment was derived from whether a mantle plume or island arc in the Cathaysian Block (Li et al. 2005, 2008b, c, 2009; Shu et al. 2011; Nie and Wang 1992; Ren et al. 1997; Wang et al. 1988; Wang and Shu 2007), and tectonic environment of Jiangnan Belt (Orogen) and the time of orogenesis. Some researchers suggested generally that the Jiangnan

Belt (Orogen) was formed during Sibao period, 1.1–0.9 Ga (Li et al. 2002, 2007a, 2008b; 2009, 2012), whereas other researches thought that the Jiangnan Belt (Orogen) may be formed during 850–820 Ma, which is later than that of the Grenville orogenic period. On account of lacking high-grade metamorphic rocks which are the characteristic in the Grenville Orogen, it is hardly affirmed that Jiangnan Belt (Orogen) was formed in the Grenvillian (Zhou et al. 2008). It is not only controversial in the time of Jiangnan orogenic collision orogeny, but also in its formation type. For its formation ways, different scholars put forward different models. Zhao and Cawood (2012) analyze all kinds of structural models about Jiangnan Belt (Orogen) in detail and considered existing three typical models of "plume-rift model" (Li et al. 2003a, 2006, Li et al. 2003; Wang and Li 2003; Wang et al. 2010a), "slab-arc model" (Zhou et al. 2002a, b, 2006a, b; Wang et al. 2004, 2006, 2007; Zhao et al. 2011; Charvet 2013), and "plate-rift model" (Zheng et al. 2007, 2008), and briefly introduced the main idea and argument for each model, and then pointed out that all these models cannot be convincingly proved or are decisively refuted because every tectonic model successfully explained some characteristic of the Jiangnan (orogenic) belt from the late Mesoproterozoic to late Neoproterozoic. On the basis of analyzing the position of junction, the time of collision and the collision patterns between Yangtze Block and Cathaysian Block, Zhao and Cawood (2012) put forward a divergent double subduction model. They suggested that the ocean between Yangtze Block and Cathaysian Block subducted to Yangtze Block and Cathaysian Block from 970 to 825 Ma, and the ocean closed from 825 to 815 Ma. Meanwhile, the margins of the two blocks gathered together. By reason of the two blocks lying in the upper position of subducted plate, extensive continental subduction and crustal thickening did not occurred. Therefore, there is only greenschist facies metamorphism. Nearly, meanwhile, oceanic lithosphere separated from the continental crust, which caused the overlying continental crust sank along orogenic belt and formed a few of sedimentary basins. In these basins, the strata of Banxi Group were uncomfortably deposited above the lower structural layers such as Sibao Group and Fanjingshan Group. This model also could be regarded as a soft-docking model.

Several basic geologic characteristics must be considered in every model. Firstly, the strata of Jiangnan Belt (Orogen) in early to middle Neoproterozoic are divided into a strong folded lower structural layer (composed by Sibao Group, Fanjingshan Group, Lengjiaxi Group, Shuangqiaoshan Group, Shuangxiwu Group, and Shangxi Group) and an upper structural layer (composed by Banxi Group, Danzhou Group, Xiajiang Group, Xiushui Group, Likou Group, and Heshangzhen Group) with the characteristics of broad and gentle fold. The relationship of these two structural layers displays uncomfortably contacts. The differences in structural features indicate that the lower structural layer underwent strong deformation. Secondly, Jiangnan Belt (Orogen) generally underwent greenschist facies metamorphism, and does not appear high-grade metamorphic rocks like the typical continental–continental collision orogenic belt. High-grade metamorphic rocks only appear in the northeast of Jiangnan Belt (Orogen) (Zhou et al. 1989, Zhou 1997; Gao J. 2001). Most of the strata of Jiangnan Belt (Orogen) solely experienced greenschist facies metamorphism. These imply that the Jiangnan Belt (Orogen) differs from typical continental–continental collision model. Furthermore, high-precision zircon dating data indicated that the northeast of Jiangnan Belt (Orogen) formed at an early time (891–926 Ma, Li et al. 2009), and there is a tendency to be younger in the southwest of Jiangnan Belt (Orogen). For example, Shuangqiaoshan Group in north-center Jiangnan Belt (Orogen) formed during 880–830 Ma (Wang et al. 2008; Gao et al. 2008a), and Lengjiaxi Group, Fanjingshan Group, and Sibao Group in the southwest of Jiangnan Belt (Orogen) formed during 855–822 Ma (Gao et al. 2011a, b; Bai et al. 2010; Zhou et al. 2009). Besides, although the relationship between the upper and lower structural layers is unconformity, their forming ages are very similar (820–780 Ma, see above). These indicate that after the lower structural layer was deformed, quickly the upper structural layer was deposited (Wang et al. 2007). Based on these geological characteristics, I suggest that the ocean between Yangtze Block and Cathaysian Block firstly closed during 891–926 Ma in the northeast of Jiangnan Belt (Orogen), together with a depth of subduction, so it reserved not only ophiolites on behalf of oceanic fragments (Zhou et al. 1989, Zhou 1997; Zhou and Zhao 1991; Zhou et al. 1990; Shen et al. 1992; Li et al. 1994a; Zhao et al. 1995), but also formed blueschists on the behalf of regional high-pressure metamorphism (Zhou et al. 1989, Zhou 1997; Gao et al. 2001). Then the subduction of oceanic slabs migrated towards the southwest. Following the subduction of oceanic slabs, Yangtze Block and Cathaysian Block merged with the soft-docking between continental crusts instead of the subduction of continental crust. The later evolution of the Orogen corresponds with the model suggested by Zhao and Cawood (2012).

3 Nanhua System and Sinian System of South China Plate

Nanhua System and Sinian System have a wide distribution and mainly distribute in the Yangtze Block and Jiangnan Belt (Orogen) of the South China Plate. Nanhua System corresponds to international Cryogenian System and Sinian System to Ediacaran System (MacGabhann 2005).

3.1 Nanhua System of South China Block

On the basis of the domestic and international developing tendency, Nanhua System was established by China National Commission of Stratigraphy at 2001 as a chronostratigraphic unit. Its original intention indicates a stratigraphic unit at system level between Qinbaikou System and Sinian System, and whose bottom boundary is the lower boundary of Doushantuo Formation, with the age of the bottom boundary is tentatively determined at 800 Ma (NCS 2001). Recently, the definition of Nanhua System has gradually been determined to the lower bound of Neoproterozoic glacial records (Zhang 2010), which approximately correspond to

Cryogenian System. The lower bound age of Cryogenian System and Nanhua System depends on how to delimit the lower boundary. At recent, Neoproterozoic stratigraphic branch of international commission on stratigraphy is working on the establishment of Cryogenian System. The Neoproterozoic stratigraphic branch of international commission on stratigraphy put forward a consultative draft about how to define the bottom boundary of Cryogenian System at the end of 2008. Through one-year consultation and a vote among committee members, it is finally put forward that the bottom of Cryogenian System ought to be under a certain oldest Neoproterozoic glacial sedimentary layer on one outcrop and must be able to define a GSSP, whereas some researchers did not agree with the definition of the oldest glacial depositions based on the oldest Neoproterozoic Kaigas Glacier which was the mountain glacier (Zhang et al. 2009). Therefore, the study about Cryogenian System's definition is remaining. At the same time, our country also carried out the study about the establishment of Nanhua System. Some researchers put forward that Yangjiaping Profile in Shimen from Hunan Province could be regarded as a representative profile of Nanhua System (Yin et al, 2004). Some others considered Zhaoxing Profile at Liping County in the southeast of Guizhou Province (Zhang and Chu 2007). There are different opinions about stratigraphic division scenario (Table 1). The difference is mainly about the division and times of the lower strata in Nanhua System. The upper Nantuo Formation is mainly composed of grayish green massive pelitic-arenaceous conglomerates. The gravels are of various sizes, poor sorting, shape diversity, and complex components. The gravels display striations and T-shaped pits, which is typical glacial outwash (Xing et al. 2000). The sedimentary thickness of this formation shows a gradually thick tendency from the Three Gorges to Guizhou Province (Peng et al. 2004; Yin et al. 2007). The mid-upper Datangpo Formation composed mainly of black-dark gray siltstones and silty shales intercalating manganiferous shale and manganiferous dolomite, which were the interstadial deposited products. However, the division of the lower part has bigger bifurcations. The lower part could be regarded as a set of glacial outwash on the whole, and is composed of pebbly sandstones, siltstones, moraineous glutenites and pebbly moraineous mudstones. There is a big divergence on Xieshuihe Formation at Shimen of Hunan Province. This formation is mainly composed of mid-fine grained to mid-coarse grained feldspar-quartz sandstone intercalating pebbly sandstones, siltstones, and slates. Some researchers thought that Xieshuihe Formation is the product of non-glacial period and should be correspond to Fulu Formation (Peng et al. 2004). However, its chemical alteration index (CIA) indicates that it has characteristics of cold environment deposition (Feng LJ et al. 2004). The phenomenon of glacial outwash such as glacier pushing structure and ice foot etching was found (Zhang et al. 2008a). The above data indicate that Xieshuihe Formation formed from cold climate environment. With the above Zhang et al. (2008a) consider that Xieshuihe Formation is only equal to Liangjiehe segment of Fulu Formation, and Fulu Formation and Changan Formation of Jiangkou Group belong to the lower glacial period of Nanhua System, which are corresponding to Sturtian glacier period. Yin et al. (2007) thought that Xieshuihe

Formation is a product of a cold event before the Gucheng glacier period, and therefore should be corresponding to Kaigas glacier period (Yin and Gao 2013).

We can find out that the bottom boundary of Nanhua System is 635 Ma from Table 2. It is consistent with the bottom age of Ediacaran System defined by International Commission on Stratigraphy. There is a big difference because it is not clear yet with respect to the definition of bottom. However, Nanhua System comprised within above Banxi Group and Danzhou Group, and these strata mainly formed at 820–780 Ma. Therefore, the bottom of Nanhua System is delimited at 780 Ma on the latest China stratigraphy table. We can also find out that all the obtained ages from the bottom of Nanhua System are younger than 760 Ma from Table 2. As a result, the age of the bottom of Nanhua System is probably younger than 780 Ma, which need to be further determined. The studies of paleomagnetism indicate that South China Plate is between 33 and 38°N (Li ZX and Powell 1996; Evans et al. 2000) and belongs to a mid-latitude region in the early Nanhua System (Zhang and Piper 1997; Zhang et al. 2009).

In period of Nanhua System, the most important geological event is the global clod event. It is also called "Snowball Earth" (Chu XL 2004). It is generally thought that there are four periods of the clod event in Neoproterozoic. Kaigas Glacier Period is the earliest one at ~ 750 Ma. Sturtian Glacier Period is the second period occurred during 720–680 Ma. Marinoan Glacier Period, which is the third period, occurred during 650–625 Ma. Gaskiers Glacier Period, namely the last period, occurred during 592–580 Ma (Huang et al. 2007; Zhang et al. 2009). It is generally thought that the first and last glacier periods were regional event, while Sturtian Period and Marinoan Glacier Period were global records of "Snowball Earth" (Zheng et al 2003; Chu 2004). The lower Gucheng Formation (Jiangkou Group) and the upper Nantuo Formation of Nanhua System in South China Plate are the sedimentary record of these both "Snowball Earth" events in China. There are different opinions about whether existing glacier records below Gucheng Formation. Some researchers suggested that Chang'an Formation lying from western Hunan Province to eastern Guizhou Province is also a sedimentary record of glacier period, named as "Changan Glacier Period". In this opinion, there are three glacier periods in South China Plate (Peng et al. 2004). Some others thought that of Chang'an Formation and Fulu Formation (including Gucheng Formation), constructing Jiangkou Group is the sedimentary record of Sturtian Glacier Period. In the opinion, there are only two sedimentary records of the global cold events in the South China Plate (Zhang and Chu 2007; Zhang et al. 2009). Although there are different understandings about the reasons of "Snowball Earth" (Hoffman et al. 1998; Hoffman and Schrag 2003; Hude et al. 2000; Schrag et al. 2002; Godderis et al. 2003), however, this special geological event displayed a significant influence on the global tectonic framework, the change of the earth environment, and the evolution of subsequent multicellular metazoan. However, it has a lot of questions to study recently. Zheng et al. (2003) summarized the questions and classified them into seven aspects for further study.

Table 2 Stratigraphic correlation table of Nanhua system in South China Plate

Inter. Comm. Stratigranhy, 2012 MacGabhann, 2005			National Comm. Stratigraphy, 2012			Yin CY et al., 2006,2007 Gao LZ., et al.2013b		Zhang QR et al.,2003,2007,2008
Neoproterozoic	Cambrian 寒武系		Neoproterozoic	Cambrian 寒武系		Shuijingtuo F. 水井沱组		
	541Ma			542Ma				
	Ediacaran 埃迪卡拉系	Gaskicrs G. (582-585Ma)		Sinian S. 震旦系	Upper	Dengying F. 灯影组	550±6Ma(4) 551±0.7Ma(4)	Sinian S. 震旦系
					550Ma			
					Lower	Doushantuo F. 陡山沱组	614±7.6Ma(8) 628±5.8Ma(5) 635±0.57Ma(5)	
	635Ma			635Ma				
	Cryogenian 成冰纪	Marinoan G. (635-660Ma)		Nanhua S. 南华系	Upper	Nantuo F. 南沱组	636±4.9Ma(10) 654.5±3.8Ma(10)	Nantuo F. 南沱组
					660Ma			
					Middle	Datangpo F. 大塘坡组	663±4.3Ma(7) 667±9.9Ma(6) 669±13Ma(11)	Datangpo F. 大塘坡组
		Sturtian G. (680-715Ma)				Gucheng F. 古城组		Jiangkou G. 江口群 / Fulu F. 富禄组 / Gucheng M. 古城段; Liangjiehe M. 两界河段
					725Ma			
		Kaigas G. (735-770Ma)			Lower	Xieshuihe F. 渫水河组	724±12Ma(9) 748±12Ma(3) 758±23Ma(2)	Jiangkou G. 江口群 / Chang'an F. 长安组
				780Ma				
				QingBaikou S. 青白口系		Banxi G. 板溪群	785±19Ma(1) 795±15Ma(1) 809±16Ma(2)	Danzhou G. 丹洲群
	850Ma							
	tonian 拉伸纪							

Annotations: the sources of chronological data in the table: (1) the top of Qingshuijiang Formation in Weng'an, Guizhou Province; the bottom of Maluping Formation in Kaiyang, Guizhou; Beiyixi Formation in Kuruktag, Xinjiang Province (after Yin et al. 2007); (2) Laoshanya Formation and Xieshuihe Formation in Yangjiaping, Shimen, Hunan Provinces (after Yin et al. 2003); (3) the lower-mid Liantuo Formation in the eastern Three Gorges, Hubei Province (after Ma et al. 1984); (4) the bottom of Dengying Formation in Jiuqunao, Zigui, Hubei Provinces (Yin et al. 2005b; Condon et al. 2005); (5) the bottom of Doushantuo Formation in Jiuqunao, Zigui, Hubei Provinces (after Yin et al. 2005a; Condon et al. 2005); (6) the lower part of Datangpo Formation in Heishuixi, Songtao, Guizhou Provinces (after Yin et al. 2006); (7) Datangpo Formation at Langgou Profile in Dongbeizhai from Guizhou Province (after Zhou et al. 2004); (8) the middle of Doushantuo Formation in Zhangcunping, Yichang, Hubei Provinces (after Liu et al. 2009); (9) the top of Liantuo Formation in the eastern Three Gorges, Hubei Province (after Gao and Zhang 2009); (10) the upper and bottom of Nantuo Formation in the eastern Three Gorges, Hubei Province (after Zhang et al. 2008c); (11) the lower Fulu Formation in Zhaoxing, Liping, Guizhou Provinces (after Yin et al. 2008)

3.2 *Sinian System of South China Plate*

Sinian System is an ancient geological terminology and has a history of 130 years in our country. Its connotation has experienced a long-term evolution (Liu 1991; Peng et al. 2012; Liu et al.2012). Today, it is generally thought that Sinian System is a late Proterozoic stratigraphy which lies above the tillite of Nantuo Formation and below Meishucun Stage (Tizhushan segment with small shelly fossils in Dengying Formation) of Early Cambrian (NCS 2002). Some investigators thought that the Sinian System of recent definition is nearly equal to Australian Ediacaran

System, and suggested using world standard "Ediacaran System" instead of the Sinian System (Peng et al. 2012). Oppositely, other researchers consider that the typical biogenic assemblage of Australian Ediacaran System is only equivalent to Miaohe Biota on the upper of Sinian System in China, and lacking lower biogenic associations; therefore, it could not be completely contrasted between the two systems. As a result, Sinian System ought to be reserved (Gao et al. 2013b). Considering long-used history of the Sinian System, this article still makes an introduction in accordance with Sinian System.

Sinian System has a wide distribution in South China Plate, such as Hubei Province, Hunan Province, Guizhou Province, Yunnan Province, Sichuan Province, Chongqing City, Guangxi Province, Jiangxi Province, Zhejiang Province, and Anhui Province. Sinian System includes Doushantuo Formation and Dengying Formation at the standard profile in the east of Three Gorges, Hubei Province. After the end of Nantuo Glacier Period, climate turned warm and ice-snow started to melt. Doushantuo Formation is the first widely transgression depositions at early Sinian System in South China Plate. At first, the grayish-white dolomite was deposited with the characteristics of strong stirring structure. It is commonly known as "cap carbonate," and has a stable horizon and widespread distribution as a sedimentary mark in early Sinian System in South China. Subsequently, gypsolite facies deposition with gypsic horizon appeared, representing a high-energy subtidal deposition in a dry climate. After this, the sea gradually deepened and deposited tabular micrite dolomite intercalating carbonaceous shale with microstratification. Micrite dolomites enriched pyrite and chert nodule, indicating a deep-sea reducing environment. In Late Doushantuo Formation, a set of massive grayish-white dolomites deposited intercalating lentoid chert beds and banded dolomite. Then a set of black silicon argillaceous shale intercalating dolomicrite lenses deposited. Dengying Formation, consisting of a set of thick shallow-deep-shallow carbonatite sedimentary sequence, is conformable overlain on Doushantuo Formation (Xing et al. 2012).

Sinian System is an important stage of biologic evolution. According to the study of the eastern Three Gorges, it could be divided into the early evolution stage of micropaleontology and the later evolution stage of macro-metazoans. In the early stage, it is characterized by extremely prosperous large spinose acritarchs. Spinose acritarchs have a large abundance and a high-degree differentiation. Besides spinose acritarchs, globular and filiform cyanobacteria and multicellular algae were also very prosperous. In the later stage, it is characterized by the appearance of Ediacaran soft-bodied macro-metazoans and extremely prosperous macro-multicellular algae. It represents a beginning of an important biologic evolution stage (Liu et al. 2012).

With the further studies of sequence stratigraphy, biostratigraphy, chemistry stratigraphy, and isotopic chronology in Sinian System, the division of Sinian System becomes more subtle. At first, Sinian System was divided into two stages of the lower Doushantuo Stage and the upper Dengyingxia Stage (Xing et al. 2000; NCS 2002). Afterwards, Wang et al. (2001) divided it into two series with four stages. The lower series of Sinian System comprises Tianjiayuanzi Stage and

Miaohe Stage, and the upper series comprises Sixi Stage and Longdengxia Stage. According to the data of sequence stratigraphy, chemistry stratigraphy, and isotopic chronology, Zhu et al. (2007) divided Sinian System into two series with five stages. The lower series called Xiadong Series comprising the first and second stages. The upper series called Yangtze Series comprising the last three stages. Based on the evolution stage of paleontological groups, Liu et al. (2012) put forward a division scenario, which also has two series with five stages, however, which exists some differences about the specific locations of series and stage between Liu's and Zhu's (2007) scenarios. Although the studies about Sinian System have made advances in the last decade, there are also some questions to further study, such as more subtle divisions of chronostratigraphy and biostratigraphy, detailed correlation about series and stages of Sinian System at different areas of South China Plate, whether Gaskiers Glacier Period has a deposition response in South China Plate, more subtle division of chemistry stratigraphy, and so on.

At the end of Sinian System, South China Plate finished the long-term Precambrian evolution and turned into a Phanerozoic geological evolution stage.

References

Bai D-Y, Jia B-H, Liu W, Chen B-H, Liu Y-R, Zhang X-Y (2010) Zircon SHRIMP U-Pb dating of the igneous rocks from Chengbu, Hunan: constraint on the Neoproterozoic tectonic evolution of the Jiangnan Orogenic Belt. Acta Geol Sinica 84(12):1715–1726 (in Chinese with English abstract)

BGMRGX (Bureau of Geology Mineral Resources of Guangxi Province) (1985) Regional Geology of Guangxi Autonomous Region. Geological Publishing House, Beijing, p 853 (in Chinese with English abstract)

BGMRGZ (Bureau of Geology Mineral Resources of Guizhou Province) (1987) Regional Geology of Guizhou Province. Geological Publishing House, Beijing, p 632 (in Chinese with English abstract)

BGMRHN (Bureau of Geology Mineral Resources of Hunan Province) (1988) Regional Geology of Hunan Province. Geological Publishing House, Beijing, p 718 (in Chinese with English abstract)

BGMRJX (Bureau of Geology Mineral Resources of Jiangxi Province) (1984) Regional Geology of Jiangxi Province. Geological Publishing House,Beijing, p 921 (in Chinese with English abstract)

BGMRSC (Bureau of Geology and Mineral Resources Sichuan Province). 1991. Regional Geology of Sichuan Province. Geological Publishing House, Beijing, p 730 (in Chinese with English abstract)

BGMRZJ (Bureau of Geology Mineral Resources of Zhejing Province) (1989) Regional geology of Zhejiang Province. Geological Publishing House, Beijing, p 688 (in Chinese with English summary)

Charvet J (2013) The Neoproterozoic-Early Paleozoic tectonic evolution of the South China Block: an overview. J Asian Earth Sci 74:198–209

Chen B-L, Dong F-X, Zhang D, Shen T-Y (2002) Analysis of the depositional environment and deformation history of Shuangqiaoshan Group: an example from Dabeiwu area Jiangxi Provinc. Resour Surv Env 23(3):157–168 (in Chinese with English abstract)

Chen W-X, Wang J, Fu X-G, Wang Z-J, Xiong G-Q (2007) Sedimentary characteristics of the Jialu Formation and its underlying granites U-Pb zircon age in southeast Guizhou. China Geol Rev 53(1):126–131 (in Chinese with English abstract)

Chen Y-L, Luo Z-H, Zhao J-X, Li Z-H, Zhang H-F, Song B (2005) Petrogenesis and dating of the Kangding complex Sichuan Province. Sci China (Series D) 48(5):622–634

Chen Z-H, Guo K-Y, Dong Y-G, Chen R, Li L-M, Liang Y-H, Li C-H, Yu X-M, Zhao L, Xing G-F (2009) Possible early Neoproterozoic magmatism associated with slab window in the Pingshui segment of the Jiangshan-Shaoxing suture zone: Evidence from zircon LA-ICP-MS U-Pb geochronology and geochemistry. Sci China: Earth Sci 52(7):925–939

Cheng Y-Q (1994) An introduction to regional geology of china. Geological Publishing House, Beijing, p 517 (in Chinese)

Chu X-L (2004) Snowball earth during the Neoproterozoic. Bull Miner Petrol Geochem 23 (3):233–238 (in Chinese with English abstract)

Condon D, Zhu M-Y, Bowring S (2005) U-Pb Ages from the Neoproterozoic Doushantuo Formation, China. Science 308:95–98

Cui J-T, Han F-L, Zhang S-H, Wang G-B, Wang B-Y, Wang X-P, Peng H-L, Wang J-A, Guo Q-M, Peng A-Y, Dui H-M (2010) Zircon SHRIMP U-Pb dating and the tectonic significance of the Xixiang Group in Southern Qinling. Geol Shaanxi 28:53–58 (in Chinese with English abstract)

Ding B-H, Shi R-D, Zhi X-C, Zheng L, Chen L (2008) Neoproterozoic (~ 850 Ma) subduction in the Jiangnan orogen: evidence from the SHRIMP U-Pb dating of the SSZ-type ophiolite in southern Anhui Province. Acta Petrol Mineral 27(5):375–388 (in Chinese with English abstract)

Dong Y-P, Liu X-M, Santosh M, Zhang X-N, Chen Q, Yang C, Yang Z (2011) Neoproterozoic subduction tectonics of the northwestern Yangtze Block in South China: constrains from zircon U-Pb geochronology and geochemistry of mafic intrusions in the Hannan Massif. Precambr Res 189:66–90

Dong Y-P, Liu X-M, Santosh M, Chen Q, Zhang X-N, Li W, He D-F, Zhang G-W (2012) Neoproterozoic accretionary tectonics along the northwestern margin of the Yangtze Block, China: constraints from zircon U-Pb geochronology and geochemistry. Precambr Res 196–197:247–274

Druschke P, Hanson AD, Yan Q-R, Wang Z-Q, Wang T (2006) Stratigraphic and U-Pb SHRIMP detrital zircon evidence for a Neoproterozoic continental arc, Central China: rodinia implications. J Geol 114:627–636

Du L-L (2010) Age and tectonic environment of Yanbian Group in Southwestern Yangtze Block. Ph.D. thesis of IGG CAS, 1–147 (in Chinese with English abstract)

Du L-L, Geng Y-S, Yang C-H, Wang X-S, Ren L-D, Zhou X-W, Shi Y-R, Yang Z-S (2005) Geochemistry and SHRIMP zircon chronology of basalts from the Yanbian Group in the Western Yangtze Block. Acta Geol Sinica 79(6):805–813 (in Chinese with English abstract)

Du L-L, Guo J-H, Geng Y-S, Yang C-H, Liu F-L, Ren L-D, Zhou X-W, Liu P-H, Meng E, Liu J-H (2013) Age and tectonic setting of Yanbian Group, southwestern Yangtze Block: constraints from clastic sedimentary rocks. Acta Petrol Sinica 29(2):641–672 (in Chinese with English abstract)

Du L-L, Guo J-H, Nutman AP, Wyman D, Geng Y-S, Yang C-H, Liu F-L, Ren L-D, Zhou X-W (2014) Implications for Rodinia reconstructions for the initiation of Neoproterozoic subduction at ~860 Ma on the western margin of the Yangtze Block: evidence from the Guandaoshan Pluton. Lithos 196–197:67–82

Evans DAD, Li Z-X, Kirschvink JL (2000) A high-quality Mid-Neoproterozoic paleomagnetic pole from South China, with implications for ice ages and the breakup configuration of Rodinia. Precambr Res 100:313–334

Fan J-L, Luo J-H, Cao Z-Y, Han W, Zhang J-Y (2010) Feature of geochemistry, isotope chronology and geological significance of Neoproterozoic granitic rocks in southeastern Guizhou China. J Northwes Univ (Nat Sci Ed) 40(4):672–678 (in Chinese with English abstract)

Feng L-J, Chu X-L, Zhang Q-R, Zhang T-G, Li H, Jiang N (2004) New evidence of deposition under cold climate for the Xieshuihe Formation of the Nanhua System in northwestern Hunan. Chin Sci Bull 49:1420–1427

Fu S-C, Ma J-Q, Lin H-C (2010) Lithostratigraphic unit re-division of medium-metamorphosed rocks in Fujian Province China. Geol Bull China 29(8):1152–1159 (in Chinese with English abstract)

Gan X-C, Li H-M, Sun D-Z, Zhuang J-M (1993) Geochronological study on the Precambrian metamorphic basement in northern Fujian. Geol Fujian 12(1):17–32 (in Chinese with English abstract)

Gao J (2001) Rock types, mineral constituents and metamorphic process of high-pressure metamorphic rocks in northeastern Jiangxi Province. Acta Petrol Mineral 20(2):134–145 (in Chinese with English abstract)

Gao L-Z, Yang M-G, Ding X-Z, Liu Y-X, Liu X, Ling H-L, Zhang C-H (2008) SHRIMP U-Pb zircon dating of tuff in the Shuangqiaoshan and Heshangzhen groups in South China: constraints on the evolution of the Jiangnan Neoproterozoic orogenic belt. Geol Bull China 27 (10):1744–1751 (in Chinese with English abstract)

Gao L-Z, Zhang C-H, Liu P-J, Ding X-Z, Wang Z-Q, Zhang Y-J (2009) Recognition of Meso- and Neoproterozoic stratigraphic framework in North and South China. Acta Geosci Sinica 30 (4):433–446 (in Chinese with English abstract)

Gao L-Z, Dai C-G, Liu Y-X, Wang M, Wang X-H, Chen J-S, Ding X-Z (2010a) Zircon SHRIMP U-Pb dating of the tuffaceous bed of Xiajiang Group in Guizhou Province and its stratigraphic implication. Geol China 37(4):1071–1080 (in Chinese with English abstract)

Gao L-Z, Dai C-G, Liu Y-X, Wang M, Wang X-H, Chen J-S, Ding X-Z, Zhang C-H, Cao Q, Liu J-H (2010b) Zircon SHRIMP U-Pb dating of tuff bed of the Sibao Group in southeastern Guizhou-northern Guangxi area, China and its stratigraphic implication. Geol Bull China 29 (9):1259–1267 (in Chinese with English abstract)

Gao L-Z, Chen J, Ding X-Z, Liu Y-R, Zhang C-H, Zhang H, Liu Y-X, Pang W-H, Zhang Y-H (2011a) Zircon SHRIMP U-Pb dating of the tuff bed of Lengjiaxi and Banxi groups, northeastern Hunan: constraints on the Wuling Movement. Geol Bull China 30(7):1001–1008 (in Chinese with English abstract)

Gao L-Z, Ding X-Z, Pang W-H, Liu Y-X, Lu S-N, Liu Y-R, Chen J, Zhang Y-H (2011b) SHRIMP zircon U-Pb dating of metamorphic tuff from the Precambrian Cangxi Complex-Group in northeastern Hunan. Geol Bull China 30(10):1479–1484 (in Chinese with English abstract)

Gao L-Z, Liu Y-X, Ding X-Z, Zhang C-H, Wang Z-Q, Chen J, Liu Y-R (2012) SHRIMP dating of Cangshuipu Group in the middle part of the Jiangnan Orogen and its implications for tectonic evolutions. Geol China 39(1):12–20 (in Chinese with English abstract)

Gao L-Z, Ding X-Z, Yin C-Y, Zhang C-H, Ettensohn FR (2013a) Qingbaikouan and Cryogenian in South China: constraints by SHRIMP zircon U-Pb dating. Acta Geol Sinica 87(6):1540–1553

Gao L-Z, Guo X-P, Ding X-Z, Zong W-M, Gao Z-J, Zhang C-H, Wang Z-Q (2013b) Nanhua glaciation event and its stratigraphic correlation in Tarim Block China. Acta Geosci Sinica 34 (1):1–19 (in Chinese with English abstract)

Gao W, Zhang C-H (2009) Zircon SHRIMP U-Pb ages of the Huangling granite and the tuff beds from Liantuo Formation in the Three Gorges area of Yangtze River, China, and its geological significance. Geol Bull China 28(1):45–56 (in Chinese with English abstract)

Ge W-C, Li X-H, Liang X-R, Wang R-C, Li Z-X, Zhou H-W (2001) Geochemistry and geological implications of mafic –ultramafic rocks with the age of ~825 Ma in Yuanbaoshan-Baotan area of northern Guizhou. Geochimica 30(2):123–130 (in Chinese with English abstract)

Geng Y-S, Yang C-H, Wang X-S, Ren L-D, Du L-L, Zhou X-W (2007) Age of crustalline basement in western margin of Yangtze terrane. Geol J China Univ 13(3):429–441 (in Chinese with English abstract)

Geng Y-S, Yang C-H, Wang X-S, Du L-L, Ren L-D, Zhou X-W (2008) Metamorphic Basement Evolution in Western Margin of Yangtze Block. Geological Publishing House, Beijing, p 215 (in Chinese)

Godderis Y, Donnadieu Y, Nedelec A, Dupre B, Dessert C, Grard A, Ramstein G, Francois LM (2003) The sturtian -snowball glaciation: fire and ice. Earth Planet Sci Lett 211:1–12

Gong S-F, Lin J-X (1987) Stratigraphic classification and correlation of Precambrian strata in Fujian Province. Geol Fujian 2:71–113 (in Chinese with English abstract)

Guo J-Q, You Z-P, Yang J, Shen W-Z, Xu S-J, Wang R-C (1998) Studing on the U-Pb dating of zircon in Tianwan and Pianlugang bodies from Shimian area, west Sichuan. J Mineral Petrol, 18 (1): 91–94 (in Chinese with English abstract)

Hoffman PF, Kaufman AJ, Halverson GP, Schrag DP (1998) A Neoproterozoic snowball Earth. Science 281:1342–1346

Hoffman PF, Schrag DP (2003) The snowball Earth hypothesis: Testing the limits of global change. Terra Nova 14:129–155

Hu X-J, Xu J-K, Tong C-X, Chen C (1991) The Precambrian geology of southwestern Zhejiang Province. Geological Publishing House, Beijing, p 278 (in Chinese with English abstract)

Hu Y-H, Gu G-M, Xu Y, Yu S-Q, Wang J-E, He Y (2011) The confirmation of the age of Caledonian Chencai Group in Zhuji area of Zhejiang province and its geological significance. Geol Bull China 30(11):1661–1670 (in Chinese with English abstract)

Huang J, Chu XL, Zhang Q-R, Feng L-J (2007) Constraints on the age of Neoproterozoic global glaciations. Earth Sci Front 14(2):249–256 (in Chinese with English abstract)

Huang X-B, Yu Z-Z, Zou G-Q (2003) Sedimentary features of the Mesoproterozoic Shuangqiaoshan Group in northwestern Jiangxi. Geol Bull China 22(1):43–49 (in Chinese with English abstract)

Hude WT, Crowley TJ, Baum SK (2000) Neoproterozoic snowball earth simulations with a coupled climate-ice-sheet model. Nature 405:425–429

Jiang X-S, Wang J, Cui X-Z, Zhuo J-W, Xiong G-Q, Lu J-Z, Liu J-H (2012) Zircon SHRIMP U-Pb geochronology of the Neoproterozoic Chengjiang Formation in central Yunnan Province (SW China) and its geological significance. Sci China: Earth Sci 55:1815–1826

Jiang Y-B, Zhang S-H, Wu H-C, Han Y-G (2005) Study on stratum characteristic and geotectonic feature of the Yanbian Group. J East China Inst Technol 28(4):347–353 (in Chinese with English abstract)

Jin S, Zhang L, Zhong Z-Q, Zhou H-W, Xiang H, Zeng W, Liu R, Lu X-Q, Li C-Z (2008) Geochemical Characteristics and Geological Implications for the Neoproterozoic Meta-Sedimentary Rocks of the Cathaysia Block in Zhejiang and Fujian Provinces. Earth Sci 33(6):764–774 (in Chinese with English abstract)

Jin W-S, Sun D-Z, Zhao F-Q (1997) Deep crustal structure and its evolution of South China. Geological Publishing House, Beijing, p 175 (in Chinese)

Kong X-S, Bao C-M, Gu G-M (1994) Discussion for main geological features and tectonic evolution of Chencai Group in Zhuji district Zhejiang province. Geol Zhejiang 10(1):15–29 (in Chinese with English abstract)

Lan Y-Q, Ye Y, Lan X, Shen Z-Y, Shen X-H (1995) Metamorphic geology of khondalite series from Chencai Group in Zhejiang province. J Zhejiang Univ, 29(3): 303–309 (in Chinese with English abstract)

Li F-H, Tan J-M, Shen Y-L, Wang F-X, Zhou G-F, Pan X-N, Li X-Z (1988) The Kangdian Region Pre-Sinian. Chongqing Publishing House, Chongqing, p 396 (in Chinese)

Li J-L, Zhang F-Q, Wang S-X (1983) Rare earth element distribution pattern of Proterozoic ophiolite, Yanbian Sichuan. Petrol Res 3:37–44 (in Chinese with English abstract)

Li J-L (1984) Eugeosyncline rock association of Yanbian Group in western Sichuan, China. Bull Chin Acad Geol Sci 9:21–34 (in Chinese with English abstract)

Li W-X, Li X-H, Li Z-X (2005) Neoproterozoic bimodal magmatism in the Cathaysia Block of South China and its tectonic significance. Precambr Res 136:51–66

Li X-H, Zhou G-Q, Zhao J-X, Fanning CW, Compston W (1994a) SHRIMP Ion Microprobe zircon U-Pb age and Sm-Nd isotopic characteristics of the NE Jiangxi ophiolite and its tectonic implications. Chin J Geochem 13(4):317–325

Li X-H, Zhou G-Q, Zhao J-X, Fanning CM, Compston W (1994b) SHRIMP Ion microprobe zircon U-Pb age of the NE Jiangxi ophiolite and its tectonic implications. Geochemica 23 (2):125–131 (in Chinese with English abstract)

Li X-H, Wang Y-X, Zhao Z-H, Chen D-F, Zhang H (1998) SHRIMP U-Pb zircon geochronology for amphibolete from the Precambrian basement in SW Zhejiang and NW Fujian provinces. Geochemica 27(4):327–334 (in Chinese with English abstract)

Li X-H, Li Z-X, Ge W-C, Zhou H-W, Li W-X, Liu Y (2001) U-Pb zircon ages of the Neoproterozoic granitoids in South China and their tectonic implications. Bull Mineral Petrol Geochem, 20(4): 271–273 (in Chinese with English abstract)

Li X-H, Zhou H-W, Li Z-X, Liu Y, Kinny PD (2001b) Zircon U-Pb age and petrochemical characteristics of the Neoproterozoic bimodal volcanics from western Yangtze block. Geochimica 30(4):315–322 (in Chinese with English abstract)

Li X-H, Li Z-X, Zhou H-W, Liu Y, Kinny PD (2002a) U-Pb zircon geochronology, geochemistry and Nd isotopic study of Neoproterozoic bimodal volcanic rocks in the Kangdian Rift of South China: implications for the initial rifting of Rodinia. Precambr Res 113:135–154

Li X-H, Li Z-X, Zhou H-W, Liu Y (2002b) SHRIPM U-Pb zircon geochronological geochemical and Nd isotopic study of the Neoproterozoic granitoids in southern Anhu. Geol Rev, 48 (Supp): 8–16 (in Chinese with English abstract)

Li X-H, Li Z-X, Ge W-C, Zhou H-W, Li W-X, Liu Y, Wingate MTD (2003a) Neoproterozoic granitoids in South China: crustal melting above a mantle plume at ca 825 Ma? Precambr Res 122:45–83

Li X-H, Li Z-X, Zhou H-W, Liu Y, Liang X-R, Li W-X (2003b) SHRIMP U-Pb zircon age, geohcemsitry and Nd isotope of the Guandaoshan pluton in SW Sichuan: Petrogenesis and tectonic significance. Science in China (Series D), 46(Supp): 73–83

Li X-H, Li Z-X, Sinclair JA, Li W-X, Carter G (2006) Revisiting the Yanbian Terrane : Implication for Neoproterozoic tectonic evolution of the western Yangtze Block, South China. Precambr Res 151:14–30

Li X-H, Li W-X, Li Z-X, Liu Y (2008a) 850-790 Ma bimodal volcanic and intrusive rocks in northern Zhejiang, South China: a major episode of continental rift magmatism during the breakup of Rodinia. Lithos 102:341–357

Li X-H, Li W-X, Li Z-X, Lo C-H, Wang J, Ye M-F, Yang Y-H (2009a) Amalgamation between the Yangtze and Cathaysia Blocks in South China: constraints from SHRIMP U-Pb zircon ages, geochemistry and Nd-Hf isotopes of the Shuangxiwu volcanic rocks. Precambr Res 174:117–128

Li X-H, Li W-X, He B (2012) Bulding of the South China Block and its relevance to assembly and breakup of Rodinia supercontinent: observations, interpretations and tests. Bull Mineral Petrol Geochem 31(6):543–559 (in Chinese with English abstract)

Li Z-X, Powell CM (1996) Break of Rodinia and Gondwanaland andassembly of Asian. Aust J Earth Sci 43(4):591–592

Li Z-X, Li X-H, Kinny P, Wang J (1999) The breakup of Rodinia: did it start with a mantle plume beneath South China. Earth Planet Sci Lett 173:171–181

Li Z-X, Li X-H, Zhou H-W, Kinny PD (2002) Grenvillian Continental collisionin south China: new SHRIMP U- Pb zircon results and implications for the configuration of Rodinia. Geology 30(2):163–166

Li Z-X, Li X-H, Kinny PD, Wang J, Zhang S, Zhou H-W (2003b) Geochronology of Neoproterozoic syn-rift magmatism in the Yangtze Craton, South China and correlations with other continents: evidence for a mantle superplume that broke up Rodinia. Precambr Res 122:85–109

Li Z-X, Wartho JA, Occhipinti S, Zhang C-L, Li X-H, Wang J, Bao C-M (2007) Early history of the eastern Sibao Orogen (South China) during the assembly of Rodinia: New mica 40Ar/39Ar dating and SHRIMP U-Pb detrital zircon provenance constraints. Precambr Res 159:79–94

Li Z-X, Li X-H, Li W-X, Ding S-J (2008b) Was Cathaysia part of Proterozoic Laurentia? —new data from Hainan Island, south China. Terra Nova 20:154–164

Li Z-X, Bogdanova SV, Collins AS, Davidson A, De Waele B, Ernst RE, Fitzsimons ICW, Fuck RA, Gladkochub DP, Jacobs J, Karlstrom KE, Lu S, Natapov LM, Pease V, Pisarevsky SA, Thrane K, Vernikovsky V (2008c) Assembly, configuration, and break-up history of Rodinia: a synthesis. Precambr Res 160:179–210

Li Z-X, Li X-H, Wartho JA, Clark C, Li W-X, Zhang C-L, Bao C (2009b) Magmatic and metamorphic events during the early Paleozoic Wuyi-Yunkai orogeny, southeastern South China: new age constraints and pressure-temperature conditions. Geol Soc Am Bull 122(5–6):772–793

Li Z-X, Li X-H, Wartho JA, Clark C, Li W-X, Zhang C-L, Bao C-M (2010) Magmatic and metamorphic events during the early Paleozoic Wuyi-Yunkai orogeny, southeastern South China: new age constraints and pressure-temperature conditions. Geol Soc Am Bull 122:772–793

Ling W-L, Zhou L, Zhang H-F, Ouyang JP (1996a) Isotopic geochronogy and crustal growth of Proterozoic basement along the northern margin of Yangtze craton: II the Huodiya group. Earth Sci 21(5):495–500 (in Chinese with English abstract)

Ling W-L, Zhang B-R, Zhang H-F, Luo T-C (1996b) An isotopic evidence for oceanic crust subduction and crust-mantle recycling in Meso-Neoproterozoic along northern margin of Yangtze craton. Earth Sci 21(3):332–336 (in Chinese with English abstract)

Ling W-L, Gao S, Zhang B-R, Li H-M, Liu Y, Cheng J-P (2003) Neoproterozoic tectonic evolution of the northwestern Yangtze craton, South China: implications for amalgamation and break-up of the Rodinia Supercontinent. Precambr Res 122:111–140

Liu H-Y (1991) The Sinian System in China. Science Press,Beijing, p 388 (in Chinese)

Liu P-J, Yin C-Y, Gao L-Z, Tang F, Chen S-M (2009a) New material of microfossils from the Ediacaran Doushantuo Formation in the Zhangcunping area, Yichang, Hubei Province and its zircon SHRIMP U-Pb age. Chin Sci Bull 54(6):1058–1064

Liu P-J, Yin C-Y, Chen S-M, Li M, Gao L-Z, Tang F (2012) Discussion on the chronostratigraphic subdivision of the Ediacaran (Sinian) in the Yangtze gorges area, south China. Acta Geol Sinica 86(6):849–866 (in Chinese with English abstract)

Liu S-W, Yan Q-R, Li Q-G, Wang Z-Q (2009b) Petrogenesis of granitoid rocks in the Kangding Comlex, western margin of the Yangtze Craton and its tectonic significance. Acta Petrol Sinica 25(8):1883–1896 (in Chinese with English abstract)

Liu S-W, Yang P-T, Wang Z-Q, Luo P, Wang Y-Q, Luo G-H, Wang W, Guo B-R (2013) LA-ICPMS zircon U-Pb ages and geochemistry of Neoproterozoic low-grade metavolcanic rocks in Wuyuan-Dexing area of northeastern Jianxi Province. Acta Petrol Sinica 29(2):581–593 (in Chinese with English abstract)

Ma G-G, Li H-Q, Zhang Z-C (1984) An investigation of the age limits of the Sinian System in South China. Bull Yichang Inst Geol mineral Resour 8:1–29 (in Chinese with English abstract)

Ma T-Q, Chen L-X, Bai D-Y, Zhou K-J, Li G, Wang X-H (2009) Zircon SHRIMP dating and geochemical characteristics of Neoproterozoic granites in southeastern Hunan. Geol Bull China 36(1):65–72 (in Chinese with English abstract)

MacGabhann (2005) Age constraints on Precambrian glaciations and the subdivision of Neoproterozoic time. IUGS Ediacarian Subcommission Circular, Aug 21

Mei H-L, Zhuang J-M, Yang C-X (1993) *P-T-t* paths and geodynamic implication of Precambrian metamorphic rocks in northern Fujian. Geol Fujian 12:182–194 (in Chinese with English abstract)

NCS (National Committee of Stratigraphy, China) (2001) Guide to stratigraphy in China and Explanation of the guide (revised edition). Geological Publishing House, Beijing, p 1–59 (in Chinese)

NCS (National Committee of Stratigraphy,China) (2002) China Commission on Stratigraphy. Explanatory notes to the regional chronostratigraphic (geochronologic) scale. Geological Publishing House, p 72 (in Chinese)

Nie G-B, Wang Y (1992) Geological features of Precambrian ophiolite in Fujian Province. Geol Fujian 11(1):48–59 (in Chinese with English abstract)

Pei X-Z (1989) The f eatures and the t ect onic significance of rock assemblage in the Bikou group, southern Qinling. J Xi'an Coll Geol 11(2):46–56 (in Chinese with English abstract)
Pei X-Z, Li Z-C, Ding S-P, Li R-B, Feng J-Y, Sun Y, Zhang Y-F, Liu Z-Q (2009) Neoproterozoic Jiaoziding peraluminous granite in the northwest margin of Yangtze Block: Zircon SHRIMP U-Pb age and geochemistry, and their tectonic significance. Earth Sci Front 16(3):231–249 (in Chinese with English abstract)
Peng S-C, Wang X-F, Xiao S-H, Tong J-N, Hua H, Zhu M-Y, Zhao Y-L (2012) A call to replace the chronostratigraphic unit Sinian System(Period) with the global Ediacaran System (Period). J Stratigr 36(1):55–59 (in Chinese with English abstract)
Peng X-J, Liu Y-R, Wu N-J, Chen J-C, Li J-Q (2004) Correlation of the Nanhuan strata on the southern margin of the Yangtze landmass. J strstigr 28(4):354–359 (in Chinese with English abstract)
Ren S-L, Li J-L, Zhou X-H (1997) Petrochemistry and mineral chemistry studies on metamorphic ultramafic rocks in Yanghou area, Zhenghe County Fujian Province. China Geochem 26 (4):13–23 (in Chinese with English abstract)
Schrag DP, Berner RA, Hoffman PF (2002) On the initiation of a snowball Earth Geochem. Geophys Geosys 3:1–21
Shen Q-H, Geng Y-S (2012) The tempo-spatial distribution, geological characteristics and genesis of blueschist belt in China. Acta Geol Sinica 86(9):1407–1446 (in Chinese with English abstract)
Shen W-Z, Zou H-B, Chu X-J, Zhou X-M, Yang J-D, Wang Y-X (1992) Nd-Sr and O isotopic study on Fuchuan ophiolites suite in Anhui Province. Sci Geol Sinica 4:33–341 (in Chinese with English abstract)
Shen W-Z, Li H-M, Xu S-J, Wang R-C (2000) U-Pb chronological study of zircons from the Huangcaoshan and Xiasuozi granites in the western margin of Yangtze plate. Geol J China Univ 6(3):412–416 (in Chinese with English abstract)
Shu L-S, Zhou G-Q, Shi Y-S, Yin J (1994) Study of the high—pressure metamorphic blueschist and its late Proterozoic age in the eastern Jiangnan belt. Chin Sci Bull 39:1200–1204
Shu L-S (2006) Pre-Devonian Tectonic evolution of South China: from Cathaysian Block to Caledonian folded orogenic belt. Geol J China Univ 12(4):418–431 (in Chinese with English abstract)
Shu L-S, Faure M, Jiang S-Y, Yang Q, Wang Y-J (2006) SHRIMP zircon U-Pb age, litho- and biostratigraphic analyses of the Huaiyu Domain in South China: evidence for a Neoproterozoic orogen, not Late Paleozoic-Early Mesozoic collision. Episodes 29:244–252
Shu L-S, Faure M, Yu J-H, Jahn B-M (2011) Geochronological and geochemical features of the Cathaysia block (South China): new evidence for the Neoproterozoic breakup of Rodinia. Precambr Res 187:263–276
Shui T (1988) Tectonic framework of the continental basement of southeast China. Sci Sinica (series .B), 31(1): 885–896
Shui T, Xu B-T, Liang R-H, Qiu Y-S (1988) Metamorphic Basement Geology of the Zhe-Min Region China. Science Press, Beijing, p 85 (in Chinese)
Sun C-M (1994) Genetic mineralogy of pyroxenes from the Yanbian Proterozoic ophiolites (Sichuan, China) and its geotectonic implications. J Mineral Petrol 14(3):1–15 (in Chinese with English anstract)
Sun W-H, Zhou M-F, Zhao J-H (2007) Geochemistry and tectonic significance of basaltic lavas in the Neoproterozoic Yanbian Group, Southern Sichuan Provance, Southwest China. Int Geol Rev 49:554–571
Sun W-H, Zhou M-F (2008) The ~ 860 Ma, Cordilleran-type Guandaoshan dioritic pluton in the Yangtze Block, SW China: implications for the origin of Neoproterozoic magmatism. J Geol 116:238–253
Sun W-H, Zhou M-F, Yan D-P, Li J-W, Ma Y-X (2008) Provenance and tectonic setting of the Neoproterozoic Yanbian Group, western Yangtze Block (SW China). Precambr Res 167:213–236

Sun W-H, Zhou M-F, Gao J-F, Yang Y-H, Zhao X-F, Zhao J-H (2009) Detrital zircon U-Pb geochronological and Lu-Hf isotopic constraints on the Precambrian magmatic and crustal evolution of the western Yangtze Block, SW China. Precambr Res 172:99–126

Tang H-S, Xiao X-D, Liu J-S (1992) The Komatiitic series and its genetic type of Sibao Group in northern Guangxi. Mineral Resour Geol 6(2):126–138 (in Chinese with English abstract)

Wan Y-S, Liu D-Y, Xu M-H, Zhuang J-M, Song B, Shi Y-R, Du L-L (2007) SHRIMP U-Pb zircon geochronology and geochemistry of metavolcanic and metasedimentary rocks in Northwestern Fujian, Cathaysia block, China: tectonic implication and the need to redefine lithostratigraphic units. Gondwana Res 12:166–183

Wang J, Li Z-X (2003) History of Neoproterozoic rift basins in South China: implications for Rodinia break-up. Precambr Res 122:141–158

Wang J, Li X-H, Duan T-Z, Liu D-Y, Song B, Li Z-X, Gao Y-H (2003) Zircon SHRIMP U-Pb dating for the Cangshuipu volcanic rocks and its implications for the lower boundary age of the Nanhua strata in South China. China Sci Bull 48:1663–1669

Wang J (2005) New advances in the study of the Nanhuaan System: with particular reference to the atratigraphic division and correlation of the Nanhuaan System South China. Geol Bull China 24(6):491–495 (in Chinese with English abstract)

Wang J, Zeng Z-G, Chen W-X, Wang Z-J, Xiong G-Q, Wang X-H (2006) The Neoproterozoic rift systems in southern China: new evidence for the sedimentary onlap and its initial age. Sediment Geol Tethyan Geol 26: 1–7 (in Chinese with English abstract)

Wang M, Shu L-S (2007) Rock geochemical characteristics of the Neoproterozoic Wuyishan ophiolite mélange. Geol China 34(4):572–583 (in Chinese with English abstract)

Wang M, Dai C-G, Wang X-H, Chen J-S, Ma H-Z (2011a) In-situ zircon geochronology and Hf isotope of muscovite-bearing leucogranites from Fanjingshan, Guizhou Province, and constraints on continental growth of the southern China block. Earth Sci Front 18(5):213–223 (in Chinese with English abstract)

Wang Q, Wyman DA, Li Z-X, Bao Z-W, Zhao Z-H, Wang Y-X, Jian P, Yang Y-H, Chen L-L (2010a) Petrology, geochronology and geochemistry of ca. 780 Ma A-type granites in South China: petrogenesis and implications for crustal growth during the breakup of the supercontinent Rodinia. Precambr Res 178:185–208

Wang X, Yang S-F, Shi J-N (1988) Discovery of collision recambin Longquan, Zhejiang province and its significance for studying collision orogenic belt in Southeastern China. J Nanjing Univ 24(3):367–378 (in Chinese with English abstract)

Wang X-C, Li Z-X, Li X-H, Li Q-L, Zhang Q-R (2011b) Geochemical and Hf-Nd isotope data of Nanhua rift sedimentary and volcaniclastic rocks indicate a Neoproterozoic continental flood basalt provenance. Lithos 127:427–440

Wang X-F, Ma D-Q, Chen X-H, Wang C-S (2001) Precambrian evolution of the Kuangling Arch, China, and its relation to the Rodinia breakup and assembly. Gondwana Res 4(4):816–817

Wang X-L, Zhou J-C, Qiu J-S, Gao J-F (2004) Geochemistry of the Meso- to Neoproterozoic basic-acid rocks from Hunan Province, South China: implications for the evolution of the western Jiangnan orogen. Precambr Res 135:79–103

Wang X-L, Zhou J-C, Qiu J-S, Zhang W-L, Liu X-M, Zhang G-L (2006) LA-ICP-MS U-Pb zircon geochronology of the Neoproterozoic igneous rocks from Northern Guangxi, South China: implications for tectonic evolution. Precambr Res 145:111–130

Wang X-L, Zhou J-C, Griffin WL, Wang R-C, Qiu J-S, O'Reilly SY, Xu X-S, Liu X-M, Zhang G-L (2007) Detrital zircon geochronology of Precambrian basement sequences in the Jiangnan orogen: dating the assembly of the Yangtze and Cathaysia Blocks. Precambr Res 159:117–131

Wang X-L, Zhao G, Zhou J-C, Liu Y, Hu J (2008) Geochronology and Hf isotopes of zircon from volcanic rocks of the Shuangqiaoshan Group, South China: implications for the Neoproterozoic tectonic evolution of the eastern Jiangnan orogen. Gondwana Res 14:355–367

Wang X-L, Shu X-J, Xing G-F, Xie S-W, Zhang C-H, Xia H (2012) LA-ICP-MS zircon U-Pb ages of the Shijiao-Huangshan intrusive rocks in Zhuji area, Zhejiang Province: implications for the petrogenesis of the ultramafic orbicular rocks. Geol Bull China 31(1):75–81 (in Chinese with English abstract)

Wang Z-J (2008) A proposal to establish the Banxi System and discussion on its foundations: based mainly on studies in eastern Guizhou area. Geol Rev 54(3):296–306 (in Chinese with English abstract)

Wang Z-J, Wang J, Duan T-Z, Xie Y, Zhou J-W, Yang P (2010) Geochronology of middle Neoproterozoic volcanic dep.osits in Yangtze Craton interior of South China and its implications to tectonic settings. Sci China Earth Sic 53(9):1307–1315

Wei C-J (1993) Epidotes from Bikou Group in the Shanxi-Gansu-Sichuan border area and their petrological signifance. Acta Petrol Mineral 12(4):332–340 (in Chinese with English abstract)

Wu R-X, Zheng Y-F, Wu Y-B (2005a) Zircon U-Pb age, element and oxygen isotope geochemisty of Neoproterozoic granites at Shiershan. Geol J China Univ 11(3):364–382 (in Chinese with English abstract)

Wu R-X, Zheng Y-F, Wu Y-B (2005b) Zircon U-Pb age, element and oxygen isotope geochemistry of granodiorite in South Anhui. Acta Petrol Sinica 21(3):587–606 (in Chinese with English abstract)

Wu R-X, Zheng Y-F, Wu Y-B, Zhao Z-F, Zhang S-B, Liu X-M, Wu F-Y (2006) Reworking of juvenile crust: element and isotope evidence from Neoproterozoic granodiorite in South China. Precambr Res 146:179–212

Wu R-X, Zheng Y-F, Wu Y-B (2007) Zircon U-Pb age and isotope geochemistry of Neoproterozoic Jingtan volcanics in South Anhui. Geol J China Univ 13(2):282–296 (in Chinese with English abstract)

Xia L-Q, Xia Z-C, Ma Z-P, Xu X-Y, Li X-M (2009) Petrogenesis of volcanic rocks from Xixiang Group in middle part of South Qinling Mountains. Northwest Geol 42:1–37 (in Chinese with English abstract)

Xia Z-C, Xia L-Q, Ren Y-X (1989) The Study on petrology of marine volcanic rocks of Bikou group f rom southern Qinling mountains. Bull Xi' Inst Geol Mineral Res CAGS 25:97–107 (in Chinese with English abstract)

Xiao L, Zhang H-F, Ni P-Z, Xiang H, Liu X-M (2007) LA-ICP-MS U-Pb zircon geochronology of early Neoproterozoic mafic-intermediat intrusions from NW margin of the Yangtze Block, South China: implication for tectonic evolution. Precambr Res 154:221–235

Xing Y-S, Gao Z-J, Wang Z-Q, Gao L-Z, Yin C-Y (2000) The Neooproterozoic Erathem. Stratigraphical Lexicon of China. Geol Publ House, Beijing, pp 295–440

Xu X-B, Zhang Y-Q, Shu L-S, Jia D, Wang R-R, Xu H-Z (2010a) Precambrian Geochronology and stratigraphy in the Wuyishan area, south China. J Stratigr 34(3):254–267 (in Chinese with English abstract)

Xu X-Y, Xiao Z-C, Xiao L-Q (2002) Volcanic cycles of the Bikou Group and their tectonic implication. Geol Bull China, p 478–495 (in Chinese with English abstract)

Xu X-Y, Chen J-L, Li X-M, Ma P-Z, Wang H-L, Xia L-Q, Xia Z-C, Li P (2010b) Geochemistry and petrogenesis of volcanic rocks from Sanlangpu Formation and Dashigou Formation. Acta Petrol Sinica 26(2):617–632 (in Chinese with English abstract)

Xu X-Y, Li T, Chen J-L, Li P, Wang H-L, Li Z-P (2011) Zircon U-Pb age and petrogenesis of intrusions from Mengzi area in the northern margin of Yangtze plate. Acta Petrol Sinica 27 (3):699–720 (in Chinese with English abstract)

Xu Y-H, Wu X-H, Lou F-S (2008) Division and correlation of the Middle Proterozoic strata in Jiangnan Oldland. Res Surces Surv Environ 29(1):1–11 (in Chinese with English abstract)

Xue H-M, Ma F, Song Y-Q, Xie Y-P (2010) Geochronology and geochemisty of the Neoproterozoic granitoid association from eastern segment of the Jiangnan orogen, China: constraints on the timing and process of amalgamation between the Yangtze and Cathaysia blocks. Acta Petrol Sinica 26(11):3215–3244 (in Chinese with English abstract)

Xue H-M, Ma F, Song Y-Q (2012) Mafic ultramafic rocks from the Fanjingshan region, southwestern margin of the Jiangnan orogenic belt: Ages, geochemical characteristics and tectonic setting. Acta Petrol Sinica 28(9):3015–3030 (in Chinese with English abstract)

Yan Q-R, Wang Z-Q, Hanson AD, Druschke PA, Yan Z, Liu D-Y, Jian P, Song B, Jiang C-F (2003) SHRIMP age and geochemistry of the Bikou volcanic terrane: implications for

Neoproterozoic tectonics on the northern margin of the Yangtze Craton. Acta Geol Sinica 77 (4):479–490

Yan Q-R, Hanson AD, Wang Z-Q, Yan Z, Druschke PA, Wang T, Liu D-Y, Song B, Jiang C-F (2004a) Geochemistry and tectonic setting of the Bikou volcanic terrane on the northern margin of the Yangtze plate. Acta Petrol Mineral 23(1):1–11 (in Chinese with English abstract)

Yan Q-R, Hanson AD, Wang Z-Q, Druschke PA, Yan Z, Wang T, Liu D-Y, Song B, Jian P, Zhou H, Jiang C-F (2004b) Neoproterozoic subduction and rifting on the northern margin of the Yangtze Plate, China: implications for Rodinia reconstruction. Int Geol Rev 46:817–832

Yang L-Z (1990) Middle proterozoic komatiite in northern Guangxi. Reg Geol China 1:14–22 (in Chinese with English abstract)

Yao J-L, Shu L-S, Santosh M, Zhao G-C (2014) Neoproterozoic arc-related mafic–ultramafic rocks and syn-collision granite from the western segment of the Jiangnan Orogen, SouthChina: constraints on the Neoproterozoic assembly of the Yangtze and Cathaysia Blocks. Precambr Res 243:39–62

Yao W-H, Li Z-X, Li W-X, Wang X-C, Li X-H, Yang J-H (2012) Post-kinematic lithospheric delamination of the Wuyi-Yunkai orogen in South China: evidence from ca. 435 Ma high-Mg basalts. Lithos 154:115–129

Ye M-F, Li X-H, Li W-X, Liu Y, Li Z-X (2007) SHRIMP zircon U-Pb geochronological and whole-rock geochemical evidence for an early Neoproterozoic Sibaoan magmatic arc along the southeastern margin of the Yangtze Block. Gondwana Res 12:144–156

Yin C-Y, Liu D-Y, Gao L-Z, Wang Z-Q, Xing Y-S, Jian P, Shi Y-R (2003) Lower boundary age of the Nanhua system and the Gucheng glacial stage: evidence from SHRIMP II dating. Chin Sci Bull 48(6):1657–1662

Yin C-Y, Gao L-Z, Xing Y-S, Wang Z-Q, Tang F (2004) Advances in the study on the Nanhua System of the Neoproterozoic and its stratotype in south China. In: Professional Papers of Stratigraphy and Palaeontology, Geological Publishing House, Beijing, p 1–10 (in Chinese with English abstract)

Yin C-Y, Tang F, Liu Y-Q, Gao L-Z, Liu P-J, Xing Y-S, Yang Z-Q, Wan Y-S (2005a) U-Pb zircon age from the base of the Ediacaran Doushantuo Formation in the Yangtze Gorges, South China: constraint on the age of Marinoan glaciation. Episodes 28(1):48–49

Yin C-Y, Tang F, Liu Y-Q, Gao L-Z, Yang Z-Q, Wang Z-Q, Liu P-J, Xing Y-S, Song B (2005b) New U-Pb zircon ages from the Ediacaran(Sinian) System in the Yangtze Gorges: constraint on the age of Miaohe biota and Marinoan glaciation. Geol Bull China 24(5):393–400 (in Chinese with English abstract)

Yin C-Y, Wang Y-G, Tang F, Wan Y-S, Wang Z-Q, Gao L-Z, Xing Y-S, Liu P-J (2006) SHRIMPIIU-Pb Zircon Date from the Nanhuan Datangpo formation in Songtao County Guizhou Province. Acta Geol Sinica 80(2):273–278 (in Chinese with English abstract)

Yin C-Y, Liu Y-Q, Gao L-Z, Wang Z-Q, Tang F, Liu P-J (2007) Phosphatized Biota in Early Sinian (Ediacaran)-Weng'an Biota and its Environment. Geological Publishing House, Beijing, p 132(in Chinese)

Yin C-Y, Liu P-J, Tang F, Gao L-Z, Yang Z-Q, Song B (2008) New U-Pb zircon age constrains on the age of the interglacial Fulu Formation in Zhaoxing, Liping, Guizhou, South China. Abstracts of the 33rd Intenational Geological Congress, Oslo

Yin C-Y, Gao L-Z (2013) Definition, time limit and stratigraphic subdivision of the Nanhuan System in China. J Stratigr 37(4):534–541 (in Chinese with English abstract)

Yu J-H, O'Reilly SY, Wang L-J, Griffin WL, Zhou M-F, Zhang M, Shu L-S (2010) Components and episodic growth of Precambrian crust in the Cathaysia Block, South China: evidence from U-Pb ages and Hf isotopes of zircons in Neoproterozoic sediments. Precambr Res 181:97–114

Zhang E-P, Niu D-Y, Huo Y-G (1993) Geological Tectonic Features of Qinling -Dabashan and Adjacent Regions. Geological Publishing House, Beijing, 1–291 (in Chinese with English abstract)

Zeng W, Zhang L, Zhou H-W, Zhong Z-Q, Xiang H, Liu R, Jin S, Lu X-Q, Li C-Z (2008) Caledonian reworking of Paleoproterozoic basement in the Cathaysia Block: constraints from zircon U-Pb dating, Hf isotopes and trace elements. Chin Sci Bull 53:895–904

Zhang D, Wu G-G, Peng R-M, Wu J-S, Di Y-J, Zhang X-X, Wang Q-F (2005a) Paleotectonic setting of the Dongyan Formation of the Mamianshan Group in Central Fujian Province. Southeast China Earth Sci Front 12(1):310–320 (in Chinese with English abstract)
Zhang K-B, Chen J-L, Lin H-C, Huang C-Q, Luo Z-X (2005b) Lithostratigraphic division and correlation of the Nanhuanan-Sinian periods in southwestern Fujian. Geol China 32(3):363–369 (in Chinese with English abstract)
Zhang Q-R, Piper JDA (1997) Palaeomagnetic study of Neoproterozoic glacial rocks of the Yangtze Block: Palaeolatitude and configuration of South China in the late Proterozoic Supercontinent. Precambr Res 85:173–199
Zhang Q-R, Chu X-L, Bahlburg H, Feng L, Dobrzinski N, Zhang T (2003) Stratigraphic architecture of the Neoproterozoic glacial rocks in the Xiang-Qian-Gui region of the central Yangtze Block, South China. Prog Nat Sci 13:783–787
Zhang Q-R, Chu X-L (2007) Problems in defining the Nanhua Perios. J Stratigr 31(3):222–228 (in Chinese with English abstract)
Zhang Q-R, Chu X-L, Feng L-J (2008a) A correlation of the Xieshuihe Formation, Nanhua system, with a discussion to its glacial sedimentary stractures. J Stratigr 32(3):246–252 (in Chinese with English abstract)
Zhang Q-R, Chu X-L, Feng L-J (2009) Discussion on the Neoproterozoic glaciations in the South China Block and their related paleolatitudes. Chin Sci Bull 54:1797–1800
Zhang Q-R (2010) The latest trends in the establishment of the Nanhua System/Period. J Stratigr 34(2):165–166 (in Chinese with English abstract)
Zhang S-B, Wu R-X, Zheng Y-F (2012) Neoproterozoic continental accretion in South China: geochemical evidence from the Fuchuan ophiolite in the Jiangnan orogen. Precambr Res 220–221:45–64
Zhang S-H, Jiang G-Q, Dong J, Han Y-G, Wu H-C (2008b) New SHRIMP U-Pb age from the Wuqiangxi Formation of Banxi Group: implications for rifting and stratigraphic erosion associated with the early Cryorogenian (sturtian) glaciation in South China. Sci China 51 (11):1537–1544
Zhang S-H, Jiang G-Q, Han Y-G (2008c) The age of the Nantuo formation and Nantuo glaciation in South China. Terra Nova 20(4):289–294
Zhang Y-Z, Wang Y-J, Geng H-Y, Zhang Y-H, Fan W-M, Zhong H (2013) Early Neoproterozoic (~850 Ma) back-arc basin in the Central Jiangnan Orogen (Eastern South China): geochronological and petrogenetic constraints from meta-basalts. Precambr Res 231:325–342
Zhao F-Q (1999) The chronotectonic framework of Precambrian basements from Cathaysia Block. Prog Precambr Res 22(2):39–46 (in Chinese with English abstract)
Zhao G-C, Sun D-Y (1994) The studies on metamorphic stages and metamorphic *pTD* path of Chencai group, southwestern Zhejiang Province. J Changchun University of Earth Sci 24 (3):246–253 (in Chinese with English abstract)
Zhao G-C, Cawood PA (1999) Tectonothermal evolution of the Mayuan assemblage in the Cathaysia Block: implications for Neoproterozoic collision-related assembly of the South China Craton. Am J Sci 299:309–339
Zhao G-C, Cawood PA (2012) Precambrian geology of China. Precambr Res 222–223:13–54
Zhao J-H, Zhou M-F (2008) Neoproterozoic adakitic plutons in the northern margin of the Yangtze block, China: partial melting of a thickened lower crust and implications for secular crustal evolution. Lithos 104:231–248
Zhao J-H, Zhou M-F, Zheng J-P, Fang S-M (2010) Neoproterozoic crustal growth and reworking of the northwestern Yangtze Block: constraints from the Xixiang dioritic intrusion, South China. Lithos 120:439–452
Zhao J-H, Zhou M-F, Yan D-P, Zheng J-P, Li J-W (2011) Reappraisal of the ages of Neoproterozoic strata in South China: no connection with the Grenvillian orogeny. Geology 39:299–302

Zhao J-X, Li X-H, McCulloch MT, Zhou G-Q, Xing F-M (1995) Petrogenesis of ophiolites from south Anhui and northwast Jiangxi, and their tectonic implications: chemical and Sm-Nd isotopic constraints. Geochimica 24(4):311–326 (in Chinese with English abstract)

Zheng Y-F (2003) Neoproterozoic magmatic activity and global change. Chin Sci Bull 48:1639–1656

Zheng Y-F, Zhang S-B, Zhao Z-F, Wu Y-B, Li X-H, Li Z-X, Wu F-Y (2007) Contrasting zircon Hf and O isotopes in the two episodes of Neoproterozoic granitoids in South China: implications for growth and reworking of continental crust. Lithos 96:127–150

Zheng Y-F, Wu R-X, Wu Y-B, Zhang S-B, Yuan H-L, Wu F-Y (2008) Rift melting of juvenile arc-derived crust: geochemical evidence from Neoproterozoic volcanic and granitic rocks in the Jiangnan Orogen, South China. Precambr Res 163:351–383

Zhu W-G, Zhong H, Li X-H, Deng H-L, He D-F, Wu K-W, Bai Z-J (2008a) SHRIMP zircon U-Pb geochronology, elemental, and Nd isotopic geochemistry of the Neoproterozoic mafic dykes in the Yanbian area, SWChina. Precambr Res 164:66–85

Zhou C-M, Tucker R, Xiao S-H, Peng Z-X, Yuan X-L, Chen Z (2004) New constraints on the ages of Neoproterozoic glaciations on south China. Geology 32(5):437–440

Zhou G-Q, Shu L-S, Wu H-L (1989) The high temperature and pressure metamorphic rocks relatied to ophiolite in northeastern Jiangxi and a discussion on the superimpurition. Acta Petrol Mineral 8(3):220–231 (in Chinese with English abstract)

Zhou G-Q, Zhao J-X (1991) Sm-Nd isotopic systematics of the NE Jiangxi ophiolite (NEJXO), SE margin of the Yangtze Craton South China. Chin Sci Bull 36(16):1374–1379

Zhou G-Q (1997a) Jadeitic rocks within high pressure metamorphic zone from northeastern Jiangxi Province and their forming and surviving conditiongs. Sci China (Series B) 40(5):477–484

Zhou J-B, Li X-H, Ge W-C, Li Z-X (2007) Age and origin of middle Neoproterozoic mafic magmatism in southern Yangtze Block and relevance to the break-up of Rodinia. Gondwana Res 12:184–197

Zhou J-C, Wang X-L, Qiu J-S (2008) Is the Jiangnan orogenic belt a Grenvillian orogenic belt: some problems about the Precanbrian geology of South China. Geol J China Univ 14(1):64–72 (in Chinese with English abstract)

Zhou J-C, Wang X-L, Qiu J-S (2009) Geochronology of Neoproterozoic mafic rocks and sandstones from northeastern Guizhou, South China: Coeval arc magmatism and sedimentation. Precambr Res 170:27–42

Zhou M-F, Yan D-P, Kennedy AK, Li Y-Q, Ding J (2002a) SHRIMP U-Pb zircon geochronological and geochemical evidence for Neoproterozoic arc-magmatism along the western margin of the Yangtze Block, South China. Earth Planet Sci Lett 196:51–67

Zhou M-F, Kennedy AK, Sun M, Malpas J, Lesher CM (2002b) Neoproterozoic arc-related mafic intrusions along the Northern Margin of South China: implications for the Accretion of Rodinia. J Geol 110:611–618

Zhou M-F, Ma Y-X, Yan D-P, Xia X-P, Zhao J-H, Sun M (2006a) The Yanbian terrane (Southern Sichuan province, SW China): a Neoproterozoic arc assemblage in the western margin of the Yangtze block. Precambr Res 144:19–38

Zhou M-F, Yan D-P, Wang C-L, Qi L, Kennedy AK (2006b) Subduction-related origin of the 750 Ma Xuelongbao adakitic complex (Sichuan Province, China): implications for the tectonic setting of the giant Neoproterozoic magmatic event in South China. Earth Planet Sci Lett 248:286–300

Zhou X-H (1997b) Study of isotopic geochemical constraint on tectonic evolution of Southeast China continental crust. In: Jinsheng Yu (ed) Study on Isotopic Geochemistry of China. Science Press, Beijing, pp 83–98 (in Chinese)

Zhou X-M, Zou H-B, Yang J-D, Wang Y-X (1990) Sm-Nd isochrononous age of Fuchuan pphiolite suite in Shexian County, Anhui Province, and its geological significance. Chin Sci Bull 35(3):208–212

Zhu M-Y, Zhang J-M, Yang A-H (2007) Integrated Ediacaran (Sinian) chronostratigraphy of South China. Palaeogeogr Palaeoclimatol Palaeoecol 254:7–61

Zhu W-G, Zhong H, Li X-H, Deng H-L, He D-F, Wu K-W, Bai Z-J (2008b) SHRIMP zircon U-Pb geochronology, elemental, and Nd isotopic geochemistry of the Neoproterozoic mafic dykes in the Yanbian area, SWChina. Precambr Res 164:66–85

Zhuang J-M, Huang Q-Z, Deng B-Z, Guo B, Jie Y-J (2000) Research of Precambrian metamorphic rock petrostrarigraphic division in Fujian. Xiamen University Press, Xiamen, p 94 (in Chinese)

Part IV
The Tarim Craton

Late Archean: Mesoproterozoic Geology of the Tarim Craton

Bei Xu

Abstract Chronological studies show that the Late Archean petrographic formation of the Tarim craton includes Tuogelakbulak complex in the northern margin and the Milan Group in the northern slope of the eastern Altun Mountains. The Tuogelakbulak complex is mainly plagioclase amphibolite, hornblende schist, gray gneiss, garnet biotite schist, actinolite mica quartz schist, biotite quartz schist, and migmatite, mixed with minor amount of marble lenticular bodies, and intruded by gneissic muscovite biotite granite. The Milan Group is composed of various types of felsic gneiss, granulite, marble, and plagioclase amphibolite, and belongs to hypometamorphic supracrustal rock series. The Neoarchean is an important episode of continental crust growth and basement cratonization of the Tarim craton, and a continental crust with the age of about 3.6 Ga might have existed in the Tarim craton basement. The Lower Proterozoic of the Tarim craton is extensively distributed and outcropped at Quruqtagh in the northern margin, Altun Mountains in the eastern margin and Yecheng area in the southwestern margin. Protoliths of the Paleoproterozoic Xingditag Group and Altun Mountains Group are mainly clastic sedimentary rock, carbonate rock, and volcanic formations, indicating a relatively stable and extensive littoral-neritic sedimentary environment. They are mainly composed of a suite of intermediate and high grade metamorphic rocks, including biotite plagioclase gneiss, muscovite biotite plagioclase gneiss, quartz schist, marble, and garnet sillimanite plagioclase quartz gneiss, resembling khondalite series. Magmatism was not significant during this episode, though the Paleoproterozoic magmatic rocks are observed in Altun Dunhuang, Tiekelike and Korla areas. During the Mesoproterozoic episode, the Tarim craton was relatively stable. The Jixianian Aierjigan Group is characterized by carbonate and clastic rock series formed in epicontinental sea. Magmatic formation consists of the Changchengian Sailajiazitage Group and the Jixianian Ailiankate Group formed in intracontinental extension setting.

B. Xu (✉)
Key Laboratory of Orogenic Belts and Crustal Evolution, School of Earth and Space Sciences, Peking University, Beijing 100871, China
e-mail: bxu@pku.edu.cn

M. Zhai (ed.), *Precambrian Geology of China*,
Springer Geology, DOI 10.1007/978-3-662-47885-1_8

Keywords Tarim craton · Late archean · Early proterozoic · Xinjiang

1 Overview

The Precambrian basement of the Tarim craton is composed of four sets of petrographic formations from bottom to top, namely, the Upper Archean, the Lower Proterozoic, the Middle Proterozoic, and the Upper Proterozoic. The Upper Archean petrographic formation is primarily distributed in the Quruqtagh and the northern margin of Altun Mountains; the Lower Proterozoic petrographic formation is mainly discovered in Quruqtagh, the southwestern Tarim and Altun Mountains; the Middle Proterozoic petrographic formation is distributed more widely and is primarily discovered in Quruqtagh, Keping, Altun Mountains, Kunlun Mountains, and the southwestern Tarim; and the Upper Proterozoic petrographic formation is mainly discovered in Quruqtagh, Aqsu and Karghilik (Fig. 1). The Precambrian stratigraphic units in various regions are shown in Table 1.

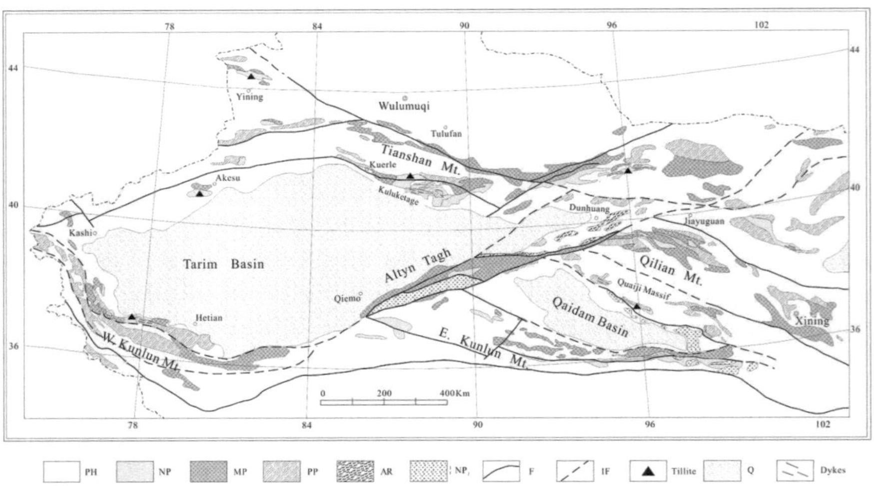

Fig. 1 The Precambrian basement distribution of the Tarim craton (after Lu et al. 2008). *PH* Phanerozoic rocks, *NP* Neoproterozoic rocks, *MP* Mesoproterozoic rocks, *PP* Palaeoproterozoic rocks, *AR* Archaean rocks, *γNP1* early Neoproterozoic granitoids, *F* faults, *IF* inferred Faults, *Tillite* Nanhua and Sinian tillite, *Q* quaternary desert and sedimentary deposits, *Dykes* mafic dyke swarm

Table 1 The Archean to Mesoproterozoic stratigraphic units in the Tarim craton

Eonothem	Erathem	System/series	North Tarim	Northwest Tarim	Southeast Tarim	Southwest Tarim
Proterozoic	Mesoproterozoic	Jixianian	Aierjigan Group	Aqsu Group		Ailiankate Group
		Changchengian	Bowam Group			Sailajiazitage Group
	Palaeoproterozoic		Xingditag Group		Altun Mountains Group	Karakax Group/Heluositan Complex
Archean	Middle-Late Archean		Tuogelakbulak complex		Milan Group/Aketashitage Complex	
	Neoarchean					

According to Lu et al. (2006), Hu and Wei (2006), Liu et al. (2010), Zhang et al. (2013)

2 The Late Archean

2.1 Petrographic Formation

Chronological studies in recent years show that the Late Archean petrographic formation of the Tarim craton includes Tuogelakbulak complex in the northern margin and the Milan Group in the northern slope of the eastern Altun Mountains (Fig. 2).

2.1.1 The Tuogelakbulak Complex

The Tuogelakbulak complex on the northern margin of the Tarim craton outcrops around Tuogelakbulak and Kalakesushuiquan at about 9 km to the south of Xinger Village. The outcropped rocks are mainly plagioclase amphibolite, hornblende schist, gray gneiss, garnet biotite schist, actinolite mica quartz schist, biotite quartz schist, and migmatite, mixed with minor amount of marble lenticular bodies, and intruded by gneissic muscovite biotite granite. The gray gneiss is the main component, and its lithogeochemical characteristic is typical Archean tonalite, trondhjemite, and granodiorite (TTG) assemblage (Hu et al. 1999). The plagioclase amphibolite generally occurs in the gray gneiss in the form of inclusion, and its protolith is continental tholeiitic basalt (Hu and Rogers 1992). To the south of Xinger Village, a relatively complete profile can be observed along the gully, with outcrop up to 800–1000 m, which is unconformably overlaid by the Paleoproterozoic Xingditag Group (Hu and Wei 2006).

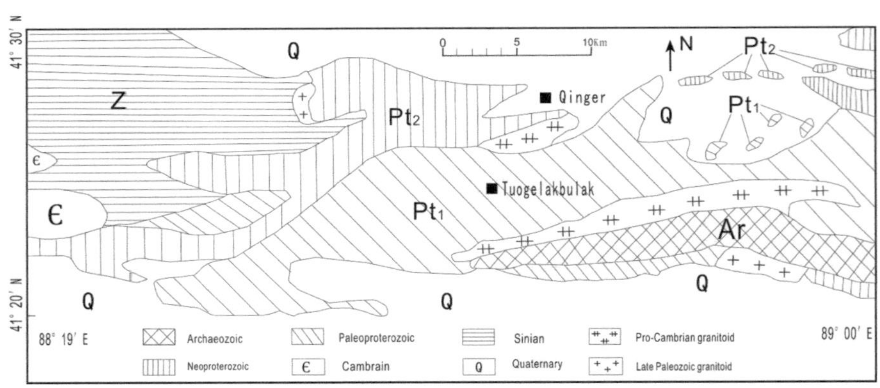

Fig. 2 Tuogelakbulak complex on the northern margin of the Tarim craton (after Hu and Wei 2006)

2.1.2 The Milan Group

The Milan rock group is distributed in the northern slope of the eastern Altun Mountains, i.e., the Dunhuang massif on the eastern margin of the Tarim craton (Fig. 3), exhibiting as EW-trending zonal distribution; and it contacts the Altun Group with a fault and extends to the Lapeiquan-Aketashige area, Qinghai Province. The previously defined Milan Group has complex lithology, mainly composed of tonalite, granulite facies pyroxene plagioclase amphibolite, and potassic migmatitic gneiss. The newly identified Neoarchean Milan Group outcrops in the Dapinggou Aketashitage Simutu region, is composed of various types of felsic gneiss, granulite, marble, and plagioclase amphibolite, and belongs to hypometamorphic supracrustal rock series (Lu et al. 2006; Liu et al. 2010). This rock group has uneven rock texture, structure, and mineral components, with banded and ptygmatic migmatization. Deformation of rocks is intense, and all stratigraphic beddings have been replaced by gneissosity (schistosity) through multiple structural events.

The main rock types are biotite plagiogneiss, hornblende plagiogneiss, biotite leptynite, sillimanite bearing garnet biotite plagiogneiss, garnet bearing plagioclase amphibolite, and plagioclase amphibolite intercalated with garnet websterite plagiogranulite, and hypersthenes biotite plagiogranulite, as well as minor amount of graphite biotite plagiogneiss, quartzite, and marble. The plagioclase amphibolites are lenticular or banded, occurring in TTG rock series and migmatitic gneiss, including plagioclase amphibolite and garnet amphibolite. The felsic gneisses include biotite garnet plagiogneiss, hornblende plagiogneiss, pyroxene hornblende plagiogneiss, and biotite plagioclase leptynite. There are granitic bands in these rocks that are primarily potassic, with width ranging from 0.5 to 5 cm, and the vein bodies and the matrix are basically distributed straightly and alternately, so it is inferred that the vein bodies are products of anatexis. There are two types of marbles: one is coarse grained marble distributed in the north of Aketashitage, including diopside marble, potash feldspar marble, and the marble with minor diagnostic metamorphic mineral; this type of marble contains lenticular bodies or

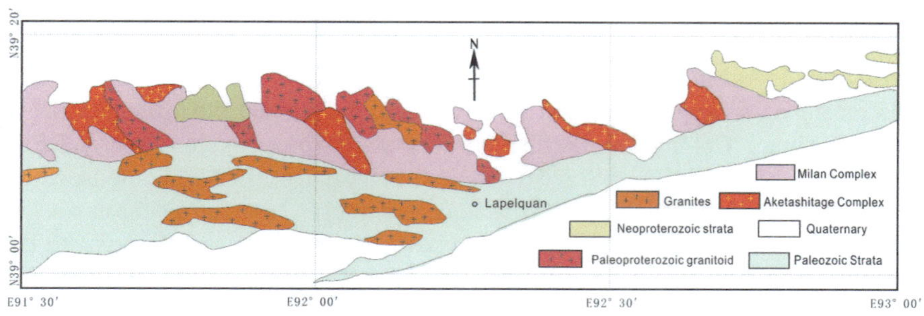

Fig. 3 The Milan Group in the Dunhuang massif on the eastern margin of the Tarim craton (modified from Long et al. 2014)

rock lumps, such as quartzite, pegmatite, garnet pyroxenite, and diabase. Protolith of the marble is likely ultrabasic carbonatite. The other is moderate to fine grained marble with blastic texture, including graphite bearing marble and phlogopite bearing marble; this type of marble is generally lenticular, banded, or laminar, occurring in quartz feldspathic gneiss (Lu et al. 2006).

The rock assemblage of the Milan rock group in the northern margin of the Altun Mountains is dominated by bimodal volcanic rock, mixed with felsic gneiss, plagioclase amphibolite, and leptynite, and locally intercalated with graphite marble and graphite bearing sillimanite garnet biotite plagiogneiss (Liu et al. 2010).

The orthogneisses consist of TTG rocks, monzogranitic, and granitic gneisses in the Dunhuang massif (Liu et al. 2010; Lu 2002). These orthogneisses are called the Aketashitage Complex. The Milan Group and Aketashitage Complex are overlain unconformably by rocks of the Neoproterozoic Annanba Group (BGMRX 1993).

2.2 *Isotopic Ages of the Neoarchean Rocks*

Some scholars have conducted isotope chronological study of the Tuogelakbulak complex using multiple methods. Hu and Rogers (1992) ever reported that the Sm–Nd whole rock age of the plagioclase amphibolite therein was 3.2 Ga, and believed, based on the Sm–Nb isochron age, 3,263 ± 126 Ma, and the Rb–Sr isochron age, 2,854 ± 594 Ma, of the amphibole in the gneiss obtained in the Quruqtagh area, that the hypometamorphic rock series in this area formed in the Early to Middle Archean. Lu (1992) obtained a U–Pb zircon age of 2,582 ± 11 Ma and a Pb–Pb zircon age of 2,488 ± 10 Ma from the granite gneiss of the Tuogelakbulak complex in this area. Hu and Wei (2006) concluded that the U–Pb age of the gray gneiss, the main component of Tuogelakbulak complex in the Xinger area, was 2,565 ± 18 Ma, indicating that these magmatic rocks with TTG components belong to the Late Neoarchean. Guo et al. (2003), based on the captured zircon ages of 2,660 ± 2 Ma and 2,782 ± 4 Ma obtained from the plagioclase amphibolite, proposed that the Neoarchean rocks might have developed in the Early Precambrian basement of the Tarim craton represented by the Quruqtagh area, but they were distributed in the Tuogelakbulak complex mainly in the form of residual inclusion, and the Early Precambrian block dominated by Tuogelakbulak gray gneiss probably formed in the Neoarchean-Paleoproterozoic.

In recent years, several isotope data have been obtained for the orthogneisses of the Aketashitage Complex in the Dunhuang massif. For example, Chen and Sun (1996) published two sets of Sm–Nd isotope isochron ages, including the isochron age of 2,792 ± 208 Ma of basic metamorphic rocks, and the isochron age of 2,787 ± 151 Ma of basic rocks metamorphosed from various types of gneiss, so the granulite formation age was the Early Archean. Mei et al. (1998) found that the TIMS U–Pb zircon age of the granite gneiss in the Dunhuang area was 2,670 ± 12 Ma. Lu et al. (2008) got a large amount of zircon isotopic ages ranging from 2,600 to 2,700 Ma in the TTG rock series around Aketashitage. Lu et al.

(2010) obtains that the SHRIMP ages of 2,567 ± 32 Ma the tonalitic gneiss "wrapping" the Milan Group and 2,705 ± 23, 2,592 ± 15, and 2,020 ± 53 Ma from the biotite plagiogneiss. Long et al. (2014) obtain SHRIMP ages of 2.7–2.8 Ga from tonalitic melanosome of the banded gneisses, and 2.51–2.53 Ga from granitic leucosome, respectively. Zircon SHRIMP U–Pb dating of metadiorites exposed in the Aketashitage Complex yielded a crystallization age of 2498 ± 10 Ma. These data show that formation age of the Milan Group is not later than the Early Archean and that it has probably experienced at least three periods of geological events.

2.3 Geochemistry

The Archean Tuogelakbulak complex outcropping to the south of Xinger in the northern margin of the Tarim craton is mainly composed of gray gneiss, and plagioclase amphibolite, as well as actinolite micaschist, migmatite, and marble. The gray gneiss is the primary component, and its lithogeochemical characteristic is typical Archean TTG assemblage. The plagioclase amphibolite generally occurs in the gray gneiss in the form of inclusion, and its protolith is continental tholeiitic basalt. Hu and Wei (2006) discussed the element geochemical characteristics of various rock types in the Tuogelakbulak complex (see the description below).

(1) Plagioclase amphibolites The plagioclase amphibolite or hornblende schist in the Tuogelakbulak complex in the northern margin of Tarim craton occurs in gray gneiss, schist,, or migmatite in the form of inclusion or relict dike, with variable outcropping area. Their strike is roughly consistent with the direction of the gneissosity. The plagioclase amphibolite is mainly composed of amphibole and plagioclase, among which the amphibole has been generally subjected to epidotization and chloritization alterations. The amphibole + epidote + chlorite account for 70–75 %; the plagioclase accounts for about 10 %, mostly subjected to saussuritization, with minor amount of quartz observed sometimes. The accessory minerals include titanite, apatite, and magnetite. As summarized by Hu et al. (1997), the plagioclase amphibolite type rocks belong to tholeiitic basalt series, in AFM discrimination diagram; and all of them fall within the range of continental basalt in the TiO_2–K_2O–P_2O_5 diagram. The plagioclase amphibolites can be divided into two groups based on element contents: the first group is characterized by high TiO_2 (1.14–1.78 %), low MgO (5.1–6.8 %), low Ni (about 40 ppm), and Cr (about 60 ppm), as well as slight LREE depletion to slight LREE enrichment; the other group contains low TiO_2 (0.59–0.63 %), high MgO (9.20–9.90 %), high Ni (about 145 ppm), and high Cr (540 ppm), with a distribution mode of LREE enrichment. This indicates that their basaltic parent rocks might have formed by fusion of two different provenances.

(2) Gray gneiss and schist Gray gneiss and schist are the rock types of the main components of the Archean complex outcropped to the south of Xinger Village, and wrapped relict masses of melanocratic amphibolite. The rocks are primarily composed of quartz and muscovite, and contain minor amount of biotite, plagioclase,

and microcline. The alteration minerals include actinolite, epidote, chlorite, zoisite, and kaolinite. The accessory minerals include zircon, titanite, and apatite, etc. Both gray gneiss and schist are orthometamorphic rock (Hu et al. 1997), having tonalite trondhjemite granodiorite petrochemical composition (TTG assemblage), in which SiO_2 content is 58–73 %, AI_2O_3 content is 13.7–17.6 % and Na_2O/K_2O is 3.8–1.0. The aluminum saturation index (ASI) ranges from 0.8 to 1.3, with most samples having ASI < 1.1 and about a half of the samples having ASI < 1. In the trace element spidergram, two trondjemitic gneiss samples exhibit negative anomalies of Nb, Sr, P and Ti, which not only indicate that they formed on the continental margin but also show the fractional effects of feldspar, apatite, titanite, and some other opaque minerals.

Lu et al. (2006) found that most granulites, plagioclase amphibolites, gneisses, and leptynites belong to the orthometamorphic rocks and their protolith might be intermediate to basic volcanic rocks, indicating that the Milan Group formed in the volcanic sedimentary rock series on the active continental margin.

(3) Metadiorite The metadiorites in the Aketashitage Complex are characterized by high Sr and Sr/Y and low Y, but their low SiO_2 and high Cr, Co, and Ni indicate similar geochemical characters of modern low-Si adakites. Their high MgO, Cr, Co, Ni, and Mg# and high εHf (t) suggest that the metadiorites were probably produced by partial melting of slab-melt metasomatized mantle peridotites. The orthogneisses in the Aketashitage Complex were generated during three periods of 2.83, 2.71–2.77, and 2.50–2.57 Ga. Except for the metamorphic rims, zircons from the Aketashitage tonalitic and monzogranitic orthogneisses have TDM c model ages ranging from 3.1 to 3.5 Ga, suggesting recycling of Paleoarchean to Mesoarchean crust. The younger TDM c model ages of 2.7–2.9 Ga for the metadioritic rocks suggest another formation period of juvenile crust in the Neoarchean (Fig. 4, Long et al. 2010).

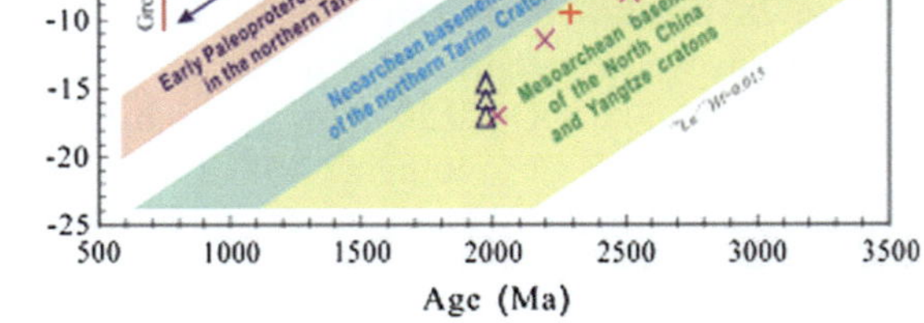

Fig. 4 εNd value and model age from the Tarim plate (after Long et al. 2014)

2.4 Growth of the Archean Primitive Continent

The outcropped plagioclase amphibolites (with a Sm–Nd isochron age of 3,263 ± 129 Ma (2σ), εNd (T)=3.2), and TTG gneiss (with a single-grain zircon U–Pb age of 2,565 ± 18 Ma of gray granodioritic gneiss) in Tuogelakbulak complex to the south of Xinger Fault in the northern margin of the Tarim craton, and blue quartzgranite (zircon U–Pb upper-intercept age is 2,582 ± 11 Ma, Liu et al. 2010) in Quruqtagh area are typical representative Archean rocks, with geochemical characteristics of primitive continental lower crustal rock, i.e., low Rb/Sr ratio (≤0.1), relatively low initial Sr ratio and Nd isotope ratio ($^{143}Nd/^{144}Nb$ is 0.5097–0.5116) and high Nd model age (TDM age = 2.8–3.5 Ga, Hu et al. 1997; Hu and Wei 2006). Zhang et al. (2012) also reported the ages of 2,641 ± 23 and 2, 602 ± 21 Ma from TTG gneiss in Quruqtagh area and based on occurrence of high Ba-Sr granite with the age of 2.53 Ga in this area, they presumed that the there was thickened lower crust with thickness of greater than 30 km, indicating the end of formation process of the Early Precambrian primitive continental crust in the Tarim craton.

Up to now, the oldest age comes from Aketashitage granitic gneiss in Milan Group north of Lapeiquan area in Ruoqiang County in the eastern piedmont of Altun Mountains. Its U–Pb upper-intercept age is 3,605 ± 43 Ma, while Nd model age of the gneiss was only 3.5 Ga (Li et al. 2001). From trondhjemite of Aketashitage area, a SHRIMP U–Pb age of 3,665 ± 15 Ma was obtained for inherited zircon, which indicates that there was the Early Archean crust in the Tarim craton.

Long et al. (2010) studied orthogneiss in Tuogelakbulak complexes from Xinger and Korla area and obtained ages of 2,516 ± 6, 2,575 ± 13 and 2,460 ± 3 as well as 2, 659 ± 15 Ma, indicating that their parent rocks formed in the Late Archean through the Early Proterozoic. The Late Archean orthogneiss has low Sr/Y ratio and low Mg#, depleted Sr, negative anomalies of Nb, Ta, and Ti, exhibiting geochemical characteristics of island-arc volcanic rock; while the Early Proterozoic gneiss is similar to adakite in terms of geochemical characteristics. The Late Archean TTG rock has low εNd value and low Hf content, and its model age is 2.9–3.3 Ga, indicating that the crustal material of Tarim basement rock comes from the Early-Middle Archean depleted mantle and underwent recycling process in the Late Archean. The Early Proterozoic orthogneiss has high εNd value and model age of 2.5–2.7 Ga, revealing the juvenile crust growth event in the Late Archean (Long et al. 2014, Fig. 4).

In summary, the Neoarchean is an important episode of continental crust growth and basement cratonization of the Tarim craton, and a continental crust with the age of about 3.6 Ga might have existed in the Tarim craton basement.

3 The Early Proterozoic

3.1 Stratigraphy

The Lower Proterozoic of the Tarim craton is extensively distributed and outcropped at Quruqtagh in the northern margin, Altun Mountains in the eastern margin and Yecheng area in the southwestern margin (Fig. 1).

The Lower Proterozoic strata of Xinger area in the northern margin of the Tarim craton is called Xingditag Group, the upper and lower parts of it are rhythmic beds composed of muscovite biotite quartz schist, biotite quartz schist, garnet biotite schist, muscovite quartz schist, and quartzite, exhibiting features of turbidite deposition; its middle part is white and yellow marble; with a total thickness of up to 5,000 m, it is unconformally covered by Yangjibulak Group (Lu et al. 2006). Its protolith is neritic clastic rock–carbonate rock. This group has no exact isotope age data.

In Altun area in the eastern margin of the Tarim craton, the Altun Mountains Group mainly consists of metamorphosed supracrustal rock composed of granulite amphibolites to high greenschist facies parametamorphic rock, and can be divided into two rock assemblages: the first rock assemblage is generally characterized by sillimanite, and its main rock types are calcite biotite quartz schist, sillimanite biotite plagioclase amphibolites schist, and garnet sillimanite biotite plagioclasegneiss, intercalated with dolomitic marble, quartzite, and garnet plagioclase amphibolites lenticular bodies; main rock types of the second rock assemblage include garnet biotite quartz schist intercalated with biotite amphibolites monzonite leptynite, metamorphic intermediate-felsic volcanic rocks, and minor amount of garnet epidote tremolite marble (Liu et al. 2010).

Another suite of the Paleoproterozoic sedimentary rock series in the southwestern margin of Tarim is Karakax Group, consisting of greenschist facies biotite quartz schist, chlorite biotite schist, and calcite quartzschist (Lu et al. 2006).

3.2 Magmatic Activities and Isotope Ages

The Paleoproterozoic magmatic rocks are observed in Altun Dunhuang, Tiekelike, and Korla areas.

Lu and Yuan (2003) reported bimodal magmatic assemblage composed of trondhjemite gneiss and amphibolites dikes in Aketashitage area east of Altun Mountains, obtained zircon U–Pb ages of 2,374 ± 10 and 2,351 ± 21 Ma, respectively. Lu et al. (2008) believed that they represent the Early Proterozoic rift setting.

In the southwestern margin of the Tarim craton, the oldest metamorphic rock series is called Heluositan Group, outcropping about 50 km southwest of Qipan Town, Yecheng County, its main rock types are banded, striated, streaky migmatite, magmaticgneiss and granitic gneiss, unconformally covered by the Mesoproterozoic epimetamorphic clastic rock. Its protolith is mainly plagioclase amphibolite, granodiorite, monzonitic granite and muscovite biotite granite (BGMRED of Xinjiang, 1993). According to Hu and Wei (2006), geochemical characteristics of the Heluositan Group granitic gneiss are similar to those of some high-K granite derived from the Paleoproterozoic collision, and its protolith is the collision-related granite. Furthermore, Zhang et al. (2007) reported that the granite of this group exhibited geochemical characteristics of A-S type granite, and the plagioclase amphibolite is featured by intraplate basalt, therefore, it should form in intracontinental extension tectonic setting. In addition, the granitic gneiss of Heluositan Group from Tiekelike uplift in the southern margin of Tarim is likely to form in the intracontinental extension tectonic setting. Some important data have been obtained from Heluositan Group granitic gneiss. The zircon U–Pb upper-intercept age of granitic gneiss is 2,199 ± 71 Ma (Hu and Wei 2006), and for the granite of this group, a zircon age of 2,426 ± 46 Ma was obtained in Qipan area while another age of 2,358 ± 10 Ma was obtained in Xuxugou area (Zhang et al. 2003). Nd model age of the Heluositan Group granitic gneiss is 2.6–2.8 Ga, indicating that the protolith of this granitic gneiss is the Archean rock (BGMRED of Xinjiang 1993), however, the zircon U–Pb age of older than 2.5 Ga has not yet obtained up to now.

3.3 Sedimentation, Intrusion, Anatexis, and Metamorphism

Sedimentation In the Paleoproterozoic episode, the first suite of cover strata, “khondalite series” as previously called, represented by Xingditag Group and Altun Mountains Group. Their protoliths are mainly clastic sedimentary rock, carbonate rock, and volcanic formations, indicating a relatively stable and extensive littoral-neritic sedimentary environment. The protolith of muscovite quartz schist in Xingditag Group is volcanic rock, and its whole rock Pb–Pb isochron age is 2,399 ± 63 Ma, representing the earliest volcanic eruption sedimentation metamorphism in the northern margin of Tarim craton during the Paleoproterozoic era (Lu et al. 2006).

Intrusion Magmatism was not significant during this episode. The restite of plagioclase amphibolites gneiss and biotite amphibolites plagioclase gneiss outcropping in Kuokesu area south of Xingdi Fault has a Sm–Nd isochron age of 2,231 ± 140 Ma and εNd (T) = +4.2, probably representing a crustal growth event (Hu et al. 1997).

Anatexis This is represented by anatexis bands occurring in Heluositan Group and Akazi pluton with a zircon age of 1,916 ± 4 Ma for the secondary growth edge, reflecting anatexis time (Lu et al. 2006).

Metamorphism Zhang et al. (2012) pointed out that, in Quruqtagh and Altun areas, Xingditag Group ,and Altun Mountains Group are mainly composed of a suite of intermediate and high grade metamorphic rocks, including biotite plagioclase gneiss, muscovite biotite plagioclase gneiss, quartz schist, marble, and garnet sillimanite plagioclase quartz gneiss, resembling khondalite series. Massive light bands, derived from light-colored metamorphic differentiation, were developed in rock. Ages of inherited zircon from this suite of metamorphic rock are mainly the Archean to the Early Paleoproterozoic, while ages of its metamorphic zircon (partially metamorphosed single grain zircon and partially edges of detrital zircon) center on 2.0–1.8 Ga (Lu et al. 2008). Zhang et al. (2007) reported that, among 2.4–2.3 Ga granite in the southwestern Tarim, zircon edge age was 1.9 Ga; while in the Mesoproterozoic Karakax Group and Ailiankate Group, massive 2.0–1.8 Ga detrital zircon occurred. Zhu et al. (2011) also reported occurrence of 1.9 Ga detrital zircon in metamorphic rock of Akesu Group. All such information demonstrated that, important tecto-metamorphic events took place in the Tarim craton during 2.0–1.8 Ga (Zhang et al. 2012).

4 The Mesoproterozoic

During the Mesoproterozoic episode as long as 0.6 Ga, the Tarim craton was relatively stable, characterized by carbonate and clastic rock series formed in epicontinental sea and sedimentary and magmatic formation in intracontinental extension setting. The former is represented by Yangjibulak Group, Bowam Group, and Aierjigan Group in the northern margin of the Tarim craton, while the latter includes the Mesoproterozoic Changchengian Sailajiazitage Group and the Ailiankate Group in the southwestern margin of the Tarim craton (Table 1).

4.1 Stratigraphy

The Bowam Group is distributed over Bowambulak area, southwest of Quruqtagh, Xinjiang, belonging to epimetamorphic rock series. Its lower part is quartzite, marble, and mica schist, with a thickness of about 2,000 m, while its bottom part has a suite of 60 m thick basal conglomerate in discordant contact with underlying the Xingditag Group. Its upper part is quartzite interbedded with epidote biotite schist and muscovite biotite schist, intercalated with marble, with a thickness of about 550 m, and in discordant contact with overlying Aierjigan Group (Gao et al. 1993; Lu et al. 2006).

The Jixianian Aierjigan Group consists of neritic carbonate rocks, including dolomitic marble, calcareous schist, silicified limestone, sericite quartz schist, sandstone, siltstone, and argillaceous limestone, with a total thickness of more than 2,000 m.

4.2 *Magmatic Activities*

The Mesoproterozoic in the southwestern margin of Tarim craton consists of the Changchengian Sailajiazitage Group and the Jixianian Ailiankate Group. The Sailajiazitage Group is a suite of epimetamorphic spilitekeratophyre intercalated with clastic rock and carbonate rock, with thickness ranging from hundreds to thousands of meters. Main rock types include spilite, quartz porphyry, quart zkeratophyre, felsites, basalt and tuff, intercalated with sandstone, shale and limestone. This suite of rock assemblage and its lithogeochemical characteristics demonstrate that this group formed in continental breakup setting and its magma originated from EMI-type enriched mantle (Lu et al. 2006).

The lower part of Ailiankate Group is a suite of bimodal volcanic rocks with thickness ranging from about 2,500 to 3,800 m, mainly composed of metarhyolite, crystalloblastic tuff, and bedded plagioclase amphibolite or plagioclase amphibolite schist intercalated with marble. Geochemical analysis of basic rocks suggests that they belong to MORB type basalt or shoshonite, respectively, representing extension tectonic setting. The zircon collected from metabasalt showed a Sm–Nd isotope isochron age of 1,200 ± 82 Ma, and is ascribed to the Jixianian. The upper part of Ailiankate Group is about 4,500 m-thick metaflysch unit that is siltstone phyllite-based turbidite series with Bouma sequence, while its top part is conglomerate (Lu et al. 2006).

The Sailajiazitage Group is a suite of epimetamorphic spilite keratophyre intercalated with clastic rock and carbonate rock assemblages, with thickness ranging from hundreds to thousands of meters. Main rock types include spilite, quartz porphyry, quartzkeratophyre, felsites, basalt and tuff intercalated with sandstone, shale, and limestone. This suite of rock assemblages and its lithogeochemical characteristic demonstrate that this group formed in continental breakup setting and its magma originated from EMI-type enriched mantle (Lu et al. 2006).

References

BGMRX (Bureau of Mineral Resources of Xinjiang Uygur Autonomous Region) (1993) Regional Geology of Xinjiang Uygur Autonomous Region, People's Republic of China, Ministry of Geology and Mineral Resources, Geological Memoirs, Series 1, No. 32. Geological Publishing house, pp 1–667 (in Chinese with English abstract)

Chen Z-C, Sun Y (1996) The age of the Altun granulite facies complex and the basement of the Tarim Basin. Reg Geol China 56:51–57 (in Chinese with English abstract)

Gao Z-J, Chen J-B, Lu S-N, Peng C-W, Qin Z-Y (1993) The Precambrian geology in northern Xinjiang. Precambrian geology, No. 6. Geological Publishing House, Beijing, pp 1–171 (in Chinese)

Guo Z-J, Zhang Z-C, Liu S-W, Li H-M (2003) U-Pb geochronological evidence for the early Precambrian complex of the Tarim Craton, NW China. Acta Petrologica Sinica 19(3):537–542

Hu A-Q, Rogers G (1992) Disovery of 3.3 Ga Archaen rocks in north Tarim Block of Xinjiang, western China. Chin Sci Bull 37(18):1546–1549

Hu A-Q, Wang Z-G, Tu G-Z (1997) Geological evolution and diagenic and metallogenetic regularity in Northern Xinjiang. Science Press, Beijing (in Chinese)

Hu A-Q, Zhang Q-F, Zhang G-X (1999) Age and crustal growth of the basement of Tianshan Orogenic Belt: constraints from Nd isotope compositions. Sci China (Series D) 29:104–122

Hu A-Q, Wei G-J (2006) On the Age of the Neo-Archean Qingir Gray Gneisses from the Northern Tarim Basin, Xinjiang, China. Acta Geologica Sinica 80(1):126–134

Li H-M, Lu S-N, Zheng J-K (2001) 3.6 Ga zircon in granitic gneiss from the eastern Altyn Tagh and its geological implication. Bull Mineral Petrol Geochem 20:259–262 (in Chinese with English abstract)

Liu Y-S, Xin H-T, Zhou S-J, Teng X-J, Yang J-Q, Lv H-Q (2010) Precambrian to Paleozoic Tectonic evolution in the Lapeiquan Area, Eastern Altyn Tagh. Geological Publishing House, pp 1–211 (in Chinese)

Long X-P, Yuan C, Sun M, Zhao G-C, Xiao W-J, Wang Y-J, Yang Y-H, Hu A-Q (2010) Archean crustal evolution of the northern Tarim craton, NW China: Zircon U-Pb and Hf isotopic constraints. Precambr Res 180:272–284

Long X-P, Cao Y, Sun M, Kröner A, Zhao G-C (2014) New geochemical and combined zircon U-Pb and Lu–Hf isotopic data of orthogneisses in the northern Altyn Tagh, northern margin of the Tibetan plateau: Implication for Archean evolution of the Dunhuang Block and crust formation in NW China. Lithos 200–201:418–431

Lu S-N (1992) Geological evolution of Proterozoic in Kuruktage, Xinjiang. Bull Tianjin Inst Geol Miner Resour 26–27:279–292 (in Chinese)

Lu S-N (2002) Preliminary Study of Precambrian Geology in the North Tibet–Qinghai Plateau. Geological Publishing House, pp 1–125 (in Chinese)

Lu S-N, Yuan G (2003) Geochronology of early Precambrian magmatic activities in Aketasdhtage, east Altyn tagh. J Acta Geol Sinica 77:61–68 (in Chinese)

Lu S-N, Yu H-F, Li H-K (2006) Research on Precambrian major problems in China. Geological Publishing Press, pp 1–206 (in Chinese)

Lu S-N, Li H-K, Zhang C-L, Liu G-H (2008) Geological and geochronological evidence for the Precambrian evolution of the Tarim Craton and surrounding continental fragments. Precambr Res 160:94–107

Mei H-L, Yu H-F, Lu S-N (1998) Archean tonalite in the Dunhuang, Gusu Province: age from the U-Pb single zircon and Nd isotope. Progr Precambr Res 21:41–45 (in Chinese with English abstract)

Zhang C-L, Zhao Y, Wang A-G, Dong Y-G (2003) Geochemistry of Meso-Proterozoic volcanites in Weatern Kunlun Mountains: evidence for the plate tectonic evolution. Acta Geol Sinica 77:180–190

Zhang C-L, Lu S-N, Yu H-F, Ye H-M (2007) Tectonic evolution of Western Orogenic belt: evidences from zircon SHRIMP and LA-ICP-MS U-Pb ages. Sci China (series D) 50:825–835

Zhang C-L, Santosh M, Li H-K, Li Z-X, Zou H-B, Wang H-Y, Ye H-M (2012) Precambrian evolution and cratonization of the Tarim Block, NW China: Petrology, geochemistry, Nd-isotopes and U-Pb zircon geochronology from Archaean gabbro-TTG-potassic granite suite and Paleoproterozoic metamorphic belt. J Asian Earth Sci 47:5–20

Zhang C-L, Li H-K, Zou H-B (2013) Tectonic framework and evolution of the Tarim Block in NW China. Gondwana Res 23(4):1306–1315

Zhu W-B, Zhang B-H, Shu L-S, Ma D-S, Wu H-L, Li Y-X (2011) Neoproterozoic tectonic evolution of the Precambrian Aksu blueschist terrane, northwestern Tarim, China: insights from LA-ICP-MS zircon U-Pb ages and geochemical data. Precambr Res 185:215–230

The Neoproterozoic Geology of the Tarim Craton

Bei Xu

Abstract The Upper Proterozoic of the Tarim craton can be divided into the upper and lower parts. The lower part belongs to the Qingbaikouan System (1.0–0.85 Ga), including epimetamorphic carbonate–clastic rock series in the northeastern and southwestern margins of the craton, and blueschists in the northwestern margin of the craton. The upper part is the Nanhuan System and the Sinian System (0.85–0.54 Ga), which occur in Aksu area of northwest Tarim, Quruqtagh area of northeast Tarim, and Yecheng area of southwest Tarim, preserving rich and varied geological records, characterized by the development of rift sedimentary formation, multiphase tillites, and magmatic activities. Four glacial periods have been recognized in Quruqtagh area, including Bayixi glaciation of 740–732 Ma, Altungal and Tereeken glaciations of 725–615 Ma, and Hankalchough glaciation of 615–542 Ma, and their international correlation has been discussed. Tarim Movement results in a stable cratonic crustal structure in the Tarim craton and is also an intense response of the Tarim craton to Rodinia supercontinent event. The Late Proterozoic magmatic activities in the Tarim craton are characterized by wide distribution and multiple episodes, and are divided into four developing phases including 820–800, 780–760, 740–730, and 650–630 Ma.

Keywords Nanhuan system · Sinian system · Glacial period · Rodinia

1 Overview

The Upper Proterozoic of the Tarim craton can be divided into the upper and lower parts (Gao and Zhu 1984). The lower part belongs to the Qingbaikouan System (1.0–0.85 Ga), including two types of formations: one is epimetamorphic carbon-

B. Xu (✉)
Key Laboratory of Orogenic Belts and Crustal Evolution, School of Earth and Space Sciences Peking University, Beijing 100871, China
e-mail: bxu@pku.edu.cn

M. Zhai (ed.), *Precambrian Geology of China*,
Springer Geology, DOI 10.1007/978-3-662-47885-1_9

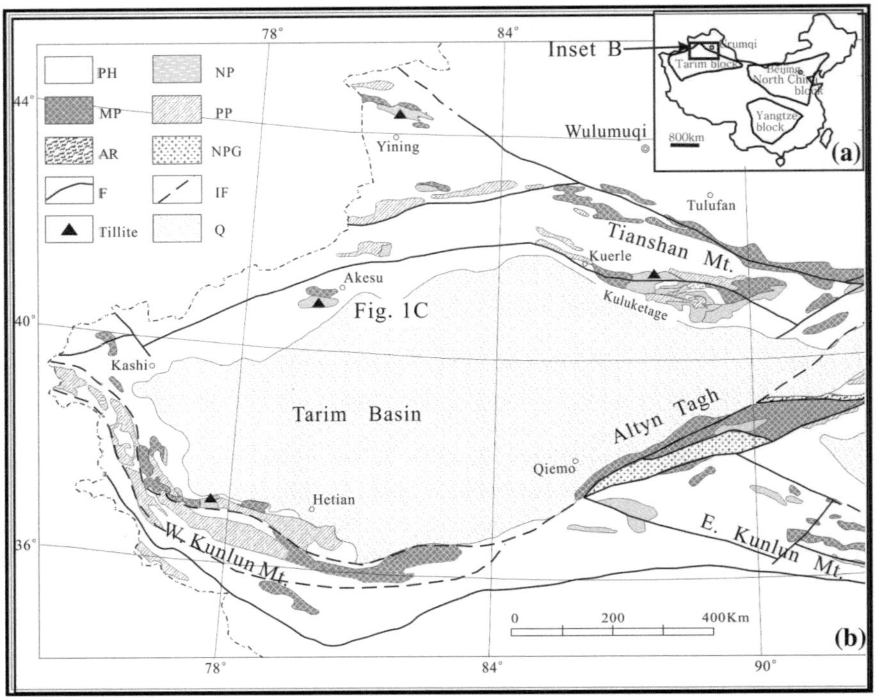

Fig. 1 Precambrian strata in the Tarim craton (after Lu et al. 2008)

ate–clastic rock series, including the Pargangtag Group in the northeastern margin of Tarim, the Silu Group in the southwestern margin of Tarim, and the Aksu Group in the northwestern margin of Tarim (Fig. 1; Table 1).

The upper part is the Nanhuan System and the Sinian System (0.85–0.54 Ga) preserving rich and varied geological records, characterized by development of rift sedimentary formation, multiphase tillites, and magmatic activities, primarily outcropping at Quruqtagh area in the northeastern margin, Aksu area in the northwestern margin, and Yecheng area in the southwestern margin of Tarim (Fig. 1).

2 Qingbaikouan System

2.1 *Stratigraphy*

2.1.1 The Pargangtag Group

The Pargangtag Group in the northeastern margin of Tarim is distributed from Pargangtag to the west of Xinger, and has huge thickness. Its lower part is clastic

Table 1 The upper Proterozoic of the Tarim craton

Eonothem	Erathem	System/series	North Tarim	Northwest Tarim		Southwest Tarim
Phanerozoic	Paleozoic	Lower Cambrian	Xishanbulake Fm	Yuertus Fm		
Proterozoic	Neoproterozoic	Sinian	Hangelchaok Fm	Wushnanshan group	Qigbulak Fm	Kezisuhumu Fm
			Shuiquan Fm		Sugetbulak Fm	Kurkak Fm
			Yukengou Fm			
			Zhamoketi Fm			
		Nanhuan	Tereeken Fm		Yourmeinak Fm	Yutang Fm
			Altungal Fm			Kelixi Fm
			Zhaobishan Fm			Bolong Fm
			Beiyixi Fm			Yalaguz Fm
		Qingbaikouan	Pargangtag group	Qiaoenbulake group		Silu group

(According to Lu et al. 2006; Hu and Wei 2006; Liu et al. 2010; Zhang et al. 2013)

Fig. 2 Unconformity between the Pargangtag Group and Nanhua system (the Pargangtag Group is below *red line*)

rock series, composed of grayish-green marine-facies fine sandstone, siltstone, and pelite; its upper part is carbonate rock, including light marble and crystalline limestone, and the top part is unconformably covered by the Nanhuan System (Fig. 2).

2.1.2 The Silu Group

The Silu Group in the southwestern margin of Tarim is divided into the Bochatetag Formation, Sumalan Formation, and Sukuluoke Formation from bottom to top, abundant in stromatolite, and formed in a setting transforming from lagoon to shallow sea. The Bochatetag Formation is primarily a suite of about 1000 m-thick magnesian dolomite with a layer of coarse clastic rock in the bottom part. The Sumalan Formation is an assemblage of bright-colored carbonate rock intercalated with fine clastic rock, as thick as about 500 m. With thickness of about 1000 m, the Sukuluoke Formation consists of a suite of fine clastic rock assemblages of marine-facies fine sandstone, siltstone, pelite, and silicalite interbeds, in parallel unconformity or slight angular unconformity contact with its overlying Nanhuan System (BGMRED of Xinjiang 1993; Lu et al. 2006).

2.2 *Tarim Movement*

During 1.0–0.9 Ga, an intense tectonic movement took place in the Tarim craton, referred to as Tarim Movement, basically represented by deformation of the Mesoproterozoic carbonate rock clastic rock series, angular unconformity between the Mesoproterozoic and Neoproterozoic, and regional metamorphism, resulting in a relatively stable cratonic crustal structure in the Tarim craton. This Movement is also an intense response of the Tarim craton to Rodinia supercontinent aggregation event (Lu et al. 2006).

Lu et al. (2006) pointed out that under the unconformities there are intense metamorphism and deformation, with developed 1.0–0.9 Ga metamorphic rock and gneissic granite, suggesting general existence of Tarim Movement. The Early Precambrian amphibolite granulite facies metamorphic rock in the northern margin of Tarim preserves evidence of 1.1–0.9 Ga metamorphic events (Zhang et al. 2012). About 1.0 Ga gneissic granite was discovered in the northern margin of Tarim (Shu et al. 2011). In the southern margin of Tarim, the Mesoproterozoic to early Neoproterozoic metavolcanic rock and sedimentary rock developed tight fold structures and foliation replacement, while amphibolite and biotite in metavolcanic rock of the Ailiankate Group in the southwestern margin of Tarim, obtained Ar/Ar plateau ages of 1,051 ± 1 Ma and 1,021 ± 1 Ma, respectively, believed to be geological record of back-arc basin subduction (Lu et al. 2006). Based on these structures, metamorphic event, and previous paleomagnetic data, Zhang et al. (2012) believed that the Tarim Movement represents the tectonic process of the Tarim craton being accreted to the northern margin of Australia and becoming part of Rodinia supercontinent.

3 The Nanhuan System and Sinian System

The Neoproterozoic Nanhuan System and Sinian System occur in Aksu area of northwest Tarim, Quruqtagh area of northeast Tarim, and Yecheng area of southwest Tarim, which are characterized by tillites, volcanic, clastic, and carbonate rocks (Fig. 1; Gao and Zhu 1984; Gao and Qian 1985; Xu et al. 2005, 2009).

3.1 *Stratigraphy in Quruqtaghh Area*

The Neoproterozoic strata crop out in Xishankou, Xinger, and Saimashan areas along the northeastern margin of the Tarim Block (Fig. 3a), which unconformably overlie the Mesoproterozoic Paergangtage Group and are unconformably overlain by the lower Cambrian siliceous rocks. The thickest and most complete Neoproterozoic sections including all eight formations occur in the Xinger area,

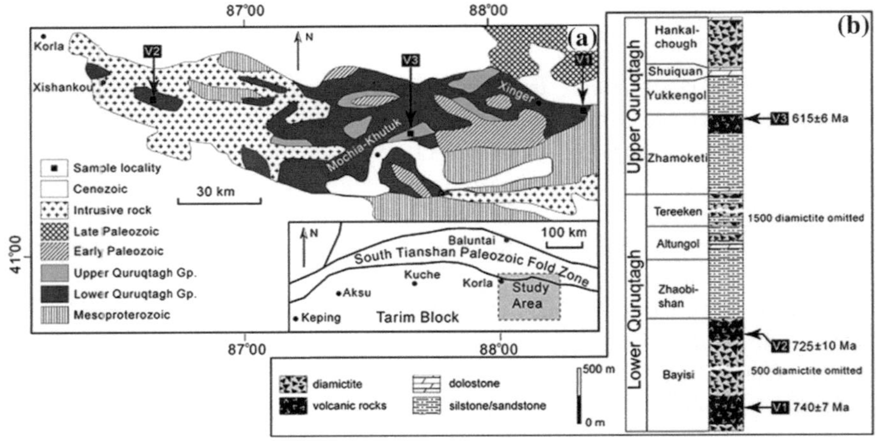

Fig. 3 Geological map of the Quruqtagh area (**a**) and stratigraphic column of the Quruqtagh Group (**b**). *Inset* in (**a**) shows major tectonic units of the Quruqtagh area. Sample localities and horizons are marked on the map and stratigraphic column (after Xu et al. 2009)

where the Neoproterozoic strata are characterized by several horizons of glacier deposits and volcanic rocks, separated by thick layers of shale, sandstone, and limestone. Four glaciations have been reported (Fig. 3b; Gao and Zhu 1984; Gao and Qian 1985; Xu et al. 2005, 2009; Kou et al. 2008).

The lower part of the Neoproterozoic succession is called the Nanhua system, composed of four formations, namely the Beiyixi, Zhaobishan, Altungal, and Tereeken formations (Table 1). The Bayisi Formation consists of phyllite-slate grade metasediments and metavolcanic rocks. The thickness of the diamictite units varies from a few meters to more than 100 m (Fig. 4a). The Bayisi diamictites are separated from the Altungol diamictite by 100–300 m-thick Zhaobishan Formation sandstones (Fig. 4b), which consists of metawackes, metarenites, calcareous siltstones, and slates. Altungol diamictites are also metamorphosed and deformed. Volcanics occur over a 50 m interval in the upper Altungol Formation at Xishankou. The overlying ~200 m-thick Tereeken Formation is less metamorphosed and consists of at least five diamictite units separated by finely laminated siltstones and mudstones (Fig. 4d). Abundant dropstones and striated clasts occur in the Tereeken diamictites (Fig. 4c, e). The Tereeken Formation is succeeded by a 3–10 m-thick cap carbonate in the basal Zhamoketi Formation (Fig. 4f).

The upper part is called the Sinian system that consists of the Zhamoketi, Yukengou, Shuiquan, and Hangelchaok formations. The Zhamoketi Formation and the overlying Yukkengol Formation consist of ~1200 m of meter-scale cycles of sandstone and siltstone, which are interpreted as deep-water turbidites. The Zhamoketi and Yukkengol formations are demarcated by 80–330 m of basaltic and andesitic volcanic rocks (V3 in Fig. 3b), which are traditionally regarded as part of the uppermost Zhamoketi Formation. The Shuiquan Formation consists of <100 m dolostone. The uppermost Sinian system is represented by the Hankalchough

Fig. 4 Filed Photos of the Quruqtagh Group in Quruqtaghh area

Formation, which consists of ca. 400 m-thick, light-gray diamictite with abundant dropstones. The Hankalchough diamictite is succeeded by a 1–5 m-thick dolostone, which is regarded as the Hankalchough cap carbonate (Fig. 4h).

3.2 Stratigraphy in Aksu Area

In the Aksu area, the Neoproterozoic strata crops out in southwest Aksu, Wushi, and Sugetbrak (Fig. 5; Xu et al. 2013).

In the sections of southwest Aksu and Wushi, the Neoproterozoic strata include the Aksu Group and the overlying Sugetbrak and Chigebrak Formations (Fig. 5). The Aksu Group is composed of pelitic, psammitic, and mafic schists (Fig. 6a). The mafic schists are characterized by greenschists and blueschists that are considered to be some of the oldest high-pressure metamorphic rocks (Fig. 6b; Liou et al. 1989, 1990; Yong et al. 2013). A tectonic evolution model for the Aksu blueschist has been proposed (Zhang et al. 2010; Zhu et al. 2011). There is a clear angular unconformity between the Aksu Group and the overlying Sugetbrak Formation in southwest Aksu and Wushi. The 400–450 m-thick Sugetbrak Formation is composed of red conglomerates (Fig. 6d), red fluvial sandstones, and gray lacustrine mudstones with three horizons of basalts. The Chigebrak Formation is characterized by thick dolostone and is overlain unconformably by the Cambrian Yuertus Formation (Fig. 6c; Gao and Qian 1985; Turner 2010; Zhu et al. 2011).

In Sugetbrak area, the Neoproterozoic shows a different succession composed of, from bottom to top, the Qiaoenbrak, Yuermeinak, Sugetbrak, and Chigebrak Formations (Figs. 7, 8a; Xu et al. 2013). The Qiaoenbrak Formation is composed of epimetamorphic sandstone and siltstone flysch, with a thickness of 1966–2094 m and belongs to the Qingbaikouan (Fig. 8f; Gao and Qian 1985). As the typical Nanhua system, the Yuermeinak Formation occurs only in the Sugetbrak area and is

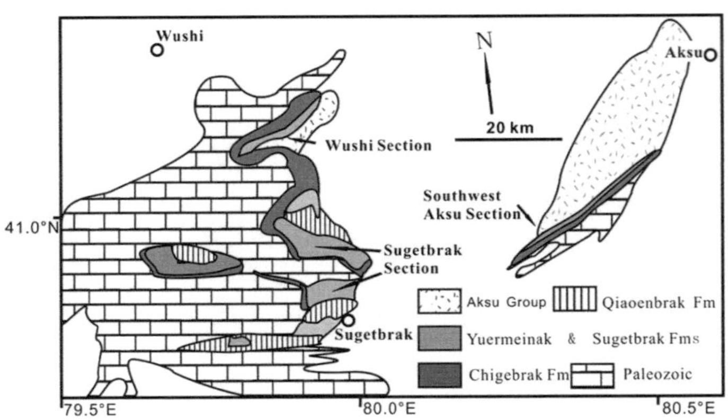

Fig. 5 Geological map of Aksu area (after Xu et al. 2013)

Fig. 6 Photos of the Neoproterozoic strata in Aksu area

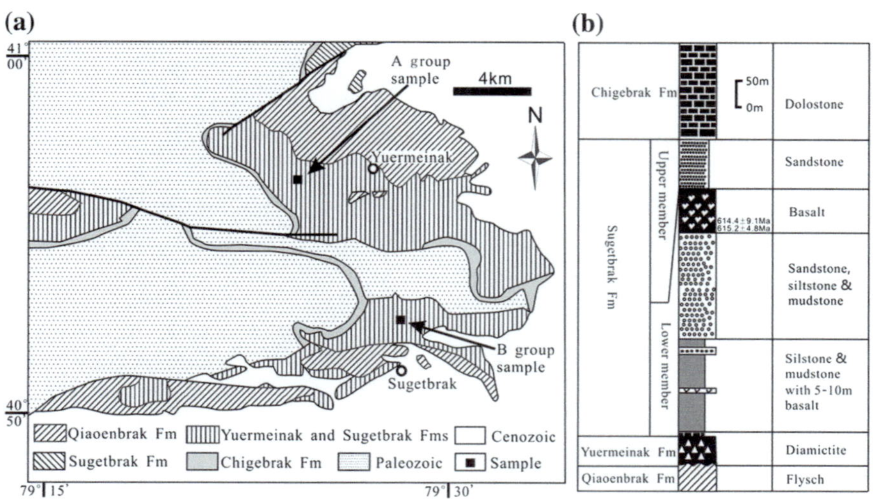

Fig. 7 Geological map and Sequence of the Neoproterozoic in Sugetbrak **a** Geological map, **b** Stratigraphic column (after Xu et al. 2013)

Fig. 8 Photos of the Neoproterozoic strata in Sugetbrak

characterized by diamictites up to 61 m thick, which are interpreted to be continental glacier deposits (Fig. 8g; Gao and Qian 1985). There is unconformity between the Qiaoenbrak and Yuermeinank Formations (Fig. 8e). The Sugetbrak Formation belongs to the Sinian system and can be divided into the Lower and Upper members in Sugetbrak area (Fig. 7a, b; Gao and Qian 1985; Zhan et al. 2007). The Lower Sugetbrak Formation is composed of thin-layered red sandstone (Fig. 8d), quartz sandstone, siltstone, and mudstone (Fig. 8b) with a discontinuous horizon of 5–10 m-thick basalt in its lower part (Fig. 8c), followed by 79 m-thick volcanic rock above. The Upper Sugetbrak Formation conformably overlies the Lower member and is conformably overlain by the Chigebrak Formation.

3.3 *Stratigraphy in Yecheng Area*

The strata of the Nanhua and Sinian systems have been divided into six formations in Yecheng area, southwest Tarim craton; from bottom to top, they are called the Yalaguz, Bolong, Kelixi, Yutang, Kuerkake, and Kezisuhum Formations. Glacial diamictite can be observed in the Bolong and Yutang Formations (Table 1, Turner 2010; Ma et al. 1989).

The Nanhua system contains four formations (Fig. 9). The lowest Yalaguz Formation is characterized by 370-m-thick purple and thick layer polymictic conglomerate with lamina mudstone and sandstone in lower part and unconformably overlies the Sukuluoke Group in Qingbaikou system. The upper part of this

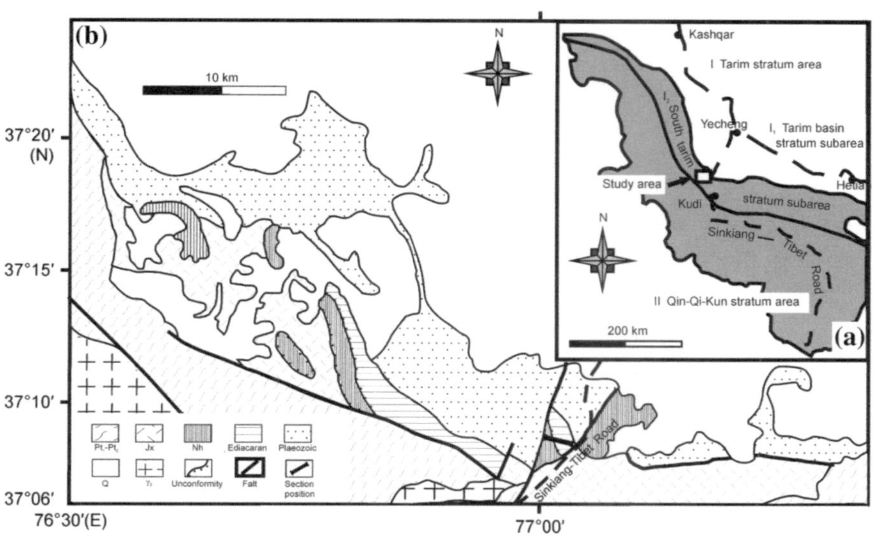

Fig. 9 **a** Precambrian stratum subareas in the southwest Tarim; **b** sketch geological map in Yecheng area, Xinjiang (after Tong et al. 2013)

Fig. 10 Photos showing Neoproterozoic sedimentary features in Yecheng area. **a** Frozen crack of moraines gravel in the Bolong Fm; **b** glacial striae; **c** olistostrome in the Yutang Fm; **d** two coarsening upward cycles in the Kelixi Fm; **e** dropstone structure; **f** convolute lamination in olistostrome; **g** diagonal bedding in the Kelixi Fm; **h** Herringbone cross-bedding in the Kuerkake Fm; **i** tepee structure in the sandstone of Kezisuhum Fm (after Tong et al. 2013)

formation consists of many cycles of grayish-green siltstone and purple lamina siltstone. The Bolong Formation mostly comprises thick layer (about 185 m) glacial diamictites (Fig. 10a). Dropstone structures occurred in the fine clastic rocks at the end of glacial diamictites (Fig. 10e). The 339-m-thick Kelixi Formation, as a whole, consists of two coarsening upward cycles (Fig. 10c, d, g), which contain siltstone, mudstone, and fine clastic rocks in the lower part and purple gritstone to conglomerate in the upper part of every cycle (Fig. 10f). The Yutang Formation consists of glacial diamictite at the bottom (Fig. 10b), thick layer gritstone in the middle part, and cycles of siltstone and mudstone in the upper part, with a total thickness of about 35 m. It unconformably underlies pebbly gritstone of the Sinian Kuerkake Formation (Tong et al. 2013).

The Sinian system contains the Kuerkake and Kezisuhum Formations. The Kuerkake Formation mainly comprises 200-m-thick coarse clastic rocks and intercalations of lamina siltstone and silty shale (Fig. 10h). The Kezisuhum Formation consists of cycles of sandstone and mudstone at the bottom, gypsum-bearing gritstone in the middle part (Fig. 10i), and thick layer mudstone in the top (Tong et al. 2013).

Six types of sedimentary facies such as alluvial fan, lacustrine, glacial, littoral, neritic, and lagoonal facies have been recognized from the Neoproterozoic in the Yecheng area by comprehensive analysis of strata features, sedimentary structures, and lithology (Fig. 11; Tong et al. 2013).

4 Geochronology of the Nanhua and Sinian Systems

The Neoproterozoic volcanic rocks occur in the Beiyixi, Altungal, Zhamoketi, and Shuiquan Formations in Quruqtagh area. Three volcanic samples for dating have been collected from three volcanic intervals (V1, V2, and V3 in Fig. 3) in the Bayisi and Zhamoketi Formations (Xu et al. 2005, 2009). Volcanic units occur at the base of the Bayisi Formation at Xinger (V1 in Fig. 3b; Fig. 4g) and the top of the Bayisi Formation at Xishankou and Xinger (V2 in Fig. 3b). Sample Xb006 came from V1 in the Xinger area, where the Bayisi Formation is 336-m thick and consists of diamictites as well as bimodal basaltic, dacitic, and felsic lavas and pyroclastic rocks, with felsic lavas being the dominant ($\sim$75 %) lithology. Sample A03112 was collected from V2 in the Xishankou area, where a unit of $\sim$50–200-m dark-gray to purple andesitic to basaltic rocks occurs near the top Bayisi Formation. Sample 2371D was collected from V3 in the uppermost Zhamoketi Formation at Mochia-Khutuk, where this volcanic interval consists of 80–330-m-thick basaltic and andesitic lavas and pyroclastic rocks. SHRIMP analysis of zircons from Xb006 (V1), A03112 (V2), and 2371D (V3) gives weighted mean ages of 740 $\pm$ 7, 725 $\pm$ 10, and 615 $\pm$ 6 Ma, respectively (Xu et al. 2005, 2009).

Zircons from two samples of the Sugetbrak basalts (SB) yield weighted mean ages of 615.2 $\pm$ 4.8 and 614.4 $\pm$ 9.1 Ma in Aksu area (Fig. 7b). These ages, interpreted as the eruption age of the SB, provide an age constraint on the timing of the Sugetbrak Formation in Sugetbrak, Aksu area, northwestern Tarim craton (Xu et al. 2013).

5 Correlation of Ice Ages

5.1 National Correlation

Few isotopic age has been obtained from the Nanhua and Sinian systems in Yecheng area up to now. Combined with the analysis of chemical index of alteration (CIA) values in the study area, Tong et al. (2013) made the further correlation (Fig. 11). The CIA values in the upper part of the Yalaguz Formation are in the range of 48–62, which are almost consistent with those CIA values (48–61) of the Altungal Formation in Quruqtagh area, and as well correspond to those CIA values (58–61) of the upper Liantuo Formation below Nantuo glacial diamictite in South

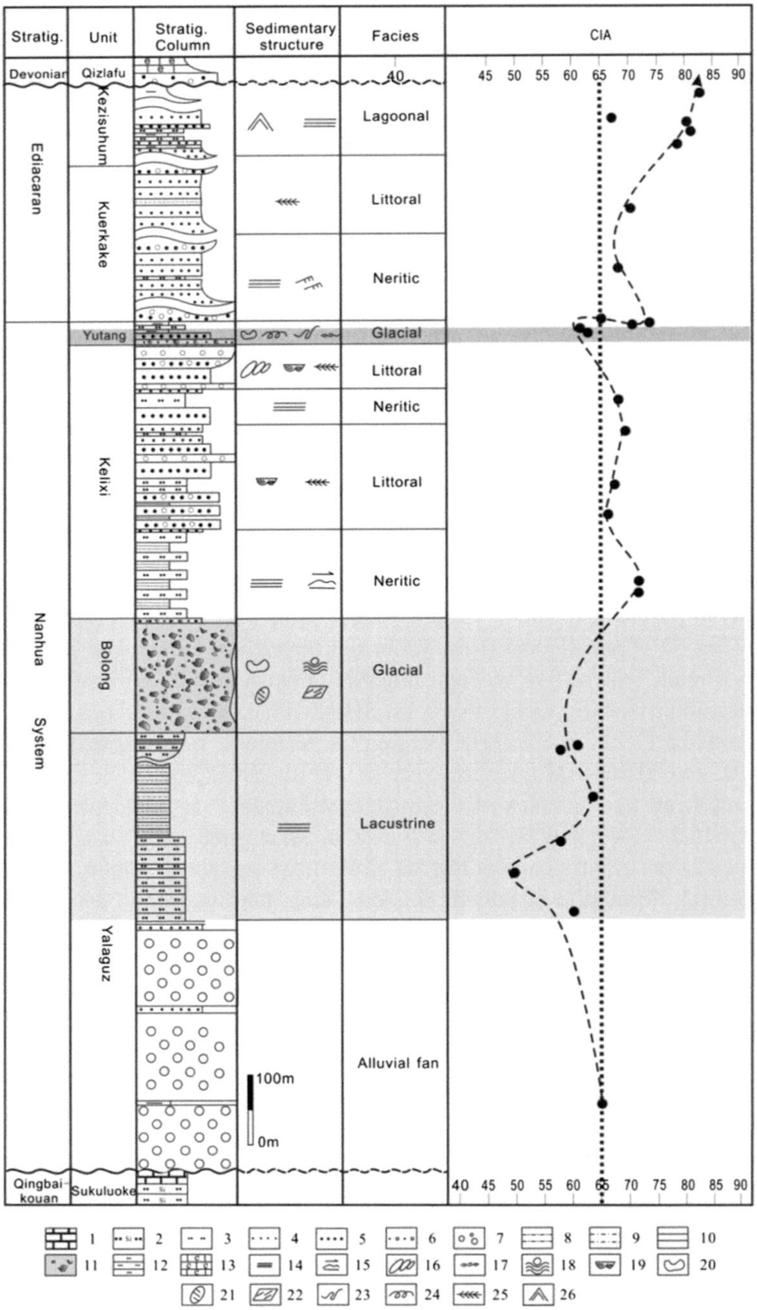
Stratig.
Unit
Stratig. Column
Sedimentary structure
Facies
CIA
Devonian
Qizlafu
Ediacaran
Kezisuhum
Kuerkake
Lagoonal
Littoral
Neritic
Glacial
Yutang
Littoral
Neritic
Kelixi
Littoral
Neritic
Nanhua System
Bolong
Glacial
Lacustrine
Yalaguz
Alluvial fan
100m
0m
Qingbaikouan
Sukuluoke
40 45 50 55 60 65 70 75 80 85 90
1 2 3 4 5 6 7 8 9 10
11 12 13 14 15 16 17 18 19 20
21 22 23 24 25 26

◀ **Fig. 11** Neoproterozoic strata and sedimentary facies in Yecheng area, Xinjing. *1* limestone;*2* siliceous siltstone;*3* siltstone;*4* medium sandstone;*5* gritstone;*6* pebbly gritstone;*7* conglomerate;*8* silty shale; *9* sandy shale;*10* shale;*11* tillite;*12* mudstone;*13* bioclastic limestone; *14* horizontal bedding;*15* metaripple;*16* imbricate structure;*17* deformed bedding;*18* dropstone structure;*19* crescent type cross-bedding;*20* saddle structure;*21* frozen crack;*22* glacial striae;*23* slump bedding;*24* convolute bedding;*25* herringbone cross-bedding;*26* tepee structure (Tong et al. 2013)

China. Besides, the CIA values in the Kelixi Formation are 66–71 (average 68.7), which can be compared with the Yukengou and Shuiquan Formations with CIA values of 67–76 (average 70.3) in Quruqtagh area. Therefore, they suggest that the Yutang glaciation can be correlated to the Hankalchough glaciation in Qurugtagh area, namely, equivalent to the worldwide Gaskers glaciation, and because the top part of the Yalaguz Formation in the study area can correlate to the Altungal Formation, the Bolong glaciation is coeval with the Tereeken glaciation.

Four glacial periods have been recognized in Quruqtagh area, including Bayixi glaciation of 740–732 Ma, Altungal and Tereeken glaciations of 725–615 Ma, and Hankalchough glaciation of 615–542 Ma (Kou et al. 2008). The Nantuo glaciation in South China was constrained between 663 and 635.4 Ma (Condon et al. 2005), which can be correlated with the Tereeken glaciation (Xu et al. 2009) and then the Bolong glaciation according to the correlation of CIA values. Based on the chemical stratigraphic correlation of carbon isotope, He et al. (2007) suggest that the upper part of Qigbulak Formation in Aksu area can be correlated to the top part of the Dengying Formation in South China, the Sugetbulak Formation can correspond to the Doushantuo Formation, and the Yulmeinak Formation can be correlated to the Nantuo Formation. Thus, the Bolong glaciation in Yecheng area also can be correlated to the Yulmeinak glaciation in Aksu area (Fig. 12).

According to the Neoproterozoic succession in the Sugetbrak section, the Lower member of Sugetbrak formation containing basalts with ages of 615 ± 5 and 614 ± 9 Ma (Xu et al. 2013) conformably overlies the Yuermeinak formation diamictite. Because there is only about 400 m of continental sedimentary succession between the basalts and Yuermeinak Formation, the occurrence of the Yuermeinak glaciation is not much earlier than the age of 614–615 Ma. This is similar to the Tereeken glaciations in the Quruqtagh area of the northeast Tarim block where there is only the Zhamoketi Formation turbidite between the Tereeken diamictite and volcanic layer with an age of 615 ± 6 Ma (Xu et al. 2009). It is therefore reasonable to correlate the Yuermeinake glaciations with the Tereeken glaciation in the Quruqtagh area. Based on a TIMS U–Pb age of 635.2 ± 0.6 Ma for the Nantuo glaciation (Condon et al. 2005) and an age of 615 ± 6 Ma for the volcanic rocks above the Tereeke glaciation, previous studies have suggested that the Tereeken glaciation can be correlated with the Nantuo glaciation in the Yangtze block and the Elatina glaciation in Australia (Xu et al. 2009). If this is correct, then the Yuermeinak glaciation should also correlate with them (Fig. 12).

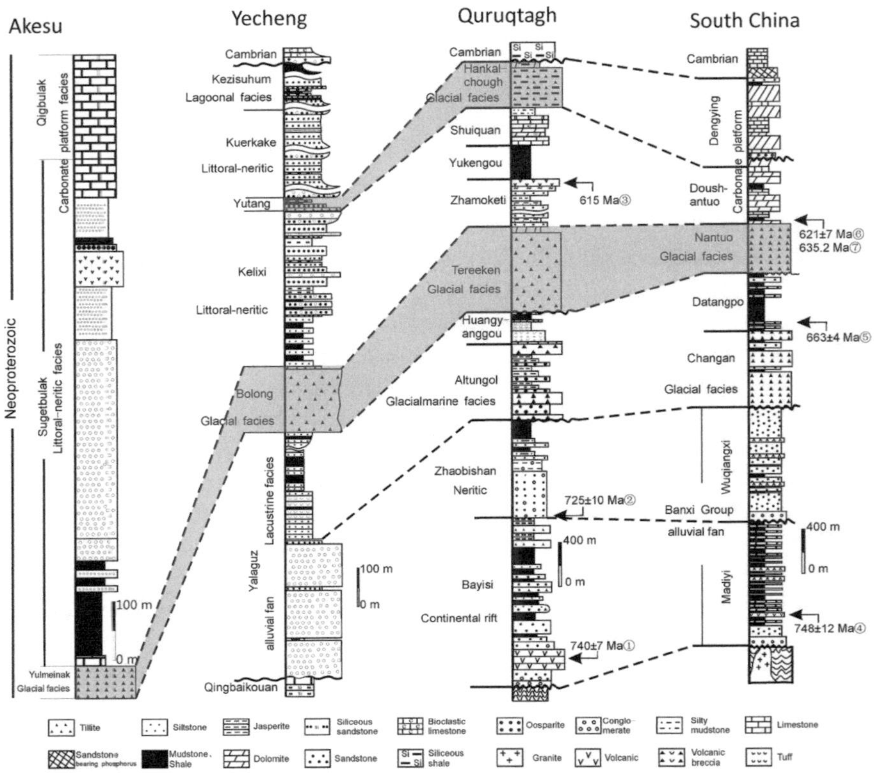

Fig. 12 Correlation of ice age between the Tarim and South China cratons. ① Xu et al. 2005; ② Xu et al. 2009; ③ Xu et al. 2013; ④ Ma et al. 1984; ⑤ Zhou et al. 2004; ⑥ Zhang et al. 2005; ⑦ Condon et al. 2005

5.2 *International Correlation*

Four glaciations have been reported in the northeast Tarim block including Bayisi (730 Ma), Altungol, Tereeken (725–615 Ma), and Hankalchough (615–542 Ma) glaciations (Kou et al. 2008; Xu et al. 2008). Xu et al. (2009) suggest an international correlation for the four ice ages (Fig. 13). They point out that "the Sturtian glaciation was believed to be ~723 Ma, on the basis of a U–Pb zircon age of 723 + 16/−10 Ma from a tuffaceous bed near the base of the Ghubrah diamictite in Oman (Brasier et al. 2000), which was regarded as an equivalent of the Australian Sturtian glaciation. So the Bayisi diamictite has been interpreted as evidence of a Sturtian age glaciation in the Quruqtagh area (Xu et al. 2009) based on the age of 725 ± 10 Ma from the Bayisi basalts. The glacial origin of the Altungol diamictite remains to be unambiguously verified. Nonetheless, its depositional age between 725 ± 10 and 615 ± 6 Ma is consistent with a correlation with either the Sturtian or Elatina glaciation. The Tereeken diamictite is unambiguously glaciogenic. Combining the Re–Os ages from

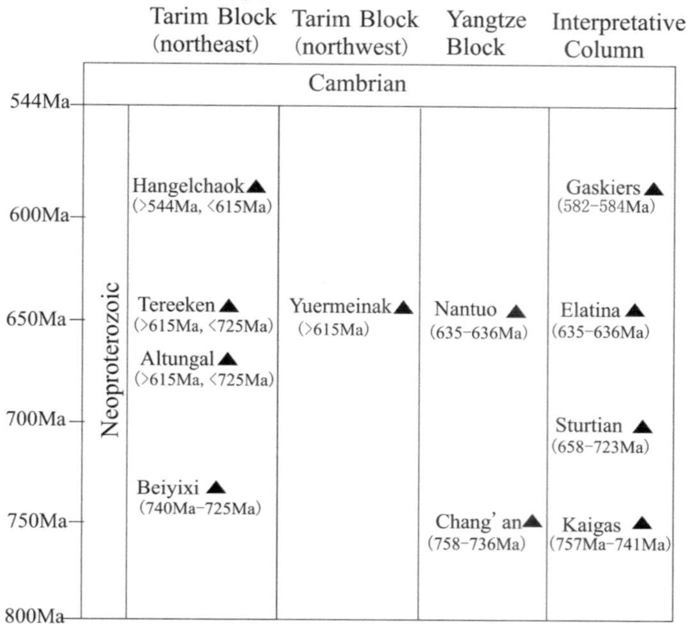

Fig. 13 Age distribution of glaciations between 780 and 544 Ma. *Triangles* denote glaciations. For stratigraphic data of the Yangtze Block, see Evans (2000). Ages of the Liantuo and Doushantuo formations are from Ma et al. (1984) and Barfod et al. (2002)

the Tindelpina Member in the Flinders Ranges and the zircon U–Pb ages from the Doushantuo cap carbonate atop the Nantuon diamictite in South China, a global Elatina age glaciation is permissible between 643.0 ± 2.4 Ma (Kendall et al., 2006) and 635.2 ± 0.6 Ma (Condon et al. 2005). Diamictites in the Tereeken (and possibly Altungol) Formation, constrained between 725 ± 10 and 615 ± 6 Ma, can thus be permissibly correlated with the Elatina glaciation. The youngest diamictite in the Quruqtagh Group, the Hankalchough diamictite, is also glaciogenic, supported by sedimentary evidence such as drop stones and striated clasts (Gao and Zhu 1984). The radiometric ages reported here further constrain the Hankalchough glaciation between 615 ± 6 and ~542 Ma and suggests that the Hankalchough glaciation may be correlated with the ~582 Ma Gaskiers glaciation in Newfoundland (Bowring et al. 2003)" (Xu et al. 2009).

6 The Late Neoproterozoic Magmatic Activities

The Late Proterozoic magmatic activities in the Tarim craton are characterized by wide distribution, multiple episodes, and being subjected to breakup of Rodinia supercontinent (Table 2). Zhang et al. (2011, 2012) subdivided magmatic activities

Table 2 Ages of the Neoproterozoic magmatic activities (after Zhang et al. 2013)

No.	Location	Rock type	Analyzed mineral	Analytical method	Age (Ma)	Reference
Neoproterozoic						
29	Xindi	Granite	Zircon	LA-ICPMS	798 ± 7	Shu et al. (2011)
30	Xindi	Granite	Zircon	LA-ICPMS	806 ± 8	Shu et al. (2011)
31	Xindi	Granite	Zircon	LA-ICPMS	775 ± 12	Shu et al. (2011)
32	Korla	Gneiss granite	Zircon	LA-ICP-MS	752 ± 13	Zhang et al. (2011)
33	South to Xindi	Granite	Zircon	LA-ICP-MS	798 ± 6	Deng et al. (2008)
34	South to Xindi	Diabase	Zircon	LA-ICP-MS	813 ± 41	Deng et al. (2008)
35	Qieganbulake	Carbonatite	Baddeleyite	TIMS	810 ± 6	Zhang et al. (2007a)
36	Qieganbulake	Gabbro	Zircon	SHRIMP	817 ± 11	Zhang et al. (2007a)
37	Taiyangdao	Trondhjemite	Zircon	SHRIMP	795 ± 9	Zhang et al. (2007a)
38	Xindi	Tonalite	Zircon	SHRIMP	820 ± 10	Zhang et al. (2007a)
39	No.2 complex	Gabbro	Zircon	SHRIMP	762 ± 6	Zhang et al. (2010a)
40	South to Xindi	Mafic dyke	Zircon	SHRIMP	824 ± 9	Zhang et al. (2009b)
41	South to Xindi	Mafic dyke	Zircon	SHRIMP	777 ± 10	Zhang et al. (2009b)
42	Askan	Mafic dyke	Baddeleyite	TIMS	773 ± 3	Zhang et al. (2009a)
43	Aksu	Mafic dyke	Zircon	SHRIMP	760 ± 7	Zhang et al. (2009a)
44	East to Xindi	Granodiorite	Zircon	LA-ICPMS	785 ± 8	Long et al. (2011)
45	East to Xindi	Granodiorite	Zircon	LA-ICPMS	798 ± 3	Long et al. (2011)
46	East to Xindi	Granodiorite	Zircon	LA-ICPMS	790 ± 3	Long et al. (2011)
47	East to Xindi	Granodiorite	Zircon	LA-ICPMS	754 ± 4	Long et al. (2011)
48	North to Korla	Mafic dyke	Zircon	SHRIMP	629 ± 7	Zhu et al. (2008)
49	North to Korla	Mafic dyke	Zircon	SHRIMP	652 ± 7	Zhu et al. (2008)
50	North to Korla	Mafic dyke	Zircon	SHRIMP	642 ± 7	Zhu et al. (2008)
51	BeiyixiFm	Tuff	Zircon	SHRIMP	740 ± 7	Xu et al. (2009)

(continued)

Table 2 (continued)

No.	Location	Rock type	Analyzed mineral	Analytical method	Age (Ma)	Reference
52	BeiyixiFm	Tuff	Zircon	SHRIMP	725 ± 10	Xu et al. (2009)
53	ZhamoketiFm	Tuff	Zircon	SHRIMP	615 ± 6	Xu et al. (2009)
54	Xuxugou	Mafic dyke	Zircon	SHRIMP	802 ± 9	Zhang et al. (2010b)
55	Kudi	A-type granite	Zircon	SHRIMP	783 ± 10	Zhang et al. (2006a)
56	Quanji	Mafic dyke	Zircon	SHRIMP	821 ± 16	Lu et al. (2008)
57	Central Tarim	Diorite	Hornblende	39Ar/40Ar	790–740	Guo et al. (2005)

after 820 Ma into four developing phases, namely, 820–800, 780–760, 740–730, and 650–630 Ma.

6.1 *Phase 1 (820–800 Ma)*

It is characterized by the formation of ultramafic mafic carbonate complex, maficdike swarm, and adakitic granite (Fig. 14; Zhang et al. 2007, 2011; Long et al. 2011; Shu et al. 2011), indicating an extension-related dynamical setting. Zhang et al. (2011) studied the granite from Dapingliang area, Saima Mountain in the northeastern margin of Tarim craton, and identified that they belong to low-Ca high-K calc-alkaline and shoshonite series and are typical I type granite, with ages ranging from 816 to 826 Ma; they are formed from partial melting of K-rich basalt in the lower crust and are products of the crust evolving from extrusion to extension. Representative examples for this phase are as follows.

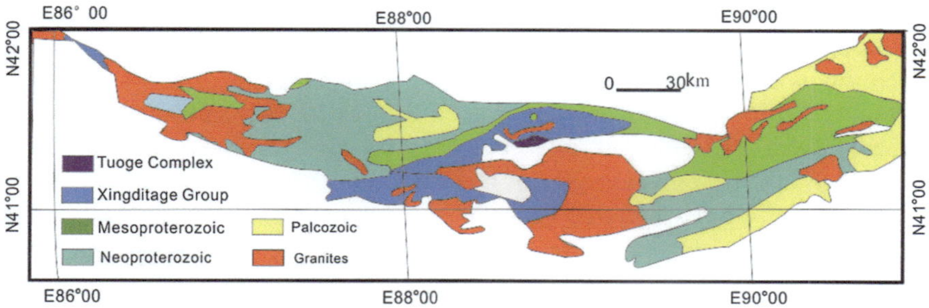

Fig. 14 Distribution of the Neoproterozoic magmatic activities in the Tarim craton (Zhang et al. 2011)

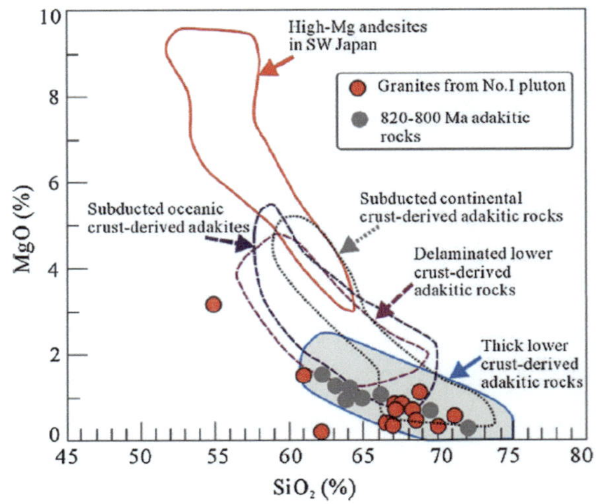

Fig. 15 Adakite characteristics of the Neoproterozoic granite in Taiyangdao, Xinger (after Zhang et al. 2007)

Zhang et al. (2007) identified ultramafic mafic complex bodies (QME), Xingdi granodiorite, and Taiyangdao granite in Quruqtagh area in the northeastern margin of Tarim (Fig. 14, location Nos. 35–37 in Table 2), and obtained their zircon and baddeleyite ages of 810 ± 6, 817 ± 11, and 795 ± 10 Ma, respectively. These granites have adakite characteristics, primarily represented by high LREE and LILE, intensely depleted HREE and HFSE. Based on low Mg# value in granite from this area and Nd–Hf isotope composition similar to that of the Archean rock in this area, these Neoproterozoic granites come from partial melting of the Archean mafic rock, with melting depth being greater than 30 km (Fig. 15; Zhang et al. 2007, 2011; Long et al. 2011).

6.2 *Phase 2 (780–760 Ma)*

Observed at Quruqtagh and Aqsuin the northern margin of Tarim, this phase is characterized by tholeiitic ultramafic mafic complex and massive dike swarms. Geochemical study demonstrates that its magma originated from metasomatized subcrustal lithospheric mantle by subducting slab fluid/melt. These extensively distributed dike swarms are unlikely to arise from island arc subduction, but more probably due to the setting of the Neoproterozoic mantle plume extension (Zhang et al. 2009, 2011). Main examples of this study are as follows.

Zhang et al. (2011) reported geochemical and isotope dating results of mafic ultramafic intrusion complex group in Xingdi area in the northeastern margin of Tarim (Fig. 14, location No. 39 and 41 in Table 2). This complex group is located about 15 km away southeastward from Xingdi, and it consists of five mafic ultramafic intrusive plutons, exhibiting EW-trending distribution. Zircon SHRIMP

U–Pb age in coarse grained gabbro from Pluton II is 762 ± 6 Ma. Geochemical analysis demonstrates that these plutons belong to tholeiitic series, and Sr–Nd isotope feature indicates that the magma comes from the enriched mantle, barely with crustal contamination, and belongs to subcrustal lithospheric mantle metasomatized by subducting slab fluid/melt.

Zhang et al. (2009) obtained a baddeleyite age of 773 ± 3 Ma and a zircon age of 759 ± 7 Ma from mafic dike swarms at Asigan and Aqsu to the south of Taiyangdao in Quruqtagh area, respectively (Fig. 14, location Nos. 42 and 43 in Table 2). According to geochemical and isotope results, it is presumed that they have the same mantle provenance as Pluton II of mafic ultramafic intrusion complex group in Xingdi area.

6.3 *Phase 3 (740–725 Ma)*

This phase is characterized by the development of bimodal volcanic rock and intrusion complexes. Main examples of this study are as follows.

Zhang et al. (2011) studied intrusion complexes I and IV at Quruqtagh and determined that they belong to bimodal intrusion complex, where five gabbro samples were obtained, with zircon ages ranging from 734 to 737 Ma (Fig. 14, complexes I and IV on the east, and west sides of location No. 40, respectively, in Table 2). Their Sr–Nd isotope characteristics demonstrate that basic rock components come from enriched mantle; trace element concentration ratio suggests that they originated from subcrustal lithospheric mantle metasomatized by the fluid that is carried by plunged slab, while incompatible element ratio characteristic indicates that they had been subjected to synergy of enriched mantle and crustal contamination. Unlike the 820–760 Ma magmatic activities which might have been subjected to synergy of mantle plume and supra-subduction zone, the magmatic activities during this phase might be related to the process of the Tarim craton separating from Rodinia supercontinent (Zhang et al. 2011, 2012).

The volcanic rock from the bottom of Beiyixi Formation in Quruqtagh area is mafic and felsic bimodal rock assemblage with ages of 740 and 725 Ma (Figs. 3 and 16, location No. 51 and 52 in Table 2), having geochemical characteristic of continental rift (Xu et al. 2005). The Beiyixi volcanic rocks consist of bimodal basalt and dacite–rhyolite with a SiO_2 gap between 55 and 65 % (Fig. 16). The mafic rocks show moderate enrichment in LILE and variable depletion in Nb, Ta, and P, resembling those of the tholeiitic basalts in continental rift. Th/La (0.12–0.13) and Nb/Ta (16.20–17.00) ratios, negative $\varepsilon_{Nd}(T)$ values (−9.9 to −10.8), higher $\sum$REE (147–158 ppm), incompatible elements (Zr, Y and Th), Mg number (50–67), and Cr (327–405 ppm) suggest that the mafic rocks were derived from partial melting of an enriched lithospheric mantle reservoir. The felsic rocks show negative $\varepsilon_{Nd}(T)$ values (−7.9 to −9.2), Nb, Ta, P, and Ti anomalies relative to the neighboring elements, higher La_N/Yb_N (62−92) and LILE, and may be generated by melting of dominant crustal materials, as a result of basalt emplacement into

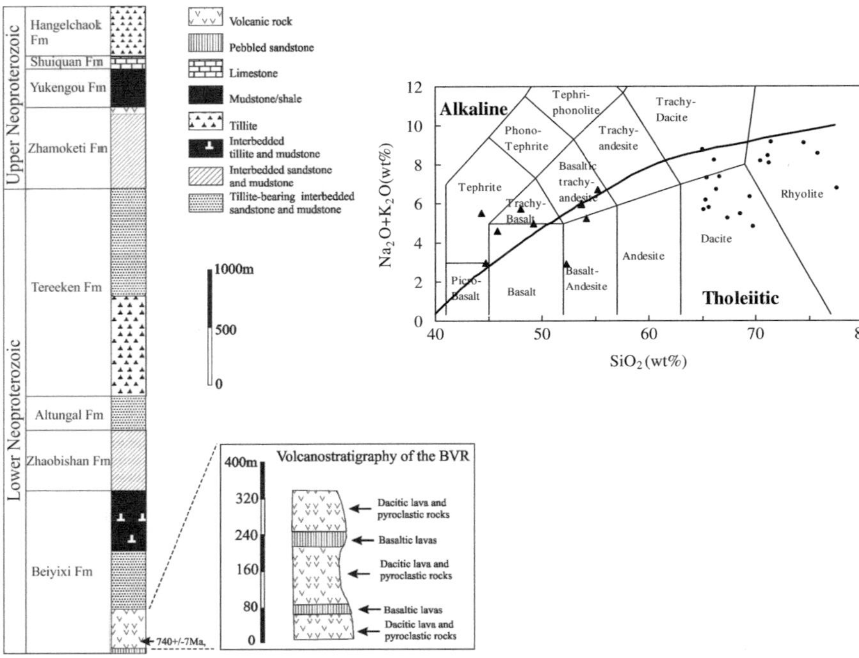

Fig. 16 Mafic and felsic bimodal rock assemblage in the Beiyixi formation in Quruqtagh area (Xu et al. 2005)

continental crust during continental rifting related with the breakup of the Rodinia supercontinent (Xu et al. 2005).

6.4 Phase 4 (650–615 Ma)

Zhu et al. (2008, 2011) studied mafic dike swarms near Korla (Fig. 14, location No. 48 and 49 in Table 2). According to geochemical analysis, such magmas originated from fractional crystallization of tholeiitic magma due to partial melting of subcrustal lithospheric mantle and had been heated by the rising mantle plume, indicating that they are situated in continental extension environment and dynamic setting of mantle plume development (Zhu et al. 2011).

Luo et al. (2011) reported an age, 647 ± 4 Ma, of biotite monzonite granite intruding into the Proterozoic basement metamorphic rock in the eastern Kuqa depression, 40 km north of Luntai County. It belongs to high-K calc-alkaline rock series, characterized by rich Si, high alkali, rich K, and low TiO_2. Geochemical characteristics demonstrate that the granite originated from partial melting of the middle- and lower crust, having characteristic of intra plate granite.

Several layers of basalts were developed in terrestrial–littoral strata among the Neoproterozoic Sugetbulak Formation in Keping area (Gao and Zhu 1984; Wang et al. 2010; Xu et al. 2013), and its geochemical characteristics are attributed to intraplate basic basalt (continental flood basalt), and its magma originated from enriched mantle and is a product of continental rift setting.

References

Barfod GH, Albarède F, Knoll AH, Xiao S, Télouk P, Frei R, Baker J (2002) New Lu–Hf and Pb–Pb age constraints on the earliest animal fossils. Earth Planet Sci Lett 201:203–212

BGMRX (Bureau of Mineral Resources of Xinjiang Uygur Autonomous Region) (1993) Regional geology of Xinjiang uygur autonomous region. People's Republic of China, Ministry of Geology and Mineral Resources, Geological Memoirs 32. Geological Publishing House, Beijing, pp 1–667 (in Chinese with English abstract)

Brasier M, McCarron G, Tucker R, Leather J, Allen P, Shields G (2000) New U-Pb zircon dates for the Neoproterozoic Ghubrah glaciation and for the top of the Huqf Supergroup. Oman. Geol. 28(2):175–178

Bowring S, Myrow P, Landing E, Ramezani J, Grotzinger J (2003) Geochronological constraints on terminal Neoproterozoic events and the rise of metazoans. Geophys Res Abs 5:13219

Condon D, Zhu M, Bowring S, Wang W, Yang A, Jin Y (2005) U-Pb ages from the Neoproterozoic Doushantuo Formation. China Sci 308:95–98

Evans DAD (2000) Stratigraphic, geochronological, and paleomagnetic constraints upon the Neoproterozoic climatic paradox. Am J Sci 300(5):347–433

Gao Z, Qian J (1985) Sinian glacial deposits in Xinjiang, Northwest China. Precambrian Res 29:143–147

Gao Z, Zhu C (1984) Precambrian geology in Xinjiang, China. Xinjiang Publishing House, Urumuqi, pp 1–151 (in Chinese with English abstract)

He J-Y, Xu B, Meng X-Y, Kou X-W, Liu B, Wang Y, Mi H (2007) Neoproterozoic sequence stratigraphy and correlation in Quruqtagh area, Xinjiang. Acta Petrologica Sinica 23(7):1645–1654 (in Chinese with English abstract)

Hu A-Q, Wei G-J (2006) On the Age of the Neo-Archean Qingir Gray Gneisses from the Northern Tarim Basin, Xinjiang, China. Acta Geologica Sinica 80 (1):126–134 (in Chinese with English abstract)

Kendall B, Creaser RA, Selby D (2006) Re–Os geochronology of postglacial black shales in Australia: constraints on the timing of "Sturtian" glaciation. Geology 34:729–732

Kou X-W, Wang Y, Wei W, He J-Y, Xu B (2008) The Neoproterozoic Altungol and Huangyanggou formations in Tarim plate: recognized newly glaciation and interglaciation? Acta Petrologica Sinica 23(12):2863–2868

Liou JG, Graham SA, Mayuyama S, Wang X, Xiao X, Carrroll AR, Chu J, Feng Y, Hendrix MS, Liang Y, Mcknight CL, Yang Y, Wang Z, Zhao M, Zhu B (1989) Proterozoic blueschist belt in western China: best-documented Precambrian blueschists in the world. Geology 17:1127–1131

Liou JG, Wang X, Xiao X, Graham A (1990) Evidence for late Proterozoic subduction from 700-Myr-old blueschists in China. Nature 346:263–265

Liu Y-S, Xin H-T, Zhou S-J, Teng X-J, Yang J-Q, Lv H-Q (2010) Precambrian to Paleozoic Tectonic evolution in the Lapeiquan area, Eastern Altyn Tagh. Geological Publishing House, Beijing, pp 1–211 (in Chinese)

Long X-P, Yuan C, Sun M, Kroner A, Zhao G-C, Wilde S, Hu A-Q (2011) Reworking of the Tarim Craton by underplating of mantle plume-derived magmas: evidence from Neoproterozoic granitoids in the Kuluketage area, NW China. Precambr Res 187:1–14

Lu S-N, Yu H-F, Li H-K (2006) Research on Precambrian major problems in China. Geological Publishing Press, Beijing, pp 1–206 (in Chinese)

Lu S-N, Li H-K, Zhang C-L, Liu G-H (2008) Geological and geochronological evidence for the Precambrian evolution of the Tarim Craton and surrounding continental fragments. Precambr Res 160:94–107

Luo J-H, Che Z-C, Zhang X-L, Han W, Zhang G-F, Nian X-Q (2011) Neo-proterozoic Granitoid Magmatism in the Kuqe Depression of Northeastern Tarim Basin and its significance. Acta Geol Sinica 85(4):467–474 (in Chinese with English abstract)

Ma G-G, Li H-Q, Zhang Z-C (1984) An investigation of the age limits of the Sinian System in South China. CAGS Bull Yichang Inst Geol Miner Res 8:1–29 (in Chinese)

Ma S-P, Wang Y-Z, Fang X-L (1989) The Sinian at North Slope Western Kunlun Mountains. Xinjiang Geol 7(4):68–79 (in Chinese)

Shu L-S, Deng XL, Zhu W-B, Ma D-S, Xiao W-J (2011) Precambrian tectonic evolution of the Tarim Block, NW China: new geochronological insights from the Quruqtagh domain. Precambr Res 42:774–790

Tong Q-L, Wei W, Xu B (2013) Neoproterozoic sedimentary facies and glacial periods in the southwest of Tarim Block. Sci China Earth Sci 56(6):901–912

Turner S (2010) Sedimentary record of Late Neoproterozoic rifting in the NW Tarim basin, China. Precambrian Res 181:85–96

Wang F, Wang B, Shu L-S (2010) Cotinental tholeiitic basalt of the Akesu area (NW China) and its implication for the Neoproterozoic rifting in the northern Tarim. Acta Petrologica Sinica 26 (2):547–558 (in Chinese with English abstract)

Xu B, Jian P, Zheng H, Zou H, Zhang L, Liu D (2005) U-Pb zircon geochronology and geochemistry of Neoproterozoic volcanic rocks in the Tarim Block of northwest China: implications for the breakup of Rodinia supercontinent and Neoproterozoic glaciations. Precambr Res 136:107–123

Xu B, Kou X-W, Song B, Wei W, Wang Y (2008) SHRIMP dating of the upper Proterozoic volcanic rocks in the Tarim Plate and constraints on the Neoproterozoic glaciation. Acta Petrologica Sinica 24(12):2857–2862 (in Chinese with English abstract)

Xu B, Xiao S-H, Zou H-B, Chen Y, Li Z-X, Song B, Liu D-Y, Zhou C-M, Yuan X-L (2009) SHRIMP zircon U-Pb age constraints on Neoproterozoic Quruqtagh diamictites in NW China. Precambr Res 168:247–258

Xu B, Zou H-B, Chen Y, He J-Y, Wang Y (2013) The Sugetbrak basalts from northwestern Tarim Block of northwest China: geochronology, geochemistry and implications for Rodinia breakup and ice age in the Late Neoproterozoic. Precambr Res 236:214–226

Yong W-J, Zhang L-F, Hall CM, Mukasa SB, Essene EJ (2013) $^{40}Ar/^{39}Ar$ and Rb–Sr chronology of the Precambrian Aksu blueschists in western China. JAES 63:197–205

Zhang C-L, Li X-H, Li Z-X, Lu S-N, Ye H-M, Li H-M (2007) Neoproterozoic ultramficmafic-carbonatite complex, granitoids in Quruqtagh of northeastern Tarim Block, western China: geochronology, geochemistry and tectonic implications. Precambr Res 152:149–169

Zhang C-L, Li Z-X, Li X-H, Ye H-M (2009) Neoproterozoic mafic dyke swarms in northern margin of the Tarim Block: age, geochemistry and implications on the breakup of Rodinia. J Asian Earth Sci 35:167–179

Zhang B-H, Zhu W-B, Jahn BM, Shu L-S, Zhang Z-Y (2010) Subducted Precambrian oceanic crust: geochemical and Sr–Nd isotopic evidence from metabasalts of the Aksu blueschist, NW China. J Geol Soc London 167:1161–1170

Zhang C-L, Li X-H, Li Z-X, Ye H-M (2011) Neoproterozoic layered ultramafic-mafic intrusive complex in Quruqtagh of the northeastern Tarim Block, NW China: two phases of mafic igneous activities with different mantle sources. Gondwana Res 19:177–190

Zhang C-L, Li H-K, Zou H-B, Wang H-Y (2012) Multiple phases of Neoproterozoic ultramafic-mafic complex in Kuruqtagh, northern margin of Tarim: interaction between plate subduction and mantle plume? Precambr Res 222–223:488–502

Zhang C-L, Li H-K, Zou H-B (2013) Tectonic framework and evolution of the Tarim Block in NW China. Gondwana Res 23(4):1306–1315
Zhang S-H, Jiang G-Q, Zhang J-M (2005) U-Pb sensitive high-resolution ion microprobe ages from the Doushantuo Formation in south China: constraints on late Neoproterozoic glaciations. Geology 33:473–476
Zhan S, Chen Y, Xu B, Wang B, Faure M (2007) Late Neoproterozoic paleomagnetic results from the Sugetbrak Formation of the Aksu area, Tarim basin (NW China) and their implications to paleogeographic reconstructions and the snowball Earth hypothesis. Precambr Res 154:143–158
Zhou C-M, Tucker R, Xiao S (2004) New constraints on the ages of Neoproterozoic glaciations in south China. Geology 32:437–440
Zhu W-B, Zhang Z-Z, Shu L-S, Lu H-F, Sun J-B, Yang W (2008) SHRIMP U-Pb zircon geochronology of Neoproterozoic Korla mafic dykes in the northern Tarim Block, NW China: implications for the long-lasting breakup process of Rodinia. J Geol Soc 165:887–890
Zhu W-B, Zhang B-H, Shu L-S, Ma D-S, Wu H-L, Li Y-X (2011) Neoproterozoic tectonic evolution of the Precambrian Aksu blueschist terrane, northwestern Tarim, China: insights from LA-ICP-MS zircon U-Pb ages and geochemical data. Precambr Res 185:215–230

The Precambrian Tectonic Evolution of the Tarim Craton

Bei Xu

Abstract The Precambrian tectonic evolution of the Tarim craton is preliminarily divided into seven important development phases, including initial crustal growth, the first extension phase of continental crust, cratonization, stable intracratonic sedimentation, formation of fold basement, mantle plume activity and subduction of oceanic crust, and continental rift activity and passive continental margin sedimentation. Isotope research indicates that significant crustal growth occurred in the Mesoarchean and Neoarchean. Four important magmatism phases are 2.36–2.55, 1.80–2.02, 0.86–1.14, and 0.68–0.84 Ga.

Keywords Tectonic evolution · Tarim craton · Precambrian

1 Tectonic Evolution

It is not yet possible to give a comprehensive outline of the Precambrian tectonic evolution of the Tarim craton from available data (Hu and Wei 2006; Lu et al. 2006, 2008; Shu et al. 2011; Zhang et al. 2013). Zhang et al. (2013) recognize four tectonic events according to magmatism and metamorphism in the Tarim craton (Fig. 1), including Columbia assemblage (2.1–1.8 Ga), Rodinia assemblage (1.0–0.9 Ga), interaction between Rodinia plume and plate subduction (820–760 Ma), and dispersing of the Rodinia continent (after 760 Ma).

The Precambrian tectonic evolution of the Tarim craton is preliminarily divided into seven important development phases.

B. Xu (✉)
Key Laboratory of Orogenic Belts and Crustal Evolution, School of Earth and Space Sciences, Peking University, Beijing 100871, China
e-mail: bxu@pku.edu.cn

M. Zhai (ed.), *Precambrian Geology of China*,
Springer Geology, DOI 10.1007/978-3-662-47885-1_10

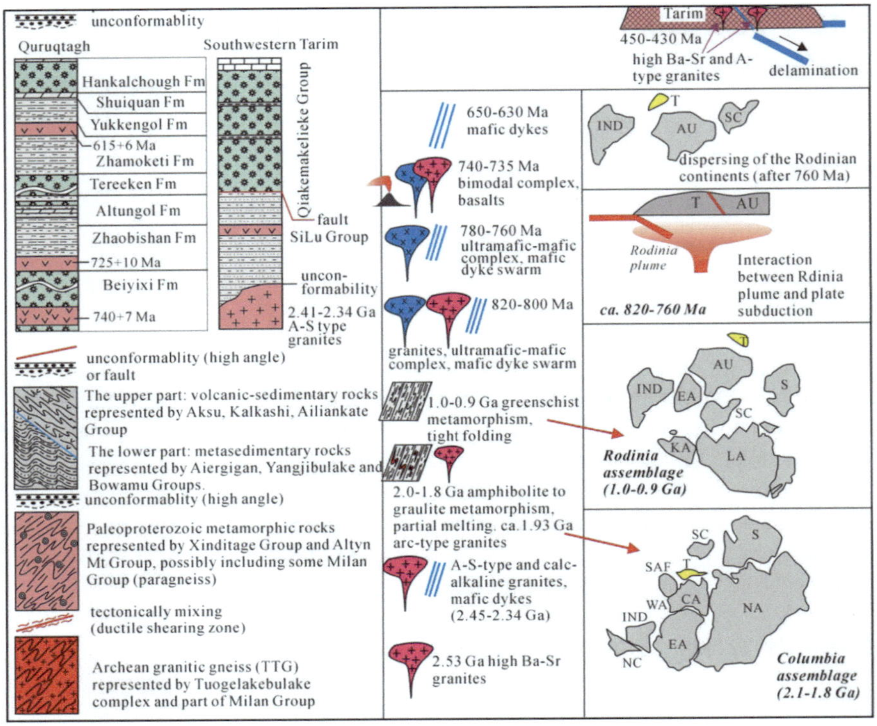

Fig. 1 Precambrian tectonic evolution of the Tarim craton (after Zhang et al. 2013)

(1) Initial crustal growth (3.2–2.5 Ga): Primitive crust might have begun to form in Tarim craton during the Early Archean, which is characterized by TTG assemblage and reached considerable size during 2.8–2.6 Ga.
(2) The first extension phase of continental crust (2.5–2.3 Ga):In both south-eastern and southwestern margins of Tarim craton, bimodal magmatic rock assemblages were developed, probably implying the first extension phase of primitive crust.
(3) Cratonization (2.3–1.8 Ga): During this phase, complex sedimentation, intrusion, anatexis, and metamorphism took place successively in the Tarim craton, among which the 2.0–1.9 Ga metamorphic event plays an important role in early cratonization of the Tarim craton, and is closely related to global aggregation event of Columbia supercontinent.
(4) Stable intracratonic sedimentation (1.8–1.0 Ga): The Tarim craton was relatively stable, characterized by sedimentary formation of carbonate rock and clastic rock series, suggesting extensive epicontinental sea and intracontinental extension environment.
(5) Formation of fold basement (1.0–0.85 Ga): This phase corresponds to the famous "Tarim Movement," and is the time when the Tarim craton responded intensely to Rodinia supercontinent aggregation, primarily represented by

geological records of epimetamorphism, intense deformation and unconformity occurring in the Mesoproterozoic carbonate and clastic rock series, and records of numerous magmatic activities and metamorphisms. These events gave rise to the fold basement of the Tarim craton.

(6) Mantle plume activity and subduction of oceanic crust in the northern margin (0.85–0.75 Ga): Dike swarms and massive intrusive rocks related to mantle plume activity were developed within the Tarim craton, while subduction of oceanic crust and high pressure metamorphism further developed in the northern margin of Tarim.

(7) Continental rift activity and passive continental margin sedimentation (7.5–5.4 Ga): Development of bimodal volcanic rocks, low-latitude tillite, and continental to littoral sedimentary formation represents continental rift and passive continental margin stage related to breakup of Rodinia supercontinent.

2 Main Growth Period of the Tarim Craton Basement

Zhang et al. (2013) have summarized the initial Nd isotopic compositions of the basement rocks and the Neoproterozoic igneous rocks (Fig. 2). The Archean mafic rocks in northern Tarim craton have positive εNd (t) values, consistent with the mantle origin of their basaltic protoliths. As for the granitic gneisses, their εNd(t) values vary from positive to negative and the data seem to follow the trend for continental or granitoid isotopic evolution (Fig. 2a). According to the Nd isotope evolution trend, most continental crust of the Tarim craton was generated in the Archean. During the whole Mesoproterozoic, almost no "juvenile crust" was added to the Tarim craton. The two phases of mantle-derived magma activities could have contributed to some crustal growth, leading to elevation of the εNd (t) values in the Neoproterozoic.

Whole-rock Nd model ages (single-stage model ages for the mafic rocks and two-stage model ages for the silicic rocks) for the basement rocks mostly range from 2.34 to 3.3 Ga and show several peaks at 2.34, 2.53, 2.74, and 3.2 Ga, indicating significant growth of the juvenile crust in the Mesoarchean and Neoarchean (Fig. 2b; Zhang et al. 2013).

Zircon Hf isotope evolution trend for the Archean, Paleoproterozoic, and Neoproterozoic igneous rocks is consistent with that of the whole-rock Nd isotope evolution (Fig. 2c; Long et al. 2010; Zhang et al. 2012). Zircon Hf model ages for the Tarim basement rocks show two peaks at ca.2.6 and ca.3.2 Ga (Long et al. 2010; Fig. 2d). Whole-rock Nd model ages and zircon Hf model ages (Figs. 2c and 2d) almost coupled with each other, especially at ca. 3.2 Ga, indicating significant crustal growth in the early Mesoarchean. Both the whole-rock Nd model ages and zircon Hf model ages suggest that the basement rocks of the Tarim craton formed later than those of the North China and Yangtze cratons.

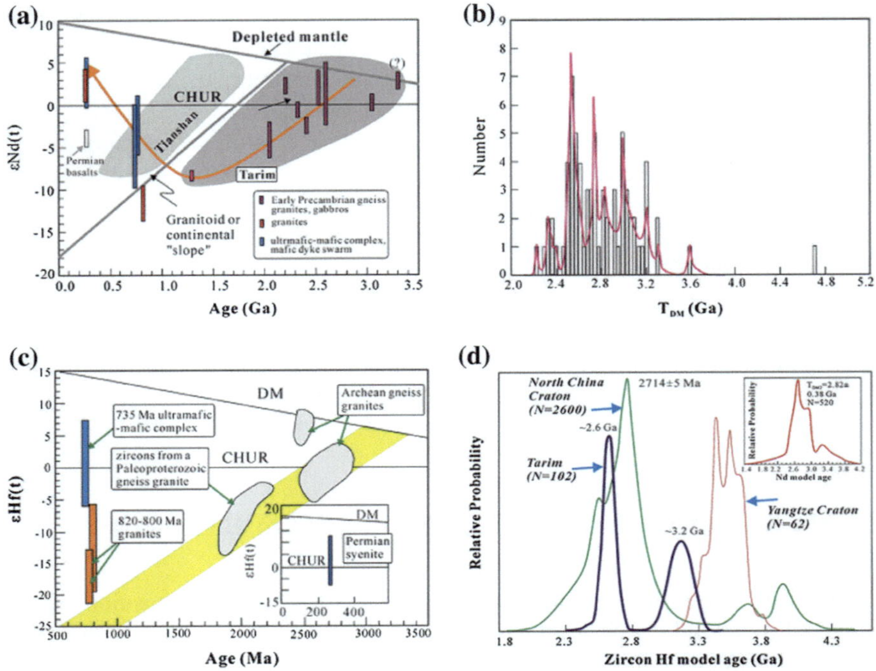

Fig. 2 Main growth period of the Tarim craton basement. (a) Whole-rock εNd(t) versus crystallizing age of the igneous rocks (with few early Precambrian gneiss samples) from Tarim craton; (b) Nd model age spectra of the basement rocks from Tarim; (c) εHf(t) versus age of the zircons from diverse-type rocks in Tarim; (d) Zircon Hf model age spectra of the Tarim, North China, and Yangtze (after Zhang et al. 2013)

3 Main Magmatic Activities and Metamorphic Events

Shu et al. (2011) systematically studied eight samples of gabbro, diorite, and granite from Quruqtagh area in the northern margin of Tarim, revealing existence of four Magmatism phases during the Precambrian, namely, 2.36–2.55, 1.80–2.02, 0.86–1.14, and 0.68–0.84, while peak episodes are 2.5, 1.87, 0.92, and 0.8 Ga, corresponding to the formation of Columbia supercontinent, formation of Rodinia supercontinent, and breakup of Rodinia supercontinent, respectively (Fig. 3).

Major Precambrian metamorphic events in the Tarim craton took place during two phases, 1.8–2.0 and 0.9–1.0 Ga, which represent aggregation events of the Columbia and Rodinia supercontinents, respectively (Fig. 4; Zhang et al. 2013).

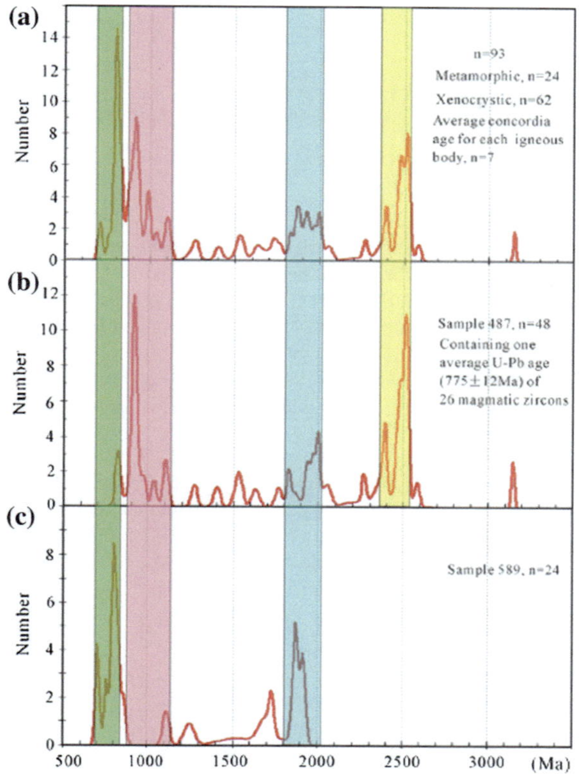

Fig. 3 Four Magmatism phases during the Precambrian in the Tarim Craton (after Shu et al. 2011)

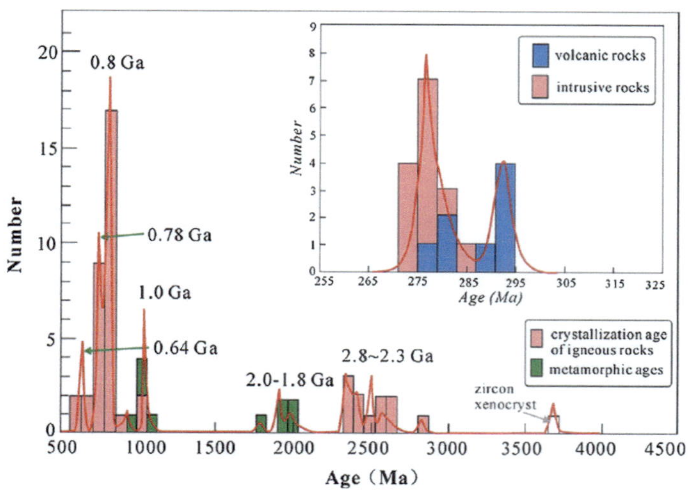

Fig. 4 Major Precambrian magmatic and metamorphic events in the Tarim craton (after Zhang et al. 2013)

References

Hu A-Q, Wei G-J (2006) On the Age of the Neo-Archean Qingir Gray Gneisses from the Northern Tarim Basin, Xinjiang. China. Acta Geologica Sinica 80(1):126–134 (in Chinese with English abstract)

Long X-P, Yuan C, Sun M, Zhao G-C, Xiao W-J, Wang Y-J, Yang Y-H, Hu A-Q (2010) Archean crustal evolution of the northern Tarim craton, NW China: Zircon U-Pb and Hf isotopic constraints. Precambr Res 180:272–284

Lu, S.-N., Yu, H.-F., Li, H.-K., 2006. Research on Precambrian Major Problems in China. Geological Publishing Press: 1-206 (in Chinese)

Lu S-N, Li H-K, Zhang C-L, Liu G-H (2008) Geological and geochronological evidence for the Precambrian evolution of the Tarim Craton and surrounding continental fragments. Precambr Res 160:94–107

Shu L-S, Deng XL, Zhu W-B, Ma D-S, Xiao W-J (2011) Precambrian tectonic evolution of the Tarim Block. NW China: New geochronological insights from the Quruqtagh domain, Precambrian Research 42:774–790

Zhang C-L, Li H-K, Santosh M, Li Z-X, Zou HB, Wang H-Y, Ye H-M (2012) Precambrian evolution and cratonization of the Tarim Block, NW China: petrology, geochemistry, Nd-isotopes and U-Pb zircon geochronology from Archaean gabbro TTG potassic granite suite and Paleoproterozoic metamorphic belt. J Asian Earth Sci 47:5–20

Zhang C-L, Li H-K, Zou H-B (2013) Tectonic framework and evolution of the Tarim Block in NW China. Gondwana Res 23(4):1306–1315

Part V
Other Chinese Paleocontinents

The Precambrian Geology of the Tibetan Plateau

Fulai Liu, Yongsheng Dong and Chaohui Liu

Abstract The Tibetan Plateau, which includes all the Tibet and Qinghai Provinces, the southern Xinjiang and Gansu Provinces and the western Sichuan and Yunnan Provinces, is a tectonically active region among the Indian, Yangtze, Tarim, and North China blocks. On the basis of the geological background, it can be divided into the Altyn-Qaidam-West Qinling orogenic system, the Kara Kunlun-Bayan Har-Chamdo-Simao orogenic system, and the Southern Qiang-Gangdise-Himalaya-Baoshan orogenic system from north to south, separated by the Kangxiwa-Southern Kunlun-Madoi-Maqên suture zone and the Longmuco-Two Lake-Lancang River suture zone, respectively. Every orogenic system can be further divided into several terranes by suture zones (melanges) consisting of deep-sea sediment residues, ophiolites (melanges), and high-pressure metamorphic rocks. This chapter describes the distribution, composition, protolith and metamorphic ages, and metamorphic characteristics of the Precambrian geologic units of the terranes and suture zones. Also included are research progress, existing problems, and an overview of different opinions from different researchers on the Precambrian history of the terranes. Additionally, a preliminary summary is given for major geological events and tectonic evolution of every Precambrian orogenic system. The Tibetan Plateau is vast in territory, but the Precambrian geological units distribute scatteredly in tectonic melanges of or between the terranes. Although abundant studies have been made about the Precambrian geology of the Tibetan Plateau, the overall level of geological research is much lower than those of other Precambrian terranes in China. Geochronologic data for each geologic unit is also of confusion, which led

F. Liu (✉) · C. Liu
Institute of Geology, Chinese Academy of Geological Sciences, Beijing, China
e-mail: lfl0225@sina.com

Y. Dong
College of Earth Sciences, Jilin University, Changchun, China

M. Zhai (ed.), *Precambrian Geology of China*,
Springer Geology, DOI 10.1007/978-3-662-47885-1_11

to divergence on their protolith and metamorphic ages, tectonic attributes, and other aspects, so the crustal evolution studies of the Precambrian Tibetan Plateau need to be further deepened.

Keywords Precambrian rocks · Protolith and metamorphic ages · Orogenic system · Precambrian terrane · Tibetan plateau

1 Distribution of the Precambrian Rocks in the Tibetan Plateau

The Tibetan Plateau is located in the southwest China, including all the Tibet and Xinjiang Provinces, the southern Gansu, and western Sichuan and Yunnan Provinces. It is a tectonically active zone among the Indian, Yangtze, Tarim blocks, and the North China Craton. Its original crustal blocks were formed by accretion, subduction, collision of the margins, or split off pieces of the peripheral continental blocks, and separated by melange belts (suture zones) consisting of deep-sea sedimentary debris, ophiolites (melanges), and high-pressure metamorphic rocks.

According to the current tectonic division scheme, the Tibetan Plateau is divided into the Altyn-Tagh-Qaidam-western Qinling orogenic system, the Karakorum-Bayanhar-Qamdo-Simao orogenic system, and the Southern Qiangtang-Gangdise-Himalaya-Baoshan orogenic system from north to south, in which the southern and western extensions of the South Qiang-Gangdise-Himalaya-Baoshan orogenic system is outside of China. The boundaries are the Kangxiwar-Muztag-Anemaqen suture zone and the Longmu Tso-Shuanghu-Lancang suture zone from north to south (Fig. 1).

The Precambrian rocks in the Tibetan Plateau are scattered and exposed within the blocks or melange belts among the blocks, and the degree of geological research is generally poor. Previous divisions and comparisons of the Precambrian geological units were mostly based on the metamorphic grade or limited and imprecise isotopic geochronology results. In recent years, a number of high precision isotope geochronology data have been achieved, which put more accurate temporal constraints on parts of the Precambrian geologic units. At the same time, some originally defined Precambrian geologic units were also questioned. From the current data, the Precambrian Tibetan Plateau has the following characteristics: (1) there is no unified Precambrian basement in the Tibetan Plateau, and most of the old basement residues have different geologic affinities; (2) the Archean ages only occur in detrital zircons from parametamorphic rocks of different ages and no Archean geological units have been found; (3) Precambrian geological units formed mainly in the Paleoproterozoic to Neoproterozoic (Nanhua). There are still the following problems: (1) due to the lack of geochronology, paleontology, and other aspects of data, parts of the geology units do not have accurate ages, which were

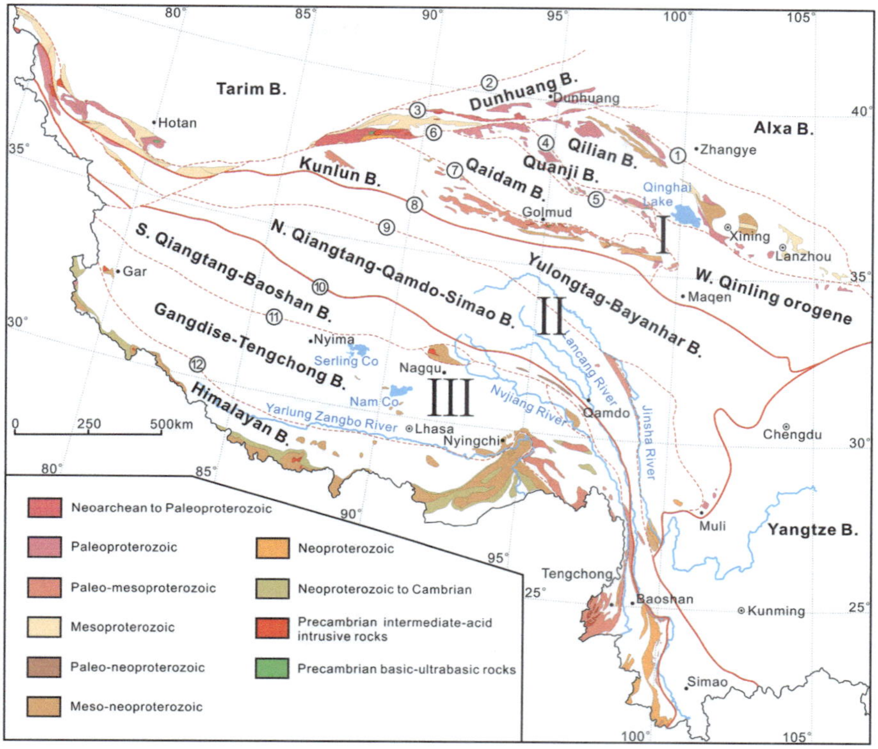

Fig. 1 Geological map showing the distribution of Precambrian rocks in the Tibetan Plateau. *I* Altyn-Tagh-Qaidam-western Qinling orogenic system; *II* Karakorum-Bayanhar-Qamdo-Simao orogenic system; *III* Southern Qiangtang-Gangdise-Himalaya-Baoshan orogenic system. ① Sunan-Tianzhu suture zone; ② Qiemo-Xingxingxia fault; ③ Hongliugou-Lapeiquan suture zone; ④ DaQaidam-Zongwulongshan fault; ⑤ North Qaidam ultrahigh-pressure metamorphic belt; ⑥ Altyn Tagh fault; ⑦ South Qaidam fault; ⑧ Kangxiwar-Muztag-Anemaqen suture zone; ⑨ Yanghu-Jinsha suture zone; ⑩ Longmu Tso-Shuanghu-Lancang suture zone; ⑪ Bangong-Nujiang suture zone; ⑫ Yarlung Zangboe suture zone. *G.* Group, *B.* Block

often placed as Paleo- to Mesoproterozoic, Meso- to Neoproterozoic, Sinian-Cambrian and other general statements, even a broadly temporal constraint as Proterozoic; (2) due to the complexity of tectonic deformation in the orogenic systems, the division is a bit confusing and there are phenomenons of homonym or homonyms; related to this, geochronology data of each geologic unit is also confused, which led to different opinions on its formation age, tectonic characteristics, and other aspects. These issues also restricted the study on the Precambrian crustal evolution of the Tibetan Plateau.

2 Precambrian Rocks in Different Orogenic System

Precambrian geological units in the Tibetan Plateau are distributed scatteredly in the Altyn-Tagh-Qaidam-western Qinling orogenic system, the Karakorum-Bayanhar-Qamdo-Simao orogenic system, and the Southern Qiangtang-Gangdise-Himalaya-Baoshan orogenic system, which have geologic affinities with the Tarim and North China blocks, Yangtze block, and the Indian block, respectively. The Precambrian geological units relatively concentrate in the northern and southern margins of the Tibetan Plateau.

2.1 Precambrian Rocks in the Altyn-Qaidam-West Qinling Orogenic System

The Altyn-Tagh-Qaidam-western Qinling orogenic system includes the Dunhuang, Qilian, Altyn, Quanji, Kunlun Mountain, and the Western Qinling blocks, and the Saishiteng-Xitieshan (North Qaidam), Kangxiwa-Muztagh-Machen suture zones (Fig. 2).

2.1.1 Neoarchean-Paleoproterozoic

Dunhuang Block The oldest metamorphic basement rocks in the Dunhuang block is the Aktas Tagg complex (formerly known as the Milan group), which outcrops EW in the southern Dunhuang block. Its southern boundary is in fault contact with the Hongliugou-Lapeiquan ophiolite melange of the northern Altyn block and unconformably overlied by the Luanshishan Formation of the Xorkol Group (Qingbaikou system) in the Inge Burak.

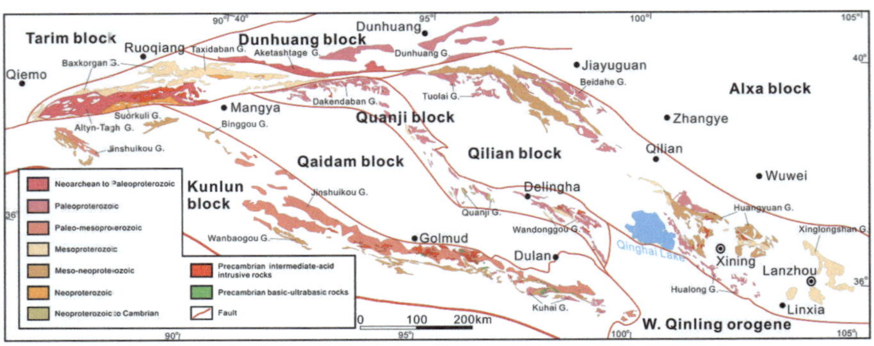

Fig. 2 Geological map showing the distribution of Precambrian rocks in Altyn-Tagh, Qaidam, and Qilian regions. *G.* Group

The Aktas Tagg complex consists of different types of mafic granulites, amphibolites, TTG granitic gneisses, and marbles, as well as charnockites. Different mineral assemblages of supracrustal rocks and charnockite indicate that they have reached the granulite-facies metamorphism (Lu et al. 2008).

Zircon U–Pb and Sm–Nd isochron ages obtained from the Aktas Tagg complex are 2460–2789 Ma (Xinjiang Uygur Autonomous Region Bureau of Geology and Mineral Resources 1993; Che and Sun 1996). Li et al. (2001), Lu et al. (2002a, b, 2006 2008) obtained zircon ages of different rock types from the Aktas Tagg complex by zircon U–Pb TIMS and SHRIMP methods. The age of magmatic zircons from tonalite gneiss is at 2604 ± 102 Ma, magmatic zircon age of adamellite is about 2830 ± 45 Ma, and magmatic zircon age of granitic gneiss is at 2396 ± 36 Ma (Lu et al. 2008), and the inheritance zircons of 3574–3665 Ma are the oldest in the northwestern China so far (Li et al. 2001; Lu et al. 2008). Meanwhile, analysis on the metamorphic zircon rims from the TTGs and metamorphic zircons from the paragneisses gave metamorphic and anatexis ages of 1978 ± 50 and 1986 ± 29 Ma. Xin et al. (2011) obtained SHRIMP U–Pb zircon ages of 2567 ± 32 and 2592 ± 15 Ma from the felsic gneisses and TTGs of the Aktas Tagg complex and suggested them to represent the continental crust accretion of the southern margin of the Tarim craton.

2.1.2 Paleoproterozoic

Dunhuang Block The Paleoproterozoic rocks in the Dunhuang block is called the Dunhuang Group, which mainly outcrops in the north of the Calatayud Alushta Georgia-Kyzy Le Tagg region (Fig. 2). Its rock assemblage has khondalite characteristic and is mainly composed of Al-rich felsic gneisses, including garnet-biotite gneisses, garnet-biotite-hornblende-plagioclase gneisses, leptites and biotite-quartz schists, marbles, graphite-bearing quartzites and garnet-amphibolites, interlayered with a small amount of biotite-plagioclase gneisses and amphibolites. The protoliths are terrigenous clastic and carbonate rocks interbedded with volcanics. The group's stratigraphic features is clear and dominated by amphibolites at the bottom, quartzites and biotite-quartz schists in the middle and marbles at the top, which has been metamorphosed at amphibolite facies.

Xin et al. (2011) obtained SHRIMP U–Pb age of 2140 ± 9 Ma from the volcanic interlayer of the Dunhuang Group and 2135 ± 10 Ma from the diorite pluton, indicating the formation age of Paleoproterozoic.

The Paleoproterozoic intrusions are distributed in the west Kumu Burak-Kalata Shi Tage area, north of the Lapeiquan. Most of the intrusions are distributed in NW or EW trending and mainly amphibolites and dioritic gneiss (gneissic diorite), quartz-dioritic gneiss (gneissic quartz diorite), gneissic hornblende syenite (K-feldspar hornblende gneisses) metamorphosed from mafic dikes. Zircon SHRIMP U–Pb ages are 2135, 2051, and 1873 Ma, respectively, representing the early subduction stage, the syn-collision stage, and post-orogenic stage (Xin et al. 2011). Zhang et al. (2012, 2013a) identified high-pressure granulites in the Dunhuang block and achieved relatively high P/T peak metamorphic conditions and clockwise P-T path, implying that these high-pressure rocks are the product of a collision event. Metamorphic ages were obtained on metamorphic zircons and restricted at 1.83–1.84 Ga, suggesting that the Dunhuang block experienced the late Paleoproterozoic (high pressure) granulite-facies metamorphic event (Zhang et al. 2012, 2013a).

Altyn Block The Neoarchean-Paleoproterozoic rocks are mainly exposed in the western part of the Altyn block and in the small area of the middle part, which are known as the Altyn group. Supracrustal rocks of the group can be divided into two types, in which the first type consists of (garnet) plagioclase amphibolites, (garnet) biotite leptynites, sillimanite-bearing biotite-plagioclase gneisses, biotite-schists, and (graphite) marbles. These rocks have experienced strong deformation and amphibolite-facies metamorphism. The second rock assemblage is high-pressure metamorphic pararocks, including kyanite-biotite-plagioclase gneisses and garnet amphibolites. The two types of rock have no contact relationship in space, but outcrop as different scale of lenticulars in the granitic gneisses. The degree of deformation and metamorphism of the supracrustal rocks are clearly different from the granitic gneisses. Besides intrusive contact, the supracrustal rocks and the granitic gneisses are in tectonic contact (ductile and brittle tectonic belt). Due to the deformation and metamorphism, supracrustal rocks only locally preserve blasto-bedding structures and geochemical composition of mafic volcanic rocks show within-plate and continental margin features.

Hu et al. (2001) obtained U–Pb zircon upper intercept age of 1820 ± 27 Ma and T_{DM} ages of 1.82–1.94 Ga from a gneiss sample of the Altyn group. Mao et al. (2000) got TIMS zircon U–Pb apparent age of 2459 ± 48 Ma from the Keke Sa biotite-feldspar gneiss. Zhang et al. (2000) obtained U–Pb zircon age of 2571 ± 340 Ma (detrital zircon), 1027 ± 10 Ma (upper intercept age), 481 ± 19 Ma (lower intercept age) from the khondalite rocks in the Tula area. Guangxi Institute of Geological Survey obtained U–Pb zircon ages of 1368 ± 49 Ma (upper intercept age) and 566 ± 14 Ma (lower intercept age) from the Qingshuiquan supracrustal rocks. Combined with the metamorphic features, the Altyn is considered as Paleoproterozoic and may experience later metamorphism.

Qilian Block The Paleoproterozoic rocks in the middle Qilian block are called the Tuolai and Beidahe groups, whereas the souther Paleoproterozoic rocks are named as the Hualong group.

The Tuolai and Beidahe groups are exposed in the Tuolaishan and Menyuan areas. The rock association is dominated by biotite-quartz schist, amphibolite, diopside marble, dolomite, quartzite, and leptynite, interbedded with biotite-amphibolite gneiss, which have experienced high amphibolite-facies metamorphism. Zhang et al. (1998) obtained Sm–Nd isochron age of 1980 ± 27 Ma from the plagioclase amphibolite, quartz-actinolitite, plagioclase-biotite schist of the Beidahe group in the north of Diaodaban, and considered it as the formation age.

However, Li et al. (2007) suggeted that the Beidahe group was deposited between 1400 Ma (the minimum detrital zircon age) and 863 Ma (the metamorphic age), not the previously believed Paleoproterozoic, based on the detrital zircon SHRIMP U–Pb results from the muscovite-quartz schist. He et al. (2010) also implied that the group was deposited between 1400 and 724 Ma based on their detrital zircon studies on the Beidahe group in the Danghe area of the Gansu Province.

The Hualong group is mainly exposed on the eastern side of the Qinghai Lake and consists of high amphibolite-facies metamorphic rocks characterized by low-pressure high-temperature metamorphic minerals, including andalusite and sillimanite (Fig. 2). Their protoliths are clastic rocks with a small amount of mafic volcanic rocks and carbonate rocks, similar to the Paleoproterozoic basement rocks in the middle Qilian block. Researcher has different opinions on the age of the Hualong group, including a series of Rb–Sr isochron age of $\sim$1.4 Ga, 2331 ± 215 Ma U–Pb upper intercept age of granitic veins in the gneisses (1:250000 Menyuan 2006), TIMS zircon U–Pb age of 750 ± 30 Ma from the potassic granite intruding the Hualong gneiss (Wan et al. 2003), zircon LA-ICP-MS upper intercept age of 875 ± 8 Ma from weakly gneissic granite vein (Xu et al. 2007). Therefore, the protolith of the Hualong group is generally believed to form at Paleo- to Mesoproterozoic and metamorphised at $\sim$1.0 Ga.

The Xinglongshan group is exposed in the southeastern Qilian block and consists of biotite-plagioclase (monzonite) gneisses, garnet amphibolites, sillimanite-garnet-biotite-quartz schists, biotite-plagioclase leptynites, quartz marbles, graphite marbles, diopside-plagioclase leptynites, and stripped migmatites. Some of the plagioclase amphibolites are mostly massive and display gradual transition relationship with mafic dikes (Chen et al. 2002).

Quanji Block The Paleoproterozoic geologic units in the Quanji block include the Delingha complex and the Dakendaban group, which are associated in space (Fig. 2).

The early Paloeproterozoic Delingha complex is dominated by monzogranitic gneiss, in which amphibolites exist as different sizes and shapes of inclusions. Single zircon U–Pb ages of 2412 ± 14 and 2366 ± 10 Ma have been obtained from the amphibolite and monzogranitic gneiss (Lu et al. 2002a).

The Dakendaban group is distributed along the southern slope of the Altyn Mountain and the northern Qaidam Basin and consists of quartzites, garnet-sillimanite-quartz schists, mica schists, amphibolite schists, and a small amount of granulites. The protoliths are dominated by major clastic rocks and minor volcanics, which suffered high amphibolite to granulite-facies metamorphism. Based on the Sm–Nd isochron age of 1791 ± 37 Ma from the mafic granulite (Zhang et al. 2001), the protoliths are suggested to form in the late Paleoproterozoic.

Kunlun Block The Kunlun block is divided into the West Kunlun and East Kunlun, separated by the Altyn fault.

In the West Kunlun, the Paleoproterozoic Kulangnagu group is well exposed in the Datong city, the Kulangnagu drainage basin and north of the Xaidula city. It is in fault contact with the surrounding stratum and is a set of middle to high metamorphosed rock, which can reach high amphibolite facies. The group can be divided into two formations, A and B. Formation A is composed of various schists (muscovite schists, biotite-quartz schists, two-mica schists, two-mica-quartz schists, amphibolite schists) and quartzites, with a small amount of marbles, gneisses (biotite-plagioclase gneisses, biotite-diopside-plagioclase gneisses) and volcanic layers. Formation B is dominated by marbles, which suffered talc and tremolite alterations. The protoliths of the group are clastic rocks at the bottom, clastic rocks interlayered with carbonate rocks in the middle, and carbonate rocks at the top. Reliable age constraints of the group are absent and the granite intrusion gave zircon U–Pb age of 480–495 Ma (1:250000 Yecheng). The metamorphic and deformation characteristics are similar with those of the Xaidula group in the west, so the group is temporarily assigned to the Paleoproterozoic.

In addition, the Paleoproterozoic Elian Carter group and Hero Stan TTG complex are exposed at the southern edge of the Tarim Basin, northern part of West Kunlun, which do not have direct contact in the region. The Elian Carter group is exposed along the Tikriklamu Ron and Boston rivers and extends south to the Hotan city. It consists of quartz-mica schists, marbles, and garnet-biotite-plagioclase (monzonite) gneisses in the north of Kangxiwar and locally unconformably overlied by the spilite keratophyres and clastic rocks of the Changcheng system Serra Jiazitage group, of which the latter is constrained as 1764 Ma, so the underlying group is placed as the Paleoproterozoic.

The Herro Stan complex scattered outcrops in the Aotulageyer and Ca Rava Schickel areas, south of the Yecheng city. The rock assemblage is TTG series granitic gneisses and unconformably overlied by the Jixian system Borchardt Tagg formation. Zhang et al. (2004) obtained zircon SHRIMP U–Pb ages of 2426 ± 46 Ma from the granitic gneiss.

In the East Kunlun, the Paleoproterozoic Jinshuikou group is exposed in the Baisha River and Jinshuikou areas and composed of various leptynites, gneisses, migmatites, migmatic gneisses, migmatic granites, dolomite marbles, olivine marbles, and amphibolites (Fig. 2). Sillimanite-andalusite schists are found in the western Kaimudu Taseer regions and sillimanite-andalusite-biotite-plagioclase gneisses are recognized in the Ren Su Haitu River and Dahuozao River areas. The metamorphic grade becomes higher in the Tiantai Mountain, where migmatites, two-pyroxene granulites and amphibolite granulites occur. Two-pyroxene granulites and leptynites occur from the Jinshuikou to Jialu River in the east. The protoliths are clastic rocks, mafic volcanic rocks, and magnesium-rich carbonate rocks, reflecting the characteristics of active continental margin tectonic setting. Overall, the group evolved from the volcanic and clastic rocks at the bottom to carbonate rocks at the top. From west to east, terrigenous clastic and volcanic rocks decrease and carbonate rocks increase, combined with decreasing of strata thickness. It is generally believed that the Baishahe group has undergone two stages of metamorphism, including the Luliang and Jinning events.

The Paleoproterozoic Kuhai group in the south section of Burhan Budai Mountain is a set of amphibolite-facies metamorphic rocks, including a variety of gneisses, plagioclase amphibolites, biotite-quartz schists, and diopside marbles. In fact, in addition to the original metamorphic supracrustal rocks (including clastic rocks, intermediate-mafic volcanic rocks, and carbonate rocks), there are a large amount of metamorphic felsic intrusions, which have experienced strong migmatization and have complex features.

The time constraints on the Kuhai group are inconsistent. Previous studies got Sm–Nd whole-rock isochron age of 2213 ± 17 Ma from the intruding mafic dykes, and obtained U–Pb zircon upper intercept age of 2330 ± 50 Ma and lower intercept age of 746 ± 6Ma from hornblende-plagioclase gneiss of the group. Wang et al. (2007) obtained U–Pb zircon age of 1441 Ma from biotite-plagioclase gneiss, Rb–Sr ages of 1167 ± 19 and 1132 ± 69 Ma, respectively, from mica-quartz schist and gneiss, and suggested that the formation age of the complex is at Paleo- to Mesoproterozoic and its final consolidation time should be the end of the Mesoproterozoic.

Western Qinling Block The Paleoproterozoic rocks of the Western Qinling block mainly occur in the Wushan-Gangu area, known as the Qinling group, which is mainly a set of high amphibolite-facies metamorphic rocks. The lower part consists of biotite-plagioclase gneisses, amphibolites, biotite-quartz schists, amphibolite schists and marbles, graphite marbles and the upper part is composed of quartz-biotite schists and quartzites, interbedded with stripped marbles and amphibolites, whose portoliths are terrigenous clastic-carbonate rocks.

Lu et al. (2006) dated the detrital zircon of pararocks of the Qinling group by U–Pb SHRIMP and LA-ICP-MS methods and found that its provenance is mainly 1.5–1.9 Ga granites and the group was most possibly deposited at the end of the Mesoproterozoic.

Saishiteng-Xitieshan Suture Zone (North Qaidam) The Mesoproterozoic rocks in the Saishiteng-Xitieshan suture zone are mainly Shaliuhe group, which mainly consists of various schists, including quartz-mica schists, banded quartz schists, plagioclase-amphibolite schists, and garnet-amphibolite-quartz schists, with eclogite enclaves. Mineral assemblage of different rocks are basically consistent with the Dakendaban group in the north. The protoliths are a set of intermediate-mafic volcanic rocks at active continental margin. The direct isotopic age constaint on the goup is absent for now. The protoliths of eclogites are ocean floor basalts, ocean island basalts, and island arc basalts, and involved continental crust materials in deep subduction (Yang et al. 2003). Based on the rutile inclusions in the kyanite and coesite inclusions in the zircon from the Beishan gneiss in the north of the Yematan (Yang et al. 2002; Song et al. 2003), supracrustal rocks experienced deep subduction and UHP metamorphism.

Single zircon U–Pb upper intercept age of 2017 ± 31 Ma and lower intercept age of 2018 ± 31 Ma have been achieved from mylonitized amphibolite (1:250000 Dulan County 2006). Upper intercept age of 2002 ± 60 Ma and lower intercept age of 219.11 Ma were obtained from the biotite-quartz shicst, showing that it was formed in the Paleoproterozoic and may be equivalent to the Dakendaban group in the north Quanji block.

Kangxiwa-Muztagh-Machen Suture Zone In the western Kangxiwa-Muztagh-Machen suture zone, the Paleoproterozoic Kangxiwa group is sandwiched in the melange as different sizes of tectonic enclaves, which consists of biotite-plagioclase gneisses, garnet-amphibolite-biotite-plagioclase gneisses, garnet-biotite-plagioclase quartzites, granet-plagioclase leptynites, garnet (sillimanite)-biotite-plagioclase leptynites, garnet quartzites, silimanite biotite-quartz schists, garnet (sillimanite)-two-mica (biotite)-quartz schists. The protoliths mainly belong to continental island arc rocks, some of which also have characteristics of oceanic island arc. The peak metamorphism reached to high amphibolite facies.

A set of middle-high metamorphosed stratum was disintegrated from the Triassic Keleqinghe group, which has reached high amphibolite metamorphism and was though to be comparable with the Ailiankete group and higher than those of the Tianshuhe, Sailatu and Sailajiazitage groups in the metamorphic grades (Shaanxi Institute of Geological Survey 2006). The geochemical compositions of metamorphic clastic rocks are similar with the Archean meta-clastic rocks, rather than the Proterozoic meta-clastic rocks. Based on regional stratigraphic correlation, the group was thought to be formed at Paleoproterozoic.

2.1.3 Mesoproterozoic

Qilian Block The Mesoproterozoic strata in the Qilian block include the Changcheng system Xinglongshan, Zhulongguan, Haiyuan groups and the Jixian system Tuolainanshan, Gaolan and Huangyuan groups, whose rock types are complex and metamorphic features change from area to area. The greenschist-facies to low amphibolite-facies metamorphic rocks and the related mineral assemblages have been found. The greenschist rocks (metabasalt) of the Haiyuan group were dated at 800 ± 40, 1002 and 1128 ± 209 Ma by Rb–Sr whole-rock method (1:50000 Nianhaowan and Dalachi 1992). The Neoproterozoic Qingbaikou system is a set of stromatolites and microfossils rich clastic-carbonate rocks, the bottom conglomerates of which unconformably overlies the metamorphic basement rocks.

The Neoproterozoic granites invading the Paleo- to Mesoproterozoic stratum have two age groups, including ~800 and ~900 Ma. ~800 Ma granitic rocks consist mainly of granodiorite and granite, whose geochemical characteristics imply rift environment. In constrast, ~900 Ma granitic rocks include quartz diorite, granodiorite, and granite, which formed in active continental margin (Tung et al. 2013). Amphibolite-facies metamorphosed mafic rocks (gabbro) in the southeastern block have crystallization ages of 905 ± 6 and 919 ± 10 Ma, whose geochemical characteristics indicate continental island arc setting (Tung et al. 2012).

Quanji Block The Mesoproterozoic metamorphic rocks distributed between the Tanjianshan and Wandonggou are called the Wandonggou group, of which the lower part is composed of phyllites, sericite phyllites interlayered with limestones and minor amphibolite schists, the central part consists of dolomites, marbles interbedded with sandstones, and the upper part is limestones, marbles, dolomite marbles sandwiched with sericite quartz schists, plagioclase amphibolite schists, and tuffaceous sandstones. Rb–Sr isochron age of 1022 ± 64 Ma has been obtained from the group (Yu et al. 1994), which represents the metamorphic age, so the group should belong to Mesoproterozoic. The late Neoproterozoic unmetamorphosed Quanji Group unconformably overlies the Wandonggou group.

The Quanji group is located between the Quanji Mountain and the Oulongbuluke Mountain in the north of the Qaidam Basin, unconformably overlies the migmatic gneisses of the Delingha complex, and is unconformable overlied by the Cambrian brachiopods-bearing stratum. The group can be divided into the following assemblagess: the bottom continental sandstones and conglomerates (Mahuanggou formation), whose basal conglomerates are dominated by siliceous gravel and scarce gravels from the underlying migmatites of the Delingha complex, implying the Quanji group was autochthonous rather than allochthonous. Bidirectional oblique beddings or parallel beddings developed in the feldspar quartz sandstones and quartz sandstones of the lower part, representing coastal to tidal flat clastic rocks (Kubaimu and Shiyingliang formations). Several hundreds of meters-thick stromatolite-rich carbonate rocks (Hongzaoshan formation) overlie the sandstones and the transition zone has three layers of volcanics, between which tuffs exist, diaplaying cycle characteristics of volcanic eruptions. Above the carbonate rocks, a set of purple and gray-green variegated sandstone (Heitupo formation) occur. Massive conglomerates (Hongtiegou formation) develop in the upper part, which were treated as tillites and can be comparable with the Hangeerqiaoke tillites of the Kuluketage group in the northern margin of Tarim Basin (Wang et al. 1980). The Cambrian brachiopods fossil-bearing phosphorus clastic rocks parallel unconformably overlie the tillites. The whole stratigraphic sequence of the group display aulacogen or passive continental margin basin characteristics. The group is almost unmetamorphosed and is dominated by brittle fractures, apparently different from the metamorphic basement, so the group should be attributed to the sedimentary cover after cratonization.

The Quanji group is the lowest part of the sedimentary covers preserved in the Oulongbuluke block, whose stratigraphic sequences show the aulacogen-like sedimentary characteristics and have sedimentary cover features. Single zircon U–Pb age of 738 ± 28 Ma has been obtained from basalt of the middle-lower Quanji group, therefore the low boundary of the group was estimated as $\sim$760 Ma. The group is important for exploring the Neoproterozoic geological evolution, the tillites of the Hongtiegou formation is a marker bed for stratigraphic correlation and is the key layer to correlate the late Neoproterozoic stratigraphic correlation of the North China Craton with the Tarim block (Lu et al. 2002b).

Altyn Block The Mesoproterozoic Changcheng System Bashkuergan and Jixian system Taxidaban groups outcrop in the northern, southern, and western margins of the Altyn Mountains, which have experienced low greenschist-facies metamorphism. The protoliths of the Bashkuergan group are basalts, andesite basalts, and dacites, of which most of basalts display tholeiitic features and a few are alkaline, showing within-plate or continental margin rift features. The protoliths of the Taxidaban group are shallow marine carbonate rocks and clastic rocks, which possibly formed by shallowing upward in filling process of rift basins.

The Neoproterozoic Qingbaikou system Xorkol group unconformably overlies the Jixian system or older strata, so its metamorphic age was estimated as ~ 1.0 Ga.

The Xorkol group (Annanba group in the Gansu Province) mainly distributes in the Annanba and Ingebulake areas and is a set of littoral-shallow marine terrigenous clastic and carbonate sedimentary rocks. The lower part (Luanshishan formation) has hundreds of meters-thick purple fluvial glutenite with good roundness at the bottom, displaying characteristics of basal conglomerate. The upper part is a set of coastal-shallow marine clastic-carbonate rocks interlayered with volcanics locally and unconformably overlies the Jixian or older stratum. It is composed of low metamorphic phyllites and greenschists, constituting the initial sedimentary cover on the block.

Kunlun Block In the West Kunlun, the Mesoproterozoic Changcheng system Sailatu group constitutes the main ridge of the western Kunlun Mountains, which consists of biotite-plagioclase gneisses, biotite schists, leptynites, quartzites interbedded with amphibolites, above which the Jixian system Sanju Tagg group unconformably overlies.

The Jixian system Sanju Tagg group is primarily exposed in the Qiaerlongnan, Akto County, and Mututage areas. It is overall a set of low to middle metamorphosed clastic and carbonate rocks, including mica or feldspar-quartz schists, mica-tremolite-calcite schists, biotite leptynites, quartzites, and marbles.

The Sinian-Cambrian Alajiaoyi group is in fault contact with the upper and lower strata, whose lower rock assemblage is biotite-quartz schist, biotite-feldspar-quartz schist, leptynite and polymictic conglomerate and the upper part is composed of dolomitic limestone, crystalline limestone, and quartz greywacke.

In the East Kunlun, the Mesoproterozoic rocks include Xiaomiaoyan formation and Jixian system Langyashan formation.

The Xiaomiaoyan formation is the upper part of the Jinshuikou group and intermittently outcrops in the Xiaojianshan, Gulamulakesayi, Ulu Akzo River, Hatu, Boluositai and Ulasitai-Naomuhunketeli-Haolaterenna around the main ridge of the Burhan Budai Mountain. It is a set of greenschist-facies metamorphic rocks, whose protoliths are coastal-shallow marine clastic-carbonate rocks interbedded with intermediate-mafic volcanics. At the bottom, it is separated from the Baishahe formation by a large amount of quartzitic rocks, but the deformation characteristics is the same with those of the Baishahe formation. At the top, the formation is parallel unconformably overlied by the Mesoproterozoic Jixian System Langyashan formation, which consists of dolomites, dolomitic limestones interlayered with quartz sandstone and mafic volcanic rocks.

The metamorphic (rim) zircon of the gneisse and leptynites from the Xiaomiaoyan formation gave SHRIMP ages of 1097 ± 30 and 969 ± 32 Ma, respectively. The Meso- to Neoproterozoic Qingbaikou system Qiujidonggou formation unconformably overlies the Langyashan formation, indicating that strong metamorphism occurred in the Mesoproterozoic. The detrital zircon evaporation Pb–Pb ages of 1901, 2053 and 2069 Ma have been achieved from garnet-mica-quartzite of the Xiaomiaoyan formation, and intrusions (two-mica plagiogneiss) gave crystallization zircon Pb–Pb ages of 913, 971, and 1011 Ma, implying that the intrusive age should be older than 1000 Ma.

The Neoproterozoic Qingbaikou system Qiujidonggou formation is overall a set of low metamorphic, low maturity terrigenous clastic rocks with mafic volcanic rocks.

The Meso- to Neoproterozoic melanges include the Changliugou-Xiangyangquan ophiolite melange in the Muztag-Xidatan ophiolite belt and the Wanbaogou-Qingshuiquan ophiolite melange in the Burhan Budai ophiolite belt. The Meso- to Neoproterozoic melange are spatially isolated blocks in the Paleozoic ophiolite melange and consist of metamorphic and deformed mafic-ultramafic rocks, mafic volcanics, clastic and carbonate rocks and cherts. The metamorphic rock assemblages are metamorphic mafic volcanic-pyroclastic rocks, sericite phyllites, epidote chlorite schists, chlorite epidote schists, muscovite chlorite schists, dolomite marbles, chlorite marbles, serpentine marbles, phlogopite marbles, and tremolite rocks. The pararocks in the Wanbaogou-Qingshuiquan ophiolite melange are called the Wanbaogou group.

Based on the whole-rock Sm–Nd isochron age of 1004.7–1372 ± 85 Ma of the mafic-ultramafic rocks (Zheng et al. 1992; Xie 1998, 1:250000 Donggicuona 2000; 1:250000 Arak Lake 2003), the Sm–Nd isochron age of 1441 ± 230 Ma of the metabasalt from the Wanbaogou group ($Pt_{2\text{-}3}$), the SHRIMP zircon U–Pb age of 1348 ± 30 Ma (1:200000 Nachitai 1981; Jiang et al. 1992; Ah et al. 2003), the protoliths were formed at the Meso- to Neoproterozoic and metamophsed at the Jinning era. However, Lu et al. (2002b) got TIMS zircon age of 522.3 ± 4.1 Ma from the Qingshuiquan gabbro, implying that the age of the Wanbaogou-Qingshuiquan ophiolite need further study.

2.1.4 The Precambrian Geological Events and Evolution of the Altyn-Qaidam-West Qinling Orogenic System

The Archean geological records are preserved very little in the Altyn-Qaidam-West Qinling orogenic system, so far the relevant age information have only been obtained from the Dunhuang block in the north. The Aketashitage complex is mainly composed of supracrustal rocks and TTG granitic rocks and the existing data suggest that the main formation age is of Archean to Paleoproterozoic (Dong et al. 2007a, b; Mei et al. 1998; Xinjiang Bureau of Geology and Mineral Resources 1993; Che and Sun 1996; Lu et al. 2002b, 2006, 2008) and the Paleoarchean age is also reported locally (Li et al. 2001). 1:25000 regional geological survey of the Xiaomiaoyan group in the eastern Kunlun found 3306 Ma detrital zircon, implying that their provenance should include Archean crust. However, due to the limited research, it is still unable to accurately determine the scope of the Archean continental nuclei.

The widespread Paleoproterozoic TTG gneisses reflect crustal vertical accretion after the initial nucleus formation, which is associated with development of clastic-carbonate rocks interbedded with mafic volcanic rocks at the active continental margin.

The widespread unconformity between the Paleoproterozoic and Mesoproterozoic strata in the Altyn, Kunlun, and the southern edge of the Tarim indicate that the areas have suffered late Paleoproterozoic tectonic events (Luliang movement), so that the Neoarchean and Paleoproterozoic strata were metamorphosed to amphibolite facies (locally granulite facies and even high-pressure granulite facies) and changed to the crystalline basement. Zhang et al. (2012) considered that the 1.85 Ga high-pressure granulite-facies metamorphism in the Tarim Craton is similar to that of the North China Craton, both of which are part of the Paleoproterozoic Columbia supercontinent.

After formation of the Paleoproterozoic crystalline basement, the Neoproterozoic sedimentary cover began to be deposited. The Changcheng system volcanics and shallow water, coastal phase passive continental margin deposition reflect the start of the Mesoproterozoic continental blocks breakup. The typical volcanics include the Sailajiazitage spilite keratophyre in the Tiekelike Mountain of the southern margin of the Tarim Basin, the within-plate basalts of the Xiaomiao group in the eastern Kunlun and the Changcheng system basalts, andesite basalts and dcaites of the Bashikuergan group in the Altyn block. The basalts of the Bashikuergan group are dominated by tholeiitic basalt with subordinate alkaline series and their geochemical characteristics show within-plate or margin rift features. The high maturity of clastic rocks of the Changchengian system represents the sedimentary response in relatively stable areas at the same time or later than the magmatic event.

The Jixianian extensional rifting continued to develop and ophiolite rocks occurred with the intensification of breakup, such as Changliuquan-Xiangyangquan ophiolite melange in the Muzitage-Xidatan of the Kangxiwa-Muztagh-Machen suture zone and the Wanbaogou-Qingshuiquan ophiolite melange in the Burhan Buda Mountain. The accompanying sedimentary response is a large amount of Jixian system passive continental margin shallow water-carbonate platform deposition in the Altyn, Kunlun and other areas, such as the Bochatataga formation in the southern edge of the Tarim, the Taxidaban group in the Altyn region, the Sangzhutage group in the West Kunlun and the Langyashan group in the East Kunlun, all of which are clastic-carbonate rocks.

The blocks gathered in the Qingbaikou area and all the Haiyuan group in the Qilian block, the Suoerkuli group in the Altyn block, and the Qiujidonggou formation in the Kunlun block unconformably overlie the underlying Jixian system or older metamorphic rocks with basal conglomerates. The Qingbaikou system geological units are comparable with the global events related to the Rodinia supercontinent formation in time and should be its response in the western China.

The Qingbaikou system deposition is a set of shallow marine clastic-carbonate rocks, whose sedimentary environments and stromatolites combination are similar to those in the North China Craton at the same period.

After the Rodinia converge event, the tillite occurred in the Nanhua-Sinian Quanji group, which is comparable with the Nantuo tillites in the Yangtze Block. This combination of early North China Craton's affinites and later Yangtze Block's affinites has two possible factors. One possibility is that these blocks were in a transition zone between the North China and Yangtze blocks; another possibility is that the blocks were parts of the North China Craton before the Nanhua era, but separated from it and connected with the Yangtze block, both of which have the Nanhua-Sinian sedimentary covers.

The Nanhua-Sinian sediments marked that the area has begun to enter a new phase of the breakup process and the blocks collaged and collided to form the current structural pattern in the early Paleozoic.

2.2 Precambrian Rocks in the Karakorum-Bayan Har-Chamdo-Simao Orogenic System

The Karakorum-Bayan Har-Chamdo-Simao orogenic system includes the Yulong Tagg-Bayan Har and North Qiang-Qamdo-Simao Blocks and the Yanghu-Jinshajiang suture zone, in which only the Proterozoic geological unit has been found (Fig. 3).

2.2.1 Paleo- to Mesoproterozoic

The Yulong Tagg-Bayan Har Block The metamorphic basement rocks in the Yulong Tagg-Bayan Har block are the Paleoproterozoic Xiacun group, which are exposed as a series of dome cores in the Jiulong-Muli region of the southern block. It is mainly composed of high greenschist to amphibolite-facies metamorphic rocks, including biotite (hornblende) plagioclase gneisses, leptynites, plagioclase amphibolites and various schists, marbles, whose protoliths are argillaceous and volcanic rocks. On the basis of regional stratigraphy and geochronological data, Yao and Ni (1990) suggested that the Xiacun group is younger than the Kangding group and older than the Huili and Yanbia groups, whose formation age is between 2000 and 1700 Ma in the Paleo- to Mesoproterozoic.

The North Qiang-Qamdo-Simao Block The Paleoproterozoic rocks distributed in the west of the Altyn fault, along the Bulunkou-Taxkorgan County, are called the Bulunkuole group, which consists of garnet and sillimanite-rich metamorphic rocks, including garnet (or hornblende)-plagioclase gneisses, garnet-sillimanite-biotite-plagioclase gneisses and marbles, interbedded with meta-felsic volcanic rocks. In the Taheman, Qiaopukalimo, Laobing, and other places also developed a set of iron formation, including layered-banded magnetite, magnetite-quartzite, and so on, forming a huge scale (up to 120 km) of sedimentary-metamorphic magnetite bed.

Zircon LA-ICP-MS U–Pb age of 2481 ± 14 Ma has been obtained from the meta-rhyolite of the Bulunkuole group in the southeastern Budaer, the Tashikuergan

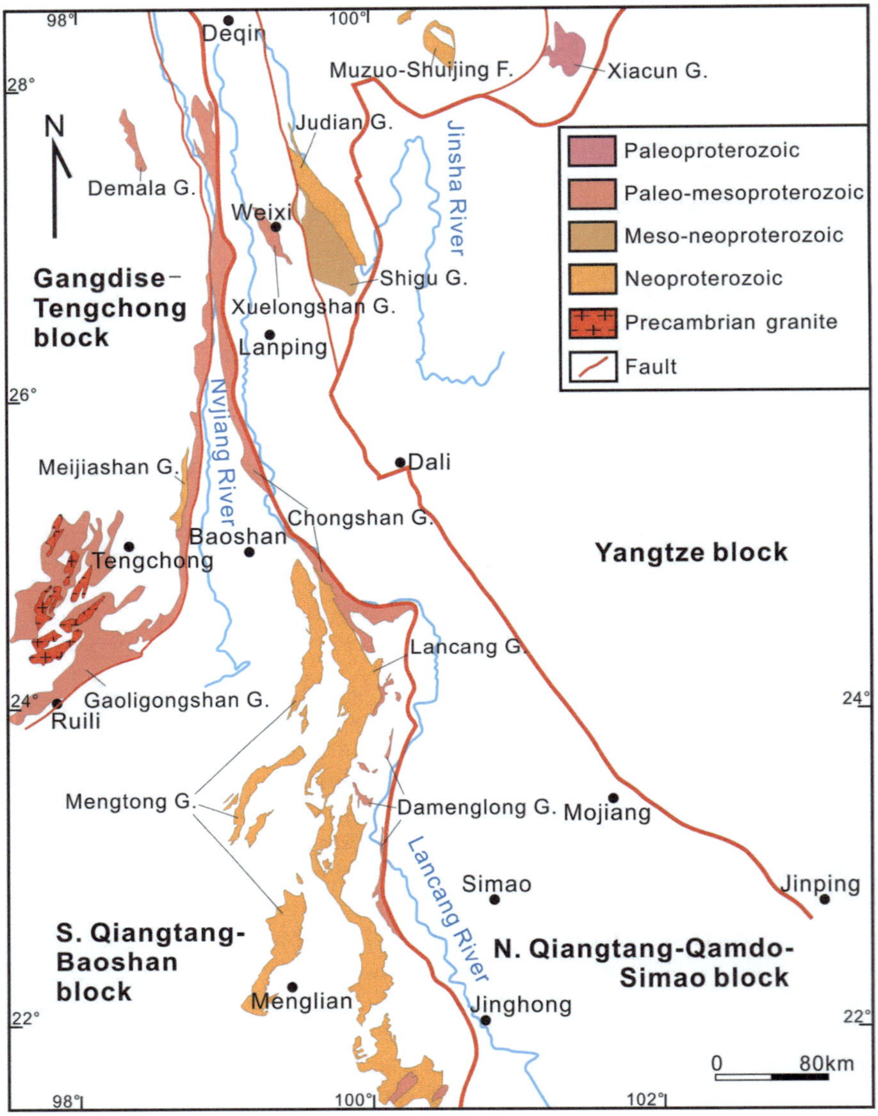

Fig. 3 Geological map showing the distribution of Precambrian rocks in western Yunnan. *F.* Formation, *G.* Group, *B.* Block

County (Ji et al. 2009), which was designated as the Paleoproterozoic.

The Paleo- to Mesoproterozoic rocks outcrop in the Xiaosumang and Jialaiduo areas of the eastern margin of the Chamdo basin and the Yanghu-Jinshajiang suture, called the Ningduo group. The main lithologies include biotite-plagioclase gneiss, quartz schist, quartzite and leptynite interbedded with amphibolite schist, hornblende schist, marble and diopsidite, which are characterized by regional schistosity and gneissosity and suffered amphibolite-facies metamorphism. Based on regional data, the protoliths were formed in the Paleoproterozoic and metamorphosed in the late Mesoproterozoic, but no reliable geochronologic data has been given.

He et al. (2011) used LA-ICP-MS zircon U–Pb method to obtain the oldest detrital zircon age of 3981 ± 9 Ma from the garnet-two-mica-quartz schist of the Ningduo group near the Yushu and also got 3505 ± 18–3127 ± 10 Ma detrital zircon ages, reflecting very old continental nucleus in the source area. Other age groups include 2600–2300, 1700–1400, 1200–850, and 700–530 Ma with peaks at ~2440, ~1532, ~982, and ~618 Ma, respectively. Among them, the main age peak of ~982 Ma is synchronous with the Grenville orogeny of the Rodinia supercontinent and the subordinate age peak of ~618 Ma is roughly simultaneous with the Pan-African movement of the Gondwana supercontinent. Thus, further study is needed for the formation age of the group.

In the Lanping area of the southern block, the Paleo- to Mesoproterozoic rocks located in the eastern side of the Lanping basin and along the Xuelong Mountian are called the Xuelongshan group. Its upper part consists of (hornblende) biotite-plagioclase leptynites, (staurolite kyanite) two-mica-quartz schists and a small amount of quartzites, amphibolite diopside marbles, amphibolite schists; the central part is composed of garnet-two-mica-quartz schists, (garnet) biotite-plagioclase leptynites, garnet-bearing biotite-plagioclase gneisses and a little biotite-quartz schists, amphibolite schists; the lower part includes biotite-plagioclase lepynites, (biotite) amphibolites, two-mica (quartz) schist and minor (amphibolite) biotite-plagioclase gneisses, marbles. The rocks are generally migmatized and metamorphosed to amphibolite facies. According to the regional data, the protolithes were formed in the Paleo- to Mesoproterozoic and metamorphosed in the late Neoproterozoic, but no reliable geochronologic data has been obtained.

2.2.2 Meso- to Neoproterozoic

The Yulong Tagg-Bayan Har Block To the west of the Altyn fault, the Meso- to Neoproterozoic Changcheng System Tianshuihe group, Jixian system Chalukou formation and Qingbaikou system Xiaoerkegudi formation are mainly a set of low metamorphic carbonate and fine-grained clastic rocks interbedded with mudrocks, including sericite (or chlorite) quartz schists, calcite schists, biotite-quartz calcite schists, chlorite-albite schists, quartzites, marbles, and so on. Metamorphic mineral assemblages reflect greenschist-facies metamorphism and the metamorphic age is of Neoproterozoic.

The Xinjiang Bureau of Geology and Mineral Resources suggest that the Tianshuihe group is formed at Changcheng era based on the *Xiayingella, Stratifera, Litia* fossiles collected from the stromatolite and regional comparison. The stromatolite fossils from the dolomites of the Xiaoerkegudi formation are tapered, cylindrical, and the laminae are very rough. According to the identification of

Researcher Feng Tang from the Institute of Geology, Chinese Academy of Sciences, they are *Tekesia* sp. *Turks* stromatolite (undetermined species), *Tungussia* sp. stromatolites (undetermined species) and its era is Qingbaikou era. Combined with the previous collected *Xiayingella, Stratifera, Litia*. Stromatolites from the formation, it is redefined as Qingbaikou system based on comprehensive analysis.

The Qingbaikou system Xiakasha formation is only distributed in the Kasha of Sichuan Muli County in the eastern block and is a set of metavolcanic rocks, pyroclastic-terrigenous clastic strata. The lower part consists of phyllites and carbonaceous slates, metamorphic felsic siltstones tuffs, albite quartz schists; the central part includes albite (quartz) leptynites, albite schists, albite quartz schists, mica schist, and phyllites; the upper part is composed of epidote-actinolite schists, chlorite-albite schists, and albite schists. Single zircon U–Pb ages of 855 ± 8 and 1083 ± 2 Ma have been obtained (Hu 1994). The overlying Nanhua System Muzuo formation is dominated by the tillites, conglomerates, and slates in the upper and lower parts and meta-siltstones and slates in the middle. The Sinian Shuijing formation consists mainly of dolomites containing algae fossils, comparable with the Doushantuo formation in the Yangtze block.

North Qiang-Qamdo-Simao Block The Neoproterozoic rocks in the Qamdo area of the eastern block are called the Caoqu formation and are distributed in the Caoqu area of the Chamdo County, in fault contact with the surrounding Upper Triassic low metamorphic rocks. The upper part consists of sericite-chlorite schists, muscovite-quartz schists, the middle and lower parts are composed of conglomerates, quartz schists, phyllites, quartzites, feldspar-quartzites and sericite-chlorite schists, interbedded with meta-mafic volcanics. The metamorphic grade is greenschist facies and the protoliths were conglomerates, muddy sandstones, calcareous shales, and mafic volcanics. Previous studies has obtained zircon U–Pb ages of 999 and 876 Ma, comparable with the Qingbaikou era.

Metamorphic strata equivalent to the Caoqu formation outcrop as “tectonic blocks” within the Yanghu-Jinshanjiang suture zone, which are known as the Judian formation. It is mainly composed of carbonaceous sericite phyllites, carbonaceous slates, sericite quartz phyllites, sericite phyllites, albite–actinolite schists, and the protoliths are intermediate-mafic volcanic rocks and volcaniclastic rocks.

Exposed in the Tacheng-Shigu area of the Lijiang city, Yunnan Pronvince, of the southern block, the Meso- to Neoproterozoic rocks are called the Shigu goup, divided into the Yangpo, Longba, and Tacheng formation from bottom to top. Among them, the Meso- to Neoproterozoic Yangpo formation ($Pt_{2\text{-}3}$) has suffered strong metamorphism and deformation and consists of biotite schists, amphibolites, sillimanite-garnet-biotite gneisses, etc., which should be basemnet of the Yangtze block. Sm–Nd model ages of 1369.8–1343.8 Ma (Zhai et al. 1990) and Rb–Sr isochron age of 996.1 ± 33.2 Ma (Zhai and Cong 1993) have been obtained from amphibolite of the Yangpo formation and the Shigui formation, respectively. The Neoproterozoic Longba and Tacheng formations (Pt_3) have low metamorphic grade, strong deformation and the metamorphic rocks are biotite schist, biotite-quartz schist, albite schist, actinolite schist, chlorite quartz schist, chlorite schist, sericite-chlorite interbedded with meta-mafic volcanic.

2.2.3 The Precambrian Geological Events and Evolution of the Karakorum-Bayan Har-Chamdo-Simao Orogenic System

The Karakorum-Bayan Har-Chamdo-Simao orogenic system is part of the Pan-Cathaysia and Yangtz blocks, but the exposed Precambrian geological units are extremely limited, so it is difficult to recover the geological evolution. This section only simply describes clues of the Neoproterozoic tectonic events.

The folded basement was formed in the early Paleo- to Neoproterozoic, including the Bulunkuole, Ningduo, Xiacun groups and the Xiakasha formation in the Muli area. The Jinning event made the fold reture and the Neoproterozoic Caoqu group may belong to molasse when the craton returns. With the regional dynamic metamorphism of greenschist facies, the crust thickened, matureded, and stabilized.

In the Nanhua-Sinian period, the system went into a stable stage of development, and fine-grained clastic rocks containing tillites comparable with those in the Yangtz block were formed.

2.3 Precambrian Rocks in the South Qiang-Gangdise-Himalaya-Baoshan Orogenic System

The South Qiang-Gangdise-Himalaya-Baoshan orogenic system includes the South Qiang-Baoshan, Gangdise-Tengchong and Himalayan blocks and Longmucuo-Shuanghu-Lancang River, Bangong-Nujiang and Yarlung Zangbo suture zones, and so on (Figs. 3 and 4).

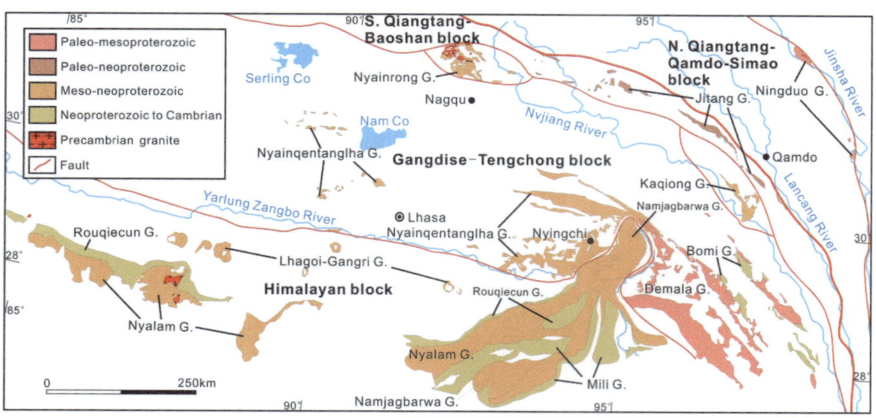

Fig. 4 Geological map showing the distribution of Precambrian rocks in Southern Qiangtang, Gangdise, and Himalayan region. *G.* Group

2.3.1 Paleo- to Mesoproterozoic

The Paleo- to Mesoproterozoic rocks in the block outcrop in the southeast, including the Chongshan and Demala groups.

South Qiang-Baoshan Block The Paleo- to Mesoproterozoic strata in the western Yunnan Province of the South Qiang-Baoshan block is called the Chongshan group, which is distributed along the Biluo and Chong Mountains. The lower part consists of the garnet-sillimanite-biotite-plagioclase gneisses, garnet-bearing sillimanite-biotite schists, quartzites interlayered with biotite-plagioclase leptynites, amphibolites, and marbles, which have generally strong migmatization. The upper part consists of biotite-quartz schists, plagioclase leptynites and hornblende leptynites, marbles, whose mineral assemblages reflect amphibolite-facies metamorphism.

Concordant U–Pb zircon age of 922 Ma has been obtained from the biotite-plagioclase leptynite of the group, the isotope model ages of amphibolites and gneiss focused on 1100–1000 and 1900–1600 Ma, respectively. The model ages of gneisses are roughly equivalent with those of the Damenglong group. Taking into account the basically same rock assemblage, deformation and metamorphism features with those of the Gaoligongshan group, both of them are crystalline basement rocks in the region. The age of 922 Ma is deemed as the metamorphisic age and the formation age is considered as the Paleo- to Mesoproterozoic.

Gangdise-Tengchong Block The Paleo- to Mesoproterozoic rocks in the block are called the De Mala group, which is exposed in the Chayu area, east of the east structural knot and is a set of high metamorphic rocks, including gneisses, schists, leptynites, migmatites and a small amount of marbles, all of which suffered amphibolite-facies metamorphism.

Sm–Nd ages of 2146–2264 and 1524–1598 Ma have been obtained from the De Mala group (1:200000 Songleng and Zhuwagen 1995), and also Sm–Nd isochron age of 2138 Ma has been obtained. Dong et al. (2011a, b) used zircon LA-ICP-MS U–Pb dating method for the orthometamorphic rocks (biotite-hornblende schist) and parametamorphic rock (biotite-quartz schist). The biotite-hornblende schist gave the age of 217 Ma, representing the crystallization age and mafic magmatism during the Late Triassic. The biotite-hornblende schist gave detrital zircon ages ranging from 188–1902 Ma, mainly in the 200–210, 520–600 and 900–1100 Ma, reflecting ages of the source area. Above data indicate that some areas of the De Mala group in the southeastern Tibet were the Paleozoic sedimentary rocks, which have undergone alteration of late Mesozoic magmatism. Although this result can not completely deny the existence of Precambrian basement metamorphic rocks, but at least reflects that the De Mala group should be determined as a set of metamorphic rocks.

2.3.2 Meso- to Neoproterozoic

South Qiang-Baoshan Block The Meso- to Neoproterozoic rocks in the eastern Tibet are mainly exposed in the Dêngqên-Riwoqe-Jitong-Bitu area and named the Jitong group in the west of the north Lancang River fault zone. The lower part consists of gneisses and migmatites, and the upper part is dominated by various schists, including the lower Enda formation and the upper Youxi formation.

Based on the current geological data, the Jitang group mainly consists of amphibolite facies biotite-plagioclase gneisses, biotite leptynites, hornblende-plagioclase gneisses, banded migmatites interbedded with quartz schists, sillimanite-garnet schist plagioclase schists, biotite-feldspar schists, and marbles. The protoliths are quartz sandstones, greywackes and argillites with a small amount of carbonate rocks and mafic volcanic rocks. Based on the metamorphic mineral assemblages, the metamorphic degree was determined as amphibolite facies.

The Jitang group was originally classified as Paleo- to Mesoproterozoic. However, He et al. (2012) used LA-ICP-MS zircon U–Pb method to date quartz chlorite schist (formerly intermediate volcanic rock) from the Youxi formation of the Jitang group along the Taniantaweng Mountain and got the formation age of 965 ± 55 Ma. Additionally, the greenschist (original intermediate-mafic volcanic rocks) was dated at 1048.2 ± 3.3 Ma, implying that the Youxi formation of the Jitamg group was formed in the Neoproterozoic and the metamorphic age is tentatively constrained in the late Neoproterozoic.

In the western Yunnan, the Meso- to Neoproterozoic rocks include the Mesoproterozoic Damenglong group and the Neoproterozoic Lancang group, which mainly outcrop in the Lincang-Lancang block and scatter as enclaves in the intrusions. The Damenglong group mainly consists of biotite-plagioclase leptynites with biotite-plagioclase gneisses; the Lancang group is dominated by quartz-mica-bearing tectonic schists with minor leptynites, phyllites, marbles and metamorphic mafic lenses.

The Gangdise-Tengchong Block The Neoproterozoic rocks in the Gangdise block are called the Nianqingtanggula group, mainly along the main ridge of the Nianqingtangula Mountain in the north of Dangxiong and in the west and north of the eastern structure knot. It is mainly composed of biotite monzonite gneisses, quartz gneisses, granitic gneisses, hornblende-plagioclase leptynites, amphibolites, magnetite quartzites, schists, marbles, and others.

About rock formation age of the group, Hu et al. (2005) have got the protolith ages of 748–787 Ma (orthometamorphic gneiss) and 782 Ma (meta-mafic rocks) from the group in the Namucuo area. Zhang et al. (2010) obtained metamorphic ages of 678–759 Ma (average age of 718 Ma) from the marbles in the Naguo area, west of Namucuo. Zhang et al. (2013b) used zircon U–Pb LA-ICP-MS method to obtain the protolith age of 742–758 Ma (Fig. 5), 666 Ma metamorphic age (Fig. 6) and 660 Ma anatexis age (Fig. 7) from the garnet amphibolites in the Yongzhu area.

Dong et al. (2009) dated detrital zircons of the group in the Nyingchi area, west of the eastern structure knot and found their age peaks mainly at around 500–600 Ma. The protoliths were formed at early Paleozoic and metamorphosed at

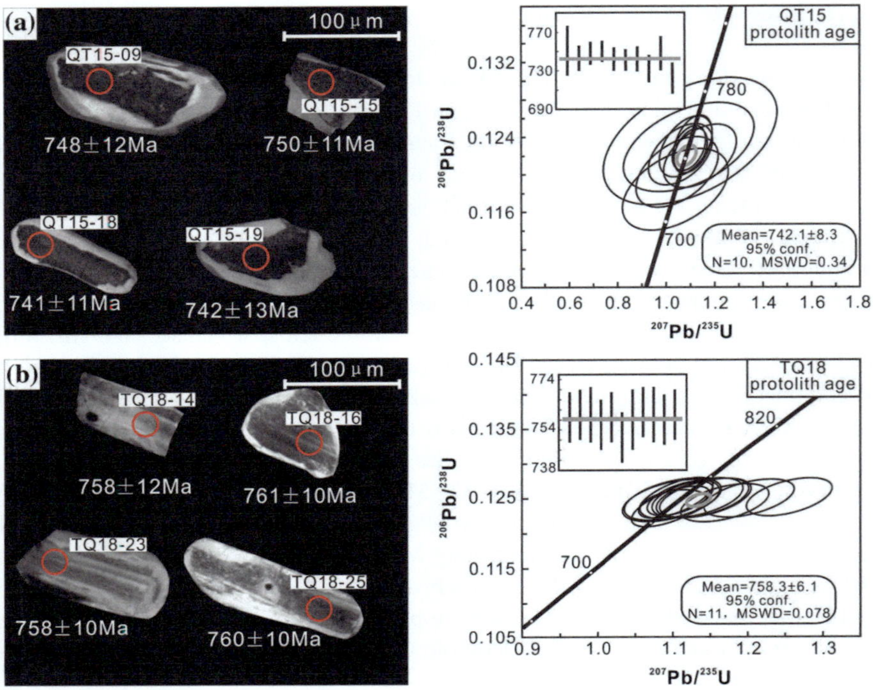

Fig. 5 Cathodoluminescence (CL) images and U–Pb concordia diagram of zircons in garnet plagioclase-amphibole gneisses (cited from Zhang et al. 2013b)

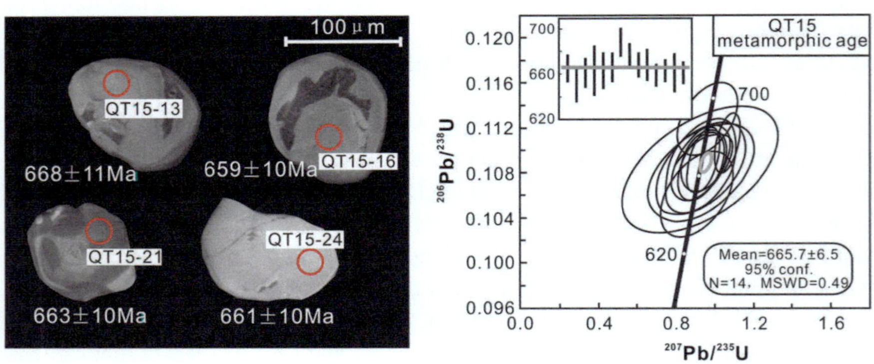

Fig. 6 Cathodoluminescence (CL) images and U–Pb concordia diagram of zircons in garnet plagioclase-amphibole gneiss (cited from Zhang et al. 2013b)

Cenozoic (about 35 Ma). Although this result cannot completely deny the existence of the original Nianqingtanggula group, but at least it should be a metamorphic complex.

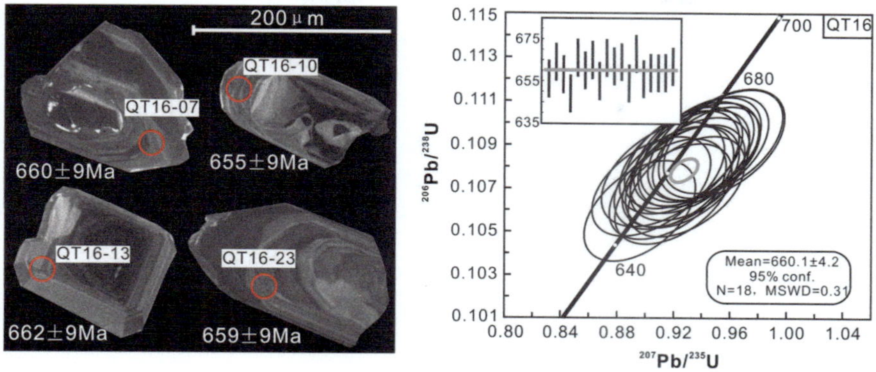

Fig. 7 Cathodoluminescence (CL) images and U–Pb concordia diagram of zircons in felsic veins within garnet plagioclase-amphibole gneiss (cited from Zhang et al. 2013b)

Small amounts of Neoproterozoic strata are exposed in the Bomi-Tongmai area of the southeastern block and called the Bomi group. It is mainly composed of two-mica-quartz schists, meta-sandstones, siltstones, intermediate-felsic volcanic rocks with minor marbles, slates, phyllites, etc. Based on the U–Pb zircon age of 564 Ma Bureau of Geology and Mineral Resources of Xizang Autonomous Region (BGMR 1993) of the quartz-mica schist in the southern Tongmai, the group was placed in the Neoproterozoic era.

The Neoproterozoic rocks in the Tengchong area of the souther block are called the Gaoligongshan group, which consists of biotite-plagioclase gneisses, diopside-hornblende-plagioclase leptynites, sillimanite-biotite schists and amphibolites and suffered amphibolite-facies metamorphism. Li et al. (2012) got magmatic zircon SHRIMP and LA-ICP-MS U–Pb ages of 497.8±7.2 and 500±14 Ma from the orthogneiss with the maximum age of 622 Ma. Therefore, the formation age of the sedimentary Gaoligongshan group should be earlier than 500 Ma. The results also show that the Gaoligongshan group suffered the Pan-African tectonic-magmatic events.

The Himalaya Block The Meso- to Neoproterozoic units in the Himalaya block include the Meso- to Neoproterozoic Laguigangri, Nanjiabarwa, and Nyalam groups and the Neoproterozoic Rouqiecun and Miri groups.

The Laguigangri group is intermittently exposed in the Laguigangri-Kangma-Lhünzê County area of the northern block as intermittent dome-shaped (structure window) outcrops. The main rock types include mica schist or mica-quartz schist with metamorphic minerals of garnet, kyanite, staurolite, biotite-plagioclase (monzonite) gneiss, staurolite-garnet-biotite-plagioclase gneiss, amphibolite plagioclase gneiss, garnet-amphibolite, marble, and quartzite. The protoliths are mainly claystone, muddy sandstone (siltstone), feldspathic sandstones, intermediate-mafic volcanic rocks, tuff, which is a set of terrigenous clastic-carbonate rocks interbedded with volcanics.

The Laguigangri and Nyalam groups are comparable in the rock assemblage and metamorphic and deformation characteristics. The orthogneiss from the Laguigangri group gave SHRIMP zircon age of 1812 ± 7 Ma (Liao et al. 2007). Xu et al. (2005) obtained SHRIMP zircon age of 528–504 Ma (the average age of 515.4 ± 9.3 Ma) in the Kangma area and considered that the metamorphic basement rocks were involved in the early Paleozoic Pan-African orogenic events; Liu et al. (2002) and Zhou et al. (2004) found the lower Ordovician basal conglomerates in the Kangma dome and suggested them as geological markers of the Pan-African event.

Given the SHRIMP age of 1812 ± 7 Ma of the orthogneiss (ancient intrusions) from the Laguigangri group, the possibility that the group extended to the Paleoproterozoic era can not be ruled out.

The Nanjiabarwa group is distributed in the Nanjiabarwa area and the Yarlung Zangbo Great bend, whose periphery was surrounded by zonal distribution of the Yarlung Zangbo suture zone ophiolite, separated by ductile shear zone. The group is divided into the Zhibaiyan formation, Duoxiongla complex and Paixiang formation, divided by two tectonic interfaces. The Zhibaiyan formation mainly consists of high-pressure granulites, garnet amphibolites, Al-rich gneisses interbedded with kyanite-garnet monzogneiss lenses and granitic gneisses. The Duoxiongla complex is composed of banded migmatites, augen migmatites, intestinal-like migmatites and leptynites, gneisses and calc-silicate rocks are locally remained. The Paixiang formation includes biotite leptynites, biotite schists, marbles, calc-silicate rocks. Except some high-pressure granulite lens in the Zhibaiyan formation, the main lithologies of the three formations are similar.

Zhang et al. (2008) accomplished petrologic and geochronological studies on the widespread felsic gneisses of the group. Except individual rocks preserving high-pressure pelitic granulite-facies mineral assemblage of garnet + kyanite + three ternary felspar, most of the gneisses have amphibolite-facies metamorphic mineral assemblages, and their protoliths include diorite and granodiorite with geochemical composition of magmatic arc granites. Zircons from the gneiss commonly have core–rim structures, and SHRIMP and LA-ICP-MS in situ analysis showed that the zircon rims gave multi-Pan-African to Cenezoic metamorphic and magmatic ages (500–10 Ma), and the Precambrian zircon cores gave ages of ~2500, ~1600, and ~1000 Ma. The analyzed zircon region has apparent magmatic oscillatory structures and high Th/U ratios, suggesting magmatic origins. The age peaks are comparable with those of the Precambrian tectonic events in the high Himalayan crystalline complex and Indian block.

The Nyalam group is distributed EW in the southern block and extend southward out of the country. The group is mainly composed of gneisses, schists, quartzites, marbles, amphibolites, and other low to intermediate metamorphic rocks, whose protoliths are mainly muddy-sandy continental margin clastic rocks with carbonates and intermediate to mafic volcanic rock, equivalent to flysch-like containing volcanics. Some of the protoliths should be granitic intrusions and (ultra) mafic intrusions.

Predecessors have obtained a lot of isotopic ages from the Nyalam group (Institute of Geochemistry, Guiyang 1973; Xu et al. 1985; Wei et al. 1989; Liu et al 1990; Hebei District Investigation Team 2004). The ages can be roughly divided into four groups: 1250–2450, 792–950, 458–664, and 42–10 Ma. The 1250–2450 Ma age should represent the formation age of the protoliths, the 792–950 Ma are representative of the Neoproterozoic (before the Pan-African) tectono-thermal events, the 458–664 Ma are coincide with the Pan-African movement; the 42–10 Ma are representative of the Himalayan tectono-thermal event. Limited to the testing methods and accuracy, further study is needed for the precise formation and metamorphic ages of the group.

The Neoproterozoic low metamorphic units adjacent to the north Nyalam group include Lower Cambrian strata at the top, which are called the Rouqiecun formation, which suffered deformation transformation and is dominated by ductile shear zone mylonite including biotite mylonites, marble mylonites, muscovite granitic mylonites and other mylonites. The upper part is composed of mica schists, quartz-mica schists, chlorite schists, slates and phyllites with crystalline limestones, meta-sandstone. The protoliths are pelitic-claystones, shale sandstones, greywackes, and carbonate rocks with a small amount of basic or acidic volcanic rocks, displaying low greenschist-facies metamorphism features. Liu et al. (1990) obtained U–Pb isochron age of 640 and 686 Ma, and Zhao et al. (2001) obtained U–Pb age of 410–515 Ma. Accordingly, the group may be formed in the Sinian-Cambrian, experienced the Pan-African orogeny main metamorphic event and the metamorphic grade is low greenschist facies (locally high greenschist facies).

The Neoproterozoic-Cambrian strata exposed in the Cona-Medog area are called the Miri group, which is composed of fine clastic rocks, dolomite, dolomitic limestone in the lower part, intermediate-mafic volcanic rocks in the middle and quartzites, feldspar sandstones, conglomerates with slates and marls in the upper part.

The Bangong-Nujiang Suture Zone The Bangong-Nujiang suture zone separates the South Qiang-Baoshan and Gangdise-Tengchong blocks and has two micro-blocks containing Precambrian geological units, called the Nierong and Kaqiong micro-blocks. The Meso- to Neoproterozoic units in the Nierong micro-block are called the Nierong group, which mainly consists of gneisses and amphibolites and the distribution range of the latter is far greater than the former. Specific rock types are biotite-plagioclase (monzonite) gneiss, hornblende-plagioclase gneiss, garnet-amphibolite diopside-plagioclase gneiss, amphibolites and pyroxene amphibolites with the metamorphic garde up to amphibolite facies.

The granitic rocks intruding the Nierong group have SHRIMP U–Pb zircon ages of 491 ± 1.15, 492 ± 111, 814 ± 18, 515 ± 14 Ma (1:250000 Anduo 2004). Their emplacement age may be the Neoproterozoic and the metamorphic age is late Neoproterozoic. The 491–530 Ma may be the geological record of the Pan-African orogeny thermal event.

The Carqiong group is located in the eastern part of the Basu County and mainly consists of sillimanite-garnet-biotite-kyanite schists, kyanite-garnet-sillimanite-biotite-monzonite gneisses, amphibolites, eclogites, granulites, and marbles, a set of medium-high metamorphic rocks. SHRIMP U–Pb zircon age of 507 ± 10 Ma has been obtained from the granitic gneiss of the Basu group in the Tongka area, which may be the geological record of the Pan-African orogeny thermal event (Li et al. 2008). In addition, retrograde eclogite xenoliths have been recognized in the group and outcropped as NW-SE beaded enclaves in the kyanite-garnet-biotite-sillimanite-monzonite gneisses. The gneissosity of the enclaves and country rocks are the same, but the age is unknown (Dong et al. 2007a, b).

2.3.3 The Precambrian Geological Events and Evolution of the South Qiang-Gangdise-Himalaya-Baoshan Orogenic System

The South Qiang-Gangdise-Himalaya-Baoshan orogenic system is distributed in the northern margin of the Indian block, whose basement metamorphic rocks are composed of metamorphic supracrustal rocks and plutons. The meta-supracrustal rocks are mainly pelitic-sandy continental margin clastic rocks with carbonates and intermediate-mafic volcanics, probably formed in the Neoproterozoic or earlier, but reliable formation or metamorphic ages of each Precambrian units are scarce.

In contrast, the Paleozoic strata in the Gangdise block to the south of the Longmucuo-Shuanghu suture zone have a large number of Precambrian detrital zircons. The early to middle Ordovician Wenquan quartzites in the south of the Longmuco-Shuanghu suture zone contain detrital zircons of 520–700, ~800, 900–1100, 1800–1900, and 2400–2500 Ma, of which the 625 and 950 Ma age peaks are most obvious. The reliable youngest detrital zircon age is 525 Ma and the oldest detrital zircon age is 3180 Ma. Therefore the source area of the Wenquan quartzites has experienced the Pan-African tectonic and magmatic events and Grenville-Jinning tectonic and magmatic events. The study of detrital zircons in the Carboniferous strata in the central Lhasa block (Leier et al. 2007) showed ages mainly in 2500, 1950–1700 Ma (peak of 1850 Ma), 1300–1050 Ma (peak of 1120 Ma), and 600–500 Ma (peak of 540 Ma).

The currently considered Precambrian strata also have such age information, such as the 1200–900 and 600–500 Ma detrital zircons from the Nianqingtanggula group in the Linzhi area (Dong et al. 2009), the 600–520 and 1100–900 Ma detrital zircons from the Demala group in the Chayu area and the 600–500 and 1100–900 Ma detrital zircon ages in the Tethyan Himalayan belt (DeCelles et al. 2000; Gehrels et al. 2003; Myrow et al. 2003).

Above age information reflects that some Precambrian geological units should be redetermined and the existence of a long-term stable source areas recording Grenville movement (~1100 Ma) and the Pan-African orogeny (~550 Ma) until the late Carboniferous in the entire South Qiang-Gangdise block.

The traditional view suggests that the Lhasa, South Qiang, and Himalaya blocks originate in the northern margin of the Indian craton before the Carboniferous-Permian era (Allègre et al. 1984; Yin and Harrison 2000). Zhang et al. (2008) implied that the Nanjiabarwa group and the high Himalayan crystalline rocks are composed of multi-episodes Archean to Neoproterozoic magmatic rocks, and experienced Rodinia and Columbia supercontinent formation and breakup process as a part of the India craton.

Another view is that the Lhasa block originated from the northern Australia (Audley-Charles 1983, 1984; Zhu et al. 2011). LA-ICP-MS dating results of the Nianqingtanggula block in the Yongzhu area indicate that the protoliths of the orthometamorphic rocks are a set of E-MORB rocks and oceanic island arc magmatic rocks and the oceanic crust was formed at 758 Ma (Zhang et al. 2013b), roughly consistent with the Rodinia supercontinent breakup, which may be a part of the nascent Neoproterozoic oceanic basin in this period of global breakup events. The 665 Ma metamorphic age and 660 Ma anatexis veins are difficult to be regionally comparable in the large range with those in the India craton (Cawood et al. 2007; Chatterjee et al. 2007a, b; Simmat and Raith 2008). However, this phase of tectonic events were recorded in the northwestern Australia (Miles orogeny, 646–671 Ma; Durocher et al. 2003; Bagas 2004), which seem to have affinities (Zhang et al. 2013b).

Combined with the regional geology and geochronology data, the South Qiang-Gangdise-Himalaya-Dianxi orogenic system suffered amphibolite-facies metamorphic at least before the formation of Neoproterozoic Rouqiecun and Miri groups, but the detailed sequence of geological events, especially in the parametamorphic rocks, are not recorded clearly.

The 590–480 Ma Pan-African tectono-thermal event made the Neoproterozoic-Cambrian strata suffered greenschist-facies metamorphism, forming the Pan-African metamorphic basement. The metamorphic strata were the Rouqiecun and Miri groups (Pt3-∈), etc., and the ancient metamorphic basement were superposed metamorphosed. The existence of a large number of the aforementioned "Pan-African" isotopic age data (664–458 Ma) in the basement metamorphic rocks may represent a period of metamorphism and the corresponding tectonic-magmatic event. Xu et al. (2005) got SHRIMP zircon ages of 530–500 Ma in the Geelong, Yadong, Nyalam, etc., and suggested that the metamorphic basement rocks were involved in the Pan-African event and continued until the early Paleozoic. The early Ordovician basal conglomerates found in the Niemula of the high Himalaya and the Kangma dome of the Tethyan Himalayan are considered as a geological mark of the Pan-African events (Zhou et al. 2004).

Acknowledgment We are grateful to the members of the Research Department of Metamorphic and Precambrian Geology, Institute of Geology, Chinese Academy of Geological Sciences for the original manuscript preparation. We also thank Mingguo Zhai, Qihan Shen, Jinghui Guo, Guochun Zhao, Peng Peng, Lingsen Zeng, Lingling Xiao, Junlai Liu, Jinjiang Zhang, Jianxin Zhang, Jing Li, Bo Hu, Lifei Zhang, Shuguang Song, and others for discussions and support in this research. We particularly thank Academician Mingguo Zhai for originating this special issue of Precambrian Geology of China. This paper is the summarized result of previous and present long-term studies of the Tibetan Plateau and adjacent areas, and supported financially by the National Natural Science Foundation of China (41372069, 41302039, 40921001), the Major State Basic Research Program of the People's Republic of China (2012CB416603), and the China Geological Survey Project (12120114023001, 1212011121276).

References

Allègre CJ, Courtillot V, Tapponnier P, Hirn A, Mattauer M, Coulon C, Jaeger JJ, Achache J, Schärer U, Marcoux J, Burg JP, Girardeau J, Armijo R, Gariépy C, Göpel C, Li T-D, Xiao X-C, Chang C-F, Li G-Q, Lin B-Y, Teng J-W, Wang N-W, Chen G-M, Han T-L, Wang X-B, Den W-M, Sheng H-B, Cao Y-G, Zhou J, Qiu H-R, Bao P-S, Wang S-C, Wang B-X, Zhou Y-X, Xu R-H (1984) Structure and evolution of the Himalaya-Tibet orogenic belt. Nature 307:17–22

An C-Y, Wang Y-Z, Ren J-Q, Bao G-P (2003) Disintegration of the wanbaogou group and discovery of early cambrian strata in the east kunlun area. Chin Geol 30(2):199–206 (in Chinese with English abstract)

Audley-Charles MG (1983) Reconstruction of eastern Gondwanaland. Nature 306:48–50

Audley-Charles MG (1984) Cold Gondwana, warm Tethys and the Tibetan Lhasa block. Nature 310:165–166

Bagas L (2004) Proterozoic evolution and tectonic setting of the northwest Paterson Orogen, Western Australia. Precambr Res 128:475–496

Bureau of Geology and Mineral Resources of XinJiang Uygur Autonomous Region (1993) Regional Geology of Xinjiang Uygur Autonomous Region. Geological Publishing House, Beijing, pp 315–318 (in Chinese)

Bureau of Geology and Mineral Resources of Xizang Autonomous Region (BGMR) (1993) Regional Geology of Xizang (Tibet) Autonomous Region. Geological Publishing House, Beijing. (in Chinese)

Chatterjee N, Mazumdar AC, Bhattacharya A, Saikia RR (2007a) Mesoproterozoic granulites of the Shillong-Meghalaya plateau: evidence of western continuation of the Prydz Bay Pan-Afircan suture into Northeastern India. Precambr Res 152:1–26

Chen J-L, Chen Y-C, Li H-P, Zhao X-S (2002) Correlation between the Longshan rock Group and the Qinling rock Group at the junction of Qilian mountain and north Qinling mountain. Geol Shaanx 20:39–49 (in Chinese)

Che Z-C, Sun Y (1996) The age of the Altun granulite facies complex and the basement of the Tarim Basin. Reg Geol China 56:51–57 (in Chinese)

Cawood PA, Johnson MRW, Nemchin AA (2007) Early Paleozoic orogenesis along the India margin of Gondw ana:tectonic response to Gondwana as sembly. Earth Planet Sci Lett 255:70–84

Chatterjee N, Mazumdar AC, Bhattacharya A, Saikia RR (2007b) Mesoproterozoic granulites of the Shillong-M eghalaya plateau: evidence of western continuation of the Prydz Bay Pan-Afircansuture into Northeastern India. Precamb rian Res 152:1–26

DeCelles PG, Gehrels GE, Quade J, LaReau B, Spurlin M (2000) Tectonic implications of U-Pb zircon ages of the Himalayan Orogenic Belt in Nepal. Science 288:497–499

Dong C-Y, Li C, Wan Y-S, Wang W, Wu Y-W, Xie H-Q, Liu D-Y (2011a) Detrital zircon age model of Ordovician Wenquan quartzite south of Lungmuco-Shuanghu Suture in the Qiangtang area, Tibet: constraint on tectonic affinity and source regions. Sci China Earth Sci 54:1034–1042

Dong G-A, Yang H-Y, Liu D-Y, Zhang J-X, Zeng J-Y, Wan Y-S (2007a) SHRIMP U-Pb geochronology of the detrital zircons from "Longshoushan Group" in the Alashan block and its tectonic significance. Chin Sci Bull 52:1414–1425 (in Chinese)

Dong X, Zhang Z-M, Wang J-L, Zhao G-C, Liu F, Wang W, Yu F (2009) Provenance and formation age of the Nyingchi Group in thesouthern Lhasa terrane, Tibetan Plateau: petrology and zircon U-Pb geochronology. Acta Petrologica Sinica 25:1678–1694 (in Chinese)

Dong Y-S, Xie Y-W, Li C, Sha S-L (2007b) Discovery of the retrometamorphic eclogite in the Baxoi area, eastern Tibet, China. Geol Bull China 26:1018–1020 (in Chinese)

Dong Y-S, Li C, Chen H, Chen W, Zhang Y (2011b) Study on geochronology and its geological significance of the Demala Group complex in Chayu area, Southeast Tibetan Plateau. Acta Petrologica Sinica 27:1198–1208 (in Chinese)

Durocher E, Kyser TK, Marlatt J, Hanly A (2003) New 40Ar/39Ar ages from the central Paterson Orogen, Western Australia. Aust J Earth Sci 50:1–10

Gehrels GE, DeCelles PG, Martin A, Ojha TP, Pinhassi G (2003) Initiation of the Himalayan Orogen as an Early Pleozoic thin-skinned thrust belt. GSA Taday 13:4–9

He S-P, Li R-S, Wang C, Yu P-S, Gu P-Y, Shi C (2010) Geochronology of gneissic plagioclase-amphibolite from Beidahe Groupcomplex in western segment of Qilian Mountains, China. Geol Bull China 29:1275–1280 (in Chinese)

He S-P, Li R-S, Wang C, Zhang H-F, Ji W-H, Yu P-S, Gu P-Y, Shi C (2011) Discovery of ~4.0 Ga detrital zircons in the Changdu Block, North Qiangtang, Tibetan Plateau. Chin Sci Bull 56:573–582 (in Chinese)

He S-P, Li R-S, Yu P-S, Gu P-Y, Wang C, Yang Y-C, Zhang W-J (2012) The Age of Youxi Formation of Jitang Rock Group in Taniantaweng Mountains, Northern Qiangtang, Qinghai-Tibet Plateau. Acta Geol Sinica 86:985–993 (in Chinese)

Hu A-Q, Zhang G-X, Chen Y-B, Zhang Q-F (2001) A model of division of the continental crust basement and the time scales of the major geological events in the Xingjiang-based on studies of isotopic geochronology and geochemistry. Xingjiang Geol 19:12–19 (in Chinese)

Hu J-C (1994) The establishment of the Qingbaikou' an lower Kasha formation in Shuilo area, Muli, Sichuan. Reg Geol China 4:366–369 (in Chinese)

Hu D-G, Wu Z-H, Jiang W, Shi Y-R, Ye P-S, Liu Q-S (2005) SHRIMP zircon U-Pb age and Nd isotopic study on the Nyainqêntanglha Group in Tibet. Sci China (Series D) 48:1377–1386 (in Chinese)

Ji W-H, Chen S-J, Zhao Z-M, Li R-S, He S-P, Wang C (2009) Discovery of the Cambiran volcanic rocks in the Xainza area, Gangdese orogenic belt, Tibet, China and its significance. Geol Bull China 28:1350–1354 (in Chinese)

Jiang C-F, Yang J-S, Feng B-G (1992) Opening and closing of the kunlun tectonic. Geological publishing house, Beijing (in Chinese)

Leier AL, Kapp P, Gehrels GE, DeCelles PG (2007) Detrital zircon geochronology of Carboniferous Cretaceous strata in the Lhasa terrane, southern Tibet. Basin Res 19:361–378

Li H-K, Lu S-N, Xiang Z-Q, Zhou H-Y, Li H-M, Liu D-Y, Song B, Zheng J-K, Gu Y (2007) SHRIMP U-Pb Geochronological Research on Detrital zircons from the Beidahe Complex-Group in the Western Segment of the North Qilian Mountains, Northwest China. Geol Rev 53:132–140

Li H-M, Lu S-N, Zheng J-K, Yu H-F, Zhao F-Q, Li H-K, Zuo Y-C (2001) Dating of 3.6 Ga zircons in granite-gneiss from the eastern Altyn Mountains and its geological significance. Bull Mineral Petrol Geochem 20:259–262 (in Chinese)

Li Z-H, Lin S-L, Cong F, Xie T, Zou G-F (2012) U-Pb ages of zircon from metamorphic rocks of the Gaoligongshan Group in western Yunnan and its tectonic significance. Acta Petrologica Sinica 28:1529–1541 (in Chinese)

Li C, Xie Y-W, Sha S-L, Dong Y-S (2008) SHRIMP U-Pb zircon dating of the Pan-African granite in Baxoi County, eastern Tibet, China. Geol Bull China 27:64–68 (in Chinese)

Liao Q-A, Li D-W, Yuan Y-M, Chu L-L, Lu L (2007) The chronological study of Paleoproterozoicgranitic gneiss and its geological significance in Dingjie from high-Himalaya and Laguigangri from north-Himalaya, Tibet. Sci China (Series D: Earth Sciences) 37(12):1579–1587 (in Chinese)

Liu G-H, Jin C-W, Wang F-B (1990) Metamorphic and magmatic rocks of Tibet. Geological publishing house, Beijing (in Chinese)

Liu W-C, Liang D-Y, Wang K-Y, Zhou Z-G, Li G-B, Zhang X-X (2002) Discovery of ordovician strata and its geological significance in Kangma, south Tibet. Earth Sci Front 9(4):247–248 (in Chinese with English abstract)

Lu S-N, Chen Z-H, Xiang Z-Q, Li H-K, Li H-M, Song B (2006) U-Pb ages of detrital zircons from the para-metamorphic rocks of the Qinling Group and their geological significance. Earth Sci Front 13:303–310 (in Chinese)
Lu S-N, Wang H-C, Li H-K, Yuan G-B, Xin H-T, Zheng J-K (2002a) Redefinition of the "Dakendaban Group" on the northern margin of the Qaidam basin. Geol Bull China 21:19–23 (in Chinese)
Lu S-N, Yu H-F, Jin W, Li H-K, Zheng J-K (2002b) Microcontinents on the eastern margin of Tarim paleocontinent. Acta Petrologica ET Mineralogica 21:317–326 (in Chinese)
Lu S-N, Li H-K, Zhang C-L, Niu G-H (2008) Geological and geochronological evidence for the Precambrian evolution of the Tarim Craton and surrounding continental fragments. Precambr Res 160:94–107
Mao D-B, Chen Z-H, Zhong C-T, Li Z-Y, Zou X-Q (2000) Re-Os isotopic dating for molybdenite of the lead-zinc deposit in Beichagoumen, North Hebei. Geochim 2:132–135 (in Chinese with English abstract)
Mei H-L, Yu H-F, Lu S-N, Li H-M, Li Q, Lin Y-X, Zuo Y-C (1998) Archean tonalite in the Dunhuang, Gansu Province: age from the U-Pb single zircon and Nd isotope. Progr Precambr Res 21:41–45 (in Chinese)
Myrow P-M, Hughes N-C, Paulsen T-S, Willlams I-S, Parcha S-K, Thompson K-R, Bowing S-A, Peng S-C, Ahluwalia A-D (2003) Integrated tectonostratigraphic analysis of the himalaya and implications tor its tectonic reconstruction. Earth Planet Sci Lett 212(3–4):433–441
Simmat R, Raith MM (2008) U-Th-Pb monazite geochronometry of the Eastern Ghats Belt, India: timing and spatial disposition of poly-metamorphism. Precambr Res 162:16–39
Song S-G, Yang J-S, Xu Z-Q, Liou JG, Shi R-D (2003) Metamorphic evolution of the coesite-bearing ultrahigh-pressure terrane in the north Qaidam, northern Tibet, NW China. J Metamorph Geol 21:631–644
Tung KA, Yang H-Y, Liu D-Y, Zhang J-X, Yang H-J, Shau YH, Tseng CY (2012) The amphibolite-facies metamorphosed mafic rocks from the Maxianshan area, Qilian block, NW China: a record of early Neoproterozoic arc magmatism. J Asian Earth Sci 46:177–189
Tung KA, Yang H-Y, Liu D-Y, Zhang J-X, Yang H-J, Shau YH, Tseng CY (2013) The Neoproterozoic granitoids from the Qilian block, NW China: evidence for a link between the Qilian and South China blocks. Precambr Res 235:163–189
Wan Y-S, Xu Z-Q, Yang J-S, Zhang J-X (2003) The precambrian high-grade basement of the Qilian terrane and neighboring areas: its ages and compositions. Acta Geosci Sin 24(4):319–324 (in Chinese with English abstract)
Wang Y-S, Zhuang Q-X, Shi C-Y (1980) Quanji group in the north margin of Tarim. In: Sinian Suberathem in China. Chengdu Science and Technology Publishing House, Tianjin (in Chinese)
Wang G-C, Wei Q-R, Jia C-X, Zhang K-X, Li D-W, Zhu Y-H, Xiang S-Y (2007) Some ideas of Precambrian geology in the East Kunlun, China. Geol Bull China 26:929–937 (in Chinese)
Wei G-Y, Shi S-Q, Mao Y-S (1989) Precambrian geological structure and metamorphism in Himalaya. Science and Technology university of Chengdu Publishing House, Chengdu (in Chinese)
Xie Y-Y (1998) Features of ophiolite with different period in the eastern sector of middle Kunlun fault and its original environments. Qinghai Geol 1:27–35 (in Chinese)
Xin H-T, Zhao F-Q, Luo Z-H, Liu Y-S, Wan Y-S, Wang S-Q (2011) Determination of the Paleoproterozoic geochronological framework in Aqtashtagh area in southeastern Tarim, China, and its geological significance. Acta Geol Sinica 85:1977–1993 (in Chinese)
Xu R-H (1985) The study of zircons from granitic rocks in the middle Tibet. Bull Miner, Petrol Geochem 2:71–72
Xu Z-Q, Yang J-S, Liang F-H, Qi X-X, Liu F-L, Zeng L-S, Liu D-Y, Li H-B, Wu C-L, Shi R-D, Chen S-Y (2005) Pan-african and early paleozoic orogenic events in the himalaya terrane: inference from SHRIMP U-Pb zircon ages. Acta Petrol Sin 21(1):1–12 (in Chinese with English abstract)

Xu W-C, Zhang H-F, Liu X-M (2007) The formation age and geological significance of high-grade metapelites in Qilian: constrains from zircon U-Pb dating. Chin Sci Bull 52 (10):1174–1180 (in Chinese)

Yao Z-D, Ni B-F (1990) The characteristics of pre-sinian metamorphic rocks in the Huili-Miyi-Yanbian area, Sichuan Province, and their age. Reg Geol China 2:166–172 (in Chinese)

Yang J-S, Xu Z-Q, Zhang J-X, Song S-G, Wu C-L, Shi R-D, Li H-B, Brunel M (2002) Early Palaeozoic North Qaidam UHP metamorphic belt on the north-eastern Tibetan plateau and a paired subduction model. Terra Nova 14:397–404

Yang J-S, Zhang J-X, Meng F-C, Shi R-D, Wu C-L, Xu Z-Q, Li H-B, Chen S-Y (2003) Ultrahigh pressure eclogites of the north Qadam and Altun mountains, NW China and their protoliths. Earth Sci Front 10(3):291–314 (in Chinese with English abstract)

Yin A, Harrison M (2000) Geologic evolution of the Himalayan-Tibetan orogen. Earth Planet Sci Lett 28:211–280

Yu, F.-C., Wei, G.-F., Sun, J.-D., 1994. Metallogenic Modal for the syntectonic gold deposits in black series: a case study of the Tanjianshan deposit. Northwest University Press, Xi'an, pp 1–130 (in Chinese)

Zhai M-G, Cong B-L (1993) The Diancangshan-Shigu metamorphic belt in west Yunnan, China: their geochemical and geochronological characteristics and division of metamorphic domains. Acta Petrologica Sinica 9:227–239 (in Chinese)

Zhai M-G, Cong B-L, Qiao G-S, Zhang R-Y (1990) Sm-Nd and Rb-Sr geochronology of metamorphic rocks from southwest Yunnan orogenic zones, China. Acta Petrologica Sinica 11:1–11 (in Chinese)

Zhang C-L, Ye H-M, Wang A-G, Dun K-Y, Dong Y-G (2004) Geochemistry of the Neoproterozoic diabase and basalt in South of Tarim plate: evidence for the Neoproterozoic breakup of the Rodinia super-continent in south of Tarim. Acta Petrologica Sinica 20:473–482 (in Chinese)

Zhang Z-M, Wang J-L, Zhao G-C, Shi C (2008) Geochronology and Precambrian tectonic evolution of the Namche Barwa complex from the eastern Himalayan syntaxis, Tibet. Acta Petrologica Sinica 24:1477–1487 (in Chinese)

Zhang Z-M, Dong X, Geng G-S, Wang W, Yu F, Liu F (2010) Precambrian metamorphism of the northern Lhasa terrane, South Tibet and its tectonic implications. Acta Geol Sinica 84:449–456 (in Chinese)

Zhang J-X, Zhang Z-M, Xu Z-Q, Yang J-S, Cui J-W (2000) Discovery of khondalite series from the western segment of Altyn Tagh and their petrological and geochronological studies. Sci China (Series D) 43:308–316

Zhang J-X, Wan Y-S, Xu Z-Q, Yang J-S, Meng F-C (2001) Discovery of basic granulite and its formation age in Delingha area, North Qaidam Mountains. Acta Petrologica Sinica 17:453–458 (in Chinese)

Zhang J-X, Gong J-H, Yu S-Y (2012) c. 1.85 Ga HP granulite-facies metamorphism in the Dunhuang block of the Tarim Craton, NW China: evidence from U-Pb zircon dating of mafic granulites. J Geol Soc 169:511–514

Zhang J-X, Yu S-Y, Gong J-H, Li H-K, Hou K-J (2013a) The latest Neoarchean-Paleoproterozoic evolution of the Dunhuang block, eastern Tarim craton, northwestern China: evidence from zircon U-Pb dating and Hf isotopic analyses. Precambr Res 226:21–42

Zhang X-Z, Dong Y-S, Li C, Xie C-M, Yang H-T, Wang M (2013b) Delineation of Meso Neoproterozoic ophiolite mélange in the northern Lhasa terrane, South Tibet and its significance. Acta Petrologica Sinica 29:674–698 (in Chinese)

Zhang Z-C, Mao J-W, Zuo G-C, Yang J-M, Wang Z-L, Zhang Z-H (1998) Geochemistry of the Proterozoic metamorphic volcanic rocks in the west part of north Qilian mountains and the tectonic setting. J Mineral Petrol 18:22–30 (in Chinese)

Zhao Z-Z, Li Y-T, Ye H-F, Zhang L-W (2001) Stratum of Qinghai-Tibet Plateau. Science Press, Beijing (in Chinese)

Zheng J-K (1992) The regional tectonic evolution of east Kunlun. Geol Qinghai 1(1):15–25 (in Chinese with English abstract)
Zhou Z-G, Liu W-C, Liang D-Y (2004) Discovery of the Ordovician and its basal conglomerate in the Kangmar area, southern Tibet-with a discussion of the relation of the sedimentary cover and unifying basement in the Himalayas. Geol Bull China 23:655–663 (in Chinese)
Zhu D-C, Zhao Z-D, Niu Y-L, Dilek Y, Mo X-X (2011) Lhasa Terrane in southern Tibet came from Australia. Geology 39:727–730

Paleocontinents in Xing'an-Mongolia Orogenic Belt (XMOB)

Bei Xu

Abstract According to new data of field observation, the youngest detrital zircon age and magmatic zircon age, four Precambrian blocks have been recognized in the Xing'an-Mongolia orogenic belt (XMOB), including the Erguna block (EB), Xing'an-Airgin Sum block (XAB), Songliao-Hunshandake block (SHB), and Jiamusi block (JB).

Keywords Xing'an-mongolia orogenic belt · Precambrian · Paleocontinents

1 Precambrian Blocks in Xing'an-Mongolia Orogenic Belt

The Central Asian Orogenic Belt (CAOB) extends from Ural area of Russia in the west, via Mongolia, to Far East area of Russia and Inner Mongolia, and Xing'an areas of China (Jahn et al. 2000; Jahn 2004; Windley et al. 2007). The CAOB takes a wide area from Xinjiang in the west, via Inner Mongolia, to northeast China in China. East segment of the CAOB extends across Inner Mongolia, Heilongjiang, Jilin, and Liaoning provinces and is called Xing'an-Mongolia orogenic belt (XMOB, Ren et al. 1980). According to new data of field observation, geochronology and geochemistry, four blocks have been recognized in the XMOB (Fig. 1, Xu et al. 2015), including the Erguna block (EB), Xing'an-Airgin Sum block (XAB), Songliao-Hunshandake block (SHB), and Jiamusi block (JB, Fig. 2).

B. Xu (✉)
Key Laboratory of Orogenic Belts and Crustal Evolution, School of Earth and Space Sciences, Peking University, Beijing 100871, China
e-mail: bxu@pku.edu.cn

M. Zhai (ed.), *Precambrian Geology of China*,
Springer Geology, DOI 10.1007/978-3-662-47885-1_12

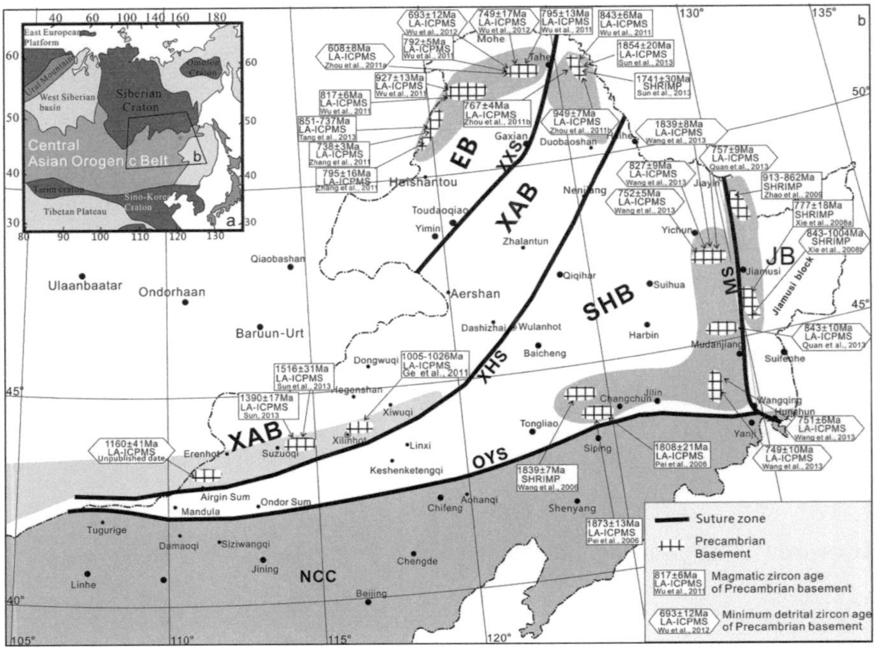

Fig. 1 Pre-Devonian blocks of the XMOB. **a** Location of the CAOB, **b** blocks in the XMOB (after Xu et al. 2015)

2 Erguna Block (EB)

The oldest Neoproterozoic intrusion in the EB includs alkali feldspar granites with ages of 792–927 Ma in Mangui and Bishui areas (Wu et al. 2011) and four stages of magmatic rocks (about 851, 792, 762 and 737 Ma) in Shiwei–Enhe area, which are characterized by syenogranites, bimodal igneous rock associations, and granodiorites that formed in an extensional environment related to the breakup of the Rodinia supercontinent (Tang et al. 2013). The Neoproterozoic Xinghuadukou Group and Ergunhe Group distribute in northern and southwestern margins of the EB, respectively. The Xinghuadukou Group is characterized by metamorphic supracrustal rocks with age of 749 ± 17 Ma and two zircon age groups of 2.0 Ga and 1000–800 Ma and has the provenance of the Paleoproterozoic and Mesoproterozoic (Wu et al. 2012). The Ergunhe Group contains dolomites, limestones, two-mica schists and sandstones, from which detrital zircon age populations around 738–1050 Ma have been reported, suggesting a Neoproterozoic sedimentary sequence (Zhang et al. 2014). Another later Precambrian sequence called the Mohe complex occurs in the northern part of the block, with the youngest zircon age of 608 ± 8 Ma from biotite plagioclase gneiss samples (Zhou et al. 2011a). These Precambrian sequences and the Neoproterozoic intrusions suggest that the EB was a later Neoproterozoic block.

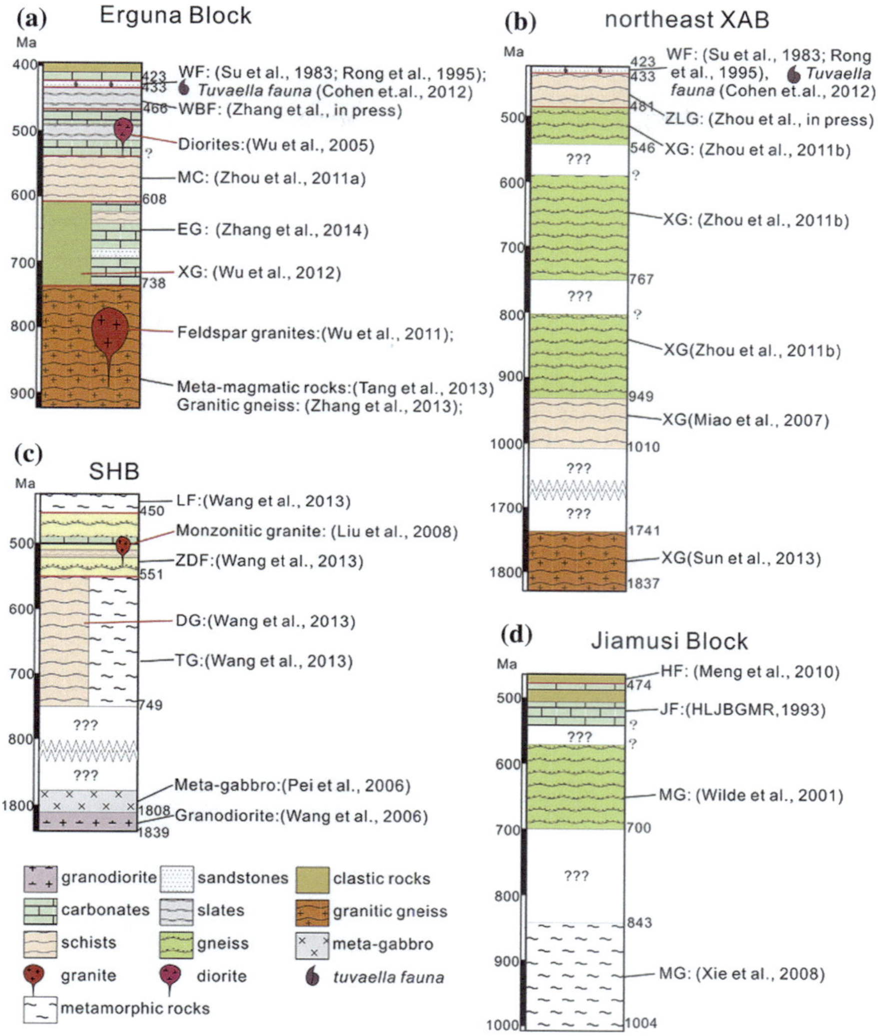

Fig. 2 Stratigraphic column for the blocks. **a** EB, **b** Northeast XAB, **c** SHB, **d** JB. See text for the ages in the right side of column. *DG* Dongfengshan Group, *EG* Ergunhe Group, *HF* Heital Formation, *JF* Jinyinku Formation, *LF* Lalagou Formation, *MC* Mohe complex, *MG* Mashan Group, *SHB* Songliao–Hunshandake block, *TG* Tadong Group, *WBF* Wubinaobao Formation, *WF* Wuduhe Formation, *XAB* Xing'an block, *XG* Xinghuadukou Group, *ZDF* Zhudundian Formation, *ZLG* Zalantun Group (after Xu et al. 2015)

3 Xing'an–Airgin Sum Block (XAB)

This block outcrops in Xing'an area in the northeast and Airgin Sum area in the west. The northeast part is called Xing'an block, which is represented by the Xinghuadukou Group in Shiqizhan area where two ages of 1837 ± 5 and 1741 ± 30 Ma from augen granitic and banded gneisses have been reported, respectively (Sun et al. 2013a, b). The Xinghuadukou Group in Hanjiayuanzi near to Shiqizhan area consists of staurolite garnet two-mica schist with a lot of detrital zircon ages older than 1010 Ma (Miao et al. 2007). The Neoproterozoic biotite plagioclase gneiss with igneous protolith ages of 767 ± 4 Ma, and garnet sillimanite gneiss with detrital zircon age peak of 949 ± 7 Ma have been reported from Xinghuadukou area (Zhou et al. 2011b).

The Precambrian basement of the west XAB includes the Airgin Sum Group in west segment, Baoyintu Group in middle segment, and Xilingol Complex in east segment. The Airgin Sum Group consists of mica quartzite with a youngest peak age of 1180 Ma, sericite quartz schists, gneisses, and gneissic granites (Fig. 3; Xu et al. 2015). In middle segment of the XAB the Baoyintu Group of the Precambrian basement is characterized by gneisses and gneissic granites with zircon ages of

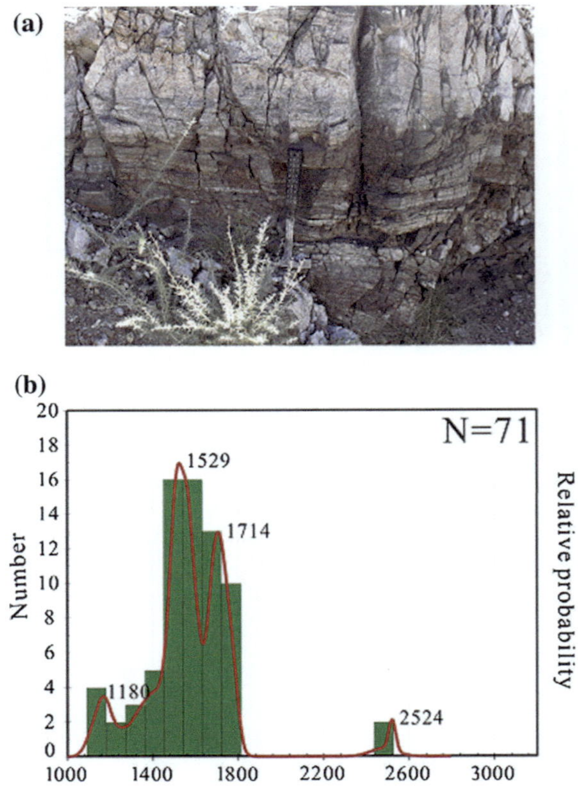

Fig. 3 Field photo and detrital zircon ages of mica quartzite in the Airgin Sum Group (after Xu et al. 2015)

1516 ± 31 and 1390 ± 17 Ma (Sun et al. 2013a, b). In the eastern segment of the west XAB, the Xilingol Complex is characterized by metamorphic core ages of detrital zircons range from 1005 to 1026 Ma in an aluminum-rich argillaceous rock, indicating that it belongs to the Precambrian basement (Ge et al. 2011).

4 Songliao–Hunshandake Block (SHB)

The oldest ages of basement rock comes from bore core in southeast margin of the block, including a SHRIMP U–Pb zircon age of 1839 ± 7 Ma from granodiorite (Wang et al. 2006) and ICP-MS U–Pb age of 1808 ± 21 Ma from metagabrro (Pei et al. 2007). The Neoproterozoic ages have been acquired from the Dongfengshan Group in the north and Tadong Group in the south along the eastern margin of the SHB, containing youngest detrital zircon age of 752 ± 5 and 751 ± 6 Ma from two-mica quartz schist and biotite quartz schist, respectively (Wang et al. 2013).

5 Jiamusi Block (JB)

The oldest rocks occur in Liumao and Ximashan areas in the southern part of the JB, which is called the Mashan Group. SHRIMP U–Pb zircon analyses from sillimanite gneiss of the Mashan Group reveal that there are two kinds of zircons: unmetamorphic zircons are characterized by high Th/U ratio and age populations from 700 to 1600 Ma. Xie et al. (2008) report magmatic zircon ages from 843 to 1004 Ma in the Mashan Group migmatite in Mulin area. These data suggest that there was a Mesoproterozoic to Neoproterozoic basement in the JB (Wilde et al. 2001; Xie et al. 2008).

References

Ge M-C, Zhou W-X, Yu Y, Sun J-J, Bao J-Q, Wang S-H (2011) Dissoluotion and supracrustal rocks dating of Xilin Gol Complex, Inner Mongolia, China. Earth Sci Frontier 18(5):182–195 (in Chinese with English abstract)

Jahn BM, Wu F-Y, Chen B (2000) Granitoids of the Central Asian Orogenic Belt and continental growth in the Phanerozoic. Trans Royal Soc Edinb: Earth Sci 91:181–193

Jahn BM (2004) The Central Asian Orogenic Belt and growth of the continental crust in the Phanerozoic. In: Malpas J, Fletcher CJN, Aitchison JC (eds) Aspects of the Tectonic Evolution of China: Geological Society London Special Publication, No. 226, pp 73–100

Miao L-C, Liu D-Y, Zhang F-Q, Fan W-M, Shi Y-R, Xie H-Q (2007) Zircon SHRIMP U-Pb ages of the "Xinghuadukou Group" in Hanjiayuanzi and Xinlin areas and the "Zhalantun Group" in Inner Mongolia, Da Hinggan Mountains. Chin Sci Bull 52(8):1112–1124

Pei F-P, Xu W-L, Yang D-B, Zhao Q-G, Liu X-M, Hu Z-C (2007) Zircon U-Pb geochronology of basement metamorphic rocks in the Songliao Basin. Chin Sci Bull 52(7):942–948

Ren J-S, Jiang C-F, Zhang Z-K, Qin D-Y (1980) The geotectonic evolution of China. Science Press, Beijing, pp 89–104 (in Chinese)

Sun L-X, Ren B-F, Zhao F-Q, Ji S-P, Geng J-Z (2013a) Late Paleoproterozoic magmatic records in the Eerguna massif: evidences from the zircon U-Pb dating of granitic gneisses. Geol Bull China 32(2/3):341–352 (in Chinese with English abstract)

Sun L-X, Ren B-F, Zhao F-Q, Li Y-H, Liu H (2013b) Zircon U-Pb dating and Hf isotopic compositions of the Mesoproterozoic granic gneiss in Xilihot Block, Inner Mongolia. Geol Bull China 32(2/3):327–340 (in Chinese with English abstract)

Tang J, Xu W-L, Wang F, Wang W, Xu M-J, Zhang Y-H (2013) Geochronology and geochemistry of Neoproterozoic magmatism in the Erguna Massif, NE China: petrogenesis and implications for the breakup of the Rodinia supercontinent. Precambr Res 224:597–611

Wang F, Xu W-L, Gao F-H, Zhang H-H, Pei F-P, Zhang L, Yang Y (2013) Precambrian terrane within the Songnen-Zhangguangcai Range Massif, NE China: evidence from U-Pb ages of detrital zircons from the Dongfengshan and Tadong group. Gondwana Res 25:402–413

Wang Y, Zhang F-Q, Zhang D-W, Miao L-C, Li T-S, Xie H-Q, Meng Q-R, Liu D-Y (2006) Zircon SHRIMP U-Pb dating of meta-diorite from the basement of the Songliao Basin and its geological significance. Chin Sci Bull 51(15):1877–1883

Wilde SA, Wu F-Y, Zhang X-Z (2001) The Mashan Complex: SHRIMP U-Pb zircon evidence for a Late Pan-African metamorphic event in NE China and its implication for global continental reconstructions. Geochimica 30(1):35–50 (in Chinese with English abstract)

Windley BF, Alexeiev D, Xiao W-J, Kröner A, Badarch G (2007) Tectonic models for accretion of the Central Asian Orogenic Belt. J Geol Soc Lond 164:31–47

Wu F-Y, De Y-S, Wen C-G, Zhang Y-B, Grant ML, Wilde SA, Jahn BM (2011) Geochronology of the Phanerozoic granitoids in northeastern China. J Asian Earth Sci 41:1–30

Wu G, Chen Y-C, Chen Y-J, Zeng Q-T (2012) Zircon U-Pb ages of the metamorphic supracrustal rocks of the Xinghuadukou Group and granitic complexes in the Argun massif of the northern Great Hinggan Range, NE China, and their tectonic implications. J Asian Earth Sci 49:214–233

Xie W-Q, Miao L-C, Chen F-K, Zhang F-Q, Liu D-Y (2008) Characteristics of the "Mashan Group" and zircon SHRIMP U-Pb dating of granite in Muling area, southeastern Heilongjiang Province, China: constraint on crustal evolution of the southernmost of Jiamusi Massif. Geol Bull China 27(12):2127–2137

Xu B, Zhao P, Wang Y-Y, Liao W, Luo Z-W, Bao Q-Z, Zhou Y-H (2015) The pre-Devonian tectonic framework of Xing'an–Mongolia orogeni belt (XMOB) in north China. JAES 97:183–196

Zhang Y-H, Xu W-L, Tang J,Wang F, Xu M-J, Wang W (2014) Age and provenance of the Ergunahe Group and the Wubinaobao Formation, northeastern Inner Mongolia, NE China: implications for tectonic setting of the Erguna Massif. Int Geol Rev 56(6):653–671

Zhou J-B, Wilde SA, Zhang X-Z, Ren S-M, Zheng C-Q (2011a) Pan-African metamorphic rocks of the Erguna Block in the Great Xing'an Range, NE China: evidence for the timing of magmatic and metamorphic events and their tectonic implications. Tectonophysics 499(1–4):105–117

Zhou J-B, Wilde SA, Zhang X-Z, Zhao G-C, Liu F-L, Qiao D-W, Ren S-M, Liu J-H (2011b) A > 1300 km late Pan-African metamorphic belt in NE China: new evidence from the Xing'an block and its tectonic implications. Tectonophiysics 509:280–292

MIX
Papier aus verantwortungsvollen Quellen
Paper from responsible sources
FSC® C105338

If you have any concerns about our products,
you can contact us on
ProductSafety@springernature.com

In case Publisher is established outside the EU,
the EU authorized representative is:
Springer Nature Customer Service Center GmbH
Europaplatz 3, 69115 Heidelberg, Germany

Printed by Libri Plureos GmbH
in Hamburg, Germany